An Annotated Catalogue of Types
of the University of Illinois
Mycological Collections (ILL)

An Annotated Catalogue of Types of the University of Illinois Mycological Collections (ILL)

J.L. Crane and Almut G. Jones

Illinois Biological Monographs 58

University of Illinois Press
Urbana and Chicago

Illinois Biological Monographs Committee
  David S. Seigler, chair
  Daniel B. Blake
  Joseph V. Maddox
  Lawrence M. Page
  Thomas Uzzell

© 1997 by the Board of Trustees of the University of Illinois
Manufactured in the United States of America
C 5 4 3 2 1

This book is printed on acid-free paper.

Library of Congress Cataloging-in-Publication Data
Crane, J.L. (Joseph Leland), 1935-
An annotated catalogue of types of the University of Illinois
mycological collections (ILL) / J.L. Crane and Almut G. Jones.
  p.   cm. — (Illinois bilogical monographs ; 58)
Includes bibliographical references (p.).
ISBN 0-252-02319-6 (alk. paper)
1. Fungi—Type specimens—Catalogs and collection—Illinois—
Urbana. 2. University of Illinois (System). Herbarium—Catalogs.
I. Jones, Almut G. II. University of Illinois (System). Herbarium.
III. Title. IV. Series.
QK600.73.U62U653   1997
589.2'074'773—dc20   96-27528
CIP

Contents

Acknowledgments vii

Introduction 1

Annotated List of Types 9

List of Lectotypes Located at ILL 349

List of Undistributed Fungus Exsiccati Holdings
  of the University of Illinois Mycological Collections 357

Literature Cited 361

Acknowledgments

We express our sincere appreciation to the curators of BPI, CUP, FH, and NY for providing literature and copies of specimen labels from collections in their charge and to the curator of LPS for lending us certain holotypes from the Spegazzini Herbarium. We gratefully acknowledge the help of Betty Nelson, who typed and printed the first three versions of the manuscript; the assistance with curating and proofing provided by Laurel McKee; and aid from the staff of the University of Illinois Map and Geography library in the search for certain locality data. A special thanks is extended to Donald P. Rogers for assistance with Latin and with the interpretation of the I.C.B.N.; to Joe Hennen for his hospitality and permission to check certain collections in the Arthur Herbarium (PUR); and to Richard P. Korf and Susan C. Gruff of CUP, who assisted us in the reconstruction of labels of specimens from C.F. Baker's Fungi Malayana. Helpful comments on the manuscript by Jack D. Rogers (Washington State University, Pullman, WA) and Roger D. Goos (University of Rhode Island, Kingston, RI) are appreciated. Financial and some clerical assistance for this project were provided by the Department of Plant Biology of the University of Illinois and the Illinois Natural History Survey.

Introduction

The herbarium of the University of Illinois' Department of Plant Biology was founded by Thomas Jonathan Burrill in 1868. Burrill was appointed Professor of Natural History in 1868 and served the university as mycologist, plant pathologist, phytobacteriologist, and administrator until 1912. His collections from the Rocky Mountains, made during the Powell expedition, and from around Champaign-Urbana, formed the nucleus of the herbarium. Specimens from the fungus disease survey of Illinois, initiated by Burrill (1884, 1885, and 1887) and largely collected by A.B. Seymour, F.S. Earle, and G.P. Clinton between the years 1868 and 1884, formed the foundation for the mycological collections. The fungus herbarium was expanded with the purchase of numerous sets of exsiccati. This institution, ILL, is regarded as the principal type depository for names published by Burrill. Exceptions are indicated in the text.

Frank Lincoln Stevens succeeded Burrill as Mycologist and Professor of Plant Pathology in 1914 and remained at the University of Illinois until his death in 1934. He was an authority on Ascomycetes, especially the Meliolales and Dothideales (Stevens, 1920, 1927a, 1928). Stevens' personal herbarium, consisting of several thousand fungus specimens, was donated to the University of Illinois (ILL). His own contributions consist of numerous specimens collected during excursions to Puerto Rico (Stevens, 1916, 1918, 1920, 1927a, 1928; see also Ryan 1924), South America (Stevens 1924; Stevens and Tehon, 1926), Central America (Stevens, 1927b), during a stay in Hawaii (Stevens, 1925), and in the Philippines (Stevens and Roldan, 1935). In addition, Stevens maintained an extensive exchange program of meliolaceous specimens, especially with H. Sydow at Berlin-Dahlem (B—most Sydow specimens located at that institution were destroyed during World War II), E.M. Doidge at Pretoria (PREM), and C.L. Spegazzini at La Plata (LPS). As a consequence, numerous duplicates and portions of types from these and other authors and collectors are deposited at ILL, which now represents one of the most extensive depositories of early 20th century collections of Meliolales from Africa, the Neotropics, the Hawaiian Islands, and the Philippines. Duplicates of Stevens' collections have been located at BISH, BPI, CUP, FH, K, LPS, MICH, NY, PUR, PREM, URM, S, and W (cf. Stafleu and Cowan, 1985).

Leland Shanor joined the botany faculty of the University of Illinois in 1934 and was appointed Curator of Fungi in 1946. He trained many prominent American mycologists until he left in 1956. Only a few of his collections were deposited at ILL. Donald P. Rogers became a faculty member of the Botany Department in 1957 and curated the mycological collections at ILL until his retirement in 1976. His contributions to the herbarium consist of several thousand specimens of resupinate Basidiomycetes from Oregon and Hawaii and also of some Hawaiian flowering plants. J. Leland Crane, Mycologist on the staff of the Illinois Natural History Survey, served as curator from 1976 to 1986. Duplicates of some of his types are at ILL. Currently, Carol A. Shearer, Professor and Head of the Department of Plant Biology, is the curator in charge of the mycological collections at ILL.

This catalogue provides information on the basionym for each type specimen, regardless of the validity or legitimacy of the name. Obligate synonyms and avowed substitute names (nomina nova) are fully cited, in chronological sequence, at the end of the paragraph under their respective basionyms; they are also cross-listed in the catalogue. Full name(s) and initials of the author(s) are given for each fungus name, as well as the place and date of publication. If the actual date of publication (i.e., of distribution in print of the published name) differs from the intended publication date or that of the journal volume, the latter is added within parentheses.

Following the example of F.C. Deighton (1968), orthographic errors of validly published fungus names were corrected in accordance with the current *International Code of Botanical Nomenclature* (Greuter et al., 1994)— abbreviated I.C.B.N. in subsequent text. We consulted the most recent edition of *Botanical Latin* (Stearn, 1992) and also several other publications dealing with orthography of scientific botanical names and epithets, e.g., Nicolson (1974, 1986, and 1987) and Nicolson and Brooks (1974). Further assistance was obtained from Latin scholars. For invalid or illegitimate names, we did not correct errors in the orthography of specific epithets but corrected spelling of the generic names where appropriate.

Type information on names based on the collections of F.L. Stevens forms the bulk of this catalogue. Since most of the taxa involved were also described by Stevens, in part in collaboration with his graduate students at the University of Illinois, the herbarium of this university (ILL) is regarded as the principal depository of his types (i.e., of the specimens he used). This assumption is supported by introductory statements in several of his publications (F.L. Stevens, 1916, 1918, 1920, 1924, and 1927b), as well as those of other authors (e.g., E. Young, 1915, and C.G. Hansford, 1961). Unless otherwise indicated, we interpret Stevens' use of the term "type" to mean "holotype," and in most cases, the specimen at ILL is assumed to be the holotype. If he cited several collections without indicating a "type," the term "syntype" is generally applied to the specimens at ILL and the term "isosyntype" to specimens deposited in other herbaria, e.g., those in the Bishop Museum (BISH). For holdings of Stevens' collections in the latter institution, the type categories used in this catalogue differ to some extent from those applied by Goos and Gowing (1992).

Following the fungus name and reference, the type citation is basically presented as published in the protologue, but standardized to the following sequence: type category, substrate or host, locality data (in descending sequence), date of collection, name of collector, collection number (not always available), and the ILL accession number, followed by an (incomplete) listing of other institutions that are known to have duplicate specimens. Complementary or contrasting information obtained from the packet label(s), explanatory notes and supplementary information, as well as the currently accepted names of host plants are inserted, within brackets, in the appropriate places. Note that we restricted use of the "≡" symbol to obligate (nomenclatural) synonyms of the fungus names; the standard "=" sign was applied to both nomenclatural and taxonomic synonyms of host plants. Herbarium abbreviations are cited according to *Index Herbariorum* (Holmgren et al., 1990). For definition of the categories of type specimens, we followed Article 9 of I.C.B.N. (Greuter et al., 1994).

References in the text give the complete (unabbreviated) names of periodicals as presented in *Botanico-Periodicum-Huntianum* (Lawrence et al., 1968), including the supplement (Bridson and Smith, 1991), and in *The World List of Scientific Periodicals* (Brown and Stratton, 1963; Porter and Koster, 1970). Titles of books and pamphlets are cited according to Stafleu and Cowan (1976–1988), and Stafleu and Mennega (1992, 1993, 1995). For a nomenclatural updating of generic names of host plants, we consulted Brummitt (1992) and Farr et al. (1979, 1986). Names of spe-

4   Introduction

cies and varieties of vascular plants, initially checked in *Index Kewensis* and the Gray Herbarium Index, were updated by consulting such works as *A Synonymized Checklist of the Vascular Flora of the United States, Canada, and Greenland* (Kartesz, 1994), *Hortus Third* (Bailey and Bailey et al., 1976), and *Flora Europaea* (Tutin et al., 1964–1980), as well as the floristic works of Backer and Bakhuizen van den Brink (1963–1968), Bond and Goldblatt (1984), Britton and Wilson (1923–1924; 1925–1930), Correll and Correll (1982), Li et al. (1975–1979), Merrill (1912, 1923–1926), Nicolson (1991), Smith (1979–1991), Standley (1937–1938), and Wagner et al. (1990). Authorities of names of vascular plant hosts are mostly abbreviated according to the standardized list of Brummitt and Powell (1992). Following convention, we did not cite authorities for names of animal hosts. Geographic information was updated to some extent following recent gazetteers and atlases.

We decided to include paratypes in our listing, even though this resulted in considerable increase in volume of the catalogue, for the following reason: Many collections host several fungus taxa, and though some of the older collections were first cited as representative specimens or paratypes under one name, they often were subsequently cited as true types (holotypes, lectotypes, or isotypes) of one or more names of other taxa. We found numerous examples of this, and we have concluded that cross-listing would provide valuable information that could aid in locating hidden types. The fact that types are not always filed under the name of the taxon to which they belong is well known to any curator dealing with older collections. Labels of older type specimens frequently give variable or incomplete information; they often were not even inscribed with the name of a new taxon found in association with or parasitic on an earlier described taxon. Isotypes were sometimes distributed before the collection was attached to or cited as the type of the name of a new taxon, or subsequent workers may have failed to recognize the importance of a specimen as a type and inscribed or filed it according to their own taxonomic decision. A case in point for problems of this kind is our recent discovery, in the general collections under the generic name *Microthyrium*, of what we assume to be the holotype of both *Monogrammia miconiae* F.L. Stevens and *Paranectria miconiae* F.L. Stevens. We are quite confident that more type material is still hidden in our general collections. Our list would probably be more substantial if specimens borrowed from ILL were always returned with proper annotations.

At this time, the annotated catalogue comprises a total of 2,507 entries, including cross-listed obligate (nomenclatural) synonyms. We recorded for ILL 820 holotypes, 45 lectotypes, 1,219 isotypes (including 63 isolectotypes and one isoneotype), and 504 syntypes (including 93 residual syntypes and 173 isosyntypes). In certain cases, we felt it was necessary to lectotypify; for instance, if Stevens had originally cited more than one collection (i.e., syntypes), we interpreted his subsequent citation of a single collection as lectotypification with the specimen located at ILL. In other instances, we interpreted citation in a monographic work of a certain institution for location of the "type" as lectotypification, e.g., in Hansford (1961). Several of our specimens were clearly lectotypified, e.g., by Chupp (1953) and Hughes (1993). The word "residual" is added to syntypes that remain after one of them was designated as the lectotype. In a few cases, we overruled lectotypification and in two cases neotypification by other authors, because we located what we assume to be the holotype or a more logical choice of lectotype. Explanations are given in all instances under the names listed. (For a separate list of lectotypes located at ILL, see p. 349.

Included in the catalogue are entries for several names published by C.L. Spegazzini (1923) that were typified with duplicates of collections made in Puerto Rico by F.L. Stevens. Some of the types were sent to us from LPS for verification of isotypes on deposit at ILL. We were unable, however, to get confirmation for what we are reasonably sure are duplicates at ILL of the specimens Spegazzini used for five of his names (*Asteridium portoricense, Micropeltidium portoricense, Scolecopeltella microcarpa, Scolecopeltella portoricensis,* and *Scolecopeltis pachyasca*), which we therefore have included, labeling them "potential isotypes." Spegazzini was often very vague in the habitat information cited; he usually did not give collection date and name of collector. Typed labels that were attached to the specimens at a later date are frequently inaccurate. Some labels are marked "leg. Spegazzini" but, as far as we know, Spegazzini did not collect in Puerto Rico, certainly not at the dates indicated, whereas Stevens was visiting the localities on those dates. The specimens we did receive perfectly match the host plants and show essentially the same locality and date as the specimens at ILL.

Generally excluded from the annotated catalogue are type specimens (mostly isotypes) in the bound exsiccati sets at ILL (see separate list of the titles on p. 357. Some of them are mentioned, however, if a separate isotype was also pulled from the general collections and placed in the type file, and consequently the name was included in the catalogue. A deliberate exception was made for an isotype of *Ravenelia opaca* P. Dietel, a species described from Illinois that is known only from the type collection.

Annotated List of Types

A

Acerbia donacina H. Rehm. Leaflets of Philippine Botany 6:2264. 1914.
Assumed isotype: Ad *Donacem cannaeformem* [= *Donax cannaeformis* (G. Forster) K. Schum.], Philippines, [Luzon, Laguna Prov., Mount Maquiling, near] Los Baños, JAN 1914, leg. C.F. Baker No. 2470b [ex Fungi Malayana No. 102], ILL 9347.

Acerbia maydis H. Rehm. Leaflets of Philippine Botany 8:2953–2954. 1916.
Assumed isotype: Ad emortuos *Zeae mays* [= *Zea mays* L.], Philippines, [Luzon, Laguna Prov., Mount Maquiling, near] Los Baños, NOV 1913, leg. M.B. Raimundo, comm. C.F. Baker No. 1993 [ex Fungi Malayana No. 100], ILL 9348.

Achaetobotrys compositarum A.C. Batista & R. Ciferri. Saccardoa 2:50–51. 1963. **Isotypes**: On leaves of undetermined Composita [= Asteraceae], Porto Rico [Puerto Rico], Luquillo Forest, 2 DEC 1913, leg. F.L. Stevens No. 5446, ILL 33528, BPI. **Holotype**: URM [I.M.U.R.] 13064. **Paratype**: On leaves of *Hibiscus* sp., Puerto Rico, Luquillo Forest, 2 DEC 1913, leg. F.L. Stevens No. 5434, ILL 33527 [also a paratype of *Antennariella californica* A.C. Batista & R. Ciferri].

Achorella attaleae F.L. Stevens. Illinois Biological Monographs 11:181. 1927. **Holotype**: On *Attalea cohune* Martius, Panama, Gamboa, 16 AUG 1927, leg. F.L. Stevens No. 1079, ILL 8431.

Achorella costaricensis F.L. Stevens. Illinois Biological Monographs 11:182, pl. VI, figs. 44, 45. 1927. **Holotype**: On *Mikania* sp., Costa Rica, Cartago, 24 JUL 1923, leg. C.H. Lancaster, [as F.L. Stevens No. 646], ILL 8432. **Paratype**: Cartago, 23 JUN 1923, leg. F.L. Stevens No. 57, ILL 8433.

Achorella guianensis F.L. Stevens. Illinois Biological Monographs 8:181–182, pl. III, figs. 24–26; pl. IV, fig. 27; pl. XII, figs. 91, 92. 1923. **Holotype**: On *Mikania* sp.(?), British Guiana [Guyana], Coverden, 8 AUG 1922, leg. F.L. Stevens No. 763, ILL 8438. **Paratypes**: Rockstone, 16 JUL 1922, leg. F.L. Stevens No. 438, ILL 8436 and 8440; Wismar, 14 JUL 1922, leg. F.L. Stevens No. 294, ILL 8439; Kartabo, 22 JUL 1922, leg. F.L. Stevens No. 563, ILL 8441.

Achorodothis poasensis H. Sydow. Annales Mycologici 24:380–383. 1926. **Isotype**: Hab. in foliis vivis *Phoebes mollicellae* [= *Phoebe mollicella* S.F. Blake] Costa Rica, in monte Poas pr. Grecia, 15 JAN 1925, leg. H. Sydow, Fungi in Itinere Costaricensi Collecti No. 32, ILL 8442.

Acladium biophilum R. Ciferri. Sydowia Annales Mycologici 10:164–165. (1956) 1957. **Isotype**: Hab. in pagina inferiore foliorum *Wissadulae* sp. [= *Wissadula* sp., Dominican Republic], Santiago Prov., Valle del Cibao, Hato del Yaque in agris, FEB 1931, leg. R. Ciferri, Mycoflora Domingensis Exsiccati No. 373, ILL 33530.

Acremonium meliolae F.L. Stevens. Botanical Gazette 65:234–235. 1918 [cited as *meliola*, corrected in accordance with Articles 32.6 & 61 of I.C.B.N.]. **Holotype**: On *Meliola paulliniae* F.L. Stevens on *Paullinia pinnata* L., Porto Rico [Puerto Rico], Vega Baja, [22 FEB 1913], leg. F.L. Stevens No. 376, ILL 14494. [also a paratype of *Meliola paulliniae* F.L. Stevens].

Acrothecium falcatum L.R. Tehon. Botanical Gazette 67:509–510, pl. XVIII, fig. 1. 1919 [as *flacatum* in the protologue but correctly spelled in the figure caption and on the packet]. **Holotype**: On *Setaria* sp., Porto Rico [Puerto Rico], leg. F.L. Stevens No. 9181, ILL 14509.

Actinodothidopsis coprosmae F.L. Stevens. Bernice P. Bishop Museum Bulletin 19:19, text fig. 4a–e, pl. II(B). 1925. **Holotype**: On *Coprosma* sp., Hawaii, Kauai, upper pipe trail, Waimea Canyon, 15 JUN [1921], leg. F.L. Stevens No. 457, ILL 8443.

Actinodothis perrottetiae F.L. Stevens. Bernice P. Bishop Museum Bulletin 19:51–52, text fig. 11a, pl. III(C,E). 1925. **Holotype**: On *Perrottetia sandwicensis* A. Gray, Hawaii, Oahu, Olympus, 24 JUN [1921], leg. F.L. Stevens No. 717, ILL 3525. **Isotypes**: BISH 499877 and 499878. **Paratypes**: Kauai, Kalalau trail, 16 JUN [1921], leg. F.L. Stevens No. 474, ILL 3529 and 3562 [F.L. Stevens No. 474 is also cited as paratype of *Amazonia perrottetiae* F.L. Stevens and, as No. 474 p.p., ILL 3562, it is the holotype of *Appendiculella kalalauensis* C.G. Hansford]; Maui, Pogue's ditch trail, 6 SEP [1921], leg. F.L. Stevens No. 1159, ILL 3528 and 3557; [Island of] Hawaii, Waimea, 30 JUL [1921], leg. F.L. Stevens No. 1055 p.p., ILL 3524 [also the holotype of *Asteridiella waimeana* C.G. Hansford and a paratype of *Appendiculella kalalauensis* C.G. Hansford]; leg. H.L. Lyon No. 68 [as No. 168 on the packet], ILL 3527. ≡ *Amazonia stevensii* C.G. Hansford. Sydowia Annales Mycologici 9:27. 1955, nom. nov., emend., Beihefte zur Sydowia Annales Mycologici 1:91. 1957.

Actinodothis suttoniae F.L. Stevens. Bernice P. Bishop Museum Bulletin 19:51, text fig. 11c, pl. III(A,B,D). 1925. **Holotype**: On *Suttonia lessertiana* (A. DC.) Mez [= *Myrsine lessertiana* A. DC.], Hawaii, Oahu, Kuliouou, 29 MAY [1921], leg. E.L. Caum, as F.L. Stevens No. 143 p.p., ILL 3540. **Isotypes**: ILL 7114, BISH 499875 and 499876 [F.L. Stevens No. 143 p.p., as ILL 7114, is also the holotype of *Asterina suttoniae* F.L. Stevens & R.W. Ryan in F.L. Stevens]. **Paratypes**: [Island of] Hawaii, Hamakua, upper ditch trail, 31 JUL [1921], leg. F.L. Stevens No. 1088, ILL 3526; Kauai, 1909, leg. C.N. Forbes-F.L. Stevens No. 267, ILL 3538; Maui, Iao Valley, 7 SEP [1921], leg. F.L. Stevens No. 1152, ILL 3535; Oahu, Puu Huluhulu, 17 JUL [1921], leg. Mrs. C.S. Judd, as F.L. Stevens No. 882, ILL 3536; [host misidentified, probably = *Myrsine sandwicensis* A. DC., Island of] Hawaii, Kealakekua, 22 JUL [1921], leg. F.L. Stevens No. 980, ILL 3537; on *Suttonia kauaiensis* (Hillebr.) Degener & Hosaka [= *Myrsine kauaiensis* Hillebr.], [Kauai], Kalalau trail, 16 JUN [1921], leg. F.L. Stevens No. 471 [erroneously as No. 47 in the protologue], ILL 3539. ≡ *Amazonia suttoniae* (F.L. Stevens) C.G. Hansford. Sydowia Annales Mycologici 9:28. 1955.

Aecidium abscedens J.C. Arthur. Mycologia 7:315. 1915. **Isotype**: On *Randia aculeata* L., Porto Rico [Puerto Rico], Aguada, 22 NOV [1913], leg. F.L. Stevens No. 5089, ILL 27120. **Holotype**: PUR 43281. **Paratypes**: Mayagüez, 2 MAY [1913], leg. F.L. Stevens No. 1125, ILL 27122; Cataño, 3 NOV [1913], leg. F.L. Stevens No. 4534, ILL 27121.

Aecidium batesianum E. Bartholomew in J.B. Ellis & B.M. Everhart. Fungi Columbiani, Century 20, No. 1901. Anno 1904. **Isotype**: On leaves and petioles of *Delphinium albescens* Rydb. [= *D. carolinianum* Walter ssp. *virescens* (Nutt.) R.E. Brooks], Nebraska, Red Cloud, 2 MAY 1903, leg. J.M. Bates, ILL 27237.

Aecidium benguetensis G.B. Cummings in F.L. Stevens. Natural and Applied Science Bulletin. University of the Philippines 2:446–447. 1932. **Isotype**: On *Smilax* sp., Philippines, [Luzon], Benguet [Prov.], Baguio, 1 JAN 1931, leg. F.L. Stevens No. 1367, ILL 27107. **Holotype**: PUR.

Aecidium crotonopsidis T.J. Burrill. Botanical Gazette 9:190. DEC 1884;

Bulletin of the Illinois State Laboratory of Natural History 2(3):237–238. AUG 1885. **Lectotype** [designated herein]: On *Crotonopsis linearis* Michaux [= *Croton michauxii* G.L. Webster], Illinois, Johnson County, [Sanburn], 16 MAY 1882, leg. A.B. Seymour [No. 4701], ILL 27716. **Isolectotypes**: PUR 11676, and J.B. Ellis & B.M. Everhart, North American Fungi No. 1824 (in bound exsiccati set at ILL). **Residual syntypes**: 13 MAY 1882, leg. A.B. Seymour [No. 4662], ILL 27714; Tunnel Hill, 12 MAY 1882, leg. A.B. Seymour [No. 4648], ILL 27715. [Collection numbers are not cited in the protologue but No. 4701 is marked "type" on the packet.].

Aecidium dicentrae T.J. Burrill. Botanical Gazette 9:189. DEC 1884; Bulletin of the Illinois State Laboratory of Natural History 2(3):223. AUG 1885, with authorship changed to W. Trelease ex T.J. Burrill. **Syntypes**: On *Dicentra cucullaria* (L.) Bernh., Illinois, [Makanda], 27 APR 1882, leg. A.B. Seymour [No. 4345], ILL 27467; 28 APR 1882, leg. A.B. Seymour [No. 4360], ILL 27466; [Twin Grove], 23 May 1882, leg. A.B. Seymour [No. 4753], ILL 27468; [Bloomington], 22 MAY 1882, leg. A.B. Seymour [No. 4731], ILL 27469; [Fountain Bluff], 21 APR 1882, leg. A.B. Seymour [No. 4211], ILL 27473; 25 APR 1882, leg. A.B. Seymour [No. 4289], ILL 27470; [Grand Tower], 20 APR 1882, leg. A.B. Seymour [No. 4195], ILL 27472; [Pine Hills], 24 APR 1882, leg. A.B. Seymour [No. 4252], ILL 27471, PUR 5629 [the latter specimen is an isosyntype].

Aecidium diodiae T.J. Burrill. Botanical Gazette 9:189–190. DEC 1884; Bulletin of the Illinois State Laboratory of Natural History 2(3):228. AUG 1885. **Lectotype** [designated herein]: On *Diodia teres* Walter, Illinois, Johnson County, [Sanburn], 13 MAY 1882, leg. A.B. Seymour [No. 4661], ILL 27474 [marked "type" on the packet]. **Isolectotype**: PUR 17153. **Residual syntype**: 16 MAY 1882, leg. A.B. Seymour [No. 4700], ILL 27475.

Aecidium eurotiae J.B. Ellis & B.M. Everhart. Journal of Mycology 6:119. 1891. **Isotype**: On *Eurotia lanata* (Pursh) Moq. [= *Krascheninnikovia lanata* (Pursh) Gueldenstaedt], Montana, Helena, [3] JUN 1889, leg. F.D. Kelsey, ex herbarium A.B. Seymour No. 20660, comm. F.W. Anderson No. 514, ILL 27544.

Aecidium favaceum J.C. Arthur. Mycologia 7:254. 1915. **Paratypes**: On *Phyllanthus nobilis* (L.f.) Muell. Arg. [= *Margaritaria nobilis* L.f.], Porto Rico [Puerto Rico], San Germán, 16 JAN [1913], leg. F.L. Stevens No. 249, ILL 27378; 12 DEC [1913], leg. F.L. Stevens No. 5832, ILL 27379 and 27380. **Holotype**: PUR 43095.

Aecidium myosotidis T.J. Burrill. Botanical Gazette 9:190. DEC 1884; Bulletin of the Illinois State Laboratory of Natural History 2(3):234. AUG

1885. **Lectotype** [designated herein]: On *Myosotis verna* Nutt., Illinois, [Union County], Cobden, 12 APR 1882, leg. A.B. Seymour [No. 4029], ILL 27258. **Isolectotype:** ILL 27264 [No. 4029 is marked "type" on both packets]. **Residual syntypes:** 12 APR 1882, leg. A.B. Seymour [No. 4026], ILL 27259; 17 APR 1882, leg. A.B. Seymour [No. 4132], ILL 27263; 26 APR 1882, leg. A.B. Seymour [No. 4306], ILL 27262.

Aecidium ocfemianum F.L. Stevens. The Philippine Agriculturist 20:87, fig. 1. 1931. **Holotype:** On *Pisonia alba* Span., Philippines, [Luzon], Tayabas, Sariaya, 9 AUG 1930, leg. G.O. Ocfemia, [as F.L. Stevens No. 198], ILL 27269.

Aecidium onobrychidis T.J. Burrill. Botanical Gazette 9:189. DEC 1884; Bulletin of the Illinois State Laboratory of Natural History 2(3):225. AUG 1885. **Holotype:** On *Psoralea onobrychis* Nutt. [= *Orbexilum onobrychis* (Nutt.) Rydb.], Illinois, La Salle County, [Seneca], 20 JUN 1882, leg. A.B. Seymour [No. 5249], ILL 27293 [the packet marked "type"]. **Isotypes:** PUR 43041, and J.B. Ellis & B.N. Everhart, North American Fungi No. 1826 [in bound exsiccati set at ILL]. ≡ *Puccinia andropogonis* L.D. v. Schweinitz var. *onobrychidis* (T.J. Burrill) J.C. Arthur. Manual of the Rusts in United States and Canada, p. 122. 1934.

Aecidium parile H. & P. Sydow. Annales Mycologici 12:197. 1914. **Assumed isotype:** Hab. in foliis *Loranthi* [= *Loranthus* sp.] ad *Goniothalamum elmeri* [= *Goniothalamus elmeri* Merrill], Philippines, [Luzon], Laguna Prov., Mount Maquiling, pr. Los Baños, JAN. 1914, leg. C.F. Baker No. 2802 [ex Fungi Malayana No. 104, the label in what appears to be H. Sydow's handwriting], ILL 27880.

Aecidium phaceliae C.H. Peck. Bulletin of the Torrey Botanical Club 11:50. 1884. **Isotype:** On living leaves of *Phacelia* sp., Utah, [Wasatch Mountains, Thistle, AUG 1833], leg. M.E. Jones s.n., ILL 27905.

Aecidium pseudo-balsameum P. Dietel & E.W.D. Holway ex E.W.D. Holway. Erythea 7:98. 1899. **Isotype:** On *Abies grandis* (Douglas ex D. Don) Lindley, [California], Eureka, 4 JUN 1896, leg. W.C. Blasdale, ILL 28251.

Aecidium sparsum J.J. Davis. Transactions of the Wisconsin Academy of Sciences, Arts and Letters 24:292–293. 1929. **Isosyntype:** On *Galium tinctorium* (L.) Scop., Wisconsin, bottom lands of the Wisconsin River opposite Sauk City, 2 JUL 1925, leg. Weber & J.J. Davis, ex Fungi Wisconsinenses Exsiccati No. 156, ILL 27875.

Aecidium trillii T.J. Burrill. Botanical Gazette 9:190. DEC 1884; Bulletin of the Illinois State Laboratory of Natural History 2(3):238. AUG 1885. **Holotype:** On *Trillium recurvatum* Beck, Illinois, Union County, Pine Hills, 24 APR 1882, leg. A.B. Seymour [No. 4251], ILL 27738 [the packet marked "type"]. **Isotype:** PUR 42973.

Aegerita webberi H.S. Fawcett. Mycologia 2:167, plates XXVIII, XXIX. 1910; Science 31:912–913. 1910. **Possible isosyntypes:** On larvae of *Aleyrodes citri* [white fly] ..., on the under surface of *Citrus* leaves, [Florida, Deland, NOV 1908, leg. H.S. Fawcett], ILL 16442; [Florida, East Palatka, 6 JUN 1909, leg. H.S. Fawcett], ILL 16443.

Ainsworthia smilacina A.C. Batista & A.F. Vital. Beihefte zur Sydowia Annales Mycologici 3:6–7, figs. 1, 2. 1962. **Holotype:** On leaves of *Smilax* sp., Hawaii, Oahu, Olympus, 24 JUN 1921, leg. F.L. Stevens [No. 981], ILL 6667 [also a syntype of *Phragmocapnias smilacina* J.M. Mendoza in F.L. Stevens and the holotype of *Trichopeltum hawaiiense* A.C. Batista & C.C.A. Costa in A.C. Batista, C.C.A. Costa & R. Ciferri (the herbarium for the latter erroneously cited as NY)].

Aleurocorticium maculatum H.S. Jackson & P.A. Lemke in P.A. Lemke. Canadian Journal of Botany 42:742–743, fig. 3. 1964. **Paratype:** On *Prunus* sp., [Canada, woods east of Long Sand Pond, east of Mainland], Ontario, (Nipissing District), Lake Timagami, 16 AUG 1939, leg. H.S. Jackson s.n., ILL 32721, TRTC 16687, DAOM. **Type specimens:** TRTC 16398, DAOM, FH, NY, NO, S.

Aleurocorticium pachysterigmatum H.S. Jackson & P.A. Lemke in P.A. Lemke. Canadian Journal of Botany 42:750–751, fig. 7. 1964. **Paratypes:** On *Thuja occidentalis* L., Ontario, York County, [swamp north of] Mt. Albert, 28 OCT 1938, leg. H.S. Jackson s.n., ILL 32714, ex TRTC 13557, NO.

Aleurodiscus abietis H.S. Jackson & P.A. Lemke in P.A. Lemke. Canadian Journal of Botany 42:225–227, fig. 1. 1964. **Paratype:** On *Abies balsamea* (L.) P. Miller, Ontario, Nipissing District, Lake Timagami, 23 AUG 1937, leg. H.S. Jackson No. 868, ILL 32716.

Aleurodiscus canadensis A.J. Skolko. Canadian Journal of Botany 22:258–260, pl. I. 1944. **Paratype:** On *Picea canadensis* (P. Miller) B.S.P. [= *Picea glauca* (Moench) Voss], Quebec, Burnet, [28] JUN 1938, leg. H.S. Jackson, s.n., ILL 32719. **Holotype:** TRTC 18147.

Aleurodiscus fruticetorum W.B. Cooke. Mycologia 35:281–282. 1943. **Isotype:** On *Arctostaphylos patula* E. Greene, California, [Siskiyou County], at 6000 ft. on Mt. Shasta, near Wagon Camp, 18 AUG 1941, leg. W.B. Cooke No. 15731, ILL 32720. **Holotype:** Probably MU.

Aleurodiscus moquiniarum A.P. Viégas. Revista de Agricultura Piracicaba 14:311–314. 1939. **Isotypes:** In corticibus *Moquiniae polymorphae* [= *Moquinia polymorpha* DC.], Brasilia [Brazil], São Paulo, pr. Santa Barbara, OCT 1938, leg. A.P. Viégas No. 2817, ILL 32723, TRTC, NO, IAC 2817.

Aleurodiscus pini H.S. Jackson. Canadian Journal of Botany 28:74–76, figs. 2, 8, 12. 1950. **Paratypes:** On *Pinus strobus* L., Massachusetts, Shirley, 11 NOV 1935, leg. D.H. Linder, comm. D.P. Rogers No. 3179, ILL 32715; New Hampshire, Jaffrey, 23 JUL 1939, leg. G.D. Darker No. 6822, ILL 33194 and 33194a. **Holotype**: TRTC 11703.

Alternaria dianthi F.L. Stevens & J.G. Hall. Botanical Gazette 47:413. figs. 3–8 [p. 412]. 1909. **Holotype**: On living leaves and stems of *Dianthus caryophyllus* L., North Carolina, Raleigh, 20 JAN 1909, leg. B.B. Higgins, as F.L. Stevens No. 1793, ILL 14519.

Alternaria sonchi J.J. Davis in J.A. Elliott. Botanical Gazette 62:414–416, fig. 1. 1916. **Isotype**: On *Sonchus asper* (L.) J. Hill, Wisconsin, Madison, [22 OCT 1914], leg. J.J. Davis, ILL 14552. **Holotype**: WIS.

Alternaria sonchi F.L. Stevens. Bernice P. Bishop Museum Bulletin 19:153–154. 1925 (nom. illegit.), non J.J. Davis in J.A. Elliott, 1916. **Syntypes** [BISH numbers cited are isosyntypes]: On *Sonchus oleraceus* L., Hawaii, Oahu, Honolulu, 19 MAY [as 9 MAY 1921 on the packet], leg. F.L. Stevens No. 6, ILL 14557, BISH 499869 and 499873; 19 MAY [as 20 MAY 1921 on the packet], leg. F.L. Stevens No. 12, ILL 14556, BISH 145650 and 499872; Tantalus, 25 MAY [1921], leg. F.L. Stevens No. 104, ILL 14553, BISH 145652 and 499874; Wahiawa, 3 JUN [1921], leg. F.L. Stevens No. 221, ILL 14555, BISH 145653 and 499871.

Amazonia acalyphae (H. Rehm) F. Theissen, 1916. **See basionym**: *Meliola acalyphae* H. Rehm, 1913.

Amazonia alyxiae C.G. Hansford. Beihefte zur Sydowia Annales Mycologici 1:89. 1957. **Holotype**: Hab. in foliis *Alyxiae olivaeformis* [= *Alyxia oliviformis* Gaudich.], Hawaii, [Oahu, Wahiawa, 3 JUN 1921], leg. F.L. Stevens No. 239, ILL 3585.

Amazonia anacardiacearum F.L. Stevens. Annales Mycologici 25:413. 1927. **Holotype**: On Anacardiaceae (?*Tapirira* sp.), British Guiana [Guyana], Wismar, 14 JUL 1922, leg. F.L. Stevens No. 277, ILL 3542. ≡ *Meliola anacardiacearum* (F.L. Stevens) C.G. Hansford. Sydowia Annales Mycologici 9:39–40. 1955.

Amazonia clermontiae C.G. Hansford. Beihefte zur Sydowia Annales Mycologici 1:89. 1957. **Holotype**: Hab. in foliis *Clermontiae multiflorae* [= *Clermontia multiflora* Hillebr.], Hawaii, [Oahu, Palolo Valley & Mt. Olympus, 10 JUN 1921], leg. F.L. Stevens No. 329, ILL 3612. **Paratype**: Leg. F.L. Stevens No. 330, ILL 3610. [The author also published the name in Sydowia Annales Mycologici 10:41–42. (1956) 1957, with a different host and type (A.A. Heller No. 2394, at K). Hansford (1961) accepted the Beihefte publication but with the later (?) published Heller type.].

Amazonia clusiae (F.L. Stevens) F.L. Stevens, 1927. **See basionym**: *Meliola clusiae* F.L. Stevens, 1916.

Amazonia ohiana F.L. Stevens. Bernice P. Bishop Museum Bulletin 19:50, text fig. 11b. 1925 [as *ohianus*, corrected by C.G. Hansford (1961:138) in accordance with currect I.C.B.N.]. **Holotype**: On *Metrosideros polymorpha* Gaudich., (Ohia lehua), Hawaii, [Island of] Hawaii, Kilauea, 14 JUL [1921], leg. F.L. Stevens No. 842, ILL 3553. **Isotypes**: BISH 499867 and 499868. **Paratypes**: [Island of] Hawaii, upper ditch trail, Hamakua, 31 JUL [1921], leg. F.L. Stevens No. 1065, ILL 3554; between Hilo and Kilauea, 10 JUL 1921, leg. F.L. Stevens No. 780, ILL 3556. ≡ *Asteridiella ohiana* (F.L. Stevens) C.G. Hansford. Beihefte zur Sydowia Annales Mycologici 2:138. 1961.

Amazonia peregrina (H. & P. Sydow) H. & P. Sydow, 1917. **See basionym**: *Meliola peregrina* H. & P. Sydow, 1913.

Amazonia perrottetiae F.L. Stevens. Bernice P. Bishop Museum Bulletin 19:47–48, text fig. 10b, pl. II(L). 1925. **Holotype**: On *Perrottetia sandwicensis* A. Gray, Hawaii, Oahu, Olympus, 24 JUN [1921], leg. F.L. Stevens No. 717a, ILL 3558. **Isotypes**: BISH 499865 and 499866. **Paratypes**: Leg. F.L. Stevens No. 702, ILL 3560; Kauai, Kalalau trail, 16 JUN [1921], leg. F.L. Stevens No. 474, ILL 3529 and 3562 [F.L. Stevens No. 474 p.p., as ILL 3562, is also the holotype of *Appendiculella kalalauensis* C.G. Hansford, and both specimens are also paratypes of *Actinodothis perrottetiae* F.L. Stevens]. ≡ *Appendiculella perrottetiae* (F.L. Stevens) C.G. Hansford. Sydowia Annales Mycologici 9:31. 1955. ≡ *Asteridiella perrottetiae* (F.L. Stevens) C.G. Hansford. Beihefte zur Sydowia Annales Mycologici 1:91. 1957.

Amazonia psychotriae (P.C. Hennings) F. Theissen var. *labordiae* C.G. Hansford. Beihefte zur Sydowia Annales Mycologici 1:89. 1957. **Holotype**: In foliis *Labordiae* spec. indet. [= *Labordia* sp. (cf. Kartesz, 1994, but = *Geniostoma* sp. fide Brummitt, 1992)], Hawaii, [Oahu, Honolulu, Tantalus, 22 JUN 1921], leg. F.L. Stevens No. 611, ILL 3587.

Amazonia psychotriae (P.C. Hennings) F. Theissen var. *straussiae* C.G. Hansford. Beihefte zur Sydowia Annales Mycologici 1:90. 1957. **Holotype**: Hab. in foliis *Straussiae marinianae* [*Straussia mariniana* (Cham. & Schlecht.) A. Gray = *Psychotria mariniana* (Cham. & Schlecht.) Fosb.], Hawaii, [Oahu, Wahiawa, 3 JUN 1921], leg. F.L. Stevens No. 244, ILL 3591. **Paratypes**: [on] *Straussia hawaiiensis* A. Gray [= *Psychotria hawaiiensis* (A. Gray) Fosb., 3 JUN 1921], leg. F.L. Stevens No. 205, ILL 3589; [on] *Straussia* sp. [= *Psychotria* sp., Kauai, Kalalau trail, 16 JUN 1921], leg. F.L. Stevens No. 530, ILL 3590; [pipe trail, Waimea Canyon, 15 JUN 1921], leg. F.L. Stevens No. 442, ILL 3588.

Amazonia scaevolae C.G. Hansford. Beihefte zur Sydowia Annales Mycologici 1:90. 1957. **Holotype**: Hab. in foliis *Scaevolae* sp. [= *Scaevola* sp.], Hawaii, [Oahu, Honolulu, Tantalus, 22 JUN 1921], leg. F.L.

Stevens No. 634, ILL 3586. **Paratype:** [On *Scaevola glabra* Hooker & Arnott, Oahu, Tantalus, 22 JUN 1921], leg. F.L. Stevens No. 640, ILL 3598.

Amazonia stevensii C.G. Hansford. Sydowia Annales Mycologici 9:27. 1955, nom. nov., emend., Beihefte zur Sydowia Annales Mycologici 1:91. 1957. **Based on:** *Actinodothis perrottetiae* F.L. Stevens, 1925, non *Amazonia perrottetiae* F.L. Stevens, 1925. **Holotype:** On *Perrottetia sandwicensis* A. Gray, Hawaii, Oahu, Olympus, 24 JUN [1921], leg. F.L. Stevens No. 717, ILL 3525. **Isotypes:** BISH 499877 and 499878.

Amazonia suttoniae (F.L. Stevens) C.G. Hansford, 1955. **See basionym:** *Actinodothis suttoniae* F.L. Stevens, 1925.

Amazonia ugandensis C.G. Hansford, Journal of the Linnean Society of London (Botany) 51:283. 1937. **Isotype:** Hab in caulis *Euphorbiae* [= *Euphorbia* sp.], Uganda, Kampala, [MAR 1930], leg. C.G. Hansford No. 1104, ILL 6405. **Holotype:** K.

Amazonia wikstroemiae C.G. Hansford. Beihefte zur Sydowia Annales Mycologici 1:93. 1957 [cited as *wilkstroemiae*, subsequently corrected by the author (Hansford, 1961:94) in accordance with the current I.C.B.N.]. **Holotype:** Hab. in foliis *Wikstroemiae foetidae* [= *Wikstroemia foetida* (L.f.) A. Gray, the packet labeled as var. *oahuensis* = *W. oahuensis* (A. Gray) Rock], Hawaii, [Oahu, Tantalus, 22 JUN 1921], leg. F.L. Stevens No. 635, ILL 3578. **Paratypes:** In foliis *Wikstroemiae phillyreifoliae* [misspelled *phillyriaefoliae* = *Wikstroemia phillyreifolia* A. Gray, Oahu, Tantalus, 22 JUN 1921], leg. F.L. Stevens No. 629, ILL 3579; [on] *Wikstroemia elongata* A. Gray [= *W. oahuensis* (A. Gray) Rock, Tantalus, 22 JUN 1921], leg. F.L. Stevens No. 610, ILL 3599; [on] *Wikstroemia* sp., [Molokai, Halawa, AUG 1912], leg. C.N. Forbes-F.L. Stevens No. 479, ILL 3580; [Oahu, Castle trail, MAR 1912], C.N. Forbes-F.L. Stevens No. 2148 [as 2148-O on the packet], ILL 3582.

Amerodothis guianensis F.L. Stevens. Illinois Biological Monographs 8:180–181, pl. II, figs. 17, 18; pl. III, fig. 19; pl. XII, fig. 90. 1923. **Holotype:** On unknown legume [= Fabaceae sp. indet.], British Guiana [Guyana], Rockstone, 16 JUL 1922, leg. F.L. Stevens No. 424, ILL 8445.

Amerosporium dolichi J.B. Ellis & B.M. Everhart. North American Fungi, Century 26, No. 2574. Anno 1891, nom. nud. **Isotype:** On *Dolichos arvensis* Vell. [a misapplied host name?; the specimen seems to belong in *Vigna unguiculata* (L.) Walp.], Mississippi, Starkville, leg. S.M. Tracy, ILL 13202.

Amphisphaeria coronata H. Rehm. Leaflets of Philippine Botany 6:2200. 1914. **Assumed isosyntype:** Ad emortuam *Gigantochloam scribnerianam* [*Gigantochloa scribneriana* Merrill = *G. levis* (Blanco) Merrill], Philippines, Luzon, Laguna Prov., [Mount Maquiling, near] Los

Baños, 1 OCT 1913, leg. M.B. Raimundo, comm. C.F. Baker [ex Fungi Malayana No. 3], ILL 9613.

Anariste poliothea H. Sydow. Annales Mycologici 25:76–78. 1927. **Isotype:** Hab. in foliis vivis vel languidis *Phoebes neurophyllae* [= *Phoebe neurophylla* Mez & Pittier], Costa Rica, Cerro de San Isidro pr. San Ramón, 9 FEB 1925, leg. H. Sydow, Fungi in Itinere Costaricensi Collecti No. 169c, ILL 6890.

Anisochora tabebuiae F.L. Stevens. Illinois Biological Monographs 8:182–183, pl. IV, figs. 28–31; pl. XIII, fig. 94. 1923. **Holotype:** On *Tabebuia* sp., Trinidad, St. Augustine, 13 AUG 1922, leg. F.L. Stevens No. 847, ILL 8446.

Anomothallus erraticus F.L. Stevens. Bernice P. Bishop Museum Bulletin 19:91–93, text fig. 22, pl. IX(J), X(A–C). 1925. **Syntypes** [additional ILL numbers and BISH numbers cited are isosyntypes]: On *Rubus hawaiensis* A. Gray, Hawaii, Maui, Pogue's ditch trail, 6 SEP [1921], leg. F.L. Stevens No. 1155, ILL 6692 and 3748; Olinda pipeline, 5 SEP [1921], leg. F.L. Stevens No. 1138, ILL 6693 and 3746, BISH 145888 and 449864.

Antennariella californica A.C. Batista & R. Ciferri. Quaderno Laboratorio Crittogamico, Istituto Botanico della Università 31:25–26. 1963. **Paratypes:** On *Hibiscus* sp., Porto Rico [Puerto Rico], Luquillo Forest, [2 DEC 1913], leg. F.L. Stevens No. 5434, ILL 33527, URM [I.M.U.R.] 13485 [also paratypes of *Achaetobotrys compositarum* A.C. Batista & R. Ciferri].

Antennellina hawaiiensis J.M. Mendoza in F.L. Stevens. Bernice P. Bishop Museum Bulletin 19:55–56, pl. IV(1–4). 1925. **Holotype:** On *Mangifera indica* L., Hawaii, Oahu, Honolulu, 14 JUN [1921], leg. F.L. Stevens No. 266, ILL 6627. **Isotypes:** BISH 499859 and 499863.

Antennellopsis mangiferae J.M. Mendoza in F.L. Stevens. Annales Mycologici 28:365. 1930. **Holotype:** On *Mangifera indica* L., British Guiana [Guyana], Vreed en Hoop, 10 AUG [1922], leg. F.L. Stevens No. 710, ILL 6629.

Anthostomella arecae H. Rehm. Leaflets of Philippine Botany 8:2938–2939. 1916. **Assumed isotype:** Ad emortuos stipites *Arecae catechu* [= *Areca catechu* L.], Philippines, [Luzon], Laguna Prov., Los Baños, APR 1914, leg. C.F. Baker No. 3068 [ex Fungi Malayana No. 106], ILL 10297.

Anthostomella coryphae H. Rehm. Leaflets of Philippine Botany 8:2940. 1916. **Assumed isotype:** Ad petiolos emortuos *Coryphae elatae* [= *Corypha elata* Roxb.], Philippines, [Luzon, Laguna Prov., Mount Maquiling, near] Los Baños, JAN 1914, leg. C.F. Baker No. 2674 [ex Fungi Malayana No. 108], ILL 10309.

Anthostomella ratibidae W.G. Solheim. Mycologia 41:623–624. 1949. **Isotype:** On leaves and stems of *Ratibida tagetes* (James) Barnhart, New

Mexico, Bernalillo County, Sandia Mountains, east of Albuquerque, Ciénaga Canyon Recreation Area, alt. 7500 ft., 24 OCT 1948, leg. W.G. & Ragnhild Solheim No. 2323 [Mycoflora Saximontanensis Exsiccata No. 428], ILL 20942.

Anthostomella sacchariferae H. Rehm. Leaflets of Philippine Botany 6:2260. 1914. **Assumed isotype**: Ad petiolos emortuos *Arengae sacchariferae* [*Arenga saccharifera* Labill. = *A. pinnata* (Wurmb.) Merrill], Philippines, [Luzon, Laguna Prov., Mount Maquiling], near Los Baños, OCT 1913, leg. S.A. Reyes, comm. C.F. Baker No. 1797 [ex Fungi Malayana No. 6], ILL 10324.

Anthostomella uberiformis H. Rehm. Leaflets of Philippine Botany 8:2937–2938. 1916. **Assumed isotype**: On rotten trunk, Philippines, [Luzon, Laguna Prov.], Mount Maquiling, near Los Baños, JUN 1914, leg. C.F. Baker No. 3411 [ex Fungi Malayana No. 111], ILL 10326.

Aphanostigme solani H. Sydow. Annales Mycologici 24:368–369. 1926. **Isotype**: Hab. in foliis *Solani* sp. [= *Solanum* sp.], Costa Rica, San Pedro de San Ramón, 2 FEB 1925, leg. H. Sydow, Fungi in Itinere Costaricensi Collecti No. 56a, ILL 8454.

Apiospora carbonacea H. Rehm. Leaflets of Philippine Botany 8:2945–2946. 1916. **Assumed isotype**: Ad *Schizostachyum* emortuum in e acumine, Philippines, [Luzon, Laguna Prov.], Mount Maquiling [near Los Baños], JUN 1914, leg. C.F. Baker No. 3427a [ex Fungi Malayana No. 114], ILL 8450.

Apiospora curvispora (C.L. Spegazzini) H. Rehm var. *rottboelliae* H. Rehm. Leaflets of Philippine Botany 6:2199. 1914. **Assumed isotype**: Ad culmos emortuos *Rottboelliae* sp. [*Rottboellia* sp.; as *R. exaltata* (L.) L.f. on the packet = *R. cochinchinensis* (Lour.) Clayton], Philippines, Luzon, Laguna Prov., [Mount Maquiling near] Los Baños, JUL 1913, leg. S.A. Reyes, comm. C.F. Baker [ex Fungi Malayana No. 7], ILL 8451.

Aposphaeria salicum P.A. Saccardo in H. & P. Sydow. Annales Mycologici 1:537–538. 1903. **Isotype**: Auf entrindeten Zweigen von *Salix viminalis* L., [Germany], Sachsen, am Elbufer bei Schmilka, 15 AUG 1903, leg. H. & P. Sydow, Mycotheca Germanica No. 87, ILL 10922.

Appendiculella adelphica H. Sydow. Annales Mycologici 24:313. 1926. **Isotype**: Hab. in foliis *Solani* spec. indet. [= *Solanum* sp., the host later identified as *S. erythrotrichum* Fernald (cf. Hansford, 1961:639)], Costa Rica, Los Angeles de San Ramón, 30 JAN 1925, leg. H. Sydow, Fungi in Itinere Costaricensi Collecti No. 55a, ILL 3732. ≡ *Irene adelphica* (H. Sydow) F.L. Stevens. Annales Mycologici 25:428. 1927. ≡ *Meliola adelphica* (H. Sydow) F. Petrak in H. Sydow & F. Petrak. Annales Mycologici 27:1. 1929. ≡ *Asteridiella adelphica* (H. Sydow) C.G. Hansford. Beihefte zur Sydowia Annales Mycologici 2:639. 1961.

Appendiculella alchorneae (F.L. Stevens & L.R. Tehon) C.G. Hansford, 1961. **See basionym:** *Irene alchorneae* F.L. Stevens & L.R. Tehon, 1926.

Appendiculella arecibensis (F.L. Stevens) R.A. Toro, 1925. **See basionym:** *Meliola arecibensis* F.L. Stevens, 1916.

Appendiculella calophylli (F.L. Stevens) R.A. Toro, 1925. **See basionym:** *Meliola calophylli* F.L. Stevens, 1916.

Appendiculella compositarum (F.S. Earle) R.A. Toro, 1925. **See basionym:** *Meliola compositarum* F.S. Earle, 1904 (1905).

Appendiculella cornu-caprae (P.C. Hennings) F. v. Höhnel, 1919. **See basionym:** *Meliola cornu-caprae* P.C. Hennings, 1904.

Appendiculella doliocarpi C.G. Hansford. Sydowia Annales Mycologici 9:29. 1955. **Isotype:** Hab. in *Doliocarpi* spec. indet. [= *Doliocarpus* sp.], Panama, France Field, 2 SEP 1924, leg. F.L. Stevens No. 233, ILL 3874 [also an isotype of *Asteridiella longipedicellata* (F.L. Stevens) C.G. Hansford var. *major* C.G. Hansford]. **Holotype:** FH. ≡ *Asteridiella doliocarpi* (C.G. Hansford) C.G. Hansford. Beihefte zur Sydowia Annales Mycologici 2:100–101. 1961.

Appendiculella echinus (P.C. Hennings) F. v. Höhnel, 1919. **See basionym:** *Meliola echinus* P.C. Hennings, 1904.

Appendiculella gloriosa (E.M. Doidge) C.G. Hansford, 1961. **See basionym:** *Meliola gloriosa* E.M. Doidge, 1920.

Appendiculella kalalauensis C.G. Hansford. Beihefte zur Sydowia Annales Mycologici 1:92. 1957. **Holotype:** Hab. in foliis *Perrottetiae sandwicensis* [= *Perrottetia sandwiciensis* A. Gray], Hawaii, [Kauai, Kalalau trail, 16 JUN 1921], leg. F.L. Stevens No. 474 p.p., ILL 3562 [also a paratype of *Amazonia perrottetiae* F.L. Stevens and of *Actinodothis perrottetiae* F.L. Stevens]. **Possible isotype:** ILL 3529. **Paratype:** [Island of] Hawaii, Waimea, 30 JUL 1921, leg. F.L. Stevens No. 1055 p.p., ILL 3524 [also the holotype of *Asteridiella waimeana* C.G. Hansford and a paratype of *Actinodothis perrottetiae* F.L. Stevens].

Appendiculella labiatarum C.G. Hansford. Beihefte zur Sydowia Annales Mycologici 2:698–699. 1961, nom. and stat. nov. **Based on:** *Irene inermis* (K. Kalchbrenner & M.C. Cooke) F. Theissen & H. Sydow var. *minor* C.G. Hansford & F.L. Stevens in C.G. Hansford, 1937. **Holotype:** Hab. in foliis Labiatarum [Lamiaceae sp.] indet., Uganda, Kampala, FEB 1930, leg. C.G. Hansford No. 1076, ILL 3750.

Appendiculella larviformis (P.C. Hennings) F. v. Höhnel var. *major* C.G. Hansford. Sydowia Annales Mycologici 9:30–31. 1955. **Paratypes:** Hab. in foliis *Acalyphae diversifoliae* [= *Acalypha diversifolia* Jacquin], Panama, [Brazos Brook Reservoir, 22 JUL 1924], leg. F.L. Stevens No. 699, ILL 3862; [Sweetwater, Fort Sherman, 6 OCT 1924], leg. F.L. Stevens No. 1070, ILL 3865; in foliis *Acalyphae* sp. [= *Acalypha* sp., Chagres Mouth, 23 AUG 1923], leg. F.L. Stevens No. 1291, ILL 3860;

Costa Rica, [Peralta, 11 JUL 1923], leg. F.L. Stevens 328, ILL 3863; [12 JUL 1923], leg. F.L. Stevens No. 366, ILL 3864; [Costa Rica, erroneously as Panama in the protologue, Siquirres, 31 JUL 1923], leg. F.L. Stevens No. 689, ILL 3861.

Appendiculella natalensis (E.M. Doidge) C.G. Hansford, 1961. **See basionym:** *Meliola natalensis* E.M. Doidge, 1917.

Appendiculella perrottetiae (F.L. Stevens) C.G. Hansford, 1955. **See basionym:** *Amazonia perrottetiae* F.L. Stevens, 1925.

Appendiculella sororcula (C.L. Spegazzini) C.G. Hansford, 1961. **See basionym:** *Meliola sororcula* C.L. Spegazzini, 1889.

Appendiculella sororcula (C.L. Spegazzini) C.G. Hansford var. *portoricensis* (F.L. Stevens) C.G. Hansford, 1961. **See basionym:** *Meliola compositarum* F.S. Earle var. *portoricensis* F.L. Stevens, 1916.

Appendiculella speciosa (E.M. Doidge) C.G. Hansford, 1961. **See basionym:** *Meliola speciosa* E.M. Doidge, 1917.

Appendiculella tonkinensis (P.A. Karsten & C. Roumeguère) R.A. Toro, 1927. **See basionym:** *Meliola tonkinensis* P.A. Karsten & C. Roumeguère, 1890.

Appendiculella tonkinensis (P.A. Karsten & C. Roumeguère) R.A. Toro var. *cecropiae* (F.L. Stevens) C.G. Hansford, 1961. **See basionym:** *Irene tonkinensis* (P.A. Karsten & C. Roumeguère) F.L. Stevens var. *cecropiae* F.L. Stevens, 1927.

Appendiculella tuberculata (F.L. Stevens) R.A. Toro, 1925. **See basionym:** *Meliola tuberculata* F.L. Stevens, 1916.

Appendiculella vernoniae (F.L. Stevens) C.G. Hansford, 1955. **See basionym:** *Irene sororcula* (C.L. Spegazzini) F.L. Stevens var. *vernoniae* F.L. Stevens, 1927.

Araneosa columellata W.H. Long. Mycologia 33:351–354, figs. 1–6. 1941. **Isotype:** Arizona, Santa Cruz County, AZ Hwy. 89, 6–8 miles from Nogales, alt. 3857 ft., 21 SEP 1934, leg. W.H. Long & V.O. Sandberg No. 7937, ILL 33529.

Argomyces insulanus J.C. Arthur. Mycologia 7:179. 1915. **Isotypes:** On *Vernonia albicaulis* Vahl ex Persoon, Porto Rico [Puerto Rico], river junction below Utuado, 30 DEC [1930], leg. F.L. Stevens No. 6596, ILL 18361 and 18364. **Holotype:** PUR. **Paratypes:** Leg. F.L. Stevens No. 6589, ILL 18362; on *Vernonia longifolia* sensu auct. non Persoon [= *V. albicaulis* Vahl ex Persoon], Villa Alba, 3 JAN [1913], leg. F.L. Stevens No. 113, ILL 18363.

Argomycetella pura H. Sydow. Annales Mycologici 23:313–314. 1925. **Isotype:** Hab. in foliis *Vernoniae patentis* [= *Vernonia patens* Kunth in H.B.K.], Costa Rica, La Caja pr. San José, 6 JAN 1925, leg. H. Sydow, Fungi in Itinere Costaricensi Collecti No. 3, ILL 18370. ≡ *Maravalia pura* (H. Sydow) E.B. Mains. Bulletin of the Torrey Botanical Club

66:178. 1939. ≡ *Uromyces purus* (H. Sydow) G.B. Cummins. Mycotaxon 5:407. 1977.

Arthrobotryum caudatum H. Sydow in E. de Wildeman. Annales du Musée du Congo Belge Série 5 (Botanique) 3:22. 1909. (Études de Systematique et de Géographie Botaniques sur la Flore du Bas et du Moyen—Congo 3:22. 1909.). **Isotype:** Parasite dans le mycelium d'un *Meliola* sp. ... sur les feuilles d'un *Randia* sp., Congo [Zaire], Kimpako, AUG 1908, leg. H. Vanderyst s.n., ILL 16368.

Arthrobotryum dieffenbachiae F.L. Stevens. Botanical Gazette 65:237, pl. V, fig. 4. 1918. **Holotype:** On *Meliola dieffenbachiae* F.L. Stevens on *Dieffenbachia seguine* (Jacquin) Schott [cited as *D. sequina*], Porto Rico [Puerto Rico], Dos Bocas below Utuado, 8 JUL 1915, leg. F.L. Stevens No. 8077, ILL 5041 [also a paratype of *Meliola dieffenbachiae* F.L. Stevens].

Arthrobotryum glabroides F.L. Stevens. Botanical Gazette 65:237–238, pl. V, figs. 1–3. 1918. **Holotype:** On *Meliola glabroides* F.L. Stevens on *Nectandra patens* (Sw.) Griseb. [= *Ocotea patens* (Sw.) Nees], Porto Rico [Puerto Rico], Mayagüez, 29 JUN 1915, leg. F.L. Stevens No. 7595, ILL 4007 [also a paratype of *Meliola glabroides* F.L. Stevens]. **Paratype:** Maricao, 20 JUL 1915, leg. F.L. Stevens No. 8867, ILL 4005 [also a paratype of *Meliola glabroides* F.L. Stevens].

Asbolisia citrina A.C. Batista & R. Ciferri. Quaderno Laboratorio Crittogamico, Istituto Botanico della Università 31:38–39, pl. I, fig. 5. 1963. **Assumed paratype:** On *Bradburya virginiana* (L.) O. Kuntze [= *Centrosema virginianum* (L.) Bentham], Porto Rico [Puerto Rico], Mayagüez, 15 JUN 1915, leg. F.L. Stevens [No. 7099], ILL 33492. **Holotype:** M. da Silva No. 5580 at URM.

Ascochyta caulicola R. Laubert. Arbeiten aus der Biologischen Abteilung für Land- und Forstwirtschaft am Kaiserlichen Gesundheitsamte 3:441–443, text figs. 1–5. 1903; also in H. & P. Sydow. Annales Mycologici 1:520. 1903. **Assumed isotype:** ...auf lebenden Stengeln, Blattstielen [und Blattnerven] von *Melilotus albus* Medik. [Germany], Brandenburg: Reichsversuchsfeld Dahlem bei Berlin, 1902 [as Aug. 1903 on the packet], leg. R. Laubert [ex H. & P. Sydow, Mycotheca Germanica No. 37], ILL 10929.

Ascomycetella quercina C.H. Peck. Bulletin of the Torrey Botanical Club 8:50. 1881. **Isosyntypes:** On living leaves of black oak, *Quercus tinctoria* Bartram (nom. nud.) [= *Q. velutina* Lam.], Illinois, Union County, [Cobden, 26 OCT 1882], leg. F.S. Earle s.n., ILL 61, 62 and 63.

Asterella aliena (J.B. Ellis & B.T. Galloway) P.A. Saccardo & A. Trotter, 1913. **See basionym:** *Asterina aliena* J.B. Ellis & B.T. Galloway in J.B. Ellis & B.M. Everhart, 1892.

Asterella prosopidis J.B. Ellis & B.M. Everhart. American Naturalist 31:340. 1897. **Isotype:** On living bark of *Prosopis dulcis* Kunth (mesquite), Mexico, near Monterey, JUL 1896, leg. B.F.G. Egeling s.n., ILL 6891. **Holotype:** NY.

Asteridiella aberrans (F.L. Stevens) C.G. Hansford, 1961. **See basionym:** *Irenina aberrans* F.L. Stevens, 1927. See also *Meliola tomentosa* H.G. Winter var. *calva* H. Rehm, 1907.

Asteridiella acalyphae (H. Rehm) C.G. Hansford ex C.G. Hansford, 1961. **See basionym:** *Meliola acalyphae* H. Rehm, 1913.

Asteridiella acervata (J.B. Ellis & B.M. Everhart) C.G. Hansford ex C.G. Hansford, 1961. **See basionym:** *Meliola acervata* J.B. Ellis & B.M. Everhart, 1897.

Asteridiella adelphica (H. Sydow) C.G. Hansford, 1961. **See basionym:** *Appendiculella adelphica* H. Sydow, 1926.

Asteridiella aibonitensis (F.L. Stevens) C.G. Hansford ex C.G. Hansford, 1961. **See basionym:** *Meliola aibonitensis* F.L. Stevens, 1916.

Asteridiella amoena (H. Sydow) C.G. Hansford ex C.G. Hansford, 1961. **See basionym:** *Irene amoena* H. Sydow, 1926.

Asteridiella anastomosans (H.G. Winter) C.G. Hansford ex C.G. Hansford, 1961. **See basionym:** *Meliola anastomosans* H.G. Winter, 1886.

Asteridiella andromedae (N.T. Patouillard) C.G. Hansford ex C.G. Hansford, 1961. **See basionym:** *Meliola andromedae* N.T. Patouillard, 1888.

Asteridiella anguriae (F.L. Stevens) C.G. Hansford ex C.G. Hansford, 1961. **See basionym:** *Irenina anguriae* F.L. Stevens, 1927.

Asteridiella angustispora F.L. Stevens & E.F. Roldan ex C.G. Hansford. Sydowia Annales Mycologici 16:312. (1962) 1963. **Based on:** *Irenina angustispora* F.L. Stevens & E.F. Roldan, Philippine Journal of Science 56:53, fig. 1f. 1935, nom. invalid. sine diagn. lat. [the latter name cited pro syn.]. **Holotype:** On Rubiaceae: *Neonauclea* sp., Philippines, Luzon, Naguilian Road, Benguet, 7 JAN 1931, leg. F.L. Stevens No. 1620, ILL 3901 [only a microscopic preparation remains].

Asteridiella atra (E.M. Doidge) C.G. Hansford ex C.G. Hansford, 1961. **See basionym:** *Meliola atra* E.M. Doidge, 1920.

Asteridiella atricha (C.L. Spegazzini) C.G. Hansford ex C.G. Hansford var. *major* C.G. Hansford ex C.G. Hansford. Beihefte zur Sydowia Annales Mycologici 2:139. 1961; originally published in Beihefte 1:95. 1957, as nom. invalid. [cf. Article 43.1 of I.C.B.N.]. **Holotype:** Hab. in foliis *Myrciae splendentis* [= *Myrcia splendens* (Sw.) DC.], Porto Rico [Puerto Rico], Aibonito, [18 JUL 1915], leg. F.L. Stevens No. 8465, ILL 5225. **Isotype:** ILL 5232.

Asteridiella aucubae (P.C. Hennings) C.G. Hansford, 1961. **See basionym:** *Meliola aucubae* P.C. Hennings, 1900 (1901).

Asteridiella barbaceniae (C.G. Hansford) C.G. Hansford ex C.G. Hansford, 1961. **See basionym**: *Irenina barbaceniae* C.G. Hansford, (1948) 1949.

Asteridiella bonii (A. Gaillard) C.G. Hansford ex C.G. Hansford, 1961. **See basionym**: *Meliola bonii* A. Gaillard, 1892.

Asteridiella brachycera (H. Sydow) C.G. Hansford ex C.G. Hansford, 1961. **See basionym**: *Meliola brachycera* H. Sydow, 1926.

Asteridiella buddlejicola (P.C. Hennings) C.G. Hansford ex C.G. Hansford, 1961. **See basionym**: *Meliola buddlejicola* P.C. Hennings, 1905.

Asteridiella callicarpae F.L. Stevens & E.F. Roldan ex C.G. Hansford. Sydowia Annales Mycologici 16:312–313. (1962) 1963. **Based on**: *Irenina callicarpae* F.L. Stevens & E.F. Roldan, Philippine Journal of Science 56:53–54. 1935, nom. invalid. sine diagn. lat. [the latter name cited pro syn.]. **Holotype**: On Verbenaceae: *Callicarpa magna* J.C. Schauer in DC., Philippines, Luzon, Naguilian Road, Benguet, 6 JAN 1931, leg. F.L. Stevens No. 1468, ILL 3957.

Asteridiella callista (H. Rehm) C.G. Hansford ex C.G. Hansford, 1961. **See basionym**: *Meliola callista* H. Rehm, 1914.

Asteridiella calva (C.L. Spegazzini) C.G. Hansford ex C.G. Hansford, 1961. **See basionym**: *Meliola calva* C.L. Spegazzini, 1889.

Asteridiella casimiroae C.G. Hansford. Sydowia Annales Mycologici 11:47. (1957) 1958 [as *Asteridella*, a typographic error later corrected (cf. C.G. Hansford, 1961)]. **Holotype**: Hab. in foliis *Casimiroae tetrameriae* [= *Casimiroa tetrameria* Millsp.], Costa Rica, El Alto, [6 JUL 1923], leg. F.L. Stevens No. 233, ILL 4002 [also a syntype of *Phyllachora casimiroae* F.L. Stevens & G.E. King].

Asteridiella cheirodendri (F.L. Stevens) C.G. Hansford, 1961. **See basionym**: *Irene cheirodendri* F.L. Stevens, 1925.

Asteridiella chloranthi F.L. Stevens ex C.G. Hansford. Beihefte zur Sydowia Annales Mycologici 1:95. 1957. **Holotype**: Hab. in foliis *Chloranthi officinalis* [*Chloranthus officinalis* Blume = *C. elatior* R. Br. ex Link], Philippines, Luzon, Laguna Prov., Mount Maquiling, [18 JAN 1931], leg. F.L. Stevens No.1851, ILL 4925. **Isotype**: CUP.

Asteridiella cleistanthi C.G. Hansford. Sydowia Annales Mycologici 11:47. (1957) 1958. **Holotype**: Hab. in foliis *Cleistanthi* sp. [= *Cleistanthus* sp.], Philippines, Mindanao [Island], Zamboanga, Malangas, [OCT–NOV 1919], leg. M. Ramos & G. Edaño, Phil. Bur. Sci. No. 36385, ILL 3952.

Asteridiella colubrinae (F.L. Stevens) C.G. Hansford ex C.G. Hansford, 1961. **See basionym**: *Irenina colubrinae* F.L. Stevens, 1927.

Asteridiella combreti (F.L. Stevens) C.G. Hansford ex C.G. Hansford, 1961. **See basionym**: *Irenina combreti* F.L. Stevens, 1927.

Asteridiella confragosa (H. & P. Sydow) C.G. Hansford ex C.G. Hansford, 1961. **See basionym**: *Meliola confragosa* H. & P. Sydow, 1912.

Asteridiella coprosmae C.G. Hansford. Beihefte zur Sydowia Annales Mycologici 1:95–96. 1957. **Holotype:** Hab. in foliis *Coprosmae* sp. [= *Coprosma* sp.], Hawaii, [Kauai, Pipe trail, Waimea Canyon, 15 JUN 1921], leg. F.L. Stevens No. 456, ILL 3570. **Paratypes:** Leg. F.L. Stevens No. 444, ILL 3571; leg. F.L. Stevens No. 437 p.p., ILL 3572; F.L. Stevens No. 458, ILL 3611; [Kalalau trail, 16 JUN 1921], leg. F.L. Stevens No. 523, ILL 3569.

Asteridiella costi (F.L. Stevens) C.G. Hansford, 1961. **See basionym:** *Irenina costi* F.L. Stevens, 1927.

Asteridiella crustacea (C.L. Spegazzini) C.G. Hansford ex C.G. Hansford, 1961. **See basionym:** *Meliola crustacea* C.L. Spegazzini, 1889.

Asteridiella cubitorum (F.L. Stevens & L.R. Tehon) C.G. Hansford ex C.G. Hansford, 1961. **See basionym:** *Irene cubitorum* F.L. Stevens & L.R. Tehon, 1926.

Asteridiella cyclopoda (F.L. Stevens) C.G. Hansford ex C.G. Hansford, 1961. **See basionym:** *Meliola cyclopoda* F.L. Stevens, 1916.

Asteridiella cyrtandrae (F.L. Stevens) C.G. Hansford, 1961. **See basionym:** *Irene cyrtandrae* F.L. Stevens, 1925 [erroneously cited as *Meliola* by Hansford].

Asteridiella doliocarpi (C.G. Hansford) C.G. Hansford, 1961. **See basionym:** *Appendiculella doliocarpi* C.G. Hansford, 1955.

Asteridiella drypeticola C.G. Hansford. Sydowia Annales Mycologici 10:53. (1956) 1957. **Holotype:** Hab. in foliis *Drypetes* sp., Porto Rico [Puerto Rico, Tanamá River, 6 JUL 1915], leg. F.L. Stevens No. 7885, ILL 4068.

Asteridiella exilis (H. & P. Sydow) C.G. Hansford ex C.G. Hansford, 1961. **See basionym:** *Meliola exilis* H. & P. Sydow, 1904.

Asteridiella glabra (M.J. Berkeley & M.A. Curtis) C.G. Hansford ex C.G. Hansford var. *isertiae* (F.L. Stevens) C.G. Hansford, 1961. **See basionym:** *Irenina isertiae* F.L. Stevens, 1927.

Asteridiella glabra (M.J. Berkeley & M.A. Curtis) C.G. Hansford ex C.G. Hansford var. *major* C.G. Hansford ex C.G. Hansford. Sydowia Annales Mycologici 16:308. (1962) 1963; originally published in Sydowia Annales Mycologici 10:55. (1956) 1957, as nom. invalid. [violation of Article 43.1 of I.C.B.N. While Hansford validated the new combination of the species name with citation of a full reference in Beihefte zur Sydowia Annales Mycologici 2:578. 1961, the varietal name was still invalid because the diagnosis was in English, and no reference was given to the earlier published Latin diagnosis (see also Articles 36.1 & 45.1 of I.C.B.N.)]. **Isotype:** Hab. in foliis *Canthii* spec. [= *Canthium* sp.] South Africa, Transvaal, Woodbush, [Zoutpansberg, Helpmakaar, 3 AUG 1911], leg. E.M. Doidge, ex PREM No. 1780, ILL 4067. **Holotype:** PREM.

Asteridiella glabriuscula (C.L. Spegazzini) C.G. Hansford ex C.G. Hansford, 1961. **See basionym:** *Meliola glabriuscula* C.L. Spegazzini, 1908.

Asteridiella glabroides (F.L. Stevens) C.G. Hansford, 1961. **See basionym:** *Meliola glabroides* F.L. Stevens, 1916.

Asteridiella gymnosporiae (H. & P. Sydow) C.G. Hansford ex C.G. Hansford, 1961. **See basionym:** *Meliola gymnosporiae* H. & P. Sydow, 1912.

Asteridiella hypelates C.G. Hansford. Beihefte zur Sydowia Annales Mycologici 1:96. 1957. **Holotype:** Hab. in foliis *Hypelates trifoliatae* [= *Hypelate trifoliata* Sw.], Cuba, leg. C. Wright, ex Plantae Cubenses Wrightianae No. 2171, ILL 3963. **Paratypes:** Porto Rico [Puerto Rico], Mona Island, [20–26 FEB 1914], leg. N.L. Britton, [J.F. Cowell & W.E. Hess] No. 1768a, ILL 3962, NY; Bahamas, New Province, [vicinity of Blue Hills, 28–29 MAY 1909], leg. P. Wilson No. 8246, ILL 3961.

Asteridiella iquitosensis (P.C. Hennings) C.G. Hansford ex C.G. Hansford, 1961. **See basionym:** *Meliola iquitosensis* P.C. Hennings, 1904.

Asteridiella irregularis (F.L. Stevens) C.G. Hansford ex C.G. Hansford, 1961. **See basionym:** *Meliola irregularis* F.L. Stevens, 1916.

Asteridiella leeicola C.G. Hansford. Sydowia Annales Mycologici 11:48. (1957) 1958. **Holotype:** Hab. in foliis *Leeae* sp. [= *Leea* sp.], Philippines, Mindanao [Island], Zamboanga, Malangas, [OCT–NOV 1919], leg. M. Ramos & G. Edaño, ex Phil. Bur. Sci. No. 36411, ILL 4124.

Asteridiella leucosykes (H.S. Yates) C.G. Hansford ex C.G. Hansford, 1961. **See basionym:** *Meliola leucosykes* H.S. Yates, (1917) 1918.

Asteridiella linocierae (H. & P. Sydow) C.G. Hansford ex C.G. Hansford, 1961. **See basionym:** *Meliola linocierae* H. & P. Sydow, 1914.

Asteridiella longipedicellata (F.L. Stevens) C.G. Hansford ex C.G. Hansford, 1961. **See basionym:** *Irenina longipedicellata* F.L. Stevens, 1927.

Asteridiella longipedicellata (F.L. Stevens) C.G. Hansford ex C.G. Hansford var. *major* CG. Hansford ex C.G. Hansford. Beihefte zur Sydowia Annales Mycologici 2:102. 1961; originally published in Sydowia Annales Mycologici 10:56. (1956) 1957, as nom. invalid. [violation of Article 43.1 of I.C.B.N.]. **Isotype:** Hab. in foliis *Doliocarpi* sp. [= *Doliocarpus* sp.], Panama, [France Field, 2 SEP 1924], leg. F.L. Stevens No. 233 p.p., ILL 3874 [also an isotype of *Appendiculella doliocarpi* C.G. Hansford]. **Holotype:** FH.

Asteridiella longipoda (A. Gaillard) C.G. Hansford ex C.G. Hansford var. *minor* C.G. Hansford ex C.G. Hansford. Sydowia Annales Mycologici 16:321. (1962) 1963; originally published in Sydowia Annales Mycologici 10:56. (1956) 1957, as nom. invalid. [violation of Article 43.1 of I.C.B.N. While Hansford's new combination of the species name was validated in Beihefte zur Sydowia Annales Mycologici 2:628–629. 1961, with full reference to the basionym, the varietal name was still invalid because the diagnosis was in English, and no reference was made to the earlier published Latin diagnosis (see also Articles 36.1 & 45.1 of I.C.B.N.)]. **Assumed holotype:** Hab. in foliis *Cordiae nitidae* [*Cordia nitida* Vahl in West = *C. laevigata* Lam.], Porto Rico [Puerto Rico, Martin Peña, 11 AUG 1915], leg. F.L. Stevens No. 9329, ILL 4127 [also a probable isotype of *Phyllachora orbicularis* C.L. Spegazzini].

Asteridiella manaosensis (P.C. Hennings) C.G. Hansford ex C.G.Hansford, 1961. **See basionym:** *Meliola manaosensis* P.C. Hennings, 1904.

Asteridiella manca (J.B. Ellis & G.W. Martin) C.G. Hansford ex C.G. Hansford, 1961. **See basionym:** *Meliola manca* J.B. Ellis & G.W. Martin, 1883.

Asteridiella megalospora (C.L. Spegazzini) C.G. Hansford, 1961. **See basionym:** *Meliola megalospora* C.L. Spegazzini, 1881.

Asteridiella meibomiae (F.L. Stevens) C.G. Hansford ex C.G. Hansford, 1961. **See basionym:** *Irenina meibomiae* F.L. Stevens, 1927.

Asteridiella melastomatacearum (C.L. Spegazzini) C.G. Hansford ex C.G. Hansford, 1961 [cited as *melastomacearum*]. **See basionym:** *Meliola melastomatacearum* C.L. Spegazzini, 1889.

Asteridiella monninae (F.L. Stevens) C.G. Hansford ex C.G. Hansford, 1961. **See basionym:** *Irenina monninae* F.L. Stevens, 1927.

Asteridiella morototonii (C.L. Spegazzini) C.G. Hansford ex C.G. Hansford, 1961. **See basionym:** *Meliola morototonii* C.L. Spegazzini, 1924.

Asteridiella naucleae (K.B. Boedijn) C.G. Hansford ex C.G. Hansford var. *libericae* C.G. Hansford ex C.G. Hansford. Beihefte zur Sydowia Annales Mycologici 2:577. 1961; orig. publ. by Hansford (1957:57) as nom. invalid. [cf. Article 43.1 of I.C.B.N.]. **Paratype:** On *Coffea robusta* Linden [= *C. canephora* Pierre ex Froehner, E. Africa], Uganda, [Kyague, SEP 1930], leg. C.G. Hansford No. 1357, ILL 3959.

Asteridiella nectandrae (C.G. Hansford) C.G. Hansford ex C.G. Hansford, 1961. **See basionym:** *Irene nectandrae* C.G. Hansford, 1955.

Asteridiella nigra (F.L. Stevens) C.G. Hansford ex C.G. Hansford, 1961. **See basionym:** *Irenina nigra* F.L. Stevens, 1927.

Asteridiella nuxiae (H. Sydow in E.M. Doidge & H. Sydow) C.G. Hansford ex C.G. Hansford, 1961. **See basionym:** *Irene nuxiae* H. Sydow in E.M. Doidge & H. Sydow, 1928.

Asteridiella obesa (C.L. Spegazzini) C.G. Hansford ex C.G. Hansford, 1961. **See basionym:** *Meliola obesa* C.L. Spegazzini, 1884.

Asteridiella obesa (C.L. Spegazzini) C.G. Hansford ex C.G. Hansford var. *obesula* (C.L. Spegazzini) C.G. Hansford, 1961. **See basionym:** *Meliola obesula* C.L. Spegazzini, 1891.

Asteridiella obscura (F.L. Stevens) C.G. Hansford ex C.G. Hansford, 1961. **See basionym:** *Irenina obscura* F.L. Stevens, 1927.

Asteridiella ohiana (F.L. Stevens) C.G. Hansford, 1961. **See basionym:** *Amazonia ohiana* F.L. Stevens, 1925.

Asteridiella olmediae C.G. Hansford. Sydowia Annales Mycologici 11:48–49. (1957) 1958. **Holotype:** Hab. in foliis *Olmediae asperae* [*Olmedia aspera* Ruiz & Pavon, recognized at this time. Perhaps = *Trophis* sp.?], Panama, Fort Sherman, [10 JUN 1924], leg. F.L. Stevens No. 1068, ILL 4050. **Paratype:** Leg. F.L. Stevens No. 1065, ILL 4015.

Asteridiella palicoureae C.G. Hansford. Beihefte zur Sydowia Annales Mycologici 1:96. 1957. **Holotype:** Hab. in foliis *Palicoureae* sp. [= *Palicourea* sp.], Porto Rico [Puerto Rico], leg. F.L. Stevens No. 1070a, ILL 4076.

Asteridiella parasitica (F.L. Stevens) C.G. Hansford ex C.G. Hansford, 1961. **See basionym:** *Irenina parasitica* F.L. Stevens, 1927.

Asteridiella peglerae (E.M. Doidge) C.G. Hansford ex C.G. Hansford, 1961. **See basionym:** *Meliola peglerae* E.M. Doidge, 1917.

Asteridiella pentaclethrae C.G. Hansford. Sydowia Annales Mycologici 10:58. (1956) 1957. **Isotype:** Hab. in foliis *Pentaclethrae* sp. [*Pentaclethra* sp., identified later as *P. macroloba* (Willd.) O. Kuntze (cf. C.G. Hansford, 1961)], British Guiana [Guyana, Demerara-Rockstone R.R., 15 JUL 1922], leg. F.L. Stevens No. 387a p.p., ILL 4961 [also the holotype of *Meliola conigera* F.L. Stevens & L.R. Tehon]. **Assumed holotype:** FH. **Paratype:** In foliis *Acaciae* sp.? [comparison with other specimens suggests that the host plant belongs in *Pentaclethra*, not *Acacia*], Costa Rica, [Sabario, 8 AUG 1923], leg. F.L. Stevens No. 787, ILL 4956 [also the lectotype of *Meliola conica* F.L. Stevens].

Asteridiella perrottetiae (F.L. Stevens) C.G. Hansford, 1957. **See basionym:** *Amazonia perrottetiae* F.L. Stevens, 1925.

Asteridiella perseae (F.L. Stevens) C.G. Hansford ex C.G. Hansford, 1961. **See basionym:** *Meliola perseae* F.L. Stevens, 1916.

Asteridiella peruviana (H. & P. Sydow) C.G. Hansford ex C.G. Hansford, 1961. **See basionym:** *Meliola peruviana* H. & P. Sydow, 1916.

Asteridiella pileae (C.G. Hansford) C.G. Hansford ex C.G. Hansford, 1961. **See basionym:** *Irenina pileae* C.G. Hansford, (1948) 1949.

Asteridiella pinicola (J. Dearness) C.G. Hansford, 1961. **See basionym:** *Meliola pinicola* J. Dearness, 1926.

Asteridiella podocarpi (E.M. Doidge) C.G. Hansford ex C.G. Hansford, 1961. **See basionym:** *Meliola podocarpi* E.M. Doidge, 1917.

Asteridiella podocarpi (E.M. Doidge) C.G. Hansford ex C.G. Hansford var. *portoricensis* C.G. Hansford. Beihefte zur Sydowia Annales Mycologici 1:96–97. 1957. **Holotype:** Hab. in foliis *Podocarpi coriacei* [= *Podocarpus coriaceus* L.C. Richard], Porto Rico [Puerto Rico], Marico, [15 JAN 1914], leg. F.L. Stevens No. 6774, ILL 4220.

Asteridiella pseudanastomosans (H. Rehm) C.G. Hansford ex C.G. Hansford, 1961. **See basionym:** *Meliola pseudanastomosans* H. Rehm, 1896.

Asteridiella pygei C.G. Hansford. Beihefte zur Sydowia Annales Mycologici 1:97. 1957. **Holotype:** Hab. in foliis *Pygei africani* [*Pygeum africanum* J.D. Hooker = *Prunus africana* (J.D. Hooker) Kalkman], South Africa, Transvaal, Woodbush, [8 MAR 1911], leg. E.M. Doidge, ex PREM No. 1761, ILL 3754. **Isotype:** PREM.

Asteridiella rinoreae (E.M. Doidge) C.G. Hansford ex C.G. Hansford, 1961. **See basionym:** *Irene rinoreae* E.M. Doidge, 1922.

Asteridiella rondeletiifolii A.C. Batista & M.L. Nascimento in A.C. Batista, M.L. Nascimento & H. da Silva Maia. Atas do Instituto de Micologia, Recife 1:16–17. 1960 [as *rondeletifolii*, corrected in accordance with Articles 60 & 61 of I.C.B.N.]. **Isotypes:** In foliis *Rondeletiae* sp. [= *Rondeletia* sp.] socius *Hormisciomycis prepusi* [= *Hormisciomyces prepusum* A.C. Batista & M.L. Nascimento] et *Hormisciellae rubi* [= *Hormisciella rubi* A.C. Batista in A.C. Batista & M.L. Nascimento], Puerto Rico, Monte Alegrillo, 14 NOV 1913, leg. F.L. Stevens No. 4511, ILL 32639, BPI. **Holotype:** URM 13391.

Asteridiella rubi C.G. Hansford. Sydowia Annales Mycologici 16:311. (1962) 1963. **Assumed holotype:** Hab. in foliis *Rubi rosaefolii* [= *Rubus rosifolius* J.E. Smith], Philippines, [Luzon, Naguilian Road, Benguet, 6 JAN 1931], leg. F.L. Stevens No. 1472, ILL 4232. [F.L. Stevens No. 1472 is also a paratype of *Irenina rubi* F.L. Stevens & E.F. Roldan (nom. invalid.) var. *angulata* F.L. Stevens & E.F. Roldan, 1935, nom. invalid. sine diagn. lat. Although Hansford possibly intended *Irenina rubi* (var. *rubi*) as the underlying basionym, his exclusion of the original type precludes recognition of F.L. Stevens & E.F. Roldan as name-bringing authors. In accordance with Article 48 of I.C.B.N., C.G. Hansford is to be cited as the sole author of *Asteridiella rubi*. See also citations of *Irenina rubi* F.L. Stevens & E.F. Roldan, 1935, and of *Irenina rubi* var. *angulata* F.L. Stevens & E.F. Roldan, 1935.].

Asteridiella saurauina C.G. Hansford. Sydowia Annales Mycologici 10:59. (1956) 1957. **Isotype:** Hab. in foliis *Saurauiae elegantis* [= *Saurauia elegans* (Choisy) Fernandez-Villar], Philippines, Luzon, [Naguilian Road], Benguet, [7 JAN 1931], leg. F.L. Stevens No. 1627, ILL 3873. **Holotype:** CUP.

Asteridiella scabra (E.M. Doidge) C.G. Hansford, 1961. **See basionym:** *Meliola scabra* E.M. Doidge, 1919.

Asteridiella schlegeliae (F.L Stevens) C.G. Hansford ex C.G Hansford, 1961. **See basionym:** *Meliola glabroides* F.L. Stevens var. *schlegeliae* F.L. Stevens, 1916.

Asteridiella sepulta (N.T. Patouillard ex F.L. Stevens) C.G. Hansford ex

C.G. Hansford, 1961. **See basionym:** *Meliola sepulta* N.T. Patouillard ex F.L. Stevens, 1916.

Asteridiella simaroubae C.G. Hansford. Beihefte zur Sydowia Annales Mycologici 1:97. 1957 [as *simarubae*, corrected in accordance with Rec. 60H.1 and Articles 60 & 61 of I.C.B.N.]. **Holotype:** Hab. in foliis *Simarubae tulae* [= *Simarouba tulae* Urban], Porto Rico [Puerto Rico, Mayagüez, 26 JUN 1915], leg. F.L. Stevens No. 7588, ILL 3965.

Asteridiella simaroubicola C.G. Hansford. Sydowia Annales Mycologici 11:49. (1957) 1958 [as *simarubicola*, corrected in accordance with Rec. 60H.1 and Articles 60 & 61 of I.C.B.N.]. **Holotype:** Hab. in foliis *Simaroubae* sp. [as *Simarubae* = *Simarouba* sp.], British Guiana [Guyana], Wismar, [14 JUL 1922], leg. F.L. Stevens No. 309, ILL 4046.

Asteridiella solanacearum (C.G. Hansford) C.G. Hansford. Sydowia Annales Mycologici 10:59. (1956) 1957 [treated as sp. nov.]; validated as comb. nov., with full reference to the basionym, in Beihefte zur Sydowia Annales Mycologici 2:638. 1961. **See basionym:** *Irene solanacearum* C.G. Hansford, 1955.

Asteridiella strophanthi (E.M. Doidge) C.G. Hansford ex C.G. Hansford, 1961. **See basionym:** *Meliola strophanthi* E.M. Doidge, 1917.

Asteridiella styracicola (C.L. Spegazzini) C.G. Hansford ex C.G. Hansford, 1961. **See basionym:** *Meliola styracicola* C.L. Spegazzini, 1912.

Asteridiella subapoda (H. & P. Sydow) C.G. Hansford ex C.G. Hansford, 1961. **See basionym:** *Meliola subapoda* H. & P. Sydow, 1914.

Asteridiella subcrustacea (C.L. Spegazzini) C.G. Hansford ex C.G. Hansford, 1961. **See basionym:** *Meliola subcrustacea* C.L. Spegazzini, 1889.

Asteridiella subglabroides C.G. Hansford. Sydowia Annales Mycologici 10:60. (1956) 1957. **Isotype:** Hab. in foliis *Annonacearum* sp. indet. [= *Annonaceae* sp. indet.], British Guiana [Guyana], Kartabo, 24 JUL 1922, leg. F.L. Stevens No. 669 [erroneously cited as No. 667 in the protologue; corrected in Hansford, 1961:31], ILL 4042. **Holotype:** K.

Asteridiella thunbergiae F.L. Stevens & E.F. Roldan ex C.G. Hansford. Sydowia Annales Mycologici 16:313. (1962) 1963. **Based on:** *Irenina thunbergiae* F.L. Stevens & E.F. Roldan, Philippine Journal of Science 56:54, fig. 1h, 1935, nom. invalid. sine diagn. lat. **Holotype:** On *Thunbergia alata* Bojer ex Sims, Philippines, Luzon, Kennon Road, Benguet, 8 JAN 1931, leg. F.L. Stevens No. 1642, ILL 4258. **Isotypes:** CALP, CUP, Phil. Bur. Sci.

Asteridiella trachylaena (H. Sydow) C.G. Hansford, 1961. **See basionym:** *Irene trachylaena* H. Sydow, 1926.

Asteridiella trematis (C.L. Spegazzini) C.G. Hansford ex A.C. Batista & H. da Silva Maia, 1957 [cited as *tremae*]. **See basionym:** *Meliola trematis* C.L. Spegazzini, 1912.

Annotated List of Types 31

Asteridiella umirayensis (H.S. Yates) C.G. Hansford ex C.G. Hansford, 1961. **See basionym**: *Meliola umirayensis* H.S. Yates, (1918) 1919.
Asteridiella uncariae (H. Rehm) C.G. Hansford ex C.G. Hansford, 1961. **See basionym**: *Meliola uncariae* H. Rehm, 1914.
Asteridiella uncariicola C.G. Hansford. Sydowia Annales Mycologici 11:49. (1957) 1958. **Holotype**: Hab. in foliis *Uncariae perrottetii* [= *Uncaria perrottetii* (A. Richard) Merrill], Philippines, Luzon, Tayabas, Quezon Forest Park, [31 NOV 1930], leg. F.L. Stevens No. 424, ILL 6408.
Asteridiella vacciniicola C.G. Hansford. Sydowia Annales Mycologici 11:49–50. (1957) 1958. **Holotype**: Hab. in foliis *Vaccinii benguetensis* [= *Vaccinium benguetense* S. Vidal], Philippines, Luzon, Benguet, Mt. San Thomas [as St. Thomas on the packet], 31 DEC 1930, leg. F.L. Stevens No. 1346 [erroneously cited as No. 1342 in the protologue], ILL 6521.
Asteridiella vegabajensis C.G. Hansford. Beihefte zur Sydowia Annales Mycologici 1:98. 1957. **Holotype**: Hab. in foliis *Psychotriae* spec. [= *Psychotria* sp.], Porto Rico [Puerto Rico], Vega Baja, [1 MAR 1913], leg. F.L. Stevens No. 516, ILL 4074.
Asteridiella viburni (H. & P. Sydow) C.G. Hansford ex C.G. Hansford, 1961. **See basionym**: *Meliola viburni* H. & P. Sydow, 1917.
Asteridiella vismiae C.G. Hansford. Sydowia Annales Mycologici 11:50. (1957) 1958. **Holotype**: Hab. in foliis *Vismiae latifoliae* [*Vismia latifolia* (probably) sensu auct. non (Aublet) Choisy = *V.* cf. *macrophylla* Kunth in H.B.K.], British Guiana [Guyana], Wismar, [14 JUL 1922], leg. F.L. Stevens No. 299 p.p., ILL 4043 p.p. [also the holotype of *Asteridiella vismiicola* C.G. Hansford].
Asteridiella vismiicola C.G. Hansford. Sydowia Annales Mycologici 11:50. (1957) 1958. **Holotype**: Hab. in foliis *Vismiae latifoliae* [*Vismia latifolia* (probably) sensu auct. non (Aublet) Choisy = *V.* cf. *macrophylla* Kunth in H.B.K.], British Guiana [Guyana], Wismar, [14 JUL 1922], leg. F.L. Stevens No. 299 p.p., ILL 4043 p.p. [also the holotype of *Asteridiella vismiae* C.G. Hansford].
Asteridiella voacangina C.G. Hansford. Beihefte zur Sydowia Annales Mycologici 1:98–99. 1957. **Holotype**: Hab. in foliis *Voacangae* spec. [= *Voacanga* sp.], Philippines, Luzon, Tayabas Prov., Mt. Binuang, [MAY 1917], leg. M. Ramos & G. Edaño, ex Phil. Bur. Sci. No. 28902, ILL 4284.
Asteridiella waimeana C.G. Hansford. Beihefte zur Sydowia Annales Mycologici 1:92–93. 1957. **Holotype**: Hab. in foliis *Perrottetiae sandwicensis* [= *Perrottetia sandwicensis* A. Gray], Hawaii, [Island of Hawaii], Waimea, [30 JUL 1921], leg. F.L. Stevens No. 1055 p.p., ILL 3524 [also a paratype of *Actinodothis perrottetiae* F.L. Stevens and of *Appendiculella kalalauensis* C.G. Hansford].

Asteridiella winteri (C.L. Spegazzini) C.G. Hansford ex C.G. Hansford, 1961. **See basionym:** *Meliola winteri* C.L. Spegazzini, 1888.

Asteridiella zeyheri (E.M. Doidge) C.G. Hansford ex C.G. Hansford, 1961. **See basionym:** *Irene zeyheri* E.M. Doidge, 1922.

Asteridiellina portoricensis (C.L. Spegazzini) F.J. Seaver & R.A. Toro in F.J. Seaver & C.E. Chardon, 1926. **See basionym:** *Asteridium portoricense* C.L. Spegazzini, 1923 (1924).

Asteridium portoricense C.L. Spegazzini. Boletín de la Academia Nacional de Ciencias en Córdoba 26:349-350. Preprint 1923 [journal part issued in 1924]. **Potential isotype** [not confirmed by LPS]: Hab. Sobre las hojas vivas de *Ocotea leucoxylon* (Sw.) de Lanessan, Porto Rico [Puerto Rico], cerca de Yajome alto [Jajome Alto, 17 JUL 1915], leg. F.L. Stevens No. 8428 p.p., ILL 4385 [also the holotype of *Meliola ocoteae* F.L. Stevens and of *Helminthosporium ocoteae* F.L. Stevens]. ≡ *Asteridiellina portoricensis* (C.L. Spegazzini) F.J. Seaver & R.A. Toro in F.J. Seaver & C.E. Chardon. New York Academy of Sciences. Scientific Survey of Porto Rico and the Virgin Islands 8(1):25. 1926 [*Asteridiellina* F.J. Seaver & R.A. Toro is a substitute name for *Asteridium* sensu C.L. Spegazzini (char. emend., 1923), non P.A. Saccardo (1891)]. ≡ *Halbaniella portoricensis* (C.L. Spegazzini) R.A. Toro in F.J. Seaver, C.E. Chardon and R.A. Toro. New York Academy of Sciences. Scientific Survey of Porto Rico and the Virgin Islands 8(2):212. 1932.

Asterina acalyphae H. Sydow. Annales Mycologici 23:395–397. 1925. **Isosyntype:** Hab. in foliis *Acalyphae macrostachyae* [as *macrostegiae*] var. *hirsutissimae* [= *Acalypha macrostachya* Jacquin var. *hirsutissima* (Willd.) Muell. Arg.], Costa Rica, San Ramón, 22 JAN 1925, leg. H. Sydow, Fungi in Itinere Costaricensi Collecti No. 206, ILL 6893 [also an isotype of *Asterostomella acalyphae* H. Sydow].

Asterina advenula H. Sydow. Annales Mycologici 25:47–48. 1927. **Isosyntype:** Hab. in foliis vivis *Rondeletiae affinis* [*Rondeletia affinis* Hemsley = *R. buddleoides* Bentham], Costa Rica, Mondongo, pr. San Ramón, 3 FEB 1925 [as 8 FEB on the packet], leg. H. Sydow, Fungi in Itinere Costaricensi Collecti No. 192, ILL 6897.

Asterina aliena J.B. Ellis & B.T. Galloway in J.B. Ellis & B.M. Everhart. The North American Pyrenomycetes. A Contribution to Mycologic Botany, p. 36. 1892. **Assumed isotype:** On leaves of pineapple (cult.) [= *Ananas comosus* (L.) Merrill, in a greenhouse], Washington, D.C., JAN 1881, leg. E.A. Southworth s.n., ILL 6900, ex NY. ≡ *Asterella aliena* (J.B. Ellis & B.T. Galloway) P.A. Saccardo & A. Trotter. Sylloge Fungorum 22: 537. 1913.

Asterina arnaudia R.W. Ryan. Mycologia 16:184. 1924. **Paratypes:** On *Passiflora multiflora* L., Porto Rico [Puerto Rico], Rio Tanamá, [7 JUL

1915], leg. F.L. Stevens No. 7943, ILL 6905; Cataño, [6 NOV 1913], leg. F.L. Stevens No. 4192, ILL 6902; on *Passiflora sexflora* Jussieu, El Alto de la Bandera, [14 JUL 1915], leg. F.L. Stevens No. 8284, ILL 6903; [15 JUL 1915], leg. F.L. Stevens No. 8642, ILL 6906; Rio Arecibo, [8 JUL 1915], leg. F.L. Stevens No. 7784, ILL 6904.

Asterina carbonacea C.L. Spegazzini var. *anacardii* R.W. Ryan. Mycologia 16:186. 1924. **Syntypes**: On *Anacardium excelsum* (Bertero & Balbis) Skeels, Porto Rico [Puerto Rico], Maricao, Indiera Fria, [8 OCT 1913], leg. F.L. Stevens No. 3369, ILL 6896; on Melastomataceae sp. indet. [as Melastomaceae], Monte de Oro, [3 DEC 1913], leg. F.L. Stevens No. 5745, ILL 6910; Luquillo Forest [misspelled "Lugiullo" in the protologue, 2 DEC 1913], leg. F.L. Stevens No. 544, ILL 6911.

Asterina celastri J.B. Ellis & W.A. Kellerman. Journal of Mycology 1:3. 1885. **Isotype**: On living leaves of *Celastrus scandens* L., Kansas, Manhattan, [5] NOV 1884, leg. W.A. Kellerman, Flora of Kansas No. 707, ILL 6916.

Asterina clausenicola E.M. Doidge. Transactions of the Royal Society of South Africa 8:263, 273–274, pl. XVI, fig. 33. 1920. **Isotype**: On leaves of *Clausena inaequalis* Bentham in Hooker, South Africa, Natal [Prov.], Hilton Road, 21 JUL 1918, leg. E.M. Doidge, ex PREM No. 11606, ILL 6933.

Asterina clermontiae F.L. Stevens & R.W. Ryan in F.L. Stevens. Bernice P. Bishop Museum Bulletin 19:73. 1925. **Lectotype** [designated herein]: On *Clermontia* sp., Hawaii, Maui, Iao valley, 7 SEP 1921, leg. F.L. Stevens No. 1154 [as No. 1154a on the packet], ILL 6881 [also an isotype of *Meliola lobeliae* F.L. Stevens and a paratype of *Trichothallus hawaiiensis* F.L. Stevens]. **Isolectotypes**: ILL 5408, BISH 499026 and 499963 [also the holotype and isotypes, respectively, of *Meliola lobeliae* F.L. Stevens], and possibly No. 1154b, ILL 6934, BISH 499858 and 499862 [also the holotype and isotypes, respectively, of *Clypeolella clermontiae* F.L Stevens & R.W. Ryan in F.L. Stevens].

Asterina consobrina H. Sydow. Annales Mycologici 25:49–51. 1927. **Isosyntypes**: Hab. in foliis *Solani* spec. [= *Solanum* sp.], Costa Rica, San Pedro de San Ramón, 2 FEB 1925, leg. H. Sydow, Fungi in Itinere Costaricensi Collecti No. 56b, ILL 6838; 5 FEB 1925, leg. H. Sydow, Fungi in Itinere Costaricensi Collecti No. 410, ILL 6937.

Asterina costaricensis H. Sydow. Annales Mycologici 25:51–52. 1927. **Isotype**: Hab. in foliis *Jacobiniae (Justiciae) tinctoriae* [*Jacobinia tinctoria* (Oersted) Hemsley = *Justicia colorifera* V. Graham], Costa Rica, Alajuela, 10 JAN 1925, leg. H. Sydow, Fungi in Itinere Costaricensi Collecti No. 220, ILL 6939.

Asterina dilabens H. Sydow var. *hilliae* R.W. Ryan. Mycologia 16:187–188.

1924. **Syntypes**: On *Hillia parasitica* Jacquin, Porto Rico [Puerto Rico], El Alto de la Bandera, [14 JUL 1915], leg. F.L. Stevens No. 8298, ILL 6958; leg. F.L. Stevens No. 8569, ILL 6956; [16 JUL 1915], leg. F.L. Stevens No. 8706, ILL 6957; Monte de Oro, [3 DEC 1917], leg. F.L. Stevens No. 5676, ILL 6954.

Asterina diplocarpa M.C. Cooke var. *cestricola* R.W. Ryan. Mycologia 16:187. 1924. **Syntypes**: On *Cestrum* sp., Porto Rico [Puerto Rico], Jajome Alto [misspelled "Ejome Alto," 17 JUL 1915], leg. F.L. Stevens No. 8397, ILL 6965; leg. F.L. Stevens No. 8384, ILL 6964; Aibonito, [as Arbonito, 16 JUL 1915], leg. F.L. Stevens No. 8463, ILL 6966; on *Cestrum macrophyllum* Vent. [= *C. laurifolium* L'Héritier, Puerto Rico, 1915], leg. F.L. Stevens s.n., ILL 6962.

Asterina diplopoda H. Sydow. Annales Mycologici 25:56–57. 1927. **Isotype**: Hab. in foliis vivis *Solani acerifolii* [= *Solanum acerifolium* Humb. & Bonpl. ex Dunal; or perhaps another species, affin. *S. capsicoides* Allioni], Costa Rica, Grecia, 19 JAN 1925, leg. H. Sydow, Fungi in Itinere Costaricensi Collecti No. 51, ILL 6967.

Asterina drypetis R.W. Ryan. Mycologia 16:180. 1924. **Holotype**: On *Drypetes* sp., Porto Rico [Puerto Rico], Rio Tanamá, [6 JUL 1915], leg. F.L. Stevens No. 7841, ILL 7162.

Asterina erebia H. Sydow. Annales Mycologici 25:59–60. 1927. **Isotype**: Hab. in foliis vivis *Palicoureae costaricensis* [= *Palicourea costaricensis* Bentham ex Oersted], Costa Rica, Piedades de San Ramón, 26 JAN 1925, leg. H. Sydow, Fungi in Itinere Costaricensi Collecti No. 135, ILL 6971.

Asterina excoecariae E.M. Doidge. Transactions of the Royal Society of South Africa 8:258, 274, pl. XV, fig. 25. 1920. **Isotype**: On leaves of *Excoecaria* sp., South Africa, Natal Prov., Winkle Spruit, 28 MAY 1915, leg. E.M. Doidge, ex PREM No. 9009, ILL 6974.

Asterina fawcettii R.W. Ryan. Mycologia 16:180–181. 1924 [as *fawcetti*]. **Holotype**: On *Eugenia buxifolia* (Sw.) Willd. [= *E. foetida* Persoon], Porto Rico [Puerto Rico], Vega Baja, [2 JUL 1915], leg. F.L. Stevens No. 7713, ILL 6981. **Paratypes**: On *Eugenia* sp., Mona Island, [20–21 DEC 1913], leg. F.L. Stevens No. 6151, ILL 6980 and 6983; leg. F.L. Stevens No. 6160, ILL 6975 and 6982; leg. F.L.Stevens No. 6163, ILL 6976 and 6979; leg. F.L. Stevens No. 6155, ILL 6977 and 6978.

Asterina fumago (G. Niessl) F. v. Höhnel, 1910. **See basionym**: *Meliola fumago* G. Niessl, 1881.

Asterina genipae R.W. Ryan. Mycologia 16:180. 1924. **Assumed holotype**: On *Genipa americana* L., Porto Rico [Puerto Rico], Mayagüez, [24 MAY 1913], leg. F.L. Stevens No. 1861 [as No. 1861a on the packet], ILL 6987.

Asterina gibbosa A. Gaillard var. *megathyria* E.M. Doidge. Transactions of the Royal Society of South Africa 8:248, 275, pl. XIII, fig. 6. 1920. **Isosyntypes:** On *Tricalysia sonderiana* Hiern in Oliver [as *Kraussia coriacea* on the packet], South Africa, Natal Prov., near Durban, 25 MAY 1897 [as 14 MAY 1897 on the packet], leg. J.M. Wood No. 6452 [ex PREM No. 9522], ILL 6996; on *Tricalysia lanceolata* (Sonder) Burtt-Davy, [Natal Prov.], Claridge, 31 MAY 1915, leg. E.M. Doidge, ex PREM No. 8992, ILL 6994; on *Alberta*? sp., Buccleuch, 20 APR 1916, leg. J.M. Sim, ex PREM No. 10151, ILL 6991. **Syntypes:** PREM.

Asterina gouldiae F.L. Stevens & R.W. Ryan in F.L. Stevens. Bernice P. Bishop Museum Bulletin 19:73. 1925. **Holotype:** On *Gouldia coriacea* (Hooker & Arnott) Hillebr. [= *Hedyotis terminalis* (Hooker & Arnott) W.L. Wagner & Herbst], Hawaii, Kauai, Kalalau trail, 16 JUN 1921, leg. F.L. Stevens No. 494, ILL 6999. **Isotypes:** BISH 499857 and 499861.

Asterina guianensis R.W. Ryan. Mycologia 16:182. 1924. **Holotype:** On *Miconia guianensis* (Aublet) Cogn. [= *M. mirabilis* (Aublet) L.O. Williams], Porto Rico [Puerto Rico], El Alto de la Bandera, [14 JUN 1915], leg. F.L. Stevens No. 8250, ILL 7003.

Asterina hamata H. Sydow. Annales Mycologici 25:61–63. 1927. **Isotype:** Hab. in foliis *Phoebes neurophyllae* [= *Phoebe neurophylla* Mez & Pittier], Costa Rica, San Pedro de San Ramón, 10 FEB 1925, leg. H. Sydow, Fungi in Itinere Costaricensi Collecti No. 389, ILL 7005.

Asterina hippocrateae R.W. Ryan. Mycologia 16:181. 1924. **Holotype:** On *Hippocratea volubilis* L., Porto Rico [Puerto Rico], Martin Peña, [11 AUG 1915], leg. F.L. Stevens No. 9296, ILL 7007. **Paratypes:** Vega Baja, [2 JUL 1915], leg. F.L. Stevens No. 7726, ILL 7008; leg. F.L. Stevens No. 9296a, ILL 7006.

Asterina isothea H. Sydow. Annales Mycologici 25:64–66. 1927. **Isotype:** Hab. in foliis vivis *Triumfettae semitrilobae* [= *Triumfetta semitriloba* Jacquin], Costa Rica, San Pedro de San Ramón, 6 FEB 1925, leg. H. Sydow, Fungi in Itinere Costaricensi Collecti No. 189, ILL 7013 [also an isotype of *Asterostomella isothea* H. Sydow].

Asterina ixorae R.W. Ryan. Mycologia 16:182. 1924 [as *ixonae*, an error corrected in accordance with Articles 60 & 61 of I.C.B.N.]. **Holotype:** On *Ixora ferrea* (Jacquin) Bentham [as *Ixona*], Porto Rico [Puerto Rico], Mayagüez [15 JUN 1915], leg. F.L. Stevens No. 7070, ILL 7016 [also a paratype of *Asterinella ixorae* R.W. Ryan]. **Isotype:** ILL 7198.

Asterina kauaiensis F.L. Stevens & R.W. Ryan in F.L. Stevens. Bernice P. Bishop Museum Bulletin 19:73. 1925. **Holotype:** On unknown host, Hawaii, Kauai, Kalalau trail, 16 JUN 1921, leg. F.L. Stevens No. 479, ILL 7017. **Isotypes:** BISH 499856 and 499860.

Asterina lobeliae F.L. Stevens & R.W. Ryan in F.L. Stevens. Bernice P. Bishop Museum Bulletin 19:74. 1925. **Holotype:** On *Lobelia* sp., Hawaii, Island of Hawaii, [erroneously cited as Kauai], Hamakua, 31 JUL 1921, leg. O.H. Swezey, as F.L. Stevens No. 1063, ILL 7019. **Isotypes:** BISH 499854 and 499855.

Asterina lophopetali H. Rehm. Leaflets of Philippine Botany 6:2228. 1914. **Assumed isotype:** Ad folia *Lophopetalae toxici* [= *Lophopetalum toxicum* Loher], Philippines, Luzon, Laguna Prov., [Mount Maquiling, near] Los Baños, AUG 1913, leg. S.A. Reyes No. 1759a, comm. C.F. Baker, [ex Fungi Malayana No. 9], ILL 7021.

Asterina macowaniana (F. v. Thümen) K. Kalchbrenner & M.C. Cooke, 1880. **See basionym:** *Meliola macowaniana* F. v. Thümen, 1876.

Asterina melastomatacearum R.W. Ryan. Mycologia 16:186–187. 1924 [cited as *melastomacearum*, corrected in accordance with current I.C.B.N. (cf. F.C. Deighton, 1968)]. **Syntypes:** On *Miconia racemosa* (Aublet) DC., Porto Rico [Puerto Rico, Mayagüez, 14 MAY 1915], leg. F.L. Stevens No. 7037 p.p., ILL 7024 [also a paratype of *Asterina racemosa* R.W. Ryan]; on *Miconia impetiolaris* (Sw.) D. Don ex DC., Jayuya [misspelled "Jayaya" in the protologue, 2 MAR 1912], leg. F.L. Stevens No. 375, ILL 7023.

Asterina mexicana J.B. Ellis & B.M. Everhart. Bulletin of the Torrey Botanical Club 27:51. 1900. **Isotype:** On *Agave mexicana* Lam., Mexico, Mexico City, MAY 1898, leg. B.F.G. Egeling, ILL 7026. **Holotype:** NY.

Asterina miconiae R.W. Ryan. Mycologia 16:181. 1924. **Holotype:** On *Miconia racemosa* (Aublet) DC., Porto Rico [Puerto Rico], Las Marias, [10 JUL 1915], leg. F.L. Stevens No. 8136, ILL 7029. **Paratypes:** Mayagüez, [24 JUN 1915], leg. F.L. Stevens No. 7417, ILL 7030; Trujillo Alto, 15 AUG 1915, leg. F.L. Stevens No. 9424, ILL 7028; on *Miconia thomasiana* DC., Sabana Grande, [13 AUG 1915], leg. F.L. Stevens No. 9373, ILL 7027.

Asterina miconiicola R.W. Ryan. Mycologia 16:182. 1924 [as *miconicola*, corrected in accordance with Rec. 60H.1 and Articles 60 & 61 of I.C.B.N.]. **Holotype:** On *Miconia racemosa* (Aublet) DC., Porto Rico [Puerto Rico], Maricao, [19 JUL 1915], leg. F.L. Stevens No. 8940, ILL 7033. **Paratypes:** El Alto de la Bandera, [as Bandera in the protologue, 14 JUL 1915], leg. F.L. Stevens No. 8292, ILL 7031; on *Palicourea* sp., Jajome Alto, [3 DEC 1913], leg. F.L. Stevens No. 5683a, ILL 7032.

Asterina myrciae R.W. Ryan. Mycologia 16:186. 1924. **Holotype:** On *Eugenia* sp., Porto Rico [Puerto Rico], Rosario, [4 AUG 1915], leg. F.L. Stevens No. 9494, ILL 7037. **Paratypes:** On *Myrcia splendens* (Sw.) DC., Maricao, [19 JUL 1915], leg. F.L. Stevens No. 8831, ILL 7036; Mayagüez Mesa, [25 JUN 1915], leg. F.L. Stevens No. 7473, ILL 7034.

Asterina natalensis E.M. Doidge. Transactions of the Royal Society of South Africa 8:248–249, 275, pl. XIII, fig. 7. 1920. **Isotype**: On *Mikania* sp., South Africa, Natal [Prov.], Winkle Spruit, 28 MAY 1915, leg. E.M. Doidge, ex PREM No. 9001, ILL 7038.

Asterina passifloricola R.W. Ryan. Mycologia 16:183. 1924. **Syntypes**: On *Passiflora rubra* L., Porto Rico [Puerto Rico], Dos Bocas, [8 JUL 1915], leg. F.L. Stevens No. 8039, ILL 7045; [on *Passiflora* sp.], Monte de Oro, [1915], leg. F.L. Stevens No. 5561, ILL 7044.

Asterina phaleriae J.M. Mendoza. Philippine Journal of Science 49:188–189, pl. 4, figs. 1–5. 1932. **Isotype**: On leaves of *Phaleria perrottetiana* (Decaisne) Fernandez-Villar, Philippines, Panay [Island], Capiz Prov., Libacao, 13 JUN 1919, leg. A. Martelino & G. Edaño, ex Phil. Bur. Sci. No. 35857, ILL 7048.

Asterina phoebes H. Sydow. Annales Mycologici 25:68–69. 1925. **Isotype**: In foliis *Phoebes costaricanae* [= *Phoebe costaricana* Mez & Pittier], Costa Rica, Piedades de San Ramón, 26 JAN 1925, leg. H. Sydow, Fungi in Itinere Costaricensi Collecti No. 167a, ILL 7050.

Asterina phyllostegiae F.L. Stevens & R.W. Ryan in F.L. Stevens. Bernice P. Bishop Museum Bulletin 19:73. 1925. **Holotype**: On *Phyllostegia* sp., Hawaii, Oahu, Olympus, 24 JUN 1921, leg. F.L. Stevens No. 718, ILL 7049. **Isotypes**: BISH 499852 and 499853.

Asterina plectroniae J.M. Mendoza. Philippine Journal of Science 49:186–187, pl. II, figs. 1–5. 1932. **Holotype**: On leaves of *Plectronia cumingii* (S. Vidal) A.D.E. Elmer [= *Canthium gynochthodes* Baillon (Rubiaceae), Philippines], Luzon, Ilocos Norte Prov., Burgos, 22 JUL 1918, leg. M. Ramos, ex Phil. Bur. Sci. No. 33399, ILL 7054.

Asterina pliniae R. Ciferri. Sydowia Annales Mycologici 10:141. (1956) 1957. **Isotype**: Hab. in foliis vivis *Pliniae* sp. [= *Plinia* sp.] (Myrtaceae), Dominican Republic, Cordillera Central, La Vega Prov., Bonao, in sylva, alt. ca. 150 m, MAR 1931, leg. R. Ciferri, Mycoflora Domingensis Exsiccata No. 367, ILL 33156.

Asterina portoricensis R.W. Ryan. Mycologia 16:185. 1924. **Holotype**: On *Solanum* sp., Porto Rico [Puerto Rico], Quebradillas, [22 NOV 1913], leg. F.L. Stevens No. 5127, ILL 7056.

Asterina psidii R.W. Ryan. Mycologia 16:185. 1924. **Holotype**: On *Psidium guajava* L., Porto Rico [Puerto Rico], Mayagüez, [30 JAN 1913], leg. F.L. Stevens No. 310, ILL 7058.

Asterina psychotriae R.W. Ryan. Mycologia 16:185. 1924. **Holotype**: On *Psychotria pubescens* Sw., Porto Rico [Puerto Rico], Mayagüez, [29 JUN 1915], leg. F.L. Stevens No. 7581, ILL 7059.

Asterina racemosae R.W. Ryan. Mycologia 16:182. 1924. **Holotype**: On *Miconia racemosa* (Aublet) DC., Porto Rico [Puerto Rico], Jajome

Alto, [17 JUL 1915], leg. F.L. Stevens No. 8402, ILL 7068. **Paratypes**: Mayagüez, [14 MAY 1915], leg. F.L. Stevens No. 7037 p.p. , ILL 7066 [also a syntype of *Asterina melastomatacearum* R.W. Ryan]; Las Piedras, [12 JUL 1915], leg. F.L. Stevens No. 9321, ILL 7069; leg. F.L. Stevens No. 9322, ILL 7064; on *Miconia impetiolaris* (Sw.) D. Don ex DC. [listed under *M. racemosa* in the protologue], Trujillo Alto, [15 AUG 1915]; leg. F.L. Stevens No. 9425, ILL 7067; Rio Piedras, [3 NOV 1913], leg. F.L. Stevens No. 5701, ILL 7071; on *Miconia sintenisii* Cogn., Rio Maricao, above Maricao, [20 SEP 1913], leg. F.L. Stevens No. 3646, ILL 7070; [18 OCT 1913], leg. F.L. Stevens No. 3887, ILL 7065; Las Marias, [22 MAR 1913], leg. F.L. Stevens No. 312, ILL 7072.

Asterina radicans J.B. Ellis. Journal of Mycology 7:276. 1893. **Isotype**: On living leaves of *Capparis cynophallophora* L., Florida, 1891, leg. C.T. Simpson No. 256, ILL 7073. **Holotype**: NY.

Asterina sidicola R.W. Ryan. Mycologia 16:181. 1924. **Holotype**: On Malvaceae, [*Sida*?], Porto Rico [Puerto Rico], Jajome Alto, [3 DEC 1913], leg. F.L. Stevens No. 5693, ILL 7099. **Paratypes**: On *Sida* sp., Rosario, leg. F.L. Stevens No. 4809, ILL 7097; Vega Baja, [21 FEB 1913], leg. F.L. Stevens No. 401, ILL 7096; on Malvaceae, [Santa Ana, 31 DEC 1913], leg. F.L. Stevens No. 6663a, ILL 7092; on *Sida carpinifolia* L.f. [= *S. acuta* N.L. Burman], Maricao, [20 JUL 1915], leg. F.L. Stevens No. 8869, ILL 7098; on *Corchorus hirtus* L., Rio Tanamá, [6 JUL 1915], leg. F.L. Stevens No. 7877, ILL 7093; on *Abutilon* sp., Jajome Alto, [3 DEC 1913], leg. F.L. Stevens No. 5642, ILL 7091.

Asterina stomatophora J.B. Ellis & G.W. Martin. Journal of Mycology 1:98–99. 1885. **Assumed isotype**: On living leaves of *Quercus laurifolia* Michaux, Florida, Green Cove Springs, MAR 1884, leg. G.W. Martin s.n., ILL 7118.

Asterina styracina H. Sydow. Annales Mycologici 25:72–74. 1927. **Isotype**: Hab. in foliis vivis *Styracis polyneurae* [= *Styrax polyneurus* Perkins], Costa Rica, in monte Poas pr. Grecia, 15 JAN 1925, leg. H. Sydow, Fungi in Itinere Costaricensi Collecti No. 156, ILL 7116.

Asterina suttoniae F.L. Stevens & R.W. Ryan in F.L. Stevens. Bernice P. Bishop Museum Bulletin 19:74. 1925. **Holotype**: On *Suttonia* sp. [= *Myrsine lessertiana* A. DC.], Hawaii, Oahu, Kuliouou, 29 MAY 1921, leg. E.L. Caum, as F.L. Stevens No. 143 p.p., ILL 7114. **Possible isotypes**: ILL 3540, BISH 499875 and 499876. [F.L. Stevens No. 143 p.p., as ILL 3540, is also the holotype of *Actinodothis suttoniae* F.L. Stevens.].

Asterina sydowiana R.W. Ryan. Mycologia 16:184–185. 1924. **Holotype**: On *Chrysophyllum* sp., Porto Rico [Puerto Rico], Monte Alegrillo, [14 SEP 1913], leg. F.L. Stevens No. 4731, ILL 7159. **Isotype**: ILL 5211. [No. 4731 is also cited as paratype of *Meliola ocoteicola* F.L. Stevens.].

Asterina tacsoniae N.T. Patouillard var. *passiflorae* R.W. Ryan. Mycologia 16:183–184. 1924. **Syntypes:** On unknown host, Porto Rico [Puerto Rico], Rio Maricao, above Maricao, [20 SEP 1913], leg. F.L. Stevens No. 3654, ILL 7164; on *Passiflora* sp., Dos Bocas, below Utuado, [30 DEC 1913], leg. F.L. Stevens No. 6575, ILL 7161; on *Passiflora sexflora* Jussieu, El Miradero, [4 AUG 1915], leg. F.L. Stevens No. 9165, ILL 7163; Dos Bocas, [8 JUL 1915], leg. F.L. Stevens No. 8040, ILL 7160.

Asterina tertia M. Raciborski var. *africana* E.M. Doidge. Transactions of the Royal Society of South Africa 8:264–265, pl. XVI, fig. 35. 1920. **Paratypes:** On *Hypoestes verticillaris* (L.f.) R. Br. ex C.B. Clarke, South Africa, Cape Prov., Kentani, 6 MAY 1915, leg. A. Pegler No. 2317, ILL 7166, PREM 9074.

Asterina tetrazygiae R.W. Ryan. Mycologia 16:183. 1924. **Holotype:** On *Tetrazygia elaeagnoides* (Sw.) DC., Porto Rico [Puerto Rico], Rio Arecibo, [8 JUL 1915], leg. F.L. Stevens No. 7778, ILL 7167. **Paratypes:** On *Tetrazygia* sp., Jajome Alto, [17 JUL 1915], leg. F.L. Stevens Nos. 8430 and 8940, ILL 7168.

Asterina theissenia R.W. Ryan. Mycologia 16:187. 1924. **Syntypes:** On Melastomataceae [as Melastomaceae in the protologue and as *Miconia* sp. on some of the packets], Porto Rico [Puerto Rico], Las Marias, [22 MAR 1913], leg. F.L. Stevens No. 451, ILL 7169 and 7170 [the latter number an isosyntype]; on *Miconia* sp., leg. F.L. Stevens No. 759a, ILL 7171.

Asterina trichiliae E.M. Doidge. Transactions of the Royal Society of South Africa 8:253–254, pl. XIV, fig. 16. 1920. **Paratype:** On *Trichilia emetica* Vahl, South Africa, Zoutpansberg [District], Louis Trichardt, 8 APR 1919, leg. V.A. Putterill, ex PREM No. 11833, ILL 7176.

Asterina versipoda R.W. Ryan. Mycologia 16:188. 1924. **Holotype:** On unknown host, Porto Rico [Puerto Rico], Utuado, [8 NOV 1913], leg. F.L. Stevens No. 4419, ILL 7057.

Asterina xylosmae J.M. Mendoza. Philippine Journal of Science 49:185–186, pl. I, figs. 1–5. 1932. **Holotype:** On leaves of *Xylosma* sp., [Philippines], Luzon, Ilocos Norte Prov., Bangui, FEB 1917, leg. M. Ramos, ex Phil. Bur. Sci. No. 27705, ILL 7179.

Asterinella capizensis J.M. Mendoza. Philippine Journal of Science 49:189–190, pl. 5, figs. 1–5. 1932. **Holotype:** On leaves of *Leucosyke capitellata* (Poiret) Weddell, Philippines, Panay [Island], Capiz Prov., Capindan [as Jamindan on the packet], San Juan Barrio, 11 APR 1918, leg. M. Ramos & G. Edaño, ex Phil. Bur. Sci. No. 32109, ILL 7182.

Asterinella hippeastri R.W. Ryan. Mycologia 16:188. 1924. **Holotype:** On *Hippeastrum* sp., Porto Rico [Puerto Rico], Mayagüez Mesa, [29 JUN 1915], leg. F.L. Stevens No. 7590, ILL 7185.

Asterinella ixorae R.W. Ryan. Mycologia 16:189. 1924 [as *ixonae*, an error corrected in accordance with Articles 60 & 61 of I.C.B.N.]. **Holotype:** On *Ixora ferrea* (Jacquin) Bentham [misspelled "*Ixona*"], Porto Rico [Puerto Rico], Mayagüez Mesa, [27 JUN 1915], leg. F.L. Stevens No. 7591, ILL 7196. **Paratypes:** [15 JUN 1915], leg. F.L. Stevens No. 7070, ILL 7016 [also the holotype of *Asterina ixorae* R.W. Ryan], ILL 7198; leg. F.L. Stevens No. 7067, ILL 7197.

Asterinella mabae F.L. Stevens & R.W. Ryan in F.L. Stevens. Bernice P. Bishop Museum Bulletin 19:75. 1925. **Syntypes** [additional ILL numbers and all BISH numbers cited are isosyntypes]: On *Maba sandwicensis* A. DC. [= *Diospyros sandwicensis* (A. DC.) Fosb.], Hawaii, Oahu, Makaleha Valley, 1914, leg. O.H. Swezey, as F.L. Stevens-C.N. Forbes No. 1995, ILL 7200 and 7202, BISH 499851; on *Maba hillebrandii* Seem. [= *Diospyros hillebrandii* (Seem.) Fosb.], 8 JUL 1922 [as JAN on the packet], leg. O.H. Swezey s.n., ILL 7199 and 7201, BISH 499849 and 499850.

Asterinella phoradendri R.W. Ryan. Mycologia 16:189. 1924. **Holotype:** On *Phoradendron* sp., Porto Rico [Puerto Rico], Maricao, [18 NOV 1913], leg. F.L. Stevens No. 4894, ILL 7207. **Paratypes:** On *Phoradendron* sp. [as *P. latifolium* (Sw.) Griseb. on the packets = *P. piperoides* (Kunth in H.B.K.) Trelease], Maricao, [1915], leg. F.L. Stevens No. 8717, ILL 7206; Las Marias, [11 JUL 1915], leg. F.L. Stevens No. 8228, ILL 7205.

Asteromyxa surigaoensis J.M. Mendoza. Philippine Journal of Science 49:190–191, pl. VI, figs. 1–7. 1932. **Holotype:** On unknown host (Annonaceae), Philippines, Mindanao [Island], Surigao Prov., Surigao, 18 MAY 1919 [as JUN on the packet], leg. M. Ramos & J. Pascasio, ex Phil. Bur. Sci. No. 35901, ILL 7210.

Asterostomella acalyphae H. Sydow. Annales Mycologici 23:415–416. 1925. **Isotype:** Hab. in foliis *Acalyphae macrostachyae* var. *hirsutissimae* [= *Acalypha macrostachya* Jacquin var. *hirsutissima* (Willd.) Muell. Arg.], Costa Rica, San Ramón, 22 JAN 1925, leg. H. Sydow, Fungi in Itinere Costaricensi Collecti No. 206, ILL 6893 [also an isosyntype of *Asterina acalyphae* H. Sydow].

Asterostomella isothea H. Sydow. Annales Mycologici 25:135. 1927. **Isotype:** Hab. in foliis vivis *Triumfettae semitrilobae* [= *Triumfetta semitriloba* Jacquin], Costa Rica, San Pedro de San Ramón, 6 FEB 1925, leg. H. Sydow, Fungi in Itinere Costaricensi Collecti No. 189, ILL 7013 [also an isotype of *Asterina isothea* H. Sydow].

Asterostroma bicolor J.B. Ellis & B.M. Everhart. Proceedings of the Academy of Natural Sciences of Philadelphia 1893:441. 1894. **Isotype:** On rotten wood, Delaware, Wilmington, [20] OCT 1893, leg. A. Commons No. 2356, ILL 32713. **Holotype:** NY.

Asterostromella ochroleuca H. Bourdot & A. Galzin. Bulletin de la Société Mycologique de France 27:266. 1911. **Isotype:** Ad frustula *Ulmi* [= debris of *Ulmus*] humum et lapides investiens [as "corpora vicina investiens" on the packet], France, Aveyron, [Beno, pr. St. Sernin, 26 APR 1911], leg. A. Galzin [as H. Bourdot No. 7921, ex herb. G. Bresadola on the packet], ILL 32589, ex NY. ≡ *Vararia ochroleuca* (H. Bourdot & A. Galzin) M.A. Donk. Nederlandsch Kruidkundig Archief 1930:79. 1930. [Superfluous later publications of this combination, based on the same type, were authored by S. Lundell in S. Lundell & J. Nannfeldt. Fungi Exsiccata Suecici Fasc. 43–44, Schedae 19. 1953, and by G.H. Cunningham. Transactions of the Royal Society of New Zealand 821:981. 1955.].

Atractobasidium corticioides G.W. Martin. Bulletin of the Torrey Botanical Club 62:340–342, fig. 2. 1935. **Isotypes:** Mexico, Vera Cruz, Jalapa, 1894–1896, leg. C.L. Smith No. 213, ILL 32817, FH, NY, BPI. **Holotype:** IA [currently at ISC].

Aulacostroma osmanthi F.L. Stevens & R.W. Ryan in F.L. Stevens. Bernice P. Bishop Museum Bulletin 19:63–64, text fig. 12, pl. VI(A,B). 1925. **Holotype:** On *Osmanthus sandwicensis* (A. Gray) Knobloch [= *Nestegis sandwicensis* (A. Gray) O. & I. Degener & L. Johnson], Hawaii, Oahu, Waialae [Maunalua on the packet], 21 JUN 1921 [as 29 MAY on the packet], leg. A.F. Judd [No 437(?)] as F.L. Stevens No. 136, ILL 7211a. **Isotype:** ILL 7221b [no type number is given in the protologue but Stevens No. 136 is cited in the caption of pl. VI and also marked "type" on the original packet; the leaf in the packet of ILL 7211a is the one illustrated in pl. VI(A)].

Aulographella baumeae F.L. Stevens & R.W. Ryan in F.L. Stevens. Bernice P. Bishop Museum Bulletin 19:77. 1925. **Holotype:** On *Baumea meyenii* Kunth [= *Machaerina mariscoides* (Gaudich.) J. Kern], Hawaii, Oahu, Waiahole ditch trail, 12 JUN 1921, leg. F.L. Stevens No. 390, ILL 7212. **Isotypes:** BISH 499847 and 499848.

Aulographum cestri R.W. Ryan. Mycologia 16:190. 1924. **Holotype:** On *Cestrum* sp., Porto Rico [Puerto Rico], Mayagüez, [4 JAN 1918], leg. F.L. Stevens No. 7576, ILL 7213 [also an isotype of *Scolecopeltis cestri* R.A. Toro].

Aulographum panici-maximi R. Ciferri. Sydowia Annales Mycologici 10:142–143. (1956) 1957. **Isotype:** Hab. in culmis siccis *Panici maximi* [= *Panicum maximum* Jacquin, Dominican Republic], Santiago Prov., Santiago, Valle del Cibao, in pratis, MAR 1930, leg. R. Ciferri, Mycoflora Domingensis Exsiccata No. 372, ILL 33103.

Auricularia ampla C.H. Persoon in C. Gaudichaud-Beaupré. Voyage autour du monde, entrepris par ordre du roi ... exécuté sur les corvettes de S.M. l'Uranie et la Physicienne, pendant les années 1817, 1818, 1819 et 1820; publié ... par M. Louis de Freycinet. Botanique par M. Charles Gaudichaud, pharmacien de la marine, Paris, Part 5, p. 177. 1827. **Isosyntype**: In insulis Marianis ..., leg. C. Gaudichaud-Beaupré, ex herb. C.H. Persoon, ILL 32814. **Syntype**(?): L.

B

Bagnisiopsis palmigena (O.A. Plunket in F.L. Stevens) F. Petrak, 1929. **See basionym**: *Coccostromopsis palmigena* O.A. Plunket in F.L. Stevens, 1923.

Bakerophoma sacchari H. Diedicke. Annales Mycologici 14:63. 1916. **Assumed isotype**: Auf der Oberseite der Blattbasis von *Saccharum officinarum* L., Philippines [Luzon, Launa Prov., Mount Maquiling, near] Los Baños, 26 DEC 1913, leg. C.F. Baker No. 2367 [ex Fungi Malayana No. 116], ILL 10992.

Balladyna melodori H. & P. Sydow. Philippine Journal of Science 9, Section C (Botany):160. 1914. **Isotype**: On leaves of *Melodorum* sp., Philippines, Palawan [Island], Taytay, MAY 1913, leg. E.D. Merrill No. 8885, ILL 6632.

Beccopycnidium palmicola F.L. Stevens. Annales Mycologici 28:369. 1930 [as *palmicolum*, corrected in accordance with Articles 32.6, 60 & 61 of I.C.B.N.]. **Holotype**: On Palmae indet. [= Arecaceae sp. indet.], British Guiana [Guyana], Demerara-Essequibo R.R., 15 JUL 1922, leg. F.L. Stevens No. 369, ILL 10993.

Beelia suttoniae F.L. Stevens & R.W. Ryan in F.L. Stevens. Bernice P. Bishop Museum Bulletin 19:71, text fig. 14b. 1925. **Holotype**: On *Suttonia lanaiensis* (Hillebr.) Mez [= *Myrsine lanaiensis* Hillebr.], Hawaii, Lanai, leg. G.C. Munro, as F.L. Stevens No. 421, ILL 6620. **Isotypes**: BISH 499845 and 499846.

Bioscypha cyatheae H. Sydow. Annales Mycologici 25:103–105. 1927. **Isotype**: Hab. in frondibus vivis *Cyatheae* spec. [= *Cyathea* sp.], Costa Rica, Piedades de San Ramón, 7 FEB 1925, leg. H. Sydow, Fungi in Itinere Costaricensi Collecti No. 63, ILL 7658.

Bisbyopeltis phoebesii A.C. Batista & A.F. Vital in A.C. Batista, C.A.A. Costa & A.F. Vital. Anais da Sociedade de Biologia de Pernambuco 15(2):403–405, figs. 2, 3. 1957. **Holotype**: In foliis vivis *Phoebes tonduzii* [= *Phoebe tonduzii* Mez], with *Chaetothyrium permixtum* H. Sydow, Costa Rica, Grecia, 19 JAN 1925, leg. H. Sydow, Fungi in Itinere Costaricensi Collecti No. 160i, ILL 6653 [also an isotype of *Chaetothyrium permixtum* H. Sydow].

Bitancourtia cassythae M.J. Thirumalachar & A.E. Jenkins. Mycologia 45:782–783, fig. 1(A–G), fig. 2. 1953. **Isotypes**: In caulibus *Cassythae filiformis* [= *Cassytha filiformis* L.], South India, [Mysore State], Bengalore, Yashavantapur, 1 JAN 1948, leg. M.J. Narasimhan [ex A.E. Jenkins and A.A. Bitancourt, Myriangiales Selecti Exsiccati, Fascicle 10, No.488], ILL 33531 and 33532, BISH 506362 and 606363, BPI 90598, HCIO, IBI 5319, K.

Bitzea ingae (H. Sydow) E.B. Mains, 1939. **See basionym**: *Maravalia ingae* H. Sydow, 1925.

Blastotrichum miconiae F.L. Stevens. Transactions of the Illinois State Academy of Science 10:202. 1917. **Syntypes**: On *Miconia laevigata* (L.) D. Don in Sweet, Porto Rico [Puerto Rico], Maricao, [18 NOV 1913], leg. F.L. Stevens No. 4822, ILL 13884; Utuado, [30 DEC 1913], leg. F.L. Stevens No. 6871, ILL 13883 [also the holotype of *Borinquenia miconiae* F.L. Stevens and a paratype of *Hyalosphaera miconiae* F.L. Stevens]; leg. F.L. Stevens No. 6862, ILL 13882 [also a paratype of *Borinquenia miconiae* F.L. Stevens and of *Hyalosphaera miconiae* F.L. Stevens]; Aguas Buenas, [9 FEB 1913], leg. F.L. Stevens No. 302, ILL 7254 [also the holotype of *Echidnodella miconiae* R.W. Ryan], and ILL 16237 [also the holotype of *Microclava miconiae* F.L. Stevens; No. 302 is also cited as a paratype of *Hyalosphaera miconiae* F.L. Stevens].

Blennoria lawsoniana P.A. Saccardo. Annales Mycologici 7:436. 1909. **Assumed isotype**: Hab. in ramulis morientibus *Chamaecyparis lawsonianae* [= *C. lawsoniana* (A. Murray) Parl., Brandenburg, Baumschulen] pr. Tamsel [now Poland, 16 JUN 1909], leg. P. Vogel No. 69 [ex H. Sydow, Mycotheca Germanica No. 832], ILL 13217.

Bolosphaera cyanomela H. Sydow. Annales Mycologici 24:335–337. 1926. **Isotype**: Hab. parasitica in mycelio *Meliolae uncitrichae* [= *Meliola uncitricha* H. Sydow] ad folia *Phoebes neurophyllae* [= *Phoebe neurophylla* Mez & Pittier], Costa Rica, Cerro de San Isidro pr. San Ramón, 9 FEB 1925, leg. H. Sydow, Fungi in Itinere Costaricensi Collecti No. 169b, ILL 9352. ≡ *Dimerium cyanomelum* (H. Sydow) C.G. Hansford. Mycological Papers. Commonwealth Mycological Institute 15:78. 1946.

Borinquenia miconiae F.L. Stevens. Transactions of the Illinois State Academy of Science 10:173–174, fig. 3. 1917. **Holotype**: On *Miconia laevigata* (L.) D. Don in Sweet, Porto Rico [Puerto Rico], Utuado, [30 DEC 1913], leg. F.L. Stevens No. 6871, ILL 13883. **Paratype**: F.L. Stevens No. 6862, ILL 13882 [F.L. Stevens Nos. 6860 and 6861 are also syntypes of *Blastotrichum miconiae* F.L. Stevens and paratypes of *Hyalosphaera miconiae* F.L. Stevens].

Botryodiplodia gossypii J.B. Ellis & E. Bartholomew. Journal of Mycology 8:175–176. 1902. **Isotype**: On dead stems of *Gossypium herbaceum* [sensu auct. non L. = *Gossypium* sp.], Alabama, Tuskegee, 29 JUL 1901, leg. G.W. Carver, ex J.B. Ellis & B.M. Everhart, Fungi Columbiani No. 1510, ILL 10995.

Botryosphaeria dasylirii (C.H. Peck) F. Theissen & H. Sydow, 1915. **See basionym**: *Dothidea dasylirii* C.H. Peck, 1882.

Botrytis epichloes J.B. Ellis & J. Dearness. Canadian Record of Science 5(5):272. 1893. **Isotype**: Parasitic on *Epichloe typhina* (C.H. Persoon:E.M. Fries) L.R. & C. Tulasne, Canada, [Ontario], London, JUL 1892, leg. J. Dearness No. 1943, ex J.B. Ellis & B.M. Everhart, North American Fungi No. 2871, ILL 13899. **Holotype**: DAOM.

Botrytis hypophylla J.B. Ellis & W.A. Kellerman. Journal of Mycology 5:143. 1889. **Isotype**: On living leaves of *Teucrium canadense* L., Kansas, Manhattan, [15] OCT 1884, leg. M.A. Carleton No. 142, ILL 13904. **Holotype**: NY.

Boudiera claussenii P.C. Hennings. Beiblatt zur Hedwigia 42:(182), figs. 1–3. 1903. **Isotype**: Auf Kaninchenkot [rabbit dung, Germany], Freiburg, (Baden), APR 1903 [as AUG on the packet], leg. P. Claussen [ex H. & P. Sydow, Mycotheca Germanica No.132], ILL 7757.

Brachysporium sphaerocolum F.E. & E.S. Clements. Cryptogamae Formationum Coloradensium, Century 1, No. 68. Anno 1906, nom. nud. **Isotype**: Colorado, Larkspur Dell, alt. 2500 m, 11 JUL 1905, leg. F.E. and E.S. Clements, ILL 14598.

Brefeldiella brasiliensis C.L. Spegazzini. Boletín de la Academia Nacional de Ciencias en Córdoba 11:558. 1889; [Fungi Puiggariani, p. 180. 1889 (as *Breffeldiella*)]. **Isotype**: Ad folia viva Bambusaceae [Poaceae trib. Bambuseae] cujusdam in uliginosis, [Brazil], pr. Apiahy [Apiai], AUG 1888, leg. C.L. Spegazzini No. 2356, ILL 6825. **Holotype**: LPS.

Brefeldiella chilensis C.L. Spegazzini. Boletín de la Academia Nacional de Ciencias en Córdoba 25:94–95, illustr. 1921. **Isotype**: Hab. sobre las hojas vivas de *Villaresia mucronata* Ruiz & Pavon [= *Citronella mucronata* (Ruiz & Pavon) D. Don, Chile?], Cerca de Los Perales, Verano 1918, leg. C.L. Spegazzini, ILL 6824 [as a microscopic preparation].

Bremia domingensis R. Ciferri. Sydowia Annales Mycologici 10:131–132. (1956) 1957. **Isotype**: Hab. in foliis *Parthenii hysterophorii* [= *Parthenium hysterophorus* L.] (Compositae), [Dominican Republic], Santiago Prov., Valle del Cibao, Hato del Yaque, in ruderatis, tempora pluviosa. Anni 1930–1931, leg. R. Ciferri, Mycoflora Domingensis Exsiccata No. 357, ILL 33169.

Bulgaria nana E.K. Cash. Beihefte zur Sydowia Annales Mycologici 1:289–

291. 1957. **Isotype**: On bark of unknown tree, Chile, Corral, DEC 1905, leg. R. Thaxter s.n., ILL 33546. **Holotype**: BPI.

Byssocallis phoebes H. Sydow. Annales Mycologici 25:14–16. 1927. **Isotype**: Hab. parasitica in mycelio *Meliolae* [= *Meliola* sp.] ad folia viva *Phoebes tonduzii* [= *Phoebe tonduzii* Mez], Costa Rica, Grecia, 19 JAN 1925, leg. H. Sydow, Fungi in Itinere Costaricensi Collecti No. 160a, ILL 8149.

C

Caeoma torreyae L. Bonar. Mycologia 43:62–64, 65, figs. 1–5. 1951. **Paratype**: On leaves of *Torreya californica* Torrey, California, Marin County, near summit of Mount Tamalpais, 1 OCT 1949, leg. L. Bonar [California Fungi No. 761], ILL 31481 [also an isotype of *Clasterosporium obclavatum* L. Bonar].

Calliospora holwayi J.C. Arthur. Botanical Gazette 39:390. 1905. **Isotype**: On *Eysenhardtia amorphoides* Kunth in H.B.K., [Mexico], Guadalajara, 28 SEP 1903, leg. E.W.D. Holway No. 5059, ILL 18373. **Paratype**: 29 SEP 1903, leg. E.W.D. Holway No. 5068, ILL 18372. **Holotype**: PUR 7349. ≡ *Uropyxis holwayi* (J.C. Arthur) J.C. Arthur. Manual of the rusts of United States and Canada, p. 77. 1934.

Calolepis congesta H. Sydow. Annales Mycologici 23:399–402, fig. 11. 1925. **Isotype**: Hab. in foliis vivis vel languidis *Mauriae biringo* [= *Mauria biringo* Tulasne], saepe in mycelio *Henningsomycetis escharoidis* [= *Henningsomyces escharoides* H. Sydow], parasitans, Costa Rica, Grecia, 17 JAN 1925, leg. H. Sydow, Fungi in Itinere Costaricensi Collecti No. 248, ILL 6660 [also an isotype of *Episoma parasiticum* H. Sydow and of *Henningsomyces escharoides* H. Sydow].

Calonectria graminicola F.L. Stevens. Botanical Gazette 65:232. 1918. **Holotype**: On *Meliola panici* F.S. Earle on *Lasiacis compacta* (Sw.) A.S. Hitchc., Porto Rico [Puerto Rico], Utuado, [8 NOV 1913], leg. F.L. Stevens No. 4663, ILL 5762a. **Isotypes**: ILL 5762b, CUP , NY. [The specimen at CUP was published as lectotype by A.Y. Rossman (Mycotaxon 8:514. 1979); she did not examine the holotype]. **Paratypes**: On *Lasiacis divaricata* (L.) A.S. Hitchc. [as *Lasiasis*], Manati, [5 NOV 1913], leg. F.L. Stevens No. 4298, ILL 5768; on *Meliola andirae* F.S. Earle on *Andira jamaicensis* (W. Wright) Urban [= *Andira inermis* (W. Wright) Kunth ex DC.], Manati, [25 NOV 1913], leg. F.L. Stevens No. 5269, ILL 4595 and 8152. ≡ *Melioliphila graminicola* (F.L. Stevens) C.L.

Spegazzini. Boletín de la Academia Nacional de Ciencias en Córdoba 26:345–346. Preprint 1923 [journal part issued in 1924].

Calopeltis acnisti H. Sydow. Annales Mycologici 23:393–395, fig. 10. 1925. **Isosyntypes:** In foliis vivis *Acnisti arborescentis* [= *Acnistus arborescens* (L.) Schlecht.], Costa Rica, San Ramón, 22 JAN 1925, leg. H. Sydow, Fungi in Itinere Costaricensi Collecti No. 223, ILL 7216; 24 JAN 1925, leg. H. Sydow, ibid. No. 352, ILL 7218; Grecia, 12 JAN 1925, leg. H. Sydow, ibid. No. 353, ILL 7217; La Caja pr. San José, 3 JAN 1925, leg. H. Sydow, ibid. No. 355, ILL 7215.

Calosphaeria (Togninia) inconspicua H. Rehm. Leaflets of Philippine Botany 6:2213. 1914. **Assumed isotype:** Ad *Gigantochloam scribnerianam* [*Gigantochloa scribneriana* Merrill = *G. levis* (Blanco) Merrill], Philippines, Luzon, Laguna Prov., [Mount Maquiling, near] Los Baños, SEP 1913, leg. M.B. Raimundo, comm. C.F. Baker No. 1698 [ex Fungi Malayana No. 13], ILL 10511.

Calospora allantospora J.B. Ellis & B.M. Everhart. Journal of Mycology 9:223. 1903. **Isosyntype**: On maple, *Acer saccharinum* L., Canada, [Ontario], London, [27] OCT 1903, leg. J. Dearness No. 2010 [ex J.B. Ellis and B.M. Everhart, Fungi Columbiani No. 1912], ILL 10489.

Calothyriella osmanthi F.L. Stevens & R.W. Ryan in F.L. Stevens. Bernice P. Bishop Museum Bulletin 19:70, pl. VI(G). 1925. **Holotype:** On *Osmanthus sandwicensis* (A. Gray) Knobloch [= *Nestegis sandwicensis* (A. Gray) O. & I. Degener & L. Johnson], Hawaii, Oahu, Maunalua, 29 MAY 1921, leg. F.L. Stevens No. 135, ILL 7219. **Isotypes:** BISH 145748 and 499844.

Calothyriopeltis clermontiae F.L. Stevens & R.W. Ryan in F.L. Stevens. Bernice P. Bishop Museum Bulletin 19:72. 1925. **Holotype:** On *Clermontia oblongifolia* Gaudich., Hawaii, Kauai, Kalalau trail, 16 JUN 1921, leg. F.L. Stevens No. 478, ILL 7225. **Isotypes:** BISH 499842 and 499843.

Calothyriopeltis metrosideri F.L. Stevens & R.W. Ryan in F.L. Stevens. Bernice P. Bishop Museum Bulletin 19:72. 1925. **Holotype:** On *Metrosideros* sp., Hawaii, Oahu, Tantalus, 22 JUN 1921, leg. F.L. Stevens No. 636, ILL 7226. **Isotypes:** BISH 499840 and 499841. **Paratypes:** On *Lobelia* sp., [Island of] Hawaii, Kealakekua [as Bishop Estate Road, Keauhou Kona on the packets, and also cited as such under *Meliola lobeliae* F.L. Stevens, op. cit. p. 29], 25 JUL 1921, leg. F.L. Stevens No. 979, ILL 5407 and 7227 [also paratypes of *Meliola lobeliae* F.L. Stevens].

Calothyriopeltis scaevolae F.L. Stevens & R.W. Ryan in F.L. Stevens. Bernice P. Bishop Museum Bulletin 19:71, text fig. 14c. 1925. **Assumed holotype:** On *Scaevola* sp. [almost certainly misidentified, the plant probably = *Psychotria* sp.], Hawaii, Kauai, Kalalau trail, 16 JUN 1921, leg. F.L. Stevens No. 476, ILL 3613. **Paratype:** Leg. F.L. Stevens No. 473, ILL 7230.

Calothyrium hippocrateae R.W. Ryan. Mycologia 16:179.1924. **Assumed holotype:** On *Hippocratea volubilis* L., Porto Rico [Puerto Rico, 14 JUN 1915], leg. F.L. Stevens [No. 7033], ILL 7220 [the protologue indicates "no data"].

Calothyrium ingae R.W. Ryan. Mycologia 16:179–180. 1924. **Assumed holotype:** On *Inga vera* Willd., Porto Rico [Puerto Rico, 1915], leg. F.L. Stevens s.n., ILL 7221 [the protologue indicates "no data"].

Calothyrium osmanthi F.L. Stevens & R.W. Ryan in F.L. Stevens. Bernice P. Bishop Museum Bulletin 19:71. 1925. **Holotype:** On *Osmanthus sandwicensis* (A. Gray) Knobloch [= *Nestegis sandwicensis* (A. Gray) O. & I. Degener & L. Johnson], Hawaii, Oahu, Ahren's ditch trail, 8 JUN 1921, leg. F.L. Stevens No. 290, ILL 7222. **Paratype:** Hawaii, Kauai, Kalalau trail, 16 JUN 1921, leg. F.L. Stevens [no number cited in the protologue, but as No. 513 on the packet], ILL 7223. [F.L. Stevens No. 513 p.p. is also cited as the type collection of *Meliola osmanthicola* C.G. Hansford and of *M. osmanthina* C.G. Hansford. Isotypes of these two names are filed as ILL 5732, 5733, and 5734; both holotypes are at S.].

Calothyrium psychotriae R.W. Ryan. Mycologia 16:179. 1924. **Holotype:** On *Psychotria* sp., Porto Rico [Puerto Rico], Preston's Ranch, [31 DEC 1913], leg. F.L. Stevens No. 6662, ILL 7224.

Calothyrium suttoniae F.L. Stevens & R.W. Ryan in F.L. Stevens. Bernice P. Bishop Museum Bulletin 19:71. 1925. **Type:** On *Suttonia sandwicensis* (A. DC.) Mez [= *Myrsine* sp., the species name uncertain], Hawaii, [Island of] Hawaii, Hamakua, upper ditch trail, 31 JUL 1921, leg. F.L. Stevens No. 143 p.p. [this number perhaps cited in error. Under ILL 7114, it is applied to the holotype of *Asterina suttoniae* F.L. Stevens & R.W. Ryan in F.L. Stevens, and under ILL 3540 to the holotype of *Actinodothis suttoniae* F.L. Stevens. The host plant in both packets, however, clearly belongs in *Myrsine (Suttonia) lessertiana* A. DC. In addition, the type locality cited in the protologue differs from that on all our packet labels of F.L. Stevens No. 143, as well as the protologues of the other two fungus names mentioned above. We have not found an original label of *Calothyrium suttoniae* attached to any specimen at ILL; nor has the name been added to any packet label originally inscribed with the other two fungus names. A collection with the locality information and the date cited in the protologue, as well as the correctly identified host (= *Myrsine sandwicensis* A. DC.), is F.L. Stevens No. 1083, ILL 7115; it is labeled *Asterina suttoniae*.].

Calycellina sadleriae (F.L. Stevens & P.A. Young in F.L. Stevens) R.W.G. Dennis, 1963. **See basionym**: *Dasyscypha sadleriae* F.L. Stevens & P.A. Young in F.L. Stevens, 1925.

Camarosporium astericola J.B. Ellis & E. Bartholomew. Journal of Mycology 8:176. 1902 [as *astericolum*, corrected in accordance with Articles 32.6, 60 & 61 of I.C.B.N.]. **Isotype**: On dead stems of *Aster multiflorus* Aiton [= *Aster ericoides* L.], Kansas, Rooks County, [3] JUN 1901, leg. E. Bartholomew No. 2884 [ex J.B. Ellis & B.M. Everhart, Fungi Columbiana No. 1512], ILL 10996.

Camarosporium coronillae (P.A. Saccardo & C.L. Spegazzini) P.A. Saccardo forma *sophorae* H. & P. Sydow. Annales Mycologici 3:420. 1905. **Isotype**: Auf Aesten von *Sophora japonica* L., Brandenburg: Baumschulen zu Tamsel [now Poland], 26 FEB 1905, leg. P. Vogel, ex H. & P. Sydow, Mycotheca Germanica No. 421, ILL 10998.

Camarosporium graminicola J.B. Ellis & B.M. Everhart. Proceedings of the Academy of Natural Sciences of Philadelphia 1893:161. 1894 [as *graminicolum*, corrected in accordance with Articles 32.6, 60 & 61 of I.C.B.N.]. **Isotype**: On dead culms of *Ammophila arenaria* (L.) Link, New York, Long Island, JUL 1892, leg. S.E. Jelliffe [ex J.B. Ellis and B.M. Everhart, North American Fungi No. 2863], ILL 10999. **Holotype**: NY.

Camptomeris calliandrae H. Sydow. Annales Mycologici 25:143–144, fig. 6. 1927. **Isotype**: Hab. in foliis *Calliandrae similis* [= *Calliandra similis* Sprague & Riley], Costa Rica, Desamparados, 30 DEC 1924, leg. H. Sydow, Fungi in Itinere Costaricensi Collecti No. 89, ILL 14600.

Camptomeris floridana E.A. Bessey. Mycologia 45:377, 379, figs. 5–7. 1953. **Isotypes**: On *Pithecellobium unguis-cati* (L.) Bentham, Florida, [Lee County], Fort Myers, 30 DEC 1950, leg. E.A. Bessey s.n., ILL 33493, BPI, FLAS, IMI, MSC, NY.

Capnodium mucronatum J.P.F.C. Montagne. Annales des Sciences Naturelles, Botanique, Series 4, 14:175–176. 1860. **Isotype**: Hab. ad ramos et folia *Weinmanniae trichospermae* [= *Weinmannia trichosperma* Cav.], Chile, leg. C. Gay s.n., ILL 6485. ≡ *Meliola mucronata* (J.P.F.C. Montagne) P.A. Saccardo. Sylloge Fungorum 1:62. 1882. ≡ *Phaeocapnias mucronata* (J.P.F.C. Montagne) R. Ciferri & A.C. Batista in A.C. Batista & R. Ciferri. Saccardoa 2:176–177. 1963. ≡ *Euantennaria mucronata* (J.P.F.C. Montagne) S.J. Hughes. New Zealand Journal of Botany 10:227. 1972.

Catacauma contractum H. Sydow. Annales Mycologici 23:365–367. 1925. **Isotype**: Hab. in foliis *Gouaniae tomentosae* [= *Gouania tomentosa* Jacquin], Costa Rica, La Caja, pr. San José, 6 JAN 1925, leg. H. Sydow, Fungi in Itinere Costaricensi Collecti No. 273, ILL 8496.

Catacauma costaricensis F.L. Stevens. Illinois Biological Monographs

11:184, pl. XV, fig. 108. 1927 [misspelled *Catacouma costaricens* in the protologue but the epithet spelled correctly in the index and figure caption]. **Holotype:** On *Myrcia costaricensis* Berg, Costa Rica, La Palma, 8 JUL 1923, leg. F.L. Stevens No. 287, ILL 8497.

Catacauma galactiae F.L. Stevens. Annales Mycologici 29:102–103. 1931. **Holotype:** On *Galactia speciosa* (DC.) Britton, Peru, Palca, 6 DEC 1924, leg. F.L. Stevens No. 36, ILL 8499. ≡ *Phyllachora nitens* (J.H. Léveillé) M.C. Cooke ssp. *isthmea* P.F. Cannon. Mycological Papers. International Mycological Institute 163:147. 1991, nom. and stat. nov., non *Phyllachora nitens* (J.H. Léveillé) M.C. Cooke ssp. *galactiae* (F.S. Earle) P.F. Cannon, 1991.

Catacauma ocoteae F.L. Stevens. Botanical Gazette 69:251. 1920. **Holotype:** On *Ocotea leucoxylon* (Sw.) de Lanessan, Porto Rico [Puerto Rico], Monte Alegrillo, [14 NOV 1913], leg. F.L. Stevens No. 4725, ILL 8506. **Isotype:** ILL 8963 [F.L. Stevens No. 4725 is also cited as paratype of *Phyllachora ocoteicola* F.L. Stevens & N.E. Dalbey]. **Paratypes:** [10 MAY 1913], leg. F.L. Stevens No. 1347, ILL 8507, [Maricao, 4 MAR 1913], leg. F.L. Stevens No. 732, ILL 8505.

Catacauma palmicola F.L. Stevens. Botanical Gazette 69:251–252, figs. 10–12. 1920. **Holotype:** On *Thrinax ponceana* O.F. Cook [= *T. morrisii* H. Wendl.], Porto Rico [Puerto Rico], Vega Baja, [2 JUL 1915], leg. F.L. Stevens No. 7716, ILL 8514a. **Isotypes:** ILL 8514b, c.

Catacauma peglerae E.M. Doidge. Bothalia 1(1):25–26. 1921. **Paratypes:** On *Eugenia capensis* (Acklon & Zeyher) Sonder, [South Africa], Natal [Prov.], Warner Beach, 1 APR 1918, leg. A.M. Bottomley, ex PREM No. 11667, ILL 8508; Scottsburgh, 5 JUL 1913, leg. I.B. Pole Evans, ex PREM No. 6841, ILL 8509.

Catacauma zanthoxyli F.L. Stevens. Illinois Biological Monographs 11:184, pl. VII, figs. 53, 54; pl. XV, fig. 109; pl. XVI, fig. 110. 1927 [the generic name misspelled *Catacouma* in the protologue]. **Holotype:** On *Zanthoxylum* sp. [as *Zanthoxylon*], Panama, Chagres [River], 3 miles from mouth, 23 AUG 1923, leg. F.L. Stevens No. 1290, ILL 8515.

Catacaumella gouaniae F.L. Stevens. Botanical Gazette 69:252, figs. 14, 15. 1920. **Holotype:** On *Gouania polygama* (Jacquin) Urban [misspelled *polygana* = *G. lupuloides* (L.) Urban], Porto Rico [Puerto Rico], Mayagüez, [31 OCT 1913], leg. F.L. Stevens No. 3923, ILL 8518. **Paratypes:** Salinas, [JAN 1914], leg. F.L. Stevens No. 6798, ILL 8521; Dos Bocas, [16 DEC 1913], leg. F.L. Stevens No. 6007, ILL 8516; [8 JUL 1915], leg. F.L. Stevens No. 8092, ILL 8523; Maricao, [10 JUL 1915], leg. F.L. Stevens No. 8953, ILL 8517; on *Gouania lupuloides* (L.) Urban, Mayagüez, [1 MAY 1913], leg. F.L. Stevens No. 1049, ILL 8522; Arecibo-Lares Road, [21 JUN 1915], leg. F.L. Stevens No. 7230, ILL 8519.

Caudella psidii R.W. Ryan. Mycologia 16:179. 1924. **Syntypes:** On *Psidium guajava* L., Porto Rico [Puerto Rico], Rio Tanamá, [6 JUL 1915], leg. F.L. Stevens No. 7834, ILL 7232; Las Marias, [10 JUL 1915], leg. F.L. Stevens No. 8128, ILL 7233; Mayagüez, [31 OCT 1913], leg. F.L. Stevens No. 3899, ILL 7236; Dos Bocas, below Utuado, [30 DEC 1913], leg. F.L. Stevens No. 6562, ILL 7234; San Sebastián, [as Sebastion, 22 NOV 1913], leg. F.L. Stevens No. 5202, ILL 7235.

Cenangium blumeanum H. Rehm. Leaflets of Philippine Botany 8:2927–2928. 1915. **Assumed isotype:** Ad *Bambusam blumeanam* [= *Bambusa blumeana* Blume ex J.H. Schultes], Philippines, [Luzon, Laguna Prov., Mount Maquiling near] Los Baños, MAR 1914, leg. M.B. Raimundo & C.F. Baker No. 2927b [ex C.F. Baker, Fungi Malayana No. 117], ILL 8064.

Cenangium yuccae F.E. & E.S. Clements. Cryptogamae Formationum Coloradensium, Century 6, No. 518. Anno 1908, nom. nud. **Isotype:** In foliis vetustis *Yuccae harrimaniae* [= *Yucca harrimaniae* Trel.], Colorado, Mesa Verde, alt. 2400 m, 6 JUL 1907, leg. F.E. & E.S. Clements, ILL 8078.

Cephalosporium bertholletianum E.R. Spencer. Botanical Gazette 72:279, fig. 3. 1921. **Holotype:** On radicle and seed of *Bertholletia nobilis* Miers & *B. excelsa* Humb. & Bonpl. ... obtained from wholesale firms in Chicago and retail grocery stores in Champaign and Urbana, Illinois, 1920, leg. E.R. Spencer s.n., ILL 13912 [a single packet holding three unmarked vials of material].

Ceratobasidium calosporum D.P. Rogers. University of Iowa Studies in Natural History 17:5, fig. 1. 1935. **Holotype:** On bark of a dead branch of *Ulmus* sp., [Iowa], Iowa City, Linder's Woods, 7 MAY 1932, leg. D.P. Rogers No. 224, ILL 32833.

Ceratobasidium obscurum D.P. Rogers. University of Iowa Studies in Natural History 17:6–7, fig. 3. 1935. **Holotype:** On lower side of a much rotted prostrate log of *Ulmus* sp., Iowa, in woods along Iowa River, east of North Liberty 11 JUN 1934, leg. D.P. Rogers No. 291, ILL 32836.

Ceratobasidium plumbeum G.W. Martin. Mycologia 31:513–514, figs. 21–27. 1939. **Isotypes:** Growing on underside of decaying log, in low forest, Panama, Canal Zone, 3 km east of Arraiján, 1 SEP 1937, leg. G.W. Martin No. 4597, ILL 32835, BPI [ex MO]. **Holotype:** IA [currently at ISC].

Ceratochaetopsis costaricensis F.L. Stevens & A.G. Weedon in F.L. Stevens. Illinois Biological Monographs 11:172–173. 1927. **Holotype:** On *Myrcia costaricensis* Berg, Costa Rica, La Palma, 8 JUL 1923, leg. F.L. Stevens No. 908, ILL 6648.

Cercoseptoria chamaesyces (F.L. Stevens & N.E. Dalbey) F. Petrak, 1925. **See basionym**: *Septoriopsis chamaesyces* F.L. Stevens & N.E. Dalbey, 1918.

Cercoseptoria piperis (F.L. Stevens & N.E. Dalbey) F. Petrak, 1925. **See basionym**: *Septoriopsis piperis* F.L. Stevens & N.E. Dalbey, 1918.

Cercospora agerati F.L. Stevens. Bernice P. Bishop Museum Bulletin 19:154. 1925. **Lectotype** [designated by C. Chupp (1953)]: On living leaves of *Ageratum conyzoides* L., Hawaii, [Island of] Hawaii, Kealakekua, 23 JUL [1921], leg. F.L. Stevens No. 944, ILL 16296. **Isolectotypes**: BISH 145751 and 499984. **Residual syntype**: Hawaii, [Island of] Hawaii, Wailuku River, 8 JUL [1921], leg. F.L. Stevens No. 750, ILL 16297. **Residual isosyntypes**: BISH 499042 and 499983. ≡ *Ragnhildiana agerati* (F.L. Stevens) F.L. Stevens & W.G. Solheim in W.G. Solheim & F.L. Stevens. Mycologia 23:402. 1931.

Cercospora arctii F.L. Stevens. Bernice P. Bishop Museum Bulletin 19:154. 1925. **Holotype**: On *Arctium lappa* L. (cult.), Hawaii, [Island of] Hawaii, Kukuihaele, leg. F.L. Stevens No. 1096, ILL 14722. **Isotypes**: BISH 499041 and 499982.

Cercospora bernardiae F.L. Stevens. Transactions of the Illinois State Academy of Science 10:213. 1917. **Holotype**: On *Bernardia bernardia* Millsp. (nom. illegit.) [= *B. dichotoma* (Willd.) Muell. Arg.], Porto Rico [Puerto Rico], Guanica, [3 FEB 1913], leg. F.L. Stevens No. 355a, ILL 14752. **Isotypes**: ILL 14750 and 14751, BPI.

Cercospora boringuensis E. Young. Mycologia 8:45. 1916. **Holotype**: On leaves of *Calopogonium orthocarpum* Urban [= *C. mucunoides* Desv.], Porto Rico [Puerto Rico], Mayagüez, [27 DEC 1913], leg. F.L. Stevens No. 6752, ILL 13949. ≡ *Cercosporina boringuensis* (E. Young) P.A. Saccardo ex A. Trotter. Sylloge Fungorum 25:905–906. 1931. ≡ *Didymaria boringuensis* (E. Young) F.L. Stevens & W.G. Solheim in W.G. Solheim & F.L. Stevens. Mycologia 23:400. 1931.

Cercospora brachypus J.B. Ellis & B.M. Everhart. Journal of Mycology 8:71–72. 1902. **Isotype**: On leaves of *Vitis rotundifolia* Michaux, Alabama, [Tuskegee, 20 SEP 1901], leg. G.W. Carver [ex J.B. Ellis and B.M. Everhart, Fungi Columbiani No. 1515], ILL 14802.

Cercospora bradburyae E. Young. Mycologia 8:46. 1916. **Holotype**: On leaves of *Bradburya pubescens* (Bentham) O. Kuntze [= *Centrosema pubescens* Bentham], Porto Rico [Puerto Rico], Rosario [15 FEB 1913], leg. F.L. Stevens No. 446, ILL 14818. **Paratypes**: Mayagüez, [25 DEC 1913], leg. F.L. Stevens No. 6296, ILL 14812 and 14813; [31 OCT 1913], leg. F.L. Stevens No. 3930, ILL 14815; [9 MAR 1913], leg. F.L. Stevens No. 479, ILL 14814; Luquillo Forest, [2 DEC 1913], leg. F.L. Stevens No. 5609, ILL14803; Dos Bocas below Utuado, [30 DEC 1913], leg. F.L. Stevens No. 6558, ILL 14810; San Germán, [8 NOV

1913], leg. F.L. Stevens No. 5785, ILL 14806; leg. F.L. Stevens No. 5796, ILL 14804; [12 DEC 1913], leg. F.L. Stevens No. 5833, ILL 14805; Guayama, [4 DEC 1913], leg. F.L. Stevens No. 5412, ILL 14807; Jayuya, [31 MAR 1913], leg. F.L. Stevens No. 446a, ILL 14817; Hormigueros, [14 JAN 1913], leg. F.L. Stevens No. 225a, ILL 14809; Cabo Rojo, [27 DEC 1913], leg. F.L. Stevens No. 6482, ILL 14808.

Cercospora calopogonii F.L. Stevens & W.G. Solheim in W.G. Solheim & F.L. Stevens. Mycologia 23:379, fig. 4. 1931. **Holotype:** On leaves of *Calopogonium* sp., Trinidad, St. Augustine, [13 AUG 1922], leg. F.L. Stevens No. 836, ILL 14992.

Cercospora carbonacea L.E. Miles. Transactions of the Illinois State Academy of Science 10:255, fig. 3. 1917. **Holotype:** On living leaves of *Dioscorea alata* L., Porto Rico [Puerto Rico], Vega Alta, [?1913], leg. F.L. Stevens No. 4178, ILL 15008. **Paratypes:** Vega Baja, [5 NOV 1913], leg. F.L. Stevens No. 4234, ILL 15009; Cabo Rojo, [27 DEC 1913], leg. F.L. Stevens No. 6469, ILL 15007; Añasco, [12 OCT 1913], leg. F.L. Stevens [& W.E. Hess] No. 3563, ILL 15010; Santa Ana, [31 DEC 1913], leg. F.L. Stevens No. 6687, ILL 15013.

Cercospora caseariae F.L. Stevens. Transactions of the Illinois State Academy of Science 10:212–213. 1917. **Lectotype** [designated by C. Chupp (1953)]: On *Casearia ramiflora* Vahl [= *C. guianensis* (Aublet) Urban], Porto Rico [Puerto Rico], Villa Alba, [3 JAN 1915, erroneously as JAN 1912 on retyped packet label], leg. F.L. Stevens No. 99, ILL 14954. **Isolectotype:** [With the original packet label, dated 3 JAN 1915], ILL 14953. **Residual syntypes and isosyntypes:** On *Casearia ramiflora* Vahl [= *C. guianensis* (Aublet) Urban], Porto Rico [Puerto Rico], Luquillo Forest, [2 DEC 1913], leg. F.L. Stevens No. 5556, ILL 14969; Quebradillas, [22 NOV 1913], leg. F.L. Stevens No. 5171, ILL 14971; Utuado, [8 NOV 1913], leg. F.L. Stevens No. 4691, ILL 14935; leg. F.L. Stevens No. 4675, ILL 14931; [as Dos Bocas on the packet, 8 JUL 1915], leg. F.L. Stevens No. 8051, ILL 14959; San Germán, [8 DEC 1913], leg. F.L. Stevens No. 4865, ILL 14929; [12 DEC 1913], leg. F.L. Stevens No. 5839, ILL 14933; Cataño, [6 NOV 1913], leg. F.L. Stevens No. 4190, ILL 14932 and 14972; Martin Peña, [11 AUG 1915], leg. F.L. Stevens No. 9306, ILL 14968; [12 AUG 1915], leg. F.L. Stevens No. 9330, ILL 14967; Mayagüez, [31 OCT 1913], leg. F.L. Stevens No. 3940, ILL 14930 and 14973; Aguada, [22 NOV 1913], leg. F.L. Stevens No. 5086, ILL 14964; Aguadilla, [25 NOV 1915], leg. F.L. Stevens No. 4858, ILL 14951; Maricao, [8 FEB 1915], leg. F.L. Stevens No. 370, ILL 14952; Vega Baja, [AUG 1915], leg. F.L. Stevens No. 9268, ILL 14962; Rio Tanamá, [7 JUL 1915], leg. F.L. Stevens No. 7925, ILL 14960; Jayuya, [31 MAR 1913], leg. F.L. Stevens No. X [s.n.], ILL 14965; Bayamon, [21 FEB 1913], leg. F.L. Stevens No.

387, ILL 14966; Preston's Ranch, [31 DEC 1913], leg. F.L. Stevens No. 6698, ILL 14963; Santa Catalina, [28 AUG 1913], leg. F.L. Stevens No. 2720, ILL 14961; on *Casearia sylvestris* Sw., Mayagüez, [15 APR 1913], leg. F.L. Stevens No. 524, ILL 14939; [31 OCT 1913], F.L. Stevens No. 3900, ILL 14936 and 14940; [27 JUL 1915], leg. F.L. Stevens No. 76, ILL 14937; [31 OCT 1913], leg. F.L. Stevens No. 3895, ILL 14938; Coamo, [20 JUN 1915], leg. F.L. Stevens No. 7275, ILL 14974, BPI; Rio Tanamá, [6 JUL 1915], leg. F.L. Stevens No. 7884, ILL 14941; leg. F.L. Stevens No. 7855, ILL 14942; Hormigueros, [23 JUN 1915], leg. F.L. Stevens No. 7364, ILL 14976; Lajas, [18 JUN 1915], leg. F.L. Stevens No. 7177, ILL 14975; Corozal, [21 FEB 1913], leg. F.L. Stevens No. 406, ILL 14956; Quebradillas, [22 NOV 1913], leg. F.L. Stevens No. 5010, ILL 14944; leg. F.L. Stevens [& W.E. Hess] No. 5004, ILL 14945; [20 JUN 1915], leg. F.L. Stevens No. 7273, ILL 14943; Ponce, [15 JUL 1915], leg. F.L. Stevens No. 8682, ILL 14955; Luquillo Forest, [2 DEC 1913], leg. F.L. Stevens No. 5431, ILL 14958; Monte de Oro, [3 DEC 1913], leg. F.L. Stevens No. 5714, ILL 14957; on *Casearia guianensis* (Aublet) Urban, Rosario, [27 OCT 1913], leg. F.L. Stevens No. 3801, ILL 14949; Corozal, [21 FEB 1915], leg. F.L. Stevens No. 420, ILL 14950; Mayaguez, [13 APR 1913], leg. F.L. Stevens No. 1386, ILL 14947 and 14948, BPI.

Cercospora cayaponiae F.L. Stevens & W.G. Solheim in W.G. Solheim & F.L. Stevens. Mycologia 23:386. 1931. **Holotype:** On leaves of *Cayaponia* sp., Porto Rico [Puerto Rico], Rosario, [27 OCT 1913], leg. F.L. Stevens No. 3777, ILL 15025. **Paratype:** Maricao, [18 NOV 1913], leg. F.L. Stevens No. 4815, ILL 15024. ≡ *Mycovellosiella cayoponiae* (F.L. Stevens & W.G. Solheim in W.G. Solheim & F.L. Stevens) M. Muntañola. Lilloa 30:205–206, text fig. 13. 1960.

Cercospora corni J.J. Davis. Transactions of the Wisconsin Academy of Sciences, Arts & Letters 18:268.1915 (1916). **Isotype:** On leaves of *Cornus paniculata* L'Héritier [= *C. racemosa* Lam.], Wisconsin, St. Croix Falls, 31 AUG 1914, leg. J.J. Davis, [ex Fungi Wisconsinenses Exsiccati No. 2], ILL 15146.

Cercospora costi F.L. Stevens. Illinois Biological Monographs 11:209. 1927. **Holotype:** On *Costus* sp., Panama, Gatun, 24 AUG 1923, leg. F.L. Stevens No. 1343, ILL 15148.

Cercospora cupaniae H. Sydow. Annales Mycologici 23:424–426. 1925. **Isotype:** Hab. in foliis *Cupaniae guatemalensis* [= *Cupania guatemalensis* Radlk.], Costa Rica, Grecia, 13 JAN 1925, leg. H. Sydow, Fungi in Itinere Costaricensi Collecti No. 351, ILL 15165.

Cercospora cylindrospora F.L. Stevens & W.G. Solheim in W.G. Solheim & F.L. Stevens. Mycologia 23:376, fig. 2. 1931. **Holotype:** On leaves of *Bradburya pubescens* (Bentham) O. Kuntze [= *Centrosema pubescens*

Bentham], Porto Rico [Puerto Rico], Cabo Rojo, [27 DEC 1913], leg. F.L. Stevens No. 6482a, ILL 15167.

Cercospora dolichi J.B. Ellis & B.M. Everhart. Journal of Mycology 5:71. 1889. **Isotype**: On leaves of *Dolichos sinensis* L. [= *Vigna unguiculata* (L.) Walp.], Mississippi, Starkville, SEP 1888, leg. S.M. Tracy [ex J.B. Ellis and B.M. Everhart, North American Fungi No. 2294], ILL 15204.

Cercospora echinochloae J.J. Davis. Transactions of the Wisconsin Academy of Sciences, Arts and Letters 18:106. 1916. **Isosyntype**: On leaves of *Echinochloa crus-galli* (L.) Beauvois, Wisconsin, [Sauk County], Devil's Lake, [9] AUG 1913, leg. J.J. Davis, [Fungi Wisconsinenses Exsiccati No. 14], ILL 15228.

Cercospora exitiosa H. & P. Sydow. Annales Mycologici 4:485–486. (1906) JAN 1907. **Isotype**: Hab. in ramis vivis vel subvivis *Tiliae platyphyllae* [= *Tilia platyphyllos* Scop., Brandenburg: Baumschulen zu] Tamsel [now Poland, JUN 1906], leg. P. Vogel, ex H. & P. Sydow, Mycotheca Germanica No. 545, ILL 15257.

Cercospora gonatoclada H. Sydow. Annales Mycologici 23:425. 1925. **Isosyntype**: Hab. in foliis *Iresines caleae* [= *Iresine calea* (Ibañez) Standley], Costa Rica, La Caja pr. San José, 5 JAN 1925, leg. H. Sydow, Fungi in Itinere Costaricensi Collecti No. 12, ILL 15294.

Cercospora guanicensis E. Young. Mycologia 8:45. 1916. **Holotype**: On leaves of *Guilandina crista* (L.) Small [= *Caesalpinia crista* L.], Porto Rico [Puerto Rico], Guanica, [29 DEC 1913], leg. F.L. Stevens No. 6840, ILL 15320. **Paratype**: Leg. F.L. Stevens No. 6845, ILL 15321. ≡ *Cercosporina guanicensis* (E. Young) P.A. Saccardo ex A. Trotter. Sylloge Fungorum 25:907. 1931.

Cercospora guianensis F.L. Stevens & W.G. Solheim in W.G. Solheim & F.L. Stevens. Mycologia 23:375, fig. 1. 1931. **Holotype**: On leaves of *Lantana* sp., British Guiana [Guyana], Rockstone, [13 JUL 1922], leg. F.L. Stevens No. 253, ILL 15322. **Isotype**: ILL 15323.

Cercospora hurae F.L. Stevens. Transactions of the Illinois State Academy of Science 10:210. 1917. **Holotype**: On *Hura crepitans* L., Porto Rico [Puerto Rico], Mayagüez, [DEC 1917], leg. F.L. Stevens No. 5830, ILL 15378 [also a syntype of *Colletotrichum curvisetum* F.L. Stevens]. **Paratypes**: [9 MAR 1913], leg. F.L. Stevens No. 478, ILL 15375 and 15377, BPI 70898; [27 JUL 1913], leg. F.L. Stevens No. 70, ILL 15379; Añasco, [19 OCT 1913], leg. F.L. Stevens No. 3594, ILL 15376 [also a syntype of *Colletotrichum curvisetum* F.L. Stevens].

Cercospora isanthi J.B. Ellis & W.A. Kellerman. Bulletin of the Torrey Botanical Club 11:115–116. 1884. **Isotype**: On leaves of *Isanthus* sp. [= *Trichostema* sp.], Kansas, [Riley County], Manhattan, [10] AUG [1884], leg. W.A. Kellerman No. 610, ILL 15405.

Cercospora ixorae W.G. Solheim. Indian Journal of Agricultural Science 3:915–916. 1933. **Assumed holotype:** On *Ixora coccinea* L., [India, Poona, Bombay, 1932], leg. B.N. Uppal No. 12, ILL 15406. **Isotype:** AMH.

Cercospora juglandis W.A. Kellerman & W.T. Swingle. Journal of Mycology 5:77. 1889. **Isotype:** On lower leaves of *Juglans nigra* L., Kansas, Manhattan, 19 AUG 1887, leg. W.A. Kellerman and W.T. Swingle No. 1079 [ex Kansas Fungi No. 34], ILL 21376.

Cercospora kalmiae J.B. Ellis & B.M. Everhart. Proceedings of the Academy of Natural Sciences of Philadelphia 1891:88. 1891. **Isotype:** On leaves of *Kalmia latifolia* L., New Jersey, Newfield, 1 JAN 1890, [leg. J.B. Ellis, ex J.B. Ellis and B.M. Everhart, North American Fungi No. 2591], ILL 15407.

Cercospora leonuri F.L. Stevens & W.G. Solheim in W.G. Solheim & F.L. Stevens. Mycologia 23:395–396, fig. 7. 1931. **Holotype:** On leaves of *Leonurus cardiaca* L., Costa Rica, Cartago, [22 JUN 1923], leg. F.L. Stevens No. 33, ILL 15411.

Cercospora malachrae E. Young. Mycologia 8:45–46. 1916, nom. illegit., non F.D. Heald & F.A. Wolf, 1911. **Holotype:** On leaves of *Malachra rotundifolia* Schrank [= *M. alceifolia* Jacquin], Porto Rico [Puerto Rico], San Sebastián, [22 NOV 1913], leg. F.L. Stevens No. 5199, ILL 14844. **Paratypes:** Ponce, [4 JAN 1913], leg. F.L. Stevens No. 5003, ILL 14840; Guanica, [10 FEB 1913], leg. F.L. Stevens No. 338a, ILL 14846; Yauco, [3 OCT 1913], leg. F.L. Stevens No. 3246, ILL 14841; San Germán, [12 DEC 1913], leg. F.L. Stevens No. 5840, ILL 14843; Vega Baja, [22 FEB 1913], leg. F.L. Stevens No. 431, ILL 14845 and 14847; leg. F.L. Stevens No. 381, ILL 14842. [= *C. malachrae* F.D. Heald & F.A. Wolf, 1911, but based on a different type (cf. C. Chupp, 1953)].

Cercospora malayensis F.L. Stevens & W.G. Solheim in W.G. Solheim & F.L. Stevens. Mycologia 23:394–395, fig. 6. 1931. **Holotype:** On leaves of *Hibiscus esculentus* L. [= *Abelmoschus esculentus* (L.) Moench], Philippines, [Luzon], Laguna Prov., Mount Maquiling near Los Baños, [JAN 1914], leg. C.F. Baker, Fungi Malayana No. 120, ILL 14824.

Cercospora maricaoensis E. Young. Mycologia 8:44–45. 1916. **Holotype:** On leaves of *Teramnus uncinatus* (L.) Sw., Porto Rico [Puerto Rico], Dos Bocas below Utuado, [30 DEC 1913], leg. F.L. Stevens No. 6554, ILL 14831. **Paratypes:** Maricao, [3 APR 1913], leg. F.L. Stevens No. 764, ILL 14832; San Germán, 8 NOV 1913], leg. F.L. Stevens No. 5815, ILL 14834; Cabo Rojo, [15 JUN 1913], leg. F.L. Stevens No. 2271, ILL 14833. ≡ *Cercosporina maricaoensis* (E. Young) P.A. Saccardo ex A. Trotter. Sylloge Fungorum 25:908. 1931.

Cercospora mikaniicola F.L. Stevens. Transactions of the Illinois State Academy of Science 10:213–214. 1917 [as *mikaniacola*, corrected in accordance with Articles 60 & 61 of I.C.B.N.]. **Holotype**: On *Mikania* sp. [misspelled "*Mikonia*" in the protologue], Porto Rico [Puerto Rico], Utuado, [7 JUL 1915], leg. F.L. Stevens No. 7923, ILL 14865. **Paratypes**: Aguada, [22 NOV 1913], leg. F.L. Stevens No. 5083, ILL 14864; Marico, [18 NOV 1915], leg. F.L. Stevens No. 4700, ILL 14863.

Cercospora negundinis J.B. Ellis & B.M. Everhart. Proceedings of the Academy of Natural Sciences of Philadelphia 1891:89. 1891. **Isotype**: On leaves of *Negundo aceroides* Moench [= *Acer negundo* L.], Nebraska, Lincoln, [14] AUG 1889, leg. R. Pound No. 37 [ex Reliquiae Seymourianae], ILL 14900.

Cercospora ocimicola F. Petrak & R. Ciferri. Annales Mycologici 30:324–325. 1932. **Isotype**: In foliis vivis *Ocimi micranthi* [*Ocimum micranthum* Willd. = *O. campechianum* P. Miller], Dominican Republic, Santiago Prov., Valle del Cibao, Santiago, Hato del Yaque, 26 NOV 1930, leg. E.L. Ekman No 3863, as R. Ciferri, Mycoflora Domingensis Exsiccata No. 359, ILL 33167.

Cercospora paspalicola F. Petrak & R. Ciferri. Annales Mycologici 30:326–327. 1932. **Isotype**: In foliis vivis *Paspali clavuliferi* [= *Paspalum clavuliferum* C. Wright, Republica Dominicana, [Dominican Republic], Valle de San Juan, Prov. de Azua, San Juan [de la Maguana], in fields, 22 AUG 1929, leg. E.L. Ekman No. 3777 p.p. [ex R. Ciferri, Mycotheca Domingensis Exsiccata No. 331], ILL 33101.

Cercospora penstemonis J.B. Ellis & W.A. Kellerman. Bulletin of the Torrey Botanical Club 11:121–122. 1884 [as *pentstemonis*; corrected in accordance with Rec. 60H.1 and Articles 60 & 61 of I.C.B.N.]. **Residual isosyntype**: On leaves of *Penstemon cobaea* Nutt. [as *Pentstemon* (orth. var.)], Kansas, [Riley County, Manhattan, 15] JUN [1884], leg. W.A. Kellerman No. 546, ILL 15444 [C. Chupp (1953) designated W.A. Kellerman No. 566 as the lectotype].

Cercospora pipturi F.L. Stevens & P.A. Glick in F.L. Stevens. Bernice P. Bishop Museum Bulletin 19:155, text fig. 33b. 1925. **Lectotype** [designated by C. Chupp (1953)]: On *Pipturus albidus* (Hooker & Arnott) A. Gray, Hawaii, Kauai, Kalalau trail, 16 JUN [1921], leg. F.L. Stevens No. 538, ILL 15471. **Isolectotypes**: ILL 15543, BISH 145766, 499909 and 500301. **Residual syntypes** [BISH numbers cited are residual isosyntypes]: Hawaii, [Island of] Hawaii, between Hilo and Kilauea, 10 JUL [1921], leg. F.L. Stevens No. 766, ILL 15469; Kapapala Ranch, 18 JUL 1919, leg. F.L. Stevens No. 894, ILL 15541, BISH 145674 and 500381; between Kona and Waimea, 27 JUL [1921], leg. F.L. Stevens No. 1020, ILL 15470; Oahu, Olympus, 24 JUN [1921], leg. F.L. Stevens No. 713, ILL 15542, BISH 145765 and 500380.

Cercospora porophylli F.L. Stevens & Moore in F.L. Stevens. Illinois Biological Monographs 11:210. 1927. **Holotype:** On *Porophyllum ruderale* (Jacquin) Cass., Costa Rica, Siquirres, 18 JUL 1923, leg. F.L. Stevens No. 554, ILL 15505.

Cercospora portoricensis F.S. Earle. Muhlenbergia 1:15–16. 1900. **Isotype:** On living leaves of *Piper aduncum* L. (with *Meliola glabra* M.J. Berkeley & M.A. Curtis on the upper surface of some leaves), Porto Rico [Puerto Rico], Mayagüez, alt. 400 ft., [23] JAN [1900], leg. A.A. Heller No. 4359 [as 4359a on the packet], ILL 4080. **Holotype:** NY.

Cercospora pulsatillae F.E. & E.S. Clements. Cryptogamae Formationum Coloradensium, Century 6, No. 514. Anno 1908, nom. nud. **Isotype:** Ad folia viva *Pulsatillae hirsutissimae* [*Pulsatilla hirsutissima* (Pursh) Britton = *P. patens* (L.) P. Miller], Colorado, Long's Peak Inn, alt. 2700 m, 5 AUG 1907, leg. F.E. & E.S. Clements, ILL 15572.

Cercospora sambuci F.L. Stevens & C.M. King in F.L. Stevens. Illinois Biological Monographs 11:211. 1927. **Holotype:** On *Sambucus mexicana* K. Presl ex DC., Costa Rica, Cartago, 7 JUL 1923, leg. F.L. Stevens No. 260 [erroneously as No. 250 in the protologue], ILL 15175.

Cercospora sechii J.A. Stevenson. Annual Report of the Insular Experiment Station of Porto Rico 1917–1918:137. 1919. **Isosyntype:** On *Sechium edule* (Jacquin) Sw., [Puerto Rico], Barceloneta, [14 MAY 1917], leg. J.A. Stevenson No. 6462, ILL 15681.

Cercospora seminalis J.B. Ellis & B.M. Everhart. Journal of Mycology 4:4. 1888. **Isotype:** In spikelets of *Buchloe dactyloides* (Nutt.) Engelm., Kansas, Manhattan, [5] JUL 1887, leg. W.T. Swingle [ex W.A. Kellerman and W.T. Swingle, Kansas Fungi No. 7], ILL 21379.

Cercospora stevensii E. Young. Mycologia 8:45. 1916. **Holotype:** On leaves of *Andira* sp., Porto Rico [Puerto Rico], Dos Bocas, below Utuado, [16 DEC 1913], leg. F.L. Stevens No. 6008, ILL 15725. **Paratype:** [30 DEC 1913], leg. F.L. Stevens No. 6549, ILL 15724.

Cercospora stromatis F.E. & E.S. Clements. Cryptogamae Formationum Coloradensium, Century 6, No. 515. Anno 1908, nom. nud. **Isotype:** Ad folia viva *Crepidis acuminatae* [= *Crepis acuminata* Nutt.], Colorado, Sulphur Springs, alt. 2400 m, 23 JUL 1907, leg. F.E. & E.S. Clements, ILL 15728.

Cercospora tectoniae F.L. Stevens. Bernice P. Bishop Museum Bulletin 19:155–156. 1925. **Holotype:** On *Tectonia grandis* L.f., Hawaii, Oahu, Honolulu, Hillebrand Gardens, 22 MAY [1921], leg. F.L. Stevens No. 52, ILL 15736. **Isotypes:** BISH 499040 and 499908.

Cercospora thaliae J.B. Ellis & A.B. Langlois. Journal of Mycology 6:36. 1891. **Isotype:** On living and dead leaves of *Thalia dealbata* Fraser ex Roscoe, Louisiana, St. Martinsville, OCT 1889 [as NOV on the packet], leg. A.B. Langlois, ex J.B. Ellis and B.M. Everhart, North American Fungi No. 2476 [erroneously as 2426 in the protologue], ILL 15744.

Cercospora thouiniae F.L. Stevens. Transactions of the Illinois State Academy of Science 10:213. 1917. **Holotype**: On *Thouinia striata* Radlk., Porto Rico [Puerto Rico], Maricao, [13 APR 1913], leg. F.L. Stevens No. 751, ILL 15747a. **Isotype**: ILL 15747b.

Cercospora trematis (F.L. Stevens & W.G. Solheim in W.G. Solheim & F.L. Stevens) C. Chupp in C.E. Chardon & R.A. Toro, 1934. **See basionym**: *Ragnhildiana trematis* F.L. Stevens & W.G. Solheim in W.G. Solheim & F.L. Stevens, 1931.

Cercospora trichophila F.L. Stevens. Transactions of the Illinois State Academy of Science 10:212. 1917. **Syntypes** plus **one isosyntype**: On *Solanum torvum* Sw., Porto Rico [Puerto Rico], Utuado, [Dos Bocas, 7 JUL 1915], leg. F.L. Stevens No. 7982, ILL 15761; Vega Baja, [20 FEB 1913], leg. F.L. Stevens No. 486, ILL 15753 and 15767; [Trujillo Alto], Rio Tanamá, [16 AUG 1915], leg. F.L. Stevens No. 9205, ILL 15764; Rio Tanamá, [6 JUL 1915], leg. F.L. Stevens No. 7832, ILL 15755; Monacillo, [12 AUG 1915], leg. F.L. Stevens No. 9339, ILL 15757; Arecibo–Lares Road, [21 JUN 1915], leg. F.L. Stevens No. 7227, ILL 15759; leg. F.L. Stevens No. 7296, ILL 15758; Lajas, [17 JUN 1915], leg. F.L. Stevens No. 7156, ILL 15763; Monati, [2 JUL 1915], leg. F.L. Stevens No. 7693, ILL 15760; Mona Island, [20–21 DEC 1913], leg. F.L. Stevens No. 6431, ILL 15766; on *Solanum verbascifolium* sensu auct. non L. [= *S.* cf. *erianthum* D. Don], El Alto de la Bandera, [14 JUL 1915], leg. F.L. Stevens No. 8260, ILL 15765.

Cercospora trichostigmatis F.L. Stevens. Transactions of the Illinois State Academy of Science 10:211. 1917 [as *trichostigmae*, corrected in accordance with Rec. 60H.1 and Articles 32.6, 60 & 61 of I.C.B.N.]. **Holotype**: On *Trichostigma octandrum* (L.) H. Walter, Porto Rico [Puerto Rico], Barceloneta, [10 AUG 1915], leg. F.L. Stevens No. 9254, ILL 15769. **Paratype**: Rio Piedras, [17 AUG 1915], leg. F.L. Stevens No. 9470, ILL 15768.

Cercospora trinidadensis F.L. Stevens & W.G. Solheim in W.G. Solheim & F.L. Stevens. Mycologia 23:376–377, fig. 3. 1931. **Holotype**: On leaves of *Croton gossypiifolius* Vahl, Trinidad, St. Augustine, [13 AUG 1922], leg. F.L. Stevens No. 839, ILL 15771.

Cercospora tuberculella J.J. Davis. Transactions of the Wisconsin Academy of Sciences, Arts and Letters 20:429–430. 1921. **Isotype**: On leaves of *Convolvulus sepium* L. [= *Calystegia sepium* (L.) R. Br.], Wisconsin, Madison, 16 SEP 1919, leg. J.J. Davis [Fungi Wisconsinenses Exsiccati No. 158], ILL 15774.

Cercospora vagnerae F.E. & E.S. Clements. Cryptogamae Formationum Coloradensium, Century 1, No. 69. Anno 1906, nom. nud. **Isotype:** In foliis languescentibus *Vagnerae stellatae* [*Vagnera stellata* (L.) Morong = *Smilacina stellata* (L.) Desf.], Colorado, Ruxton Park, alt. 2800 m, 1 SEP 1904, leg. F.E. & E.S. Clements, ILL 15783.

Cercospora verruculosa F.L. Stevens & W.G. Solheim in W.G. Solheim & F.L. Stevens. Mycologia 23:397, fig. 8. 1931. **Holotype:** On leaves of *Caladium* sp., Trinidad, St. Augustine, [13 AUG 1922], leg. F.L. Stevens No. 829, ILL 15793.

Cercospora vismiae H. Sydow. Annales Mycologici 23:427. 1925. **Isotype:** Hab. in foliis *Vismiae ferrugineae* [*Vismia ferruginea* Kunth in H.B.K. = *V. baccifera* (L.) Triana & Planchon ssp. *ferruginea* (Kunth in H.B.K.) Ewan], Costa Rica, pr. San Ramón, 22 JAN 1925, leg. H. Sydow, Fungi in Itinere Costaricensi Collecti No. 334, ILL 15796.

Cercosporella californica L. Bonar. Mycologia 34:188. 1942. **Paratype:** On *Rhus diversiloba* Torrey & Gray [= *Toxicodendron diversilobum* (Torrey & Gray) E. Greene, California, Monterrey County, [Pineridge trail, one mile east of Big Sur], 13 AUG 1937, leg. L. Bonar [California Fungi No. 606], ILL 13916.

Cercosporella elvirae H. Sydow. Annales Mycologici 25:140–141. 1927. **Isosyntypes:** Hab. in foliis *Elvirae biflorae* [*Elvira biflora* (L.) Cass. = *Delilia biflora* (L.) O. Kuntze], Costa Rica, La Caja pr. San José, 24 DEC 1924, leg. H. Sydow, Fungi in Itinere Costaricensi Collecti No.98, ILL 13928; Desamparados, 30 DEC 1924, leg. H. Sydow, Fungi in Itinere Costaricensi Collecti No. 399, ILL 13929.

Cercosporella trichophila J.J. Davis. Transactions of the Wisconsin Academy of Sciences, Arts and Letters 18:266–267. 1916. **Isotype:** On leaves of *Fraxinus pennsylvanica* Marshall, Wisconsin, Bridgeport, [4] AUG 1914, leg. J.J. Davis [Fungi Wisconsinenses Exsiccati No. 73], ILL 13938.

Cercosporina boringuensis (E. Young) P.A. Saccardo ex A. Trotter, 1931. **See basionym:** *Cercospora boringuensis* E. Young, 1916.

Cercosporina carthami H. & P. Sydow. Annales Mycologici 11:406. 1913. **Isotype:** Hab. in foliis *Carthami tinctorii* [= *Carthamus tinctorius* L.], Philippines, [Luzon], Laguna Prov., Los Baños, 12 JUN 1913, leg. C.F. Baker No. 1248 [ex Fungi Malayana No. 125], ILL 15829.

Cercosporina guanicensis (E. Young) P.A. Saccardo ex A. Trotter, 1931. **See basionym:** *Cercospora guanicensis* E. Young, 1916.

Cercosporina josensis H. Sydow. Annales Mycologici 23:427–428. 1925. **Isotype:** Hab. in foliis *Crotalariae guatemalensis* [*Crotalaria guatemalensis* Bentham ex Oersted = *C. vitellina* Ker Gawler], Costa Rica, San José, 22 DEC 1924, leg. H. Sydow, Fungi in Itinere Costaricensi No. 99, ILL 15830.

Cercosporina maricaoensis (E. Young) P.A. Saccardo ex A. Trotter, 1931. **See basionym**: *Cercospora maricaoensis* E. Young, 1916.

Cerinomyces pallidus G.W. Martin. Mycologia 41:83–85, figs. 5, 6, 10–13. 1949. **Paratypes**: On rotten oak [*Quercus* sp.], Iowa, Iowa City, [1 NOV 1940], leg. G.W. Martin No. 5180, ILL 32846; [on rotten] apple [= *Malus* sp. wood], Iowa City, [18 AUG 1939], leg. G.W. Martin No. 3914, ILL 32845. **Holotype**: IA [currently at ISC].

Ceuthocarpon talaumae H. Rehm. Leaflets of Philippine Botany 8:2953. 1916. **Assumed isotype**: Ad folia emortua *Talauma villariana* Rolfe [= *Magnolia villariana* (Rolfe) D.C.S. Raju & M.P. Nayar], Philippines, [Luzon, Laguna Prov., Mount Maquiling near] Los Baños, FEB 1914, leg. P.M.B. Raimundo, comm. C.F. Baker No. 2843 [ex Fungi Malayana No. 126], ILL 9359.

Chaconia ingae (H. Sydow) G.B. Cummins, 1956. **See basionym**: *Maravalia ingae* H. Sydow, 1925.

Chaetasbolisia falcata V.M. Miller & L. Bonar. University of California Publications in Botany 19:413–414, pl. 67, fig. 10. 1941. **Isotype**: On living leaves and stems of *Lithocarpus densiflorus* (Hooker & Arnott) Rehder, California, Marin County, Inverness Ridge, 21 FEB 1932, leg. L. Bonar [California Fungi No. 607], ILL 33496. **Holotype**: UC 498797. **Paratypes**: On *Umbellularia californica* (Hooker & Arnott) Nutt., [14] MAR 1931, leg. H.E. Parks No. 3577, [ex L. Bonar, California Fungi No. 609], ILL 33495; on *Sequoia sempervirens* (Lambert ex D. Don) Endl., Monterey County, Big Sur, 14 AUG 1937, leg. L. Bonar [California Fungi No. 608], ILL 33494 [also a paratype of *Phaeosaccardinula dematia* V.M. Miller & L. Bonar].

Chaetodiplodia aragalli F.E. & E.S. Clements. Cryptogamae Formationum Coloradensium, Century 5, No. 485. Anno 1908, nom. nud. [as *aragali*]. **Isotype**: In caulibus emortuis *Aragalli lamberti* [*Aragallus lambertii* (Pursh) E. Greene = *Oxytropis lambertii* Pursh], Colorado, La Veta, alt. 2100 m, 20 JUL 1907, leg. F.E. & E.S. Clements, ILL 11008.

Chaetomium dolichotrichum L.M. Ames in G.A. Greathouse & L.M. Ames. Mycologia 37:145, fig. 5(8–10) & table I. 1945. **Isotype**: Tennessee, the Great Smoky Mountains, Cades Cove, [1 MAY 1945], leg. L.M. Ames [No. 1044.7], ILL 9331, ex L.M. Ames herb.

Chaetomium pachypodioides L.M. Ames in G.A. Greathouse & L.M. Ames. Mycologia 37:145–146, fig. 7(1–3) & table I. 1945. **Isotype**: Tennessee, the Great Smoky Mountains, Cades Cove, [1 MAY 1945], leg. L.M. Ames [No. 1044.3], ILL 9339, ex L.M. Ames herb.

Chaetophoma dugaldeae F.E. & E.S. Clements. Cryptogamae Formationum Coloradensium, Century 3, No. 246. Anno 1907, nom. nud. **Isotype**: In foliis vetustis *Dugaldeae hoopesii* [= *Dugaldia hoopesii* (A.

Gray) Rydb.], Colorado, Beaver Park, alt. 2700 m, 25 AUG 1906, leg. F.E. & E.S. Clements, ILL 11009.
Chaetoplaca memecyli H. & P. Sydow. Annales Mycologici 15:232–234, text fig. I, subfigs. 1–7. 1917. **Assumed isotype**: Hab. in foliis *Memecyli* sp. [= *Memecylon* sp., Philippines, Luzon], Bataan Prov., DEC 1915, leg. M. Ramos, ex Phil. Bur. Sci. No. 23986, ILL 6694, [only microscopic preparations remain; the Phil. Bur. Sci. No. is omitted from the packet label].
Chaetothyriopsis panamensis F.L. Stevens & M. Dorman. Mycologia 19:237. 1927. **Holotype**: On upper surface of leaves of *Oncoba laurina* Warb., Panama, Darien, 10 SEP 1924, leg. F.L. Stevens No. 411, ILL 6649.
Chaetothyrium ceibae F. Petrak & R. Ciferri in R. Ciferri. Sydowia Annales Mycologici 10:132. (1956) 1957. **Isotype**: Hab. in foliis vivis *Ceibae pentandrae* (cult.) [= *Ceiba pentandra* (L.) Gaertner, Dominican Republic], Santiago Prov., Santiago, Valle del Cibao, Hato del Yaque, in sylva, 28 JAN 1931, leg. R. Ciferri, ex Mycoflora Domingensis Exsiccata No. 335, ILL 33115.
Chaetothyrium dominicanum R. Ciferri. Sydowia Annales Mycologici 10:133. (1956) 1957. **Isotype**: Hab. in foliis vivis *Eupatorii odorati* [= *Eupatorium odoratum* L.] (Compositae), [Dominican Republic], Santiago Prov., Santiago, Valle del Cibao, Palmerejo, 21 NOV 1930, leg. R. Ciferri & E.L. Ekman, ex Mycoflora Domingensis Exsiccata No. 358, ILL 33168.
Chaetothyrium hawaiiense J.M. Mendoza in F.L. Stevens. Bernice P. Bishop Museum Bulletin 19:56–57, pl. IV(24–27). 1925. **Holotype**: On *Morinda citrifolia* L., Hawaii, Oahu, Hakipuu, on Mr. A.F. Judd's property, 19 JUN [1921], leg. F.L. Stevens No. 577, ILL 6651. **Isotypes**: BISH 499014 and 499918.
Chaetothyrium mangiferae J.M. Mendoza in F.L. Stevens. Bernice P. Bishop Museum Bulletin 19:57, pl. IV(28–33). 1925 [as *magniferae*, corrected in accordance with Articles 60 & 61 of I.C.B.N.]. **Holotype**: On *Mangifera indica* L., Hawaii, Oahu, Honolulu, 6 JUN [1921], leg. F.L. Stevens No. 267, ILL 6652 [also the holotype of *Deslandesia honoluluensis* A.C. Batista & A.F. Vital in A.C. Batista & R. Ciferri]. **Isotypes**: BISH 499013 and 499917.
Chaetothyrium permixtum H. Sydow. Annales Mycologici 24:348–352. 1926. **Isotype**: Hab. in foliis vivis *Phoebes tonduzii* [= *Phoebe tonduzii* Mez], Costa Rica, Grecia, 19 JAN 1925, leg. H. Sydow, Fungi in Itinere Costaricensi Collecti No. 160i, ILL 6653 [also the holotype of *Bisbyopeltis phoebesii* A.C. Batista & A.F. Vital]. **Paratypes**: In foliis *Ingae verae* (varietas) [= *Inga vera* Willd.], San Pedro de San Ramón, 23 JAN 1925, leg. H. Sydow, Fungi in Itinere Costaricensi Collecti

No.108c, ILL 6654 [also the holotype of *Epistigme ingae* A.C. Batista & A.F. Vital]; in foliis *Ocoteae veraguensis* [= *Ocotea veraguensis* (Klotzsch ex Meisner) Mez], San Pedro de San Ramón, 22 JAN 1925, leg. H. Sydow, Fungi in Itinere Costaricensi Collecti No. 171c, ILL 6655.

Chaetothyrium pongamiae H.A. Harris in F.L. Stevens & A.S. Peirce. The Indian Journal of Agricultural Science 3:912. 1933. **Assumed holotype**: On *Pongamia glabra* Vent., [India, Poona, Bombay, leg. B.N. Uppal No. 37 in 1932], ILL 6656. **Isotype**: BSI.

Chaetothyrium straussiae J.M. Mendoza in F.L. Stevens. Bernice P. Bishop Museum Bulletin 19:56, pl. IV(20–23). 1925. **Holotype**: On *Straussia mariniana* (Cham. & Schlecht.) A. Gray [= *Psychotria mariniana* (Cham. & Schlecht) Fosb.], Hawaii, Oahu, Wahiawa, 31 MAY [1921], leg. F.L. Stevens No.157, ILL 6658. **Isotypes**: BISH 499012 and 499916.

Chaetotrichum macrosporum L. Bonar. Mycologia 34:189, fig. 1(F,G). 1942. **Isotype**: On *Gaultheria shallon* Pursh, California, Humboldt County, Trinidad, 10 MAY 1931, leg. H.E. Parks [ex L. Bonar, California Fungi No. 611], ILL 15831. **Holotype**: UC 653859. **Paratype**: On *Physocarpus capitatus* (Pursh) O. Kuntze, Humboldt County, Bishop Pine Lodge, 5 APR 1932, leg. H.E. Parks No. 4033 [ex L. Bonar, California Fungi No. 612], ILL 15832.

Chalara cibotii (O.A. Plunkett in F.L. Stevens) T.R. Nag Raj & B. Kendrick, 1975. **See basionym**: *Excioconidium cibotii* O.A. Plunkett in F.L. Stevens, 1925.

Chionosphaera apobasidialis D.E. Cox. Mycologia 68:503–505. 1976. **Holotype**: In ramis corticatis *Quercus macrocarpae* [= *Q. macrocarpa* Michaux, the packet marked "scrub oak prairie on limbs of tops of bur oak"], Illinois, Effingham County, [back of Cassidy's house], 28 MAR 1971, leg. D.E. Cox s.n., ILL 33563 [the packet marked holotype]. **Isotypes**: BPI, NY, FH, IA, and in D.E. Cox's personal herbarium. **Paratypes**: [On] *Quercus stellata* Wang. [as post oak, Cassidy's woods on the packet], 30 DEC 1972, leg. D.E. Cox s.n., ILL 33565; [on] *Carpinus caroliniana* Walter, Champaign County, Trelease Woods, 9 NOV 1974, leg. D.E. Cox s.n., ILL 33564.

Chorostate sydowiana P.A. Saccardo. Annales Mycologici 6:561. 1908. **Isotype**: Hab. in ramis morientibus *Sorbi aucupariae* [= *Sorbus aucuparia* L., Germany], Bayern: Birgsau bei Oberstdorf im Allgäu, [6] JUL 1906, leg. P. Sydow, [ex H. Sydow, Mycotheca Germanica No. 688], ILL 9361. ≡ *Diaporthe sydowiana* (P.A. Saccardo) P.A. Saccardo. Sylloge Fungorum 22:377. 1913.

Cicinnobella asperula H. Sydow. Annales Mycologici 23:412–413. 1925. **Isotype**: Hab. parasitica in mycelio et thyriotheciis *Asterinae acalyphae*

[= *Asterina acalyphae* H. Sydow] in foliis *Acalyphae macrostachyae* var. *hirsustissimae* [= *Acalypha macrostachya* Jacquin var. *hirsutissima* (Willd.) Muell. Arg.], Costa Rica, San Ramón, 22 JAN 1925, leg. H. Sydow, Fungi in Itinere Costaricensi Collecti No. 349, ILL 6610 [also an isotype of *Phaeodimeriella asperula* H. Sydow].

Cicinnobella consimilis H. Sydow. Annales Mycologici 25:105. 1927. **Isotype** : Hab. parasitica in mycelio *Irenes escharoides* [*Irene escharoides* H. Sydow ≡ *Meliola tabernaemontanae* C.L. Spegazzini var. *escharoides* (H. Sydow) C.G. Hansford], in mycelio et stromatibus *Neostomellae tabernaemontanae* [= *Neostomella tabernaemontanae* H. Sydow], ad folia *Tabernaemontanae sananho* [= *Tabernaemontana sananho* Ruiz & Pavon]; Costa Rica, Cerro de San Isidro pr. San Ramón, 9 FEB 1925, leg. H. Sydow, Fungi in Itinere Costaricensi Collecti No. 395 p.p., ILL 3658 [also an isotype of *Dimerium consimile* H. Sydow, and an isosyntype of *Neostomella tabernaemontanae* H. Sydow].

Cicinnobella exigua H. Sydow. Annales Mycologici 24:411–412. 1926. **Isotype**: Hab. parasitica in mycelio *Asterinae* spec. indet. [= *Asterina* sp.], ad folia *Roupalae veraguensis* [= *Roupala veraguensis* Klotzsch ex Meisner in Martius], Costa Rica, Mondongo pr. San Ramón, 3 FEB 1925, leg. H. Sydow, Fungi in Itinere Costaricensi Collecti No. 229b, ILL 6612 [also an isotype of *Phaeodimeriella exigua* H. Sydow].

Ciliochorella bambusarum L. Shanor. Mycologia 38:335–336. 1946. **Paratypes**: On *Arthrostylidium racemiflorum* Steudel, Mexico, 5 JAN–6 FEB 1892, leg. E. Palmer No. 1914, ILL 31385, ex US 1021482; Panama Canal Zone, JUL 1923, leg. H. Johnson No. 17, ILL 31383, ex US 1167472.

Cladosporium calotropidis F.L. Stevens. Transactions of the Illinois State Academy of Science 10:207. 1917. **Holotype**: On *Calotropis procera* (Aiton) Aiton f., Porto Rico [Puerto Rico], Guayanilla, [JUL 1915], leg. F.L. Stevens No. 9130, ILL 15840. **Isotypes**: ILL 15842 and 15843, BPI 70873. **Paratype**: F.L. Stevens No. 291, ILL 15841.

Cladosporium guanicensis F.L. Stevens. Transactions of the Illinois State Academy of Science 10:207–208. 1917. **Holotype**: On *Argemone mexicana* L., Porto Rico [Puerto Rico], Guanica, [3 FEB 1913], leg. F.L. Stevens No. 347a, ILL 15873. **Paratype**: Coamo, [6 APR 1913] leg. F.L. Stevens No. 620, ILL 15872.

Cladosporium humile J.J. Davis. Transactions of the Wisconsin Academy of Sciences, Arts and Letters 19(2):702. 1919. **Isotype**: On leaves of *Acer rubrum* L., Wisconsin, Luck, 25 AUG 1916, leg. J.J. Davis [Fungi Wisconsinenses Exsiccati No. 64], ILL 15887.

Cladosporium lysimachiae E.F. Guba. Rhodora 41:513. 1939. **Isotype**: On living leaves, rarely on stems, of *Lysimachia vulgaris* L., [Massachusetts, Nantucket County], Nantucket, in waste places near the water front east of Main Street, 15 AUG 1936, leg. E.F. Guba No. 115, ILL 21101.

Cladosporium mikaniae F.L. Stevens. Transactions of the Illinois State Academy of Science 10:208. 1917. **Holotype:** On *Mikania* sp., Porto Rico [Puerto Rico], Las Marias, 22 MAR 1913, leg. F.L. Stevens No. 314, ILL 15889. ≡ *Mycovellosiella mikaniae* (F.L. Stevens) F.C. Deighton. Mycological Papers. Commonwealth Mycological Institute 137:45–47. 1974.

Cladosporium nervale J.B. Ellis & J. Dearness in J.B. Ellis and B.M. Everhart. Fungi Columbiani, Century 21, No. 2010. Anno 1905. **Isotype:** On living leaves of *Rhus typhina* L., Canada, [Ontario], London, JUL & AUG 1904, leg. J. Dearness s.n., ILL 15890.

Clasterosporium obclavatum L. Bonar. Mycologia 43:64, 66, fig. 1. 1951. **Isotype:** On leaves of *Torreya californica* Torrey, California, Marin County, [summit of] Mt. Tamalpais, 1 OCT 1949, leg. L. Bonar [California Fungi No. 761], ILL 31481 [also a paratype of *Caeoma torreyae* L. Bonar].

Clasterosporium pulchrum J.B. Ellis & B.M. Everhart. Proceedings of the Academy of Natural Sciences of Philadelphia 1893:169. 1894 [cited as *Clasterisporium*, corrected in accordance with Articles 60 & 61 of I.C.B.N.]; originally distributed as *Clasterisporium pulcherrimum* J.B. Ellis & B.M. Everhart. North American Fungi, Century 29, No. 2877. Anno 1893, nom. nud. **Isotype:** On bark of dead *Carpinus americana* Michaux [= *C. caroliniana* Walter], Canada, [Ontario], London, JUN 1892 [as JUL 1892 on the packet], leg. J. Dearness No. 1900, ILL 15923.

Cleistothecopsis circinans F.L. Stevens & E. Young True. University of Illinois Agriculture Experiment Station Bulletin 220:530, figs. 17–19. 1919. **Assumed holotype:** On onion (= *Allium* sp.), ILL 3628 [as microscopic preparations. No type is specified in the protologue, but the only authentic material is at ILL.].

Clithris pandani L.R. Tehon. Botanical Gazette 65:555. 1918. **Holotype:** On dead leaves of cultivated *Pandanus* sp., Porto Rico [Puerto Rico], San Juan, 16 NOV 1913, leg. F.L. Stevens No. 4090 [as 4090A on the packet], ILL 7426.

Clypeodiplodina baccharidis F.L. Stevens. Mycologia 19:235–237, pl. 18, figs. 3–7; pl. 20; pl. 21, figs. 2–4. 1927. **Holotype:** [Parasitic on living leaves of] *Baccharis floribunda* Kunth in H.B.K., Ecuador, Guapulo, 12 NOV 1924, leg. F.L. Stevens No. 267, ILL 11023. **Isotype:** ILL 11024.

Clypeolella clermontiae F.L. Stevens & R.W. Ryan in F.L. Stevens. Bernice P. Bishop Museum Bulletin 19:72. 1925. **Holotype:** On *Clermontia* sp.,

Hawaii, Maui, Iao Valley, 7 SEP 1921, leg. F.L. Stevens No. 1154b, ILL 6934 [also an isolectotype of *Asterina clermontiae* F.L. Stevens & R.W. Ryan in F.L. Stevens and an isotype of *Meliola lobeliae* F.L. Stevens]. **Isotypes:** BISH 499858 and 499862.

Clypeolum exiguum H. Sydow. Annales Mycologici 25:89–90. 1927. **Isotype:** Hab. in foliis *Phoebes tonduzii* [= *Phoebe tonduzii* Mez], Costa Rica, Grecia, 19 JAN 1925, leg. H. Sydow, Fungi in Itinere Costaricensi Collecti No. 160h, ILL 6695.

Clypeoseptoria rockii F.L. Stevens & P.A. Young in F.L. Stevens. Bernice P. Bishop Museum Bulletin 19:141, text fig. 29. 1925. **Holotype:** On living leaves of *Platydesma campanulata* H. Mann [= *P. spathulata* (A. Gray) B. Stone], Hawaii, Maui, Honomanu, MAY 1911, leg. J.F. Rock, as H.L. Lyon No. 286, ILL 11025.

Clypeosphaeria bakeriana H. Rehm. Leaflets of Philippine Botany 8:2948–2949. 1916. **Assumed isotype:** Ad ramulos *Eugeniae bataanensis* [= *Eugenia bataanensis* Merrill], Philippines, [Luzon, Laguna Prov.], Mount Maquiling, [near Los Baños], MAY 1914, leg. C.F. Baker No. 3481a [ex Fungi Malayana No.127a (in exsiccati set as No. 127a, b on the same packet label)], ILL 9362 [also, under 127b, an assumed isotype of *Massaria bataanensis* H. Rehm].

Coccomyces dubius H. Rehm. Leaflets of Philippine Botany 8:2926. 1915. **Assumed isotype:** Ad folia *Fici minahassae* [= *Ficus minahassae* Miq.], Philippines, [Luzon, Laguna Prov., Mount Maquiling, near] Los Baños, MAY 1914, leg. S.A. Reyes, comm. C.F. Baker No. 3480 [ex Fungi Malayana No. 128], ILL 7444.

Coccomyces quadratus (A.W.F. Schmidt & O. Kuntze) P.A. Karsten var. *philippinus* H. Rehm. Leaflets of Philippine Botany 8:2926. 1915. **Assumed isotype:** Ad folia *Neolitseae* sp. [= *Neolitsea* sp.], Philippines, [Luzon, Laguna Prov.], Mount Maquiling, [near Los Baños], JUN 1914, leg. C.F. Baker No. 3446 [ex Fungi Malayana No.129], ILL 7446.

Cocconia palmae F.L. Stevens. Illinois Biological Monographs 11:175, pl. III, fig. 24. 1927. **Holotype:** On palm [Arecaceae], Costa Rica, Peralta, 13 JUL 1923, leg. F.L. Stevens No. 432, ILL 7451.

Coccostromopsis palmigena O.A. Plunkett in F.L. Stevens. Illinois Biological Monographs 8:176–177, pl. I, figs. 5, 6. 1923. **Holotype:** On leaves of palm [Arecaceae], species indet., Trinidad, Cumuto, 16 AUG 1922, leg. F.L. Stevens No. 1001, ILL 8527. ≡ *Bagnisiopsis palmigena* (O.A. Plunkett in F.L. Stevens) F. Petrak. Annales Mycologici 27:332. 1929.

Coleosporium paraphysatum P. Dietel & E.W.D. Holway. Botanical Gazette 31:337. 1901. **Isotype:** On *Liabum discolor* (W.J. Hooker & Arnott) Bentham & J.D. Hooker ex Hemsley, [Mexico], Chapala, 23 SEP 1899, leg. E.W.D. Holway No. 3483, ILL 27089.

Coleosporium pini B.T. Galloway. Journal of Mycology 7:44. 1891. **Isotype:** On leaves of *Pinus inops* Aiton [= *P. virginiana* P. Miller, Maryland], near Washington, D.C., [Garrett Park, 26 MAY 1891], leg. B.T. Galloway No. 1342, ILL 27095. **Holotype:** BPI.

Colletotrichum catenulatum F.L. Stevens. Annales Mycologici 28:370. 1930. **Holotype:** On *Agave angustifolia* Haw. cult. *marginata*, British Guiana [Guyana], Tumatumari, 8 JUL 1922, leg. F.L. Stevens No. 32, ILL 13224.

Colletotrichum cereale T.F. Manns in A.D. Selby & T.F. Manns. Bulletin of the Ohio Agricultural Experiment Station 203:207, pl. I–X, text figs. 2, 3. 1909. **Isosyntypes:** On culms of orchard grass, *Dactylis glomerata* L., [Ohio, Wooster, Ohio Experiment Station, 1908], leg. A.D. Selby s.n., ILL 13228; on culms of timothy, *Phleum pratense* L., [Ohio Experiment Station, JUL 1908], leg. A.D. Selby s.n., ILL 13229; on culms of volunteer chess, *Bromus secalinus* L., [Sharpsburg, 4 JUL 1908], leg. A.D. Selby s.n., ILL 13225; on culms of valley wheat, *Triticum vulgare* Vill. [= *T. aestivum* L., Sharpsburg, 23 JUN 1908], leg. A.D. Selby s.n., ILL 13226.

Colletotrichum curvisetum F.L. Stevens. Transactions of the Illinois State Academy of Science 10:199. 1917. **Syntypes:** On *Hura crepitans* L., Porto Rico [Puerto Rico], Añasco, [19 OCT 1913], leg. F.L. Stevens No. 3594, ILL 15376 [also a paratype of *Cercospora hurae* F.L. Stevens.]; Mayagüez, [DEC 1917], leg. F.L. Stevens No. 5830, ILL 15378 [also the holotype of *Cercospora hurae* F.L. Stevens].

Colletotrichum dianellae F.L. Stevens & P.A. Young in F.L. Stevens. Bernice P. Bishop Museum Bulletin 19:145. 1925 [the epithet misspelled "*dianallae*" in the protologue but correctly spelled in the published indices and on the packet labels]. **Holotype:** On languid leaves of *Dianella odorata* sensu Hillebr. non Blume [= *D. sandwicensis* Hooker & Arnott], Hawaii, Kauai, Waimea Canyon, 15 JUN [1921], leg. F.L. Stevens No. 447, ILL 13235. **Isotypes:** BISH 499011 and 499914.

Colletotrichum iresines F.L. Stevens. Illinois Biological Monographs 11:206–207. 1927. **Holotype:** On *Iresine calea* (Ibañez) Standley, Costa Rica, Desamparados, 24 JUN 1923, leg. F.L. Stevens No. 139, ILL 13284.

Colletotrichum lobeliae F.L. Stevens. Transactions of the Illinois Academy of Science 10:198. 1917. **Holotype:** On *Lobelia assurgens* L. var. *portoricensis* (A. DC.) Urban [= *L. robusta* Graham var. *portoricensis* (A. DC.) McVaugh], Porto Rico [Puerto Rico], Maricao, [3 APR 1913], leg. F.L. Stevens No. 776, ILL 13309. **Paratype:** Añasco, [12 OCT 1913], leg. F.L. Stevens No. 3513, ILL 13308.

Colletotrichum lycopersici J.B. Ellis & B.M. Everhart. North American

Fungi, Century 29, No. 2868. Anno 1893, nom. nud. **Isotype:** On fruit of *Solanum lycopersicum* L. [= *Lycopersicon esculentum* P. Miller], Delaware, Newark, OCT 1891, leg. F.D. Chester, ILL 13310.

Colletotrichum passiflorae F.L. Stevens & P.A. Young in F.L. Stevens. Bernice P. Bishop Museum Bulletin 19:146. 1925. **Syntypes** [BISH numbers cited are isosyntypes]: On fruits of *Passiflora laurifolia* L., Hawaii, [Island of] Hawaii, Kealakekua, 21 JUL [1921], leg. F.L. Stevens No. 914, ILL 13317, BISH 499010 and 508404; on living leaves of *Passiflora edulis* Sims, Kauai, Pipe trail, [upper Waimea Canyon], 15 JUN [1921], leg. F.L. Stevens No. 465, ILL 13318, BISH 145782 and 499913.

Colletotrichum piperis F.L. Stevens. Transactions of the Illinois State Academy of Science 10:198–199. 1917. **Holotype:** On *Piper umbellatum* L. [= *Lepianthes umbellata* (L.) Raf.], Porto Rico [Puerto Rico], Caguas, [9 FEB 1913], leg. F.L. Stevens No. 288, ILL 13324. **Paratype:** Leg. F.L. Stevens No. 291a, ILL 13325.

Colletotrichum sordidum J.J. Davis. Transactions of the Wisconsin Academy of Sciences, Arts and Letters 18:265. 1916. **Isotype:** On leaves of *Menispermum canadense* L., Wisconsin, Grant County, river bottom opposite Bridgeport, 31 JUL 1914, leg. J.J. Davis [Fungi Wisconsinenses Exsiccati No. 5], ILL 13328.

Colletotrichum toluiferae F.L. Stevens & W.G. Solheim in F.L. Stevens. Annales Mycologici 28:370. 1930. **Holotype:** On *Toluifera* sp. [nom. rej. = *Myroxylon* L.f., nom. cons. (Fabaceae)], Trinidad, Port of Spain, 8 AUG 1922, leg. F.L. Stevens No. 820, ILL 13330.

Comatricha mirabilis R.K. Benjamin & A.W. Poitras. Mycologia 42:515–517, figs. 1–4. 1950. **Holotype:** On goat dung, Illinois, Champaign County, Urbana, NOV 1947, leg. R.K. Benjamin and A.W. Poitras, ILL 21003. **Isotypes:** PBI, FH, NY. IA [currently at ISC].

Comoclathris ipomoeae F.E. Clements. Minnesota Botanical Studies 4:185–186. 1911; originally distributed as *Pyrenophora ipomoeae* F.E. & E.S. Clements. Cryptogamae Formationum Coloradensium Centuries 5–6, No. 450. Anno 1908, nom. nud., the latter cited pro syn. under the new name. **Isotype:** In caulibus vetustis *Ipomoeae leptophyllae* [= *Ipomoea leptophylla* Torrey], Colorado, Wray, alt. 1100 m, [25] AUG 1907, leg. F.E. & E.S. Clements, ILL 10189.

Comoclathris lanata F.E. & E.S. Clements ex F.E. Clements. Minnesota Botanical Studies 4:185, pl. XXV, fig.1. 1911; originally distributed as an invalid name: ex Cryptogamae Formationum Coloradensium, Centuries 5–6, No. 444. Anno 1908, nom. nud. **Isotype:** In caulibus vetustis *Leptotaeniae multifidae* [*Leptotaenia multifida* Nutt. ex Torrey & Gray = *Lomatium dissectum* (Nutt. ex Torrey & Gray) Mathias

& Constance var. *multifidum* (Nutt. ex Torrey & Gray) Mathias & Constance], Colorado, Silverton, alt. 2800 m, 8 JUN 1907, leg. F.E. & E.S. Clements, ILL 9366.

Coniophora cyanospora D.P. Rogers. University of Iowa Studies in Natural History 17:25–26, fig. 12. 1935. **Holotype:** On the underside of prostrate log of some broad-leaved species, Iowa, Iowa City, Linder's Woods, 31 OCT 1934, leg. D.P. Rogers No. 330, ILL 32546.

Coniophora vaga E.A. Burt. Annals of Missouri Botanical Garden 4:251, fig. 8. 1917. **Isotype:** On bark of old log of *Ulmus americana* L., New York, [north of] Hudson Falls, 1 SEP 1915, leg. S.H. Burnham No. 2 [comm. W.L. White No. 2882], ILL 32637, ex MO 54498. **Holotype:** BPI [ex MO].

Coniosporium arundinellae J.B. Ellis & S.M. Tracy ex J.B. Ellis & B.M. Everhart. North American Fungi, Century 28, No. 2795. Anno 1892, nom. nud. **Isotype:** On *Arundinaria tecta* (Walter) Muhl. [erroneously as *Arundinella* in the protologue = *Arundinaria gigantea* (Walter) Muhl. ssp. *tecta* (Walter) McClure], Alabama, Auburn?, MAR 1889, leg. F.L. Scribner, ILL 15926.

Coniothyrium dracaenae F.L. Stevens & A.G. Weedon in F.L. Stevens. Bernice P. Bishop Museum Bulletin 19:135. 1925. **Holotype:** On *Dracaena aurea* H. Mann [= *Pleomele aurea* (H. Mann) N.E. Brown], Hawaii, Kauai, Pipe trail, upper Waimea Canyon, 15 JUN [1921], leg. F.L. Stevens No. 419a, ILL 11032. **Isotype:** ILL 10006.

Coniothyrium glabroides F.L. Stevens. Botanical Gazette 65:234. 1918. **Holotype:** On *Meliola glabroides* F.L. Stevens on *Piper aduncum* L., Porto Rico [Puerto Rico], Maricao, 18 NOV 1913, leg. F.L. Stevens No. 4802, ILL 4066. [also a paratype of *Helminthosporium glabroides* F.L. Stevens and of *Meliola glabroides* F.L. Stevens]. **Paratypes:** On *Meliola tortuosa* H.G. Winter on *Piper umbellatum* L. [= *Lepianthes umbellata* (L.) Raf.], Maricao, [Indiera Fria on the packet], 8 OCT 1913, leg. F.L. Stevens No. 3379, ILL 4461; on *Meliola compositarum* F.S. Earle var. *portoricensis* F.L. Stevens, on *Eupatorium portoricense* Urban, river junction, [Dos Bocas] below Utuado, [16 DEC 1913], leg. F.L. Stevens No. 6032, ILL 3777 and 11033 [also the holotype and an isotype, respectively, of *Perisporium meliolae* F.L. Stevens, as well as paratypes of *Helminthosporium glabroides* F.L. Stevens and of *Meliola compositarum* F.S. Earle var. *portoricensis* F.L. Stevens].

Coniothyrium marisci L.R. Tehon. Botanical Gazette 67:508, pl. XVIII, fig. 6. 1919. **Holotype:** On *Mariscus jamaicensis* (Crantz) Britton [= *Cladium mariscus* (L.) Pohl ssp. *jamaicense* (Crantz) Kükenth.], Porto Rico [Puerto Rico], leg. F.L. Stevens No. 124, ILL 11034.

Coniothyrium melanconieum P.A. Saccardo. Annales Mycologici 7:436.

1909. **Isotype:** Hab. in ramulis languidis v. emortuis *Ribis grossulariae* [= *Ribes grossularia* L., Brandenburg: Baumschulen], pr. Tamsel [now Poland], 27 SEP 1908, leg. P. Vogel [ex H. Sydow, Mycotheca Germanica No. 815], ILL 11035.

Cornuella lemnae W.A. Setchell. Proceedings of the American Academy of Arts and Sciences (= Daedalus) 26:19. MAY 1891. **Isosyntype:** In the fronds of *Lemna* (*Spirodela*) *polyrrhiza* L. [= *Spirodela polyrrhiza* (L.) Schleiden], Massachusetts, Cambridge, [30 JUN 1889, leg. W.A. Setchell No. 32706], ILL 16743. ≡ *Tracya lemnae* (W.A. Setchell) H. & P. Sydow. Hedwigia 40: (3). 1901. [*Tracya* H. & P. Sydow, 1901, is a substitute name for *Cornuella* W.A. Setchell, May 1891 (nom. illegit.), non J.B.L. Pierre, JAN 1891.].

Corticium bambusae E.A. Burt. Annals of the Missouri Botanical Garden 13:218. 1926. **Isotype:** ["very common on bamboo" = *Bambusa* sp.], Trinidad, [Port of Spain], Maravel [Valley, 23 JAN 1914], leg. R. Thaxter, comm. W.G. Farlow No. 19, ILL 32718. **Holotype:** FH.

Corticium bicolor C.H. Peck. Bulletin of the Buffalo Society of Natural Sciences 1:62. 1873. **Isotype:** On rotten wood, [New York], Center, October, leg. C.H. Peck s.n., ILL 32550, ex NYS.

Corticium botryoideum L.O. Overholts. Mycologia 26:510, pl. 55, fig. 10. 1934. **Isotype:** On dead *Alnus rugosa* (Du Roi) Sprengel [= *A. incana* (L.) Moench ssp. *rugosa* (Du Roi) Clausen], Pennsylvania, Adams County, Biglerville, 26 JUL 1932, [leg. W.L. White No. 1070] as L.O. Overholts No. 14503, ILL 32624. **Holotype:** PAC.

Corticium calceum E.M. Fries. Epicrisis Systematis Mycologici seu Synopsis Hymenomycetum, p. 562. 1838. **Authentic material:** Ad ligna pinea commune [Norvegia: Christiania], leg. M.N. Blytt s.n., ex E.M. Fries herb., ILL 33547, UPS [no type designated].

Corticium canadense E.A. Burt. Annals of the Missouri Botanical Garden 13:290–291. 1926. **Paratype:** [On pine log = *Pinus* sp.], New Hampshire, Chocorua, [30 JUL 1921], leg. E.A. Burt s.n., ILL 32551, ex FH.

Corticium chlamydosporum E.A. Burt in C.H. Peck. Annual Report. New York State Museum of Natural History 54:154–155. 1901. **Isotype:** On bark of *Ulmus americana* L., New York, Westport, SEP [1900], leg. C.H. Peck s.n., ILL 32512. **Holotype:** FH.

Corticium confluens E.M. Fries var. *subcalceum* P.A. Karsten. Revue Mycologique 10:74. 1888. **Isotype:** Supra corticem et lignum *Betulae* [= *Betula* sp., Finland], circa Mustiala, leg. P.A. Karsten, ?comm. J.B. Ellis, ILL 32620, ex IA.

Corticium conigenum C.L. Shear & R.W. Davidson. Mycologia 36:296, figs. 1, 2. 1944. **Possible isotype:** On stump of *Quercus* sp. [as bark of decayed oak on the packet], Florida, Weikiwa Spa, 10 MAR 1942, leg.

C.L. Shear No. 1405 [as 5562 on the packet], ILL 32536. **Holotype:** BPI.

Corticium consimile G. Bresadola. Mycologia 17:68. 1925. **Isotype:** Hab. ad truncos decorticatos *Laricis occidentalis* [= *Larix occidentalis* Nutt.], Idaho, [Bonner County, Priest River], leg. J.R. Weir No. 10808 [as 16808 on the packet, 16 OCT 1920], ILL 32538, ex BPI. **Holotype:** BPI.

Corticium crustulinum E.A. Burt. Annals of the Missouri Botanical Garden 13:209–210. 1926. **Isotype:** On very rotten frondose wood, Porto Rico [Puerto Rico], Rio Piedras, [26] JUL [1915], leg. J.A. Stevenson No. 2914, ILL 32539, ex MO 3130. **Holotype:** BPI [transferred from MO].

Corticium delectabile H.S. Jackson. Canadian Journal of Research, Section C (Botanical Sciences) 26:145, fig. 1. 1948. **Isotype:** On partly decayed, decorticated wood, [Canada], Ontario, York County, woods west of Maple, 1 OCT 1938, leg. R.F. Cain, ex TRTC 13684, ILL 32540.

Corticium effuscatum M.C. Cooke & J.B. Ellis. Grevillea 9:103. 1881. **Isotype:** On rotten log [in a swamp], New Jersey, [Newfield, NOV 1880], leg. J.B. Ellis & H.W. Harkness No. 1704, publ. as M.C. Cooke No. 3401, ILL 32585. **Holotype:** NY.

Corticium electum H.S. Jackson. Canadian Journal of Research, Section C (Botanical Sciences) 26:146. 1948. **Isotype:** On decorticated wood of *Pinus* sp., Canada, Ontario, Lake Timagami, Bear Island, 23 AUG 1939 [as 1938 on the packet], leg. H.S. Jackson, ex TRTC 14982, ILL 33557.

Corticium epigaeum J.B. Ellis & B.M. Everhart. Journal of Mycology 1:88. 1885. **Isotype:** On bare soil, [Oregon], JUL 1884, leg. W.C. Carpenter No. 100, ILL 32541. **Holotype:** NY.

Corticium ermineum E.A. Burt. Annals of the Missouri Botanical Garden 13:182–183. 1926. **Isotypes:** On decorticated, very rotten wood of logs of *Thuja plicata* Donn ex D. Don, Idaho, Priest River, [2 AUG 1919], leg. E.E. Hubert, comm. J.R. Weir No. 12026, ex MO 63379, ILL 32717, ex TRTC.

Corticium eximum H.S. Jackson. Canadian Journal of Research, Section C (Botanical Sciences) 26:152–154, fig. 6. 1948. **Isotype:** On bark of lower dead branches of *Pinus strobus* L., [Canada], Ontario, York County, woods near Holland River Marsh, 5 miles west of Aurora, 29 MAY 1937, leg. H.S. Jackson No. 12916, ILL 32542.

Corticium ferax J.B. Ellis & B.M. Everhart. The American Naturalist 31:339–340. 1897. **Isotype:** On dead wood, Canada, 10 SEP 1896, leg. Macoun, ILL 32804. **Holotype:** NY.

Corticium furfuraceum G. Bresadola. Mycologia 17:69. 1925. **Isosyntype:** Ad ligna *Abietis grandis* [= *Abies grandis* (Douglas ex D. Don) Lindley], Idaho, [13 AUG 1920], leg. J.R. Weir, ILL 32602 [a note in the packet indicates this material was sent by USDA as part of the type]. **Syntype:** BPI.

Corticium hinnuleum G. Bresadola. Hedwigia 56:303. 1915. **Isotype:** Hab. ad truncos *Bambusae* [= *Bambusa* sp.], Philippines, Luzon, Benguet [APR 1904], leg. A.D.E. Elmer No. 6215, ILL 32603, ex NY.

Corticium incanum E.A. Burt. Annals of the Missouri Botanical Garden 13:205. 1926. **Paratypes:** On poplar wood [= *Populus* sp.], North Carolina, Chapel Hill, [Raleigh Road, 1 APR 1920], leg. J.N. Couch No. 4225, comm. W.C. Coker under the name of *C. ochraceo-niveum*, nom. nud., ILL 32595 [ex MO 57412], BPI.

Corticium involucrum E.A. Burt. Annals of the Missouri Botanical Garden 13:271–272. 1926. **Isotype:** Vermont, Middlebury, [27 NOV 1897], leg. E.A. Burt, ILL 32831. **Holotype:** FH. **Paratypes:** Beneath charred hardwood log in banana grove, Cuba, Ceballos, 17 DEC 1914, leg. C.J. Humphrey No. 2793, ILL 32830 [ex MO 20200], BPI.

Corticium jamaicense E.A. Burt. Annals of the Missouri Botanical Garden 13:273. 1926. **Isotypes:** On decaying wood, [mountainous region, 4500–5200 ft.], Jamaica, Cinchona, [25 DEC 1908–8 JAN 1909], leg. W.A. & E.L. Murrill No. 456, ILL 32531 and 32588, ex TRTC. **Holotype:** NY.

Corticium maculare E.D. Lair. Journal of the Elisha Mitchell Scientific Society 62:216–218, figs. 3–12. 1946. **Isosyntypes:** Hab. ad truncos vivibus *Querci albi* [= *Quercus alba* L., North Carolina, Durham, 25 MAY 1946], leg. E.D. Lair s.n., ILL 32532, ex TRTC, FH, NY.

Corticium maculatum V. Litschauer. Oesterreichische Botanische Zeitschrift 77:124–126, fig. 3A–C. 1928. **Isosyntype:** Hab. in cortice et in ligno truncorum putridorum *Piceae excelsae* [*Picea excelsa* (Lam.) Link = *P. abies* (L.) H. Karsten, Austria], Innsbruck, Kranebittenklamm, 25 AUG 1926, leg. V. Litschauer s.n., ILL 32513, ex TRTC.

Corticium murrillii E.A. Burt. Annals of the Missouri Botanical Garden 13:289. 1926 [as murrilli, corrected in accordance with Articles 32.6 & 61 of I.C.B.N.]. **Isotypes:** On bark of decaying log, Mexico, Jalapa, [12–20] DEC [1909], leg. W.A. & E.L. Murrill No. 182, ILL 32605, ex NY, MO 44967 [currently at BPI]. **Holotype:** NY.

Corticium notabile H.S. Jackson. Canadian Journal of Research, Section C (Botanical Sciences) 26:156–157, fig. 10. 1948. **Paratypes:** On coniferous wood, Canada, Ontario, Lake Timagami, Bear Island, 16 AUG 1935, leg. R. Biggs No. 394, ILL 32533, TRTC 8215. **Holotype:** TRTC 14911.

Corticium permodicum H.S. Jackson. Canadian Journal of Research, Section C (Botanical Sciences) 28:721–722, fig. 5. 1950. **Isotype:** On wood of deciduous trees, [Canada], Ontario, Lake Timagami, Bear Island, 18 JUL 1936, leg. H.S. Jackson No. 16705, ILL 32577, ex TRTC. **Holotype**: TRTC.

Corticium pilosum E.A. Burt. Annals of the Missouri Botanical Garden 13:262–263. 1926. **Isotype**: On bark of fallen limbs of *Alnus rugosa* (Du Roi) Sprengel [= *A. incana* (L.) Moench ssp. *rugosa* (Du Roi) Clausen], Georgia, Atlanta, [9 OCT 1924], leg. E. Bartholomew No. 8982, ex MO 63463, ILL 32619. **Holotype**: BPI [transferred from MO].

Corticium praestans H.S. Jackson. Canadian Journal of Research, Section C (Botanical Sciences) 26:148–149, fig. 3. 1948. **Isotype**: On bark of decaying branches of *Quercus alba* L. on ground, [Canada], Ontario, Chalk River, Petawawa Forest Reserve, 12 SEP 1939, leg. H.S. Jackson s.n., ex TRTC 14121, ILL 32521.

Corticium pruni L.O. Overholts. Mycologia 21:282, pl. 23, fig. 6; pl. 25, figs. 11, 12. 1929. **Isotype** : On bark of dead *Prunus* sp., New Hampshire, North Conway, 18 AUG 1918, leg. L.O. Overholts No. 5111, ILL 32601. **Holotype**: PAC.

Corticium racemosum E.A. Burt. Annals of the Missouri Botanical Garden 13:287–288. 1926. **Isotype**: On bark and wood of decaying logs of *Thuja plicata* Donn ex D. Don, Idaho, Priest River, [JUL–SEP 1912], leg. J.R. Weir No. 39, ILL 32590. **Holotype**: FH.

Corticium rigescens M.J. Berkeley & M.A. Curtis in M.C. Cooke. Grevillea 20:12. 1891. **Assumed isosyntype**: On wood, etc., Venezuela, leg. A. Fendler s.n., ILL 32607, ex NY.

Corticium roseo-pallens E.A. Burt in G.R. Lyman. Proceedings of the Boston Society of Natural History 33:173–176, pl. 20, figs. 56–73. 1907. **Assumed isosyntype**: On bark of beech log [= *Fagus* sp.], Vermont, Middlebury, [Ripton], Chapman's Mill, 4 NOV 1896, leg. E.A. Burt s.n., ILL 32530, ex FH. **Syntype**: FH.

Corticium separatum H.S. Jackson & E.R. Dearden. Canadian Journal of Research, Section C (Botanical Sciences) 27:154, fig. 7. 1949. **Isotypes**: On *Abies grandis* (Douglas ex D. Don) Lindley, [Canada], British Columbia, Vancouver Island, Cadboro Bay, 21 NOV 1940, leg. J.E. Bier, OTBF 10111 [currently DAOM], ILL 32523, ex TRTC, DAOM.

Corticium subapiculatum G. Bresadola. Mycologia 17:69. 1925. **Assumed isotype**: Hab. ad truncos *Pini* [= *Pinus* sp.], Idaho, [Clarkia], leg. J.R. Weir No. 16928 [as H.S. Rhoades on the packet], 23 JUN 1920, ILL 32524. **Holotype**: BPI.

Corticium sublilascens V. Litschauer. Oesterreichische Botanische Zeitschrift 77:122. 1928. **Isotype**: In cortice putrida *Pini silvestris* [= *Pinus sylvestris* L., Austria], Gnadenwald apud Hall, Tiroliae septentrionalis, 12 APR 1924 [as 1922 on the packet], leg. V. Litschauer s.n., ILL 32525, ex TRTC.

Corticium vinoso-scabens E.A. Burt. Annals of the Missouri Botanical Garden 13:267–268. 1926. **Isotype:** On bark of fallen trunk of *Abies rubra* Poiret (p.p.) [= *Picea* cf. *mariana* (P. Miller) B.S.P.], Vermont, Ripton, Little Notch, [9 NOV 1902], leg. E.A. Burt s.n., ILL 32514, ex TRTC. **Holotype:** FH.

Corynelia clavata (L.) P.A. Saccardo var. *portoricensis* F.L. Stevens. Transactions of the Illinois State Academy of Science 10:181, continued on p. 179 [printer's error], fig. 5 [p. 178]. 1917. **Lectotype** [designated herein]: On *Podocarpus coriaceus* L.C. Richard, Porto Rico [Puerto Rico, Rio Maricao], near Maricao, [20 OCT 1913], leg. F.L. Stevens No. 784, ILL 9632 [lectotypification is necessary because Stevens (1917) did not indicate a holotype, and the collection was subdivided]. **Isolectotypes:** ILL 9638a, b, FH, K. ≡ *Corynelia portoricensis* (F.L. Stevens) H.M. Fitzpatrick. Mycologia 12:259. 1920.

Corynelia portoricensis (F.L. Stevens) H.M. Fitzpatrick, 1920. **See basionym:** *Corynelia clavata* (L.) P.A. Saccardo var. *portoricensis* F.L. Stevens, 1917.

Corynelia pteridicola F.L. Stevens. Transactions of the Illinois State Academy of Science 10:179, fig. 6. 1917. **Holotype:** On *Campyloneurum* sp., Porto Rico [Puerto Rico], Añasco, [12 OCT 1913], leg. F.L. Stevens No. 3551, ILL 9647.

Corynespora colebrookeae W.G. Solheim in F.L. Stevens & A.S. Peirce. Indian Journal of Agricultural Science 3:916. 1933 [as *colebrookiae*, corrected in accordance with Rec. 60H.1 and Articles 60 & 61 of I.C.B.N.]. **Assumed holotype:** On *Colebrookea oppositifolia* J.E. Smith, [India, Poona, Bombay, 1932], leg. B.N. Uppal No. 10, ILL 13943. **Isotype:** AMH.

Coryneum vogelianum P.A. Saccardo. Annales Mycologici 3:514. FEB 1906. **Assumed isotype:** Hab. ad ramulos emortuos corticatos *Aceris campestris* [= *Acer campestre* L.], Brandenburg [now Poland, Baumschulen zu] Tamsel, [24] MAY 1905, leg. P. Vogel [ex H. & P. Sydow, Mycotheca Germanica No. 437], ILL 13346.

Crepidotus rhizomorphus E.A. Burt. Annals of the Missouri Botanical Garden 10:181. 1923. **Holotype:** On culm of an undetermined grass [Poaceae sp. indet.], Hawaii, [Island of Hawaii, Kona, Keauhou, 23 JUL 1921], leg. F.L. Stevens No. 940, ILL 30029.

Cribropeltis citrullina L.R. Tehon. Mycologia 25:252. 1933. **Isotype:** On *Citrullus vulgaris* Schrader [= *C. lanatus* (Thunb.) Matsumura & Nakai], Illinois, Woodford County, Spring Bay, 30 SEP 1921, leg. G.L. Stout s.n., ILL 33550. **Holotype:** ILLS 22882.

Crossopsora stevensii H. Sydow. Mycologia 17:255–256. 1925. **Syntypes**: On species of Apocynaceae [as Asclepiadaceae in the protologue; *Mandevilla scabra* (Hoffmgg. ex Roemer & J.A. Schultes) K. Schum. on the packet], British Guiana [Guyana], Rockstone, 17 JUL 1922, leg. F.L. Stevens Nos. 490 and 491 [in one packet], ILL 18031; on *Echites tomentosa* Vahl [= *Mandevilla hirsuta* (L.C. Richard) K. Schum.], Trinidad, Cumuto, 16 AUG 1922, leg. F.L. Stevens No. 933, ILL 18030 [two packets, both labeled *Crossospora*].

Cryptosporium ludwigii P.A. Saccardo. Annales Mycologici 11:18–19. 1913. **Isotype**: Hab. in ramis ramulisque corticatis emortuis *Sarothamni scoparii* [*Sarothamnus scoparius* (L.) Wimmer ex K. Koch = *Cytisus scoparius* (L.) Link, France], Lotharingiae, Forbach, [30 DEC 1911], leg. A. Ludwig [No. 498d], comm. H. Sydow, ILL 13349.

Curreya sandwicensis J.B. Ellis & B.M. Everhart. Bulletin of the Torrey Botanical Club 24:135. 1897 [as *sandicensis*, printer's error]. **Isotypes**: On living leaves of *Alphitonia ponderosa* Hillebr., Sandwich Islands [Hawaii], (Kauai), 1895, leg. A.A. Heller No. 2758, ILL 8612, ex MO 17828 [as microscopic preparations], BPI. **Holotype**: NY.

Cyclostomella oncophora H. Sydow. Annales Mycologici 25:30–32. 1927. **Isosyntype**: In foliis vivis *Ocoteae veraguensis* [= *Ocotea veraguensis* (Klotzsch ex Meisner) Mez], Costa Rica, San Pedro de San Ramón, 6 FEB 1925, leg. H. Sydow, Fungi in Itinere Costaricenci Collecti No. 405, ILL 8532.

Cylindrosporium corni W.G. Solheim. Mycologia 41:630. 1949. **Isotype**: In foliis vivis *Corni stoloniferae* [*Cornus stolonifera* Michaux. = *C. sericea* L.], Wyoming, Carbon County, Medicine Bow Mountains, Six Mile Gap, Platte River, 7 SEP 1948, leg. W.G. Solheim No. 2224, ex Mycoflora Saximontanensis Exsiccata No. 488, ILL 20882.

Cylindrosporium heraclei J.B. Ellis & B.M. Everhart. Journal of Mycology 4: 52. 1888. **Isotype**: On leaves of *Heracleum lanatum* Michaux, with *Phyllachora heraclei* (E.M. Fries:E.M. Fries) L. Fuckel, Utah, Ogden, 1 AUG 1887, leg. S.M Tracy [as Tracy & Evans No. 590 on the packet], ILL 8850.

Cylindrosporium saximontanense W.G. Solheim. Mycologia 41:631. 1949. **Isotype**: In foliis vivis *Populi angustifoliae* [= *Populus angustifolia* James], Colorado, Ouray County, Ouray Picnic Grounds, 12 OCT 1948, leg. W.G. & R. Solheim No. 2258, ex W.G. Solheim, Mycoflora Saximontanensis Exsiccata No. 490, ILL 20880.

Cytospora actinidiae H. & P. Sydow. Annales Mycologici 4:485. 1907. **Isotype**: Hab. in ramis *Actinidiae argutae* [= *Actinidia arguta* (Sieb. & Zucc.) Franchet & Savatier ex Miquel, Germany, Brandenburg: Späth'sche Baumschulen zu] Rixdorf pr. Berlin, [1 OCT 1905], leg. H. & P. Sydow, Mycotheca Germanica No. 519, ILL 11044.

Cytospora annulata J.B. Ellis & B.M. Everhart. Proceedings of the Academy of Natural Sciences of Philadelphia 1893:160. 1893. **Isotype**: On dead limbs of *Negundo aceroides* Moench [= *Acer negundo* L.], with *Sphaeropsis albescens* J.B. Ellis & B.M. Everhart, South Dakota, Brookings, OCT 1891, leg. T.A. Williams [ex J.B. Ellis and B.M. Everhart, North American Fungi No. 2770], ILL 11046.

Cytospora ostryae H. Sydow. Annales Mycologici 8:492. 1910. **Isotype**: Hab. in ramis *Ostryae virginicae* [= *Ostrya virginiana* (P. Miller) K. Koch, Brandenburg: Baumschulen zu] Tamsel [now Poland, in Gesellschaft von *Diplodia ostryae* H. Sydow], 10 NOV 1909 [as 19 NOV on the packet], leg. P. Vogel, ex H. Sydow, Mycotheca Germanica No. 920, ILL 11076.

Cytospora prunorum P.A. Saccardo & H. Sydow in H. & P. Sydow. Annales Mycologici 2:191–192. 1904. **Isotype**: Hab. in ramis *Pruni avium* [= *Prunus avium* L., Germany, Brandenburg]: Späth'sche Baumschulen [Rixdorf] pr. Berolinum [Berlin, 28 OCT 1903], leg. H. & P. Sydow, Mycotheca Germanica No. 136, ILL 11080.

Cytospora pulchella P.A. Saccardo in H. & P. Sydow. Annales Mycologici 1:538. 1903. **Isotype**: Hab. in ramis *Fraxini excelsioris* [= *Fraxinus ecelsior* L.] in summo apice montis, Germaniae [Germany], Saxoniae, "Grosser Winterberg" [bei Schmilka, 8 AUG 1903], leg. H. & P. Sydow, Mycotheca Germanica No. 89, ILL 11082.

Cytospora querna P.A. Saccardo. Annales Mycologici 6:561–562. 1908. **Isotype**: Hab. in ramis emortuis *Quercus pedunculatae* [*Quercus pedunculata* Ehrh. = *Q. robur* L., Germany], Biesenthal pr. Bernau, [14] APR 1907, leg. H. Sydow, Mycotheca Germanica No. 717, ILL 11085.

Cytospora spinescens P.A. Saccardo in H. & P. Sydow. Annales Mycologici 2:192. 1904. **Isotype**: Hab. in ramis corticatis *Betulae papyraceae* [*Betula papyracea* Aiton = *B. papyrifera* Marshall, Germany, Brandenburg]: Späth'sche Baumschulen, [Rixdorf] pr. Berolinum [Berlin, 25 SEP 1903], leg. H. & P. Sydow, Mycotheca Germanica No. 137, ILL 11089.

Cytospora tamaricella H. & P. Sydow. Annales Mycologici 2:192. 1904. **Isotype**: Hab. in ramis emortuis vel subemortius *Tamaricis anglicae* [*Tamarix anglica* Webb in Hooker = *T. gallica* L., meist in Gesellschaft von *Coniothyrium caespitulosum* P.A. Saccardo, Germany, Brandenburg]: Späth'sche Baumschulen [Rixdorf], pr. Berolinum [Berlin, 25 SEP 1903], leg. H. & P. Sydow, Mycotheca Germanica No. 138, ILL 11090.

Cytosporina sorbi C. Oudemans var. *macrospora* F.E. & E.S. Clements. Cryptogamae Formationum Coloradensium, Century 3, No. 248. Anno 1907, nom. nud. **Isotype**: Colorado, Minnehaha, alt. 2600 m, 7 JUL 1906, leg. F.E. & E.S. Clements, ILL 11100.

D

Dacrymyces punctiformis W. Neuhoff. Arkiv för Botanik 28A:45–47, pl. VII, text fig. 1d. 1936. **Paratype**: An *Pinus sylvestris* L. [as *silvestris*], Sweden, Småland, Lidhult [Parish], am Südufer des Askakesjöns [Lake], 1 JUL 1929, leg. J.A. Nannfeldt No. 2195 [ex Fungi Exsiccati Suecici Praesertim Upsalienses No. 474], ILL 32847.

Dasyscypha cyatheae H. Rehm. Leaflets of Philippine Botany 6:2280. 1914. **Assumed isotype**: Ad emortuam rhachidem *Cyatheae caudatae* [= *Cyathea caudata* (J.E. Smith) Copeland, Philippines, [Luzon, Laguna Prov.], Mount Maquiling, pr. Los Baños, JAN 1914, leg. C.F. Baker No. 2727 [ex Fungi Malayana No. 23], ILL 7782.

Dasyscypha nigra F.E. & E.S. Clements. Cryptogamae Formationum Coloradensium, Century 1, No. 83. Anno 1906, nom. nud. **Isotype**: Saprophilus gregarius ad ramos dejectos *Holodisci dumosae* [= *Holodiscus dumosus* (Nutt. ex Hooker) A.A. Heller], Colorado, Minnehaha, alt. 2600 m, 12 AUG 1905, leg. F.E. & E.S. Clements, ILL 7786.

Dasyscypha pulverulenta (M.A. Libert) P.A. Saccardo var. *conicola* H. Rehm ex H. Sydow. Annales Mycologici 8:492. 1910. **Isotype**: [Auf Fruchtzapfen von *Pinus sylvestris* L., Germany, Brandenburg: Kupferhammer bei Mixdorf bei Mullrose, 3 AUG 1909], leg. P. Sydow, ex H. Sydow, Mycotheca Germanica No. 907, ILL 7790.

Dasyscypha sadleriae F.L. Stevens & P.A. Young in F.L. Stevens. Bernice P. Bishop Museum Bulletin 19:11, pl. III(G). 1925. **Holotype**: On living leaves of *Sadleria* sp., Hawaii, [Island of] Hawaii, Hamakua, 31 JUL [1921], leg. F.L. Stevens No. 1078, ILL 7793. **Isotypes**: BISH 499000 and 499903. ≡ *Calycellina sadleriae* (F.L. Stevens & P.A. Young in F.L. Stevens) R.W.G. Dennis. Kew Bulletin 17:363. 1963.

Dendrodochium epistroma F. v. Höhnel ex F. v. Höhnel. Sitzungsberichte der Kaiserl. Akademie der Wissenschaften in Wien. Mathematisch-Naturwissenschaftliche Klasse. Abt. 1, 118:424. 1909; Fragmente zur Mykologie, Mitteilung VI, No. 284. **Isosyntypes**: Auf den Stromaten von *Diatripella favacea* (E.M. Fries:E.M. Fries) V. Cesati & G. de Notaris an Birkenzweigen [= *Betula* sp.] schmarotzend und dieselben oft ganz überziehend, [Poland], Brandenburg, [Schmidt's Grund], Tamsel bei [near] Cüstrin, 24 FEB 1906, leg. P. Vogel, ex H. Sydow, Mycotheca Germanica No. 648 [sub *Hymenula epistroma*], ILL 16505 and 16506. **Syn.**: *Hymenula epistroma* F. v. Höhnel in H. Sydow, 1907, nom. nud. [Both ILL numbers are also isosyntypes of the latter, invalid name.].

Dendrodochium gouaniae H. Sydow. Annales Mycologici 25:155–156. 1927.

Isotype: Hab. in foliis vivis vel languidis *Gouaniae tomentosae* [= *Gouania tomentosa* Jacquin], Costa Rica, ad fluv. Rio Poas inter Sabanilla de Alajuela et San Pedro, 10 JAN 1925, leg. H. Sydow, Fungi in Itinere Costaricensi Collecti No. 392 p.p., ILL 8399 [also an isosyntype of *Phacellula gouaniae* H. Sydow].

Dendroecia farlowiana (P. Dietel) J.C. Arthur, 1906. **See basionym**: *Ravenelia farlowiana* P. Dietel, 1894.

Dendroecia opaca (P. Dietel) J.C. Arthur, 1907. **See basionym**: *Ravenelia opaca* P. Dietel, 1895.

Dendrophoma gouldiae F.L. Stevens & O.A. Plunkett in F.L. Stevens. Bernice P. Bishop Museum Bulletin 19:135. 1925. **Holotype**: On living leaves of *Gouldia coriacea* (Hooker & Arnott) Hillebr. [= *Hedyotis terminalis* (Hooker & Arnott) W.L. Wagner & Herbst], Hawaii, Kauai, Kalalau trail, 16 JUN [1921], leg. F.L. Stevens No. 499, ILL 11171. **Isotypes**: BISH 486023 and 499007.

Densocarpa shanori H.M. Gilkey. North American Flora, Series 2, Part 1:16. 1954. **Holotype**: Illinois, Urbana, 14 JUN 1953, leg. L. Shanor, det.: H.M. Gilkey No. 749a, ILL 33534.

Deslandesia ficina (H. Sydow) A.C. Batista var. *microspora* A.C. Batista & A.F. Vital. Beihefte zur Sydowia Annales Mycologici 3:41–42, fig. 29. 1962, nom. invalid. [in violation of Articles 37.1 & 43.1 of I.C.B.N.]. **Holotype**: With *Limaciniella psidii* J.M. Mendoza in F.L. Stevens; on *Psidium guajava* L. [cited as *guayava*], Hawaii, Kauai, Waimea [misspelled "Wannea"], 16 JUN 1921, leg. F.L. Stevens No. 542, ILL 6661 [also the holotype of *Limaciniella psidii* J.M. Mendoza in F.L. Stevens].

Deslandesia honoluluensis A.C. Batista & A.F. Vital. Beihefte zur Sydowia Annales Mycologici 3:42–43, fig. 30. 1962, nom. invalid. [in violation of Articles 37.1 & 43.1 of I.C.B.N.]. **Holotype**: On leaves of *Mangifera indica* L., associated with *Chaetothyrium mangiferae* J.M. Mendoza in F.L. Stevens and *Microxyphium columnatum* A.C. Batista, R. Ciferri & M.L. Nascimento, Hawaii, Oahu, Honolulu, Olympus, 6 JUN 1921, leg. F.L. Stevens No. 267, ILL 6652 [also the holotype of *Chaetothyrium mangiferae* J.M. Mendoza in F.L. Stevens]. **Isotypes**: BISH 499013 and 499917.

Desmotascus portoricensis F.L. Stevens. Botanical Gazette 68:476, pl. 30. 1919. **Holotype**: On *Bromelia pinguin* L., Porto Rico [Puerto Rico], Mayagüez, [Jayuda], 31 MAR 1913, leg. F.L. Stevens No. 964, ILL 8533.

Detonia constellatio F.E. & E.S. Clements var. *aurantiaca* F.E. & E.S. Clements. Cryptogamae Formationum Coloradensium, Century 2, No. 114. Anno 1906, nom. nud. **Isotype**: ...ad terram..., Colorado, Minnehaha, alt. 2600 m, 6 SEP 1904, leg. F.E. & E.S. Clements, ILL 7660.

Detonia dictyospora F.E. & E.S. Clements. Cryptogamae Formationum

Coloradensium, Century 2, No. 116. Anno 1906, nom. nud. **Isotype:** ...ad terram muscosam ..., Colorado, Beaver Dam, alt. 2700 m, 18 AUG 1904, leg. F.E. & E.S. Clements, ILL 7661.

Dexteria pulchella F.L. Stevens. Transactions of the Illinois State Academy of Science 10:174–175, fig. 4. 1917. **Holotype:** On *Paullinia pinnata* L., Porto Rico [Puerto Rico], Mayagüez, [4 MAY 1913], leg. F.L. Stevens No. 1207, ILL 8249 [also an isotype of *Perisporium paulliniae* F.L. Stevens]. **Isotype:** ILL 6222 [also the holotype of *Perisporium paulliniae* F.L. Stevens].

Diaporthe albocarnis J.B. Ellis & B.M. Everhart. Proceedings of the Academy of Natural Sciences of Philadelphia 1893:140. 1893. **Isosyntype:** On smaller dead limbs of *Cornus* sp., Canada, [Ontario], London, MAY 1891, leg. J. Dearness s.n. [ex J.B. Ellis and B.M. Everhart, North American Fungi No. 2820], ILL 10353.

Diaporthe caryigena J.B. Ellis & B.M. Everhart. Journal of Mycology 9:223. 1903. **Isotype:** On dead hickory limbs [as *Hicoria minima* (Marshall) Britton = *Carya cordiformis* (Wang.) K. Koch on the packet], Canada, [Ontario], London, OCT 1903, leg. J. Dearness No. 2863 [ex J.B. Ellis and B.M. Everhart, Fungi Columbiani No. 1919], ILL 10356.

Diaporthe confusa J.B. Ellis & B.M. Everhart. Fungi Columbiani, Century 16, No. 1528. Anno 1901, nom. nud. **Isotype:** On fire-killed *Rhus venenata* DC. [= *Toxicodendron vernix* (L.) O. Kuntze], New Jersey, Newfield, 23 OCT 1900, leg. J.B. Ellis, ILL 10357.

Diaporthe crinigera J.B. Ellis & B.M. Everhart. Proceedings of the Academy of Natural Sciences of Philadelphia 1890:234. 1890. **Isotype:** On dead oak limbs [= *Quercus* sp.], Canada, [Ontario], London, MAR 1890, leg. J. Dearness No. 1347B [ex J.B. Ellis and B.M. Everhart, North American Fungi No. 2533], ILL 10359.

Diaporthe leucosarca J.B. Ellis & B.M. Everhart. Proceedings of the Academy of Natural Sciences of Philadelphia 1890:233–234. 1890. **Isotype:** On dead limbs of *Carpinus americana* Michaux [= *C. caroliniana* Walter], Canada, [Ontario], London, MAY 1890, leg. J. Dearness No. 1696 [ex J.B. Ellis and B.M. Everhart, North American Fungi No. 2743], ILL 10365.

Diaporthe melonis L. Beraha & M.J. O'Brien. Phytopathologische Zeitschrift 94(3):205–206. 1979. **Isotypes:** In caulibus *Glottidii* [= *Glottidium* stems inoculated in vitro with mycelium from isolated conidia of the fungus], leg. L. Beraha s.n., ILL 33551, NY, IMI. **Holotype:** BPI 71906. Status conidicus: (?)*Phomopsis cucurbitae* C.D. McKeen [original host of the anamorph: *Cucumis melo* L. (Cucurbitaceae) from Texas; the teleomorph was developed in culture; the host, *Glottidium* sp., is a member of Fabaceae].

Diaporthe nivosa J.B. Ellis & E.W.D. Holway in J.B. Ellis & B.M. Everhart. Proceedings of the Academy of Natural Sciences of Philadelphia 1890:222. 1890. **Isotype**: On dead alder [= *Alnus* sp.], Canada, Lake Superior, Isle Royale, JUL 1889, leg. E.W.D. Holway s.n. [ex J.B. Ellis and B.M. Everhart, North American Fungi No. 2535], ILL 10372.

Diaporthe ostryigena J.B. Ellis & J. Dearness in J.B. Ellis and B.M. Everhart. Fungi Columbiani, Century 21, No. 2019. Anno 1905. **Isotype**: On dead trunks and branches of *Ostrya virginica* [= *Ostrya virginiana* (P. Miller) K. Koch], Canada, [Ontario], London, 28 JUN 1904, leg. J. Dearness s.n., ILL 10375.

Diaporthe pruni J.B. Ellis & B.M. Everhart. Proceedings of the Academy of Natural Sciences of Philadelphia 1893:141–142. 1893. **Assumed isosyntype**: On dead limbs of *Prunus virginiana* L., Canada, [Ontario], London, APR 1892, leg. J. Dearness No. 1695 [ex J.B. Ellis and B.M. Everhart, North American Fungi No. 2822], ILL 10378.

Diaporthe subcongrua J.B. Ellis & B.M. Everhart. The North American Pyrenomycetes. A Contribution to Mycologic Botany, pp. 425–426. 1892. **Isotype**: On dead maple limbs [*Acer saccharinum* L.], Canada, [Ontario], London, [autumn and winter of 1891–1892], leg. J. Dearness s.n. [ex J.B. Ellis and B.M. Everhart, North American Fungi No. 2744], ILL 10382.

Diaporthe sydowiana (P.A. Saccardo) P.A. Saccardo, 1913. **See basionym**: *Chorostate sydowiana* P.A. Saccardo, 1908.

Dichomera prunicola J.B. Ellis & J. Dearness in P.A. Saccardo. Sylloge Fungorum 22:1085. 1913. **Isotype**: Hab. in ramis emortuis *Pruni virginianae* [= *Prunus virginiana* L.], Canada, [Ontario], London, [13 OCT 1903], leg. J. Dearness s.n. [ex J.B. Ellis and B.M. Everhart, Fungi Columbiani No. 2021], ILL 11176.

Dicoccum cupaniae H. Sydow. Annales Mycologici 23:421–422. 1925. **Isosyntype**: Hab. in foliis vivis *Cupaniae guatemalensis* [= *Cupania guatemalensis* Radlk.], Costa Rica, Alajuela, 9 JAN 1925, leg. H. Sydow, Fungi in Itinere Costaricensi Collecti No. 35, ILL 15938.

Dictyochorella andropogonis E.M. Doidge. Bothalia 1(2):66. 1922. **Isotype**: Hab. in foliis *Andropogonis nardi* [= *Andropogon nardus* L.] South Africa, Natal [Prov.], Tugela Valley, [near Goodoo], 16 MAY 1920, leg. E.M. Doidge, ex PREM No. 14104, ILL 8546.

Dictyothyriella guianensis F.L. Stevens & H.W. Manter. Botanical Gazette 79:274, figs. 10–13. 1925. **Syntypes**: On *Costus* sp., British Guiana [Guyana], Rockstone, 16 JUL [1922], leg. F.L. Stevens No. 1005, ILL 6697; on *Posoqueria latifolia* (Rudge) Roemer & J.A. Schultes [misspelled "*Pesequeria*"], Kartabo, 22 JUL [1922], leg. F.L. Stevens No. 534, ILL 6701; on *Mauritia* sp. [misspelled "*Maurita*"], Trinidad [as

British Guiana on the packet], Cumuto, 16 AUG [1922], leg. F.L. Stevens No. 1007, ILL 6700; on *Tabernaemontana* sp., British Guiana [Guyana], Rockstone, 17 JUL [1922], leg. F.L. Stevens No. 473, ILL 6707; on unknown member of the Marantaceae, Kartabo, 21 JUL [1922], leg. F.L. Stevens No. 505, ILL 6705; on unknown member of the Apocynaceae, Tumatumari, 11 JUL [1922], leg. F.L. Stevens No. 163, ILL 6708; on unknown host, Kartabo, 23 JUL [1922], leg. F.L. Stevens No. 608, ILL 6706; leg. F.L. Stevens No. 606, ILL 6699; 22 JUL [1922], leg. F.L. Stevens No. 530 [as 530-1 on the packet], ILL 6702; Rockstone, 17 JUL [1922], leg. F.L. Stevens No. 469, ILL 6710; Tumatumari, 8 JUL [1922], leg. F.L. Stevens No. 155, ILL 6698; leg. F.L. Stevens No. 993, ILL 6704; Wismar, 14 JUL [1922], leg. F.L. Stevens No. 275, ILL 6696.

Dictyothyriella inaequiseptata R. Ciferri. Sydowia Annales Mycologici 10:143–144. (1956) 1957. **Isotype:** Hab. in foliis *Inga ? laurina* (Sw.) Willd. [misspelled "*laurima*," Dominican Republic], Cordillera Central, Santo Domingo Prov., La Cumbre, alt. ca. 250 m, JUL 1927, leg. R. Ciferri, Mycoflora Domingensis Exsiccata No. 378, ILL 33065.

Dictyothyriella philodendri F.L. Stevens & H.W. Manter. Botanical Gazette 79:274–275, figs. 24–26. 1925. **Syntype:** On *Philodendron* sp., British Guiana [Guyana], Kartabo, 23 JUL [1922], leg. F.L. Stevens No. 1006, ILL 6727.

Dictyothyriella vismiae F.L. Stevens & H.W. Manter. Botanical Gazette 79:276, figs. 27–29. 1925. **Syntypes:** On *Vismia* sp., British Guiana [Guyana], Rockstone R. Ry., 15 JUL [1922], leg. F.L. Stevens No. 1014, ILL 6729; on *Coccoloba* sp., Rockstone, 17 JUL [1922], leg. F.L. Stevens No. 476, ILL 6728; on unknown member of the Loranthaceae, Kartabo, 21 JUL [1922], leg. F.L. Stevens No. 540, ILL 6731; on unknown host, Kartabo, 23 JUL [1922], leg. F.L. Stevens No. 591, ILL 6730.

Dictyothyrium disporum F.L. Stevens & H.W. Manter. Botanical Gazette 79:272, fig. 8. 1925. **Holotype:** On unknown member of the Menispermaceae, British Guiana [Guyana], Rockstone, 16 JUL [1922], leg. F.L. Stevens No. 435, ILL 6732.

Diderma diadematum J.D. Schoknecht & J.L. Crane. Transactions of the British Mycological Society 70:146–150, figs. 1–17. 1978. **Isotypes:** On submerged, decayed leaves of angiosperms (*Acer* sp.) and *Taxodium distichum* (L.) L.C. Richard, Illinois, Johnson County, Elvira Cypress Swamp (Deer Pond), northwest of Vienna, 9 MAR 1977, leg. J.D. Schoknecht and J.L. Crane, ILL 33532 and 33552, K. **Holotype:** ILLS 36664.

Didymaria boringuensis (E. Young) F.L. Stevens & W.G. Solheim in W.G. Solheim & F.L. Stevens, 1931. **See basionym:** *Cercospora boringuensis* E. Young, 1916.

Didymaria conjugans F.L. Stevens & W.G. Solheim in W.G. Solheim & F.L. Stevens. Mycologia 23:401–402, fig. 10. 1931. **Holotype:** On leaves of an unknown legume [= Fabaceae sp. indet.], British Guiana [Guyana], Tumatumari, [8 JUL 1922], leg. F.L. Stevens No. 54, ILL 13951.

Didymella clibadii F.L. Stevens. Annales Mycologici 29:105–106. 1931. **Syntypes:** On *Clibadium* sp., Peru, Hda. Chalhuapuquio, 6 DEC 1924, leg. F.L. Stevens No. 119, ILL 9962; Ecuador, Terecita, 30 OCT 1924 [as SEP. on the packet], leg. F.L. Stevens No. 130, ILL 9961.

Didymella cocos A.G. Weedon. Mycologia 18:219. 1926. **Holotype:** On leaves of *Cocos alphonsei* [probably = *C. nucifera* L. cult.], Florida, St. Petersburg, 15 FEB 1923, A.G. Weedon No. 6, ILL 9963.

Didymella eupatorii F.L. Stevens. Illinois Biological Monographs 11:200–201, pl. IX, fig. 76; pl. X, fig. 77. 1927. **Holotype:** On *Eupatorium* sp., Costa Rica, La Palma, 8 JUL 1923, leg. F.L. Stevens No. 307, ILL 9965.

Didymella eusticha F.E. & E.S. Clements. Cryptogamae Formationum Coloradensium, Century 5, No. 428. Anno 1908, nom. nud. **Isotype:** In ramuli emortuis corticatisque *Ribis longifolii* [*Ribes longifolium*, nom. illegit. = *R. aureum* Pursh], Colorado, Long's Peak Inn, alt. 2700 m, 10 AUG 1907, leg. F.E. & E.S. Clements, ILL 9964.

Didymella eutypoides H. Rehm. Leaflets of Philippine Botany 8:2943. 1916. **Assumed isotype:** Ad *Bambusam* emortuam [= *Bambusa vulgaris* Schrader ex J.C. Wendl.], Philippines, [Luzon, Laguna Prov., Mount Maquiling, near] Los Baños, OCT 1913, leg. S.A. Reyes, comm. C.F. Baker No. 1915c [ex Fungi Malayana No. 139c, (in exsiccati set as 139 a,b,c, on the same packet label)], ILL 9668 [also, under No. 139a, an assumed isosyntype of *Guignardia bambusina* H. Rehm and, under No. 139b, an assumed isotype of *Massarinula bambusincola* H. Rehm].

Didymosphaeria aphysa F.E. & E.S. Clements. Cryptogamae Formationum Coloradensium, Century 3, No. 233. Anno 1907, nom. nud. **Isotype:** Saprophilus gregarius in foliis putridis *Betulae occidentalis* [= *Betula occidentalis* Hooker], Colorado, Minnehaha, alt. 2600 m, 26 JUN 1906, leg. F.E. & E.S. Clements, ILL 9968.

Didymosphaeria inconspicua H. Rehm. Leaflets of Philippine Botany 8:2948. 1916. **Assumed isotype:** Ad ramum emortuum *Premnae odoratae* [= *Premna odorata* Blanco], Philippines, [Luzon, Laguna Prov., Mount Maquiling], near Los Baños, OCT 1913, leg. C.F. Baker No. 2110b, [ex Fungi Malayana No. 132], ILL 9978.

Didymosphaeria manitobiensis J.B. Ellis & B.M. Everhart. The North American Pyrenomycetes. A Contribution to Mycologic Botany, p. 732. 1892. **Isotype:** On raspberry leaves [*Rubus* sp., Canada], Manitoba, banks of the Little Saskatchewan [River], 3 OCT 1891, leg. J. Dearness s.n. [ex J.B.Ellis and B.M. Everhart, North American Fungi No. 2761], ILL 9979.

Didymosphaeria rehmii J. Kunze. Fungi Selecti Exsiccati, Century 1, No. 90. Anno 1876, nom. nud. **Isotype:** Sub epidermide in *Verbenae officinalis* [= *Verbena officinalis* L.] caulibus vivis vel emortuis, [Germany], (Sax. Bor.), Kloster-Mansfeld ad Islebiam [Eisleben], SEP 1875, leg. J. Kunze, ILL 9981.

Didymosphaeria saccata F.E. & E.S. Clements. Cryptogamae Formationum Coloradensium, Century 1, No. 21. Anno 1906, nom. nud. **Isotype:** ... in foliis vetustis *Iridis missouriensis* [= *Iris missouriensis* Nutt.], Colorado, Larkspur Dell, alt. 2500 m, 11 JUL 1905, leg. F.E. & E.S. Clements, ILL 9982.

Diedickea piceae L. Bonar. Mycologia 34:185–187, fig. 2b. 1942. **Isotype:** On needles of *Picea sitchensis* (Bong.) Carrière, California, Humboldt County, Trinidad, MAY 1931, leg. H.E. Parks No. 3890 [ex California Fungi No. 624], ILL 13114. **Holotype:** UC 653851.

Dielsiella ciferriana F. Petrak in F. Petrak & R. Ciferri. Annales Mycologici 30:183–185. 1932. **Isotype:** In foliis vivis *Capparidis flexuosae* [= *Capparis flexuosa* (L.) L.], in thickets, Dominican Republic, Santiago Prov., Valle del Cibao, Hato del Yaque, 31 DEC 1930, leg. E.L. Ekman No. 4046 [ex R. Ciferri, Mycoflora Domingensis Exsiccata No. 321], ILL 33127.

Dimeriella erigeronicola F.L. Stevens. Transactions of the Illinois State Academy of Science 10:166–167. 1917. **Holotype:** On *Erigeron spathulatus* Vahl in H. West [= *Conyza apurensis* Kunth], Porto Rico [Puerto Rico], Quebradillas, [17 JAN 1914], leg. F.L. Stevens No. 6821, ILL 3652. **Paratypes:** Maunabo, [23 JUN 1913], leg. F.L. Stevens No. 2453, ILL 3651; Yauco, [3 OCT 1913], leg. F.L. Stevens No. 3240, ILL 3653; El Gigante, [16 JUL 1915], leg. F.L. Stevens No. 8522, ILL 3647; Maricao, [20 JUL 1915], leg. F.L. Stevens No. 8935, ILL 3648; on *Erigeron pusillus* Nutt. [= *Conyza canadensis* (L.) Cronq. var. *pusilla* (Nutt.) Cronq.], Maricao, leg. F.L. Stevens No. 8805, ILL 3646.

Dimeriella olyrae F.L. Stevens. Transactions of the Illinois State Academy of Science 10:167. 1917. **Holotype:** On *Olyra latifolia* L., Porto Rico [Puerto Rico], Preston's Ranch, [31 DEC 1913], leg. F.L. Stevens No. 6770, ILL 3643. **Paratypes:** [Rio Maricao above] Maricao, [20 SEP 1913], leg. F.L. Stevens No. 3639, ILL 3642; Maricao, [Indiera Fria, 8 OCT 1913], leg. F.L. Stevens No. 3472, ILL 3644; Maricao, [1 OCT 1912], leg. F.L. Stevens No. 190, ILL 3640; [19 JUL 1915], leg. F.L. Stevens No. 8942, ILL 3638; [20 JUL 1915], leg. F.L. Stevens No. 8959, ILL 5833 and 5737; Mayagüez [as Mayagüez Mesa, 25 JUN 1915], leg. F.L. Stevens No. 7486, ILL 5834; [29 JUN 1915], leg. F.L. Stevens No. 7587, ILL 5837.

Dimeriella pseudotsugae V.M. Miller & L. Bonar. University of California Publications in Botany 19:405, pl. 67, fig. 1. 1941. **Isotype:** On lower side of leaves of young *Pseudotsuga taxifolia* (Lambert) Britton [= *P. menziesii* (Mirbel) Franco], California, Humboldt County, Trinidad, Mill Creek, MAY 1931, leg. J.P. Tracy & H.E. Parks No. 3680 [ex California Fungi No. 626], ILL 3634. **Holotype:** UC 498795.

Dimeriellopsis costaricensis F.L. Stevens. Illinois Biological Monographs 11:170, pl. III, fig. 19; pl. XIII, fig. 99. 1927. **Holotype:** On *Canavalia* sp. [as *Phaseolus* sp. on the packet], Costa Rica, Swamp Mouth, 8 AUG 1923, leg. F.L. Stevens No. 783, ILL 3655.

Dimerina dodonaeae F.L. Stevens. Illinois Biological Monographs 11:171, pl. III, fig. 20; pl. XIII, fig. 100. 1927. **Holotype:** On *Dodonaea viscosa* (L.) Jacquin, Costa Rica, Swamp Mouth, 8 AUG 1923, leg. F.L. Stevens No. 780, ILL 3659. ≡ *Episphaerella dodonaeae* (F.L. Stevens) M.L. Farr, J.D. Schoknecht & J.L. Crane, Canadian Journal of Botany 63:1983–1986. 1985.

Dimerina jacquiniae P. Garman. Mycologia 7:337. 1915. **Holotype:** On leaves of *Jacquinia barbasco* (Loefl.) Mez [nom. illegit. = *J. arborea* Vahl], Porto Rico [Puerto Rico], Mona Island, 20–21 DEC 1913, leg. F.L. Stevens No. 6087, ILL 3680.

Dimeriopsis arthrostylidiicola F.L. Stevens. Transactions of the Illinois State Academy of Science 10:171–172, fig. 2. 1917 [as *arthrostylidicola*, corrected in accordance with Articles 32.6, 60 & 61 of I.C.B.N.]. **Holotype:** On *Arthrostylidium sarmentosum* Pilger in Urban, Porto Rico [Puerto Rico], Monte Alegrillo, [14 NOV 1913], leg. F.L. Stevens No. 4772, ILL 3657.

Dimerium consimile H. Sydow. Annales Mycologici 24:324–325. 1926. **Isotype:** Hab. parasiticum in mycelio *Irenes escharoides* [*Irene escharoides* H. Sydow ≡ *Meliola tabernaemontanae* C.L. Spegazzini var. *escharoides* (H. Sydow) C.G. Hansford], in mycelio et stromatibus *Neostomella tabernaemontanae* H. Sydow ad folia *Tabernaemontanae sananho* [= *Tabernaemontana sananho* Ruiz & Pavon], Costa Rica, Cerro de San Isidro pr. San Ramón, 9 FEB 1925, leg. H. Sydow, Fungi in Itinere Costaricensi Collecti No. 395 p.p., ILL 3658 [also an isotype of *Cicinnobella consimilis* H. Sydow and an isosyntype of *Neostomella tabernaemontanae* H. Sydow].

Dimerium cyanomelum (H. Sydow) C.G. Hansford, 1946. **See basionym:** *Bolosphaera cyanomela* H. Sydow, 1926.

Dimerium fumago (G. Niessl) P.A. & D. Saccardo & P. Sydow in P.A. & D. Saccardo, 1905. **See basionym:** *Meliola fumago* G. Niessl, 1881.

Dimerium guianense F.L. Stevens. Illinois Biological Monographs 8:197, pl. IX, fig. 77; pl. XVII, fig. 107. 1923. **Holotype:** On unknown rosaceous host, British Guiana [Guyana], Kartabo, 24 JUL 1922, leg. F.L. Stevens No. 656, ILL 3679.

Dimerium macowanianum (F. v. Thümen) E.M. Doidge, 1917. **See basionym**: *Meliola macowaniana* F. v. Thümen, 1876.

Dimerium psilostomatis (F. v. Thümen) P.A. & D. Saccardo & P. Sydow in P.A. & D. Saccardo, 1905; emend. [as *D. psilostomae*] by F. v. Höhnel, 1910. **See basionym**: *Meliola psilostomatis* F. v. Thümen, 1877.

Dimerium stevensii P. Garman. Mycologia 7:337. 1915. **Holotype**: On leaves of *Cordia corymbosa* (L.) G. Don [= *C. polycephala* (Lam.) I.M. Johnston], Porto Rico [Puerto Rico], Mayagüez, Quebradillos, College grounds, [30 APR 1913], leg. F.L. Stevens No. 934, ILL 3709. **Paratype**: Maricao, [18 NOV 1913], leg. F.L. Stevens No. 4816, ILL 3704.

Dimerosporina dinochloae H. & P. Sydow. Philippine Journal of Science, Section C (Botany) 9:161. 1914. **Isosyntype**: On *Dinochloa scandens* (Blume) O. Kuntze, [Philippines], Palawan [Island], Taytay, APR 1913, leg. E.D. Merrill No. 8736, ILL 3716.

Dimerosporium cordiae P.C. Hennings. Hedwigia 48:4. 1908. **Isotype**: In foliis *Cordiae* sp. [= *Cordia* sp., Brazil], São Paulo, FEB 1903, leg. A. Puttemans No. 640, ILL 3718.

Dimerosporium fumago (G. Niessl) P.A. Saccardo, 1882. **See basionym**: *Meliola fumago* G. Niessl, 1881.

Dimerosporium macowanianum (F. v. Thümen) P.A. Saccardo, 1882. **See basionym**: *Meliola macowaniana* F. v. Thümen, 1876.

Dimerosporium psilostomatis (F. v. Thümen) P.A. Saccardo, 1882. **See basionym**: *Meliola psilostomatis* F. v. Thümen, 1877.

Diplocladium cylindrosporum J.B. Ellis & B.M. Everhart. Bulletin of the Torrey Botanical Club 27:58. 1900. **Isotypes**: On dead leaves [or broken limb] of *Asimina triloba* (L.) Dunal, West Virginia, Fayette County, Nuttallburg, 31 JUL 1899, leg. L.W. Nuttall, ex J.B. Ellis No. 962, ILL 33533 and 33558, ex NY. **Holotype**: NY.

Diplodia ostryae H. Sydow. Annales Mycologici 8:493. 1910. **Isotype**: Hab. in ramis *Ostryae virginicae* [= *Ostrya virginiana* (P. Miller) K. Koch, Brandenburg: Baumschulen], Tamsel [now Poland; in Gesellschaft von *Cytospora ostryae* H. Sydow], 10 NOV 1909, leg. P. Vogel, ex H. Sydow, Mycotheca Germanica No. 922, ILL 11220.

Diplodia platanicola P.A. Saccardo. Annales Mycologici 6:562. 1908. **Isotype**: Hab. in ramis emortuis *Platani orientalis* [= *Platanus orientalis* L., Brandenburg: Baumschulen], Tamsel [now Poland, 25 FEB 1907], leg. P. Vogel, ex H. Sydow, Mycotheca Germanica No. 720, ILL 11224.

Diplodia veratri F.E. & E.S. Clements. Cryptogamae Formationum Coloradensium, Century 5, No. 483. Anno 1908, nom. nud. **Isotype**: Saprophilus gregarius in foliis *Veratri speciosi* [*Veratrum speciosum* Rydb. = *V. californicum* Durand], Colorado, Sierra Blanca, alt. 2800 m, 21 JUN 1907, leg. F.E. & E.S. Clements, ILL 11245.

Diplodia viciae F.E. & E.S. Clements. Cryptogamae Formationum Coloradensium, Century 5, No. 484. Anno 1908, nom. nud. **Isotype:** ... in foliis vivis et exsiccatis *Viciae linearis* [*Vicia linearis* (Nutt.) E. Greene = *V. americana* Muhl. ex Willd. ssp. *minor* (Hooker) C.R. Gunn], Colorado, LaVeta, alt. 2100 m, 19 JUN 1907, leg. F.E. & E.S. Clements, ILL 11246.

Diplodina equiseti P.A. Saccardo in H. & P. Sydow. Annales Mycologici 3:233. 1905. **Isotype:** Hab. in caulibus emortuis *Equiseti limosi* [= *Equisetum limosum* L., gelegentlich in Gesellschaft von *Phoma equiseti* J.B.H.J. Desmazières, Germany, Brandenburg]: Hennigsdorf pr. Berolinum [Berlin, 28 AUG 1904], leg. H. Sydow, ex H. & P. Sydow, Mycotheca Germanica No. 336, ILL 11266.

Discella populina P.A. Saccardo. Annales Mycologici 6:562, tab. 24, fig. 4. 1908. **Isotype:** Hab. in ramis subvivis *Populi albae* var. *bolleanae* [= *Populus alba* L. var. *bolleana* Lauche, [Brandenburg: Baumschulen], Tamsel [now Polland, 10] MAY 1908, leg. P. Vogel [ex H. Sydow, Mycotheca Germanica No. 726], ILL 13208.

Ditopella asclepiadea L. Bonar. Mycologia 57:382–383. 1965. **Isotype:** On dry stems of *Asclepias mexicana* sensu auct. non Cav. [= *A. fascicularis* Decaisne], California, San Bernardino County, roadside, 5 miles north of Lake Arrowhead, 14 MAR 1957, leg. L. Bonar s.n., ILL 33537. **Holotype:** UC 1272180.

Doratospora guianensis J.M. Mendoza in F.L. Stevens. Annales Mycologici 28:366–367. 1930. **Holotype:** On *Alchornea cordata* (A. Jussieu) Muell. Arg. in DC., non Bentham in Hooker [= *Aparisthmium cordatum* (A. Jussieu) Baillon], British Guiana [Guyana], Tumatumari, 10 JUL [1922], leg. F.L. Stevens No. 153, ILL 6659.

Dothichiza exigua P.A. Saccardo. Annales Mycologici 6:562, tab. 24, fig. 7. 1908. **Isotype:** Hab. in acuibus *Pini strobi* morientibus [on needles of dying *Pinus strobus* L., Brandenburg: Im Forst], Tamsel [now Poland], 31 JAN 1908, leg. P. Vogel [ex H. Sydow, Mycotheca Germanica No. 725], ILL 13209.

Dothidea dasylirii C.H. Peck. Botanical Gazette 7:57. 1882. **Isotypes:** Leaves of some species of *Dasylirion*, probably *D. wheeleri* S. Watson [as *D. wheeleri* on the packet], Arizona, MAY [1881], leg. C.G. Pringle s.n., ex herb. W.G. Farlow, ILL 8555 and 8556. ≡ *Phyllachora dasylirii* (C.H. Peck) P.A. Saccardo, Sylloge Fungorum 2:606. 1883. ≡ *Botryosphaeria dasylirii* (C.H. Peck) F. Theissen & H. Sydow. Annales Mycologici 13:663. 1915. **Holotype:** FH.

Dothidella andiricola C.L. Spegazzini. Boletín de la Academia Nacional de Ciencias en Córdoba 26:359–360. illustr. Preprint 1923 [journal part issued in 1924]. **Probable isotypes:** Hab. En las hojas vivas de *Andira* *jamaicensis* (W. Wright) Urban [= *Andira inermis* (W. Wright) Kunth ex DC.], Porto Rico [Puerto Rico], cerca de Maricao, [20 SEP 1913], leg. F.L. Stevens No. 3628, ILL 10102 and 10108 [also paratypes of *Physalospora andirae* F.L. Stevens]. **Holotype:** LPS 314.

Dothidella flava F.L. Stevens. Botanical Gazette 69:250–251, text figs. 2, 3. 1920. **Holotype:** On *Lithachne pauciflora* (Sw.) Beauvois ex Poiret, Porto Rico [Puerto Rico], Florida Adentro, [1 JUL 1915], leg. F.L. Stevens No. 7665, ILL 8576. **Paratypes:** Trujillo Alto, [15 AUG 1915], F.L. Stevens No. 7394 [erroneously as 9394 in the protologue], ILL 8579; Trujillo Alto [as Florida Adentro on the packet, 1 JUL 1915], leg. F.L. Stevens No. 7654, ILL 8581; Mayagüez, [1 MAY 1913], leg. F.L. Stevens No. 1062, ILL 8580; [27 JUN 1915], leg. F.L. Stevens No. 7432, ILL 8578.

Dothidella portoricensis F.L. Stevens. Botanical Gazette 69:249–250, pl. 13, figs. 8, 9 and text fig. I. 1920. **Holotype:** On *Gleichenia* sp., Las Marias, [22 MAR 1913], leg. F.L. Stevens No. 355I, X.62, ILL 8586.

Dothidina amadelpha H. Sydow. Annales Mycologici 23:387–389. 1925. **Isotype:** Hab. in foliis *Miconiae furfuraceae* [= *Miconia furfuracea* (Vahl) Griseb.], Costa Rica, Mondongo pr. San Ramón, 3 FEB 1925, leg. H. Sydow, Fungi in Itinere Costaricensi Collecti No. 149, ILL 8591.

Dothidina costaricensis F.L. Stevens. Illinois Biological Monographs 11:180–181, pl. VI, figs. 41–43. 1927. **Holotype:** On palm (*Astrocaryum*, *Acrocomia* or *Bactris*) [the latter cited as *Bactria*], Costa Rica, El Roble, 25 JUL 1923, leg. F.L. Stevens No. 622, ILL 8588.

Duportella velutina N.T. Patouillard. Philippine Journal of Science, Section C (Botany) 10:87. 1915. **Paratype:** Sur les branches mortes *Gliricidia sepium* (Jacquin) Kunth ex Walpers [misspelled "*Gliciridia*"], Philippines, [Luzon], Laguna Prov., Mount Maquiling, Evaristo, [near Los Baños], DEC 1913, leg. C.F. Baker No. 1183 [ex Fungi Malayana No. 25], ILL 28479.

E

Echidnodella cocculi F.L. Stevens & R.W. Ryan in F.L. Stevens. Bernice P. Bishop Museum Bulletin 19:76, pl. VII(A). 1925. **Syntypes** [BISH numbers cited are isosyntypes]: On *Cocculus ferrandianus* Gaudich. [= *C. trilobus* (Thunb.) DC.], Hawaii, [Island of] Hawaii, Kealakekua, [Keauhou Kona], 25 JUL [1921], leg. F.L. Stevens No. 989, ILL 7240,

BISH 145811 and 499910; 21 JUL [1922], leg. F.L. Stevens No. 998a [illustrated], ILL 7239, BISH 499937; Hilo to Kilauea, 10 JUL [1921], leg. F.L. Stevens No. 767, ILL 7241; Oahu, Nuuanu Valley, JAN 1912, leg. C.N. Forbes No. 1729, ILL 7242, BISH 145812 and 499912. ≡ *Lembosina cocculi* (F.L. Stevens & R.W. Ryan in F.L. Stevens) J.A. v. Arx. Beiträge zur Kryptogamenflora der Schweiz 11(2):121. 1962.

Echidnodella furcraeae R.W. Ryan. Mycologia 16:195. 1924 [as *fourcroyae*, corrected in accordance with Rec. 60H.1 and Articles 60 & 61 of I.C.B.N.]. **Syntypes**: On *Furcraea* sp. [cited as *Fourcroya*], Porto Rico [Puerto Rico], Maricao, [14 JUL 1915], leg. F.L. Stevens No. 8771, ILL 7247; [19 JUL 1915], leg. F.L. Stevens No. 8822, ILL 7248; Calia Baja, [29 JUL 1915], leg. F.L. Stevens No. 9081, ILL 7249; Rosario, [4 AUG 1915], leg. F.L. Stevens No. 9499, ILL 7246; on *Furcraea hexapetala* (Jacquin) Urban [cited as *Fourcroya*], Maricao, [8 OCT 1913], leg. F.L. Stevens No. 3496 [as 3496a on the packet], ILL 7250.

Echidnodella mabae F.L. Stevens & R.W. Ryan in F.L. Stevens. Bernice P. Bishop Museum Bulletin 19:76. 1925. **Holotype**: On *Maba sandwicensis* A. DC. [= *Diospyros sandwicensis* (A. DC.) Fosb.], Hawaii, Oahu, Makaleha Valley, 8 JAN 1922, leg. O.H. Swezey s.n., ILL 7251. **Isotypes**: ILL 7252 and 7253, BISH 145813, 499938, 499939 and 499940.

Echidnodella melastomatacearum R.W. Ryan. Mycologia 16:195–196. 1924 [cited as *melastomacearum*, corrected in accordance with current I.C.B.N. (cf. F.C. Deighton, 1968)]. **Lectotype** [designated herein]: On *Miconia* sp., Porto Rico [Puerto Rico], Las Marias, [10 JUL 1915], leg. F.L. Stevens No. 8160, ILL 6930. **Isolectotypes**: ILL 4411, 7326 and 7345. [Stevens No. 8160 is also cited as paratype of *Morenoella miconiae* R.W. Ryan and *Meliola miconiae* F.L. Stevens, but only the packet of ILL 4411 is marked with the latter name.]. **Paratypes**: Las Marias, [22 MAR 1913], leg. F.L. Stevens No. 812, ILL 6932; on *Miconia rubiginosa* (Bonpl.) DC. [the epithet misspelled "*religinosa*" in the protologue], Maricao, [18 NOV 1913], leg. F.L. Stevens No. 4796, ILL 6925, 6926, 6929, and 6931; on Melastomataceae sp. indet. [cited as Melastomaceae], Ponce, [8 NOV 1913], leg. F.L. Stevens No. 4375, ILL 6927; Venezuela, Caracas, [15 JUL 1913], leg. F.L. Stevens No. 2993, ILL 6924; leg. F.L. Stevens No. 2997, ILL 6928.

Echidnodella miconiae R.W. Ryan. Mycologia 16:195. 1924. **Holotype**: On *Miconia laevigata* (L.) D. Don in Sweet, Porto Rico [Puerto Rico], Aguas Buenas, [9 FEB 1913], leg. F.L. Stevens No. 302, ILL 7254. **Isotype**: ILL 16237 [this number is also the holotype of *Microclava miconiae* F.L. Stevens (ILL 7254 an isotype). Furthermore, F.L. Stevens No. 302 is cited as a syntype of *Blastotrichum miconiae* F.L. Stevens].

Echidnodella myrciae R.W. Ryan. Mycologia 16:195. 1924. **Holotype**: On *Myrcia splendens* (Sw.) DC., Porto Rico [Puerto Rico], Jajome Alto, [17 JUL 1915], leg. F.L. Stevens No. 8413 [erroneously as 8431 in the protologue], ILL 7255.

Echidnodella raillardiae F.L. Stevens & R.W. Ryan in F.L. Stevens. Bernice P. Bishop Museum Bulletin 19:76. 1925. **Holotype**: On *Raillardia* sp. [= *Dubautia* sp.], Hawaii, [Island of] Hawaii, Kilauea, 15 JUL 1921 [as 16 JUL on the packet], leg. F.L. Stevens No. 853, ILL 7258.

Echidnodella rondeletiae R.W. Ryan. Mycologia 16:195. 1924. **Holotype**: On *Rondeletia* sp., Porto Rico [Puerto Rico], Monte Alegrillo, [14 NOV 1913], leg. F.L. Stevens No. 4505, ILL 7259.

Echidnodes bromeliae R.W. Ryan. Mycologia 16:194. 1924. **Holotype**: On *Bromelia* sp., Porto Rico [Puerto Rico], Rio Tanamá, [6 JUL 1915], leg. F.L. Stevens No. 7913, ILL 7261. **Paratypes**: Maricao, [19 JUL 1915], leg. F.L. Stevens No. 8853, ILL 7262; Jajome Alto, [17 JUL 1915], leg. F.L. Stevens No. 8425, ILL 7260.

Echidnodes mammeae R.W. Ryan. Mycologia 16:194–195. 1924. **Holotype**: On *Mammea americana* L., Porto Rico [Puerto Rico], El Miradero, [4 JUN 1915], leg. F.L. Stevens No. 9139, ILL 7263.

Echidnodes pisoniae F.L. Stevens & R.W. Ryan in F.L. Stevens. Bernice P. Bishop Museum Bulletin 19:76–77, text fig. 14d, pl. VI(D). 1925. **Holotype**: On *Pisonia sandwicensis* Hillebr., Hawaii, Oahu, Tantalus, 22 JUN 1921, leg. F.L. Stevens No. 651, ILL 7269. **Isotypes**: ILL 7264, BISH 499941 and 499942. **Paratypes**: On *Pisonia umbellifera* (J.R. & G. Forster) Seem., [Island of] Hawaii, mountain near Kilauea, OCT 1916, leg. C.N. Forbes No. 537-H, ILL 7268; Oahu, Ahren's ditch trail, 1917, leg. C.N. Forbes No. 2494-O, ILL 7267; 8 JUN 1921, leg. F.L. Stevens No. 288, ILL 7265 [the host erroneously as *Eugenia malaccensis* L. on the packet label].

Echidnodes rhoina E.M. Doidge. Transactions of the Royal Society of South Africa 8:269–270, pl. 18, fig. 43. 1920. **Isosyntypes**: On leaves of *Rhus lucida* L., South Africa, Cape Prov., Van Staden's Pass, 13 NOV 1917, leg. E.M. Doidge, ex PREM No. 10887, ILL 7270; Grahamstown, Howiesons Poort, [17 NOV 1917], leg: E.M. Doidge, ex PREM No. 10957, ILL 7271.

Elachopeltis phoebes H. Sydow. Annales Mycologici 25:121–123, fig. 2. 1927. **Isotype**: In foliis vivis *Phoebes costaricanae* [= *Phoebe costaricana* Mez & Pittier], Costa Rica, Piedades de San Ramón, 26 JAN 1925, leg. H. Sydow, Fungi in Itinere Costaricensi Collecti No. 167e, ILL 6734.

Elasmomyces russuloides W.A. Setchell. Journal of Mycology 13:240–241, pl. 107, figs. 1–3. 1907. **Isotype**: Sub foliis *Heteromeles arbutifoliae* [=

H. arbutifolia (Lindley) M. Roemer] et *Quercus agrifoliae* [= *Q. agrifolia* Née], California, Berkeley, [29 APR 1905], leg. N.L. Gardner & W.A. Setchell No. 220, ILL 30620. **Holotype:** UC.

Ellisiella portoricensis F.L. Stevens. Transactions of the Illinois State Academy of Science 10:203–204, fig. 10. 1917. **Holotype:** On dead leaves of *Clusia rosea* Jacquin, Porto Rico [Puerto Rico], Arecibo, [17 JAN 1914], leg. F.L. Stevens No. 6809, ILL 15941. **Paratypes:** Lajas, [17 JUN 1915], leg. F.L. Stevens No. 7136a, ILL 15940; Hormigueros, [23 JUN 1915], leg. F.L. Stevens No. 7348, ILL 15939.

Elsinoe corni A.E. Jenkins & A.A. Bitancourt. Journal of the Washington Academy of Sciences 38:362–364, figs. 1, 2. 1948. **Paratypes:** On *Cornus florida* L., Maryland, Helen, 6 AUG 1948, leg. R.A. Jehle s.n., ILL 32641, BPI 90594; Denton, 21 JUL 1948, leg. R.A. Jehle s.n., ILL 32640, BPI 90583. **Holotype:** BPI.

Elsinoe ilicis A.G. Plakidas. Mycologia 46:351–352, fig. 2. 1954. **Isotypes:** In foliis caulibus baccisque *Ilicis cornutae* [= *Ilex cornuta* Lindley & Paxton], Louisiana, Hammond, [16 DEC 1948], leg. A.G. Plakidas [ex A.E. Jenkins and A.A. Bitancourt, Myriangiales Selecti Exsiccati No. 472], ILL 32643, BPI [as USM 90804 on the packet], IB 5354, LSUM. **Paratypes:** Myriangiales Selecti Exsiccati No. 471, ILL 32642 and 32644, IB 5353.

Endodothella tapirirae F.L. Stevens. Illinois Biological Monographs 8:190–191, pl. VII, figs. 57–59. 1923 [as *tapirae*, corrected in accordance with Articles 60 & 61 of I.C.B.N.]. **Holotype:** On *Tapirira* sp. [misspelled "*Tapira*"], British Guiana [Guyana], Kartabo, 22 JUL 1922, leg. F.L. Stevens No. 525, ILL 32646 [also the holotype of *Myrianginella tapirirae* F.L. Stevens & A.G. Weedon]. **Isotype:** ILL 8600 [as a microscopic preparation].

Endophyllum lacus-regis D.B. Savile & J.A. Parmelee. Mycologia 48:577–579, figs. 1, 2. 1956. **Isotypes:** On *Claytonia caroliniana* Michaux, moist beech-maple woods, Canada, Quebec, Gatineau County, Kingsmere Road, between Kingsmere and Old Chelsea, 13 MAY 1955, leg. D.B. Savile & J.A. Parmelee [No. 3236], ILL 33538, DAOM 46405.

Englerula macowaniana (F. v. Thümen) F. v. Höhnel, 1910, nom. illegit. [incorrect name, cited in discussion only and not accepted by the author]. **See basionym:** *Meliola macowaniana* F. v. Thümen, 1876.

Englerulaster macowaniana (F. v. Thümen) G. Arnaud, 1918. **See basionym:** *Meliola macowaniana* F. v. Thümen, 1876.

Englerulaster popowiae E.M. Doidge. Transactions of the Royal Society of South Africa 8:243, pl. XIII, fig. 1. 1920. **Isotype:** On leaves of *Popowia caffra* J.D. Hooker & Thomson, South Africa, Natal Prov., Buccleuch, 24 MAR 1916, leg. E.M. Doidge, ex PREM No. 9714, ILL 7279.

Paratype: East London, 24 NOV 1917, leg. E.M. Doidge, ex PREM No. 10917, ILL 7280.

Entyloma browalliae H. Sydow. Annales Mycologici 23:326. 1925. **Isosyntype:** Hab. in foliis *Browalliae demissae* [= *Browallia demissa* L.], Costa Rica, Aserri, 26 DEC 1924, leg. H. Sydow, Fungi in Itinere Costaricensi Collecti No. 356, ILL 17450.

Epicyta ampliata H. Sydow. Annales Mycologici 24:414. 1926. **Isotype:** In foliis vivis *Phoebes neurophyllae* [= *Phoebe neurophylla* Mez & Pittier], Costa Rica, Cerro de San Isidro pr. San Ramón, 9 FEB 1925, leg. H. Sydow, Fungi in Itinere Costaricensi Collecti No. 169a, ILL 11276.

Episoma parasiticum H. Sydow. Annales Mycologici 23:329–331. 1925. **Isotype:** Hab. parasiticum in mycelio *Henningsomycetes escharoidis* [= *Henningsomyces escharoides* H. Sydow] ad folia *Mauriae biringo* [= *Mauria biringo* Tulasne], Costa Rica, Grecia, 17 JAN 1925, leg. H. Sydow, Fungi in Itinere Costaricensi Collecti No. 248, ILL 6660 [also an isotype of *Henningsomyces escharoides* H. Sydow and of *Calolepis congesta* H. Sydow].

Episphaerella dodonaeae (F.L. Stevens) M.L. Farr, J.D. Schoknecht & J.L. Crane, 1985. **See basionym:** *Dimerina dodonaeae* F.L. Stevens, 1927.

Epistigme ingae A.C. Batista & A.F. Vital. Anais da Sociedade de Biologia de Pernambuco 15(2):417–419, fig. 4. 1957. **Holotype:** In foliis *Ingae* sp. [= *Inga* sp., Costa Rica], San Pedro de San Ramón, 23 JAN 1925, leg. H. Sydow, [Fungi in Itinere Costariensi Collecti No. 108c], ILL 6654 [also a paratype of *Chaetothyrium permixtum* H. Sydow].

Euantennaria mucronata (J.P.F.C. Montagne) S.J. Hughes, 1972. **See basionym:** *Capnodium mucronatum* J.P.F.C. Montagne, 1860.

Eumela chiococcae H. Sydow. Annales Mycologici 23:335–337. 1925. **Isosyntype:** Hab. in foliis vivis *Chiococcae racemosae* [*Chiococca racemosa* L. = *C. alba* (L.) A.S. Hitchcock], Costa Rica, San Pedro de San Ramón, 23 JAN 1925, leg. H. Sydow, Fungi in Itinere Costaricensi Collecti No. 70, ILL 9374.

Eutypa petiolaris R. Ciferri. Sydowia Annales Mycologici 10:145. (1956) 1957. **Isotype:** Hab. in petiolis foliorum vivis *Nectandrae coriaceae* [*Nectandra coriacea* (Sw.) Griseb. = *Ocotea coriacea* (Sw.) Britton], (Lauraceae), [Dominican Republic], Cordillera Central, La Vega Prov., Bonao, in sylva humida, 1930–1931, leg. R. Ciferri, Mycoflora Domingensis Exsiccata No. 369, ILL 33157.

Excioconidium cibotii O.A. Plunkett in F.L. Stevens. Bernice P. Bishop Museum Bulletin 19:156–158, text fig. 34. 1925 [misspelled "*cibotti*" in the protologue but correctly spelled on the packet; corrected in accordance with Articles 60 & 61 of I.C.B.N.]. **Assumed holotype:** Saprophytic on stems of *Cibotium chamissoi* Kaulfuss, Hawaii, [Island of] Hawaii, Kilauea, 13 JUL [1921], leg. F.L. Stevens No. 810 [as 811 on the packet], ILL 15942. **Isotypes**: BISH 486050 and 499005. ≡ *Chalara cibotii* (O.A. Plunkett in F.L. Stevens) T.R. Nag Raj & B. Kendrick. A Monograph of *Chalara* and Allied Genera, p. 101. 1975.

Exophoma magnoliae A.G. Weedon. Mycologia 18:221–222, figs. 1, 2. 1926. **Holotype**: On *Magnolia grandiflora* L., Florida, St. Petersburg, 15 FEB 1923, leg. A.G. Weedon No. 5, ILL 11277.

Exosporium leucaenae F.L. Stevens & N.E. Dalbey. Mycologia 11(1):5, pl. 2, figs. 5–7. DEC 1918. **Holotype**: On *Leucaena glauca* sensu auct. non Bentham [= *L. leucocephala* (Lam.) de Wit], Porto Rico [Puerto Rico], Arecibo, [17 JAN 1914], leg. F.L. Stevens No. 6792, ILL 16469. **Paratypes**: Vega Baja, [5 NOV 1913], leg. F.L. Stevens No. 4295, ILL 16473; Manati, [20 NOV 1915], leg. F.L. Stevens No. 5265, ILL 16470; Quebradillas, [22 NOV 1915], leg. F.L. Stevens No. 5122, ILL 16472; Aguada, [22 NOV 1915], leg. F.L. Stevens No. 5076, ILL 16471.

F

Fusarium meliolicola F.L. Stevens. Botanical Gazette 65:245. 1918 [cited as *meliolicolum*, corrected in accordance with Articles 32.6, 60 & 61 of I.C.B.N.]. **Holotype**: On *Meliola paulliniae* F.L. Stevens on *Casearia sylvestris* Sw., Porto Rico [Puerto Rico], Mayagüez, [1 May 1913], leg. F.L. Stevens No. 1051, ILL 5884 [also the holotype of *Nectria meliolicola* F.L. Stevens and a paratype of *Meliola paulliniae* F.L. Stevens].

Fusarium uredinum J.B. Ellis & B.M. Everhart. North American Fungi, Century 28, No. 2799. Anno 1892, nom. nud. **Isotype**: Parasitic on Uredo of *Melampsora salicina* J.H. Léveillé, on leaves of *Salix* sp., Wisconsin, Racine, SEP 1890, leg. J.J. Davis s.n., ILL 16498.

Fusicladium angelicae J.B. Ellis & B.M. Everhart. Proceedings of the Academy of Natural Sciences of Philadelphia 1891:87 1891. **Isotype**: On leaves of *Angelica atropurpurea* L., Wisconsin, Racine, Sep 1890, leg. J.J. Davis No. 9035 [ex J.B. Ellis and B.M. Everhart, North American Fungi No. 2790], ILL 15952.

Fusicladium crataegi R. Aderhold. Berichte der Deutschen Botanischen Gesellschaft 20:200. 1902. **Isotype**: Auf schwarzen Flecken der Früchte [*Crataegus* sp.] im Herbst bis Frühjahr, leg. H. Diedicke [ex H. & P. Sydow, Mycotheca Germanica No. 45], ILL 15955 [this is the anamorph of *Venturia crataegi* R. Aderhold].

Fusicladium gnaphaliatum L. Bonar. Mycologia 57:392–393. 1965.

Paratypes: Parasitic on leaves of *Gnaphalium chilense* Sprengel [= *G. stramineum* Kunth in H.B.K.], California, Humboldt County, Trinity River Valley, [opposite Willow Creek], 15 JUL 1950, leg. J.P. Tracy No. 19045, ILL 33502, UC 1272182. **Holotype:** UC 52330.

Fusicladium pirinum (M.A. Libert) L. Fuckel var. *cladophilum* J.B. Ellis & B.M. Everhart. North American Fungi, Century 28, No. 2791. Anno 1892, nom. nud. **Isotype:** On living shoots of pear trees [= *Pyrus* sp.], Wisconsin, Pewaukee, MAY 1890, leg. G.P. Pfeiffer, ILL 15980.

G

Galera viscosa F.E. & E.S. Clements. Cryptogamae Formationum Coloradensium, Century 4, No. 380. Anno 1907, nom. nud. **Isotype:** Saprophilus copiosus ad fimum vaccinum udumque, etc., Colorado, Cameron Glen, alt. 2800 m, 25 AUG 1906, leg. F.E. & E.S. Clements, ILL 30020.

Galzinia cymosa D.P. Rogers. Mycologia 36:100–101, fig. 12. 1944. **Holotype:** On firm fallen wood of *Pinus rigida* P. Miller, Massachusetts, Springfield, woods east of Watershops Pond, 17 AUG 1943, leg. D.P. Rogers No. 1026, ILL 32515.

Galzinia occidentalis D.P. Rogers. Mycologia 36:102–103, fig. 14. 1944. **Holotype:** On wood of *Pseudotsuga mucronata* (Raf.) Sudw. in Holz. [= *P. menziesii* (Mirbel) Franco], Oregon, Linn County, near Lebanon, 9 APR 1937, leg. D.P. Rogers No. 371, ILL 32516.

Gibberidea turfosa H. Sydow. Annales Mycologici 6:480. 1908. **Isotype:** Hab. in ramis emortuis *Vaccinii uliginosi* [= *Vaccinium uliginosum* L., Germany, Hessen-Nassau]: Rotes Moor bei Gersfeld im Rhöngebirge, [6 AUG 1907], leg. H. Sydow, Mycotheca Germanica No. 690, ILL 9615.

Gillotiella late-maculans H. Rehm. Leaflets of Philippine Botany 6:2278. 1914. **Assumed isotype:** Ad *Arengam* [= *Arenga* sp.], Philippines, [Luzon], Laguna Prov., [Mount Maquiling] near Los Baños, AUG 1913, leg, S.A. Reyes, comm. C.F. Baker No. 1433a [ex Fungi Malayana No. 29], ILL 7281.

Gloeocystidium caliciferum V. Litschauer. Oesterreichische Botanische Zeitschrift 77:126–128, fig. 4. 1928. **Holotype:** Ad truncos cariosor *Alni incanae* [= *Alnus incana* (L.) Moench, Austria], Tirol, Innsbruck, 20 OCT 1926, leg.V. Litschauer s.n., ILL 32916.

Gloeocystidium ochroleucum G. Bresadola in H. Bourdot and A. Galzin.

Bulletin de la Société Mycologique de France 28:365. 1912. **Isosyntype**: Sur boi carié, trom creux de châtaignier, [France], Aveyron, [12 FEB 1912], leg. A. Galzin No. 10801 [renumbered by H. Bourdot: No. 8645], ILL 32606, ex NY.

Gloeocystidium polygonium (C. Persoon:E.M. Fries) G. Bresadola var. *fulvescens* G. Bresadola. Mycologia 17:69. 1925. **Isosyntype**: Hab. ad corticem *Populi trichocarpae* [*Populus trichocarpa* Torrey & Gray = *P. balsamifera* L. ssp. *trichocarpa* (Torrey & Gray) Brayshaw], Idaho, [Bonner County, Priest River, 12 OCT 1920], leg. J.R. Weir No. 16824, ILL 32520. **Syntype**: BPI.

Gloeocystidium sernanderi V. Litschauer. Svensk Botanisk Tidskrift 25:437, fig. 1. 1931. **Isosyntype**: [An der Rinde eines morschen *Betula* Stammes], Suecia [Sweden], Uplandia:Vårdsätra, Uppsala, 11 JAN 1930, leg. S. Lundell No. 809, ex herb. V. Litschauer, ILL 32598, ex TRTC.

Gloeocystidium triste V. Litschauer & S. Lundell. Svensk Botanisk Tidskrift 32:292, fig. 5. 1938. **Isotype**: [Underside of prostrate coniferous] log, Suecia [Sweden], Småland, "Dullaberget", Femsjö, 6 AUG 1937, leg. S. Lundell No. 1470, ex Fungi Exsiccati Suecici No. 744, ILL 32527, ex TRTC. **Holotype**: UPS.

Gloeosporium barringtoniae F.L. Stevens & P.A. Young in F.L. Stevens. Bernice P. Bishop Museum Bulletin 19:144. 1925. **Holotype**: On living leaves of *Barringtonia asiatica* (L.) Kurz, Hawaii, Oahu, Honolulu, Hillebrand Gardens, 18 JUN [1921], leg. F.L. Stevens No. 42, ILL 13368.

Gloeosporium bicolor J.J. Davis. Transactions of the Wisconsin Academy of Sciences, Arts and Letters 20:427. 1921. **Isotype**: On leaves of *Quercus bicolor* Willd., Wisconsin, Chippewa Falls, 14 SEP 1918, leg. J.J. Davis [Fungi Wisconsinenses Exsiccati No. 116], ILL 13380.

Gloeosporium (*Marsonia*) *brunneum* J.B. Ellis & B.M. Everhart. Journal of Mycology 5:154. 1889. **Isotype**: On leaves of *Populus candicans* Aiton [= *P. balsamifera* L.], New Jersey, Newfield, AUG 1889, leg. J.B. Ellis [ex J.B. Ellis and B.M. Everhart, North American Fungi No. 2444], ILL 13381.

Gloeosporium cladosporioides J.B. Ellis & B.D. Halsted. Journal of Mycology 6:34. 1890. **Isotype**: On living stems and leaves of *Hypericum mutilum* L., New Jersey, Metuchen, [Terra Cotta Bank], JUL 1889, leg. B.D. Halsted [ex J.B. Ellis and B.M. Everhart, North American Fungi No. 2438], ILL 13400.

Gloeosporium lagenarium (G. Passerini) P.A. Saccardo & C. Roumeguère var. *foliicolum* J.B. Ellis & B.M. Everhart. North American Fungi, Century 25, No. 2448. Anno 1890, nom. nud. **Isotype**: On leaves of

(a) *Cucumis sativus* L. and (b) *Citrullus vulgaris* Schrader [= *C. lanatus* (Thunb.) Matsumura & Nakai, in a single packet], New Jersey, Newfield, SEP 1889, leg. J.B. Ellis, ILL 13537.

Gloeosporium peleae F.L. Stevens. Bernice P. Bishop Museum Bulletin 19:144–145. 1925. **Holotype**: On galls caused by the psyllid *Hevaheva perkini* on *Pelea* sp. [= *Melicope* sp.], Hawaii, Oahu, Tantalus, 22 JUN [1921], leg. F.L. Stevens No. 632, ILL 13574. **Isotype**: BISH 499004.

Gloeosporium ramosum J.B. Ellis & B.M. Everhart. Journal of Mycology 5:154. 1889. **Isotype**: On leaves and stems of *Polygala polygama* Walter, New Jersey, Newfield, JUN 1889, leg. J.B. Ellis [ex J.B. Ellis and B.M. Everhart, North American Fungi No. 2440], ILL 13597.

Gloeosporium revolutum J.B. Ellis & B.M. Everhart. Journal of Mycology 5:153. 1889. **Isotype**: On living leaves of *Robinia pseudoacacia* L. [as *pseudacacia*], New Jersey, Newfield, [North Vineland], AUG 1889, leg. J.B. Ellis [ex J.B. Ellis and B.M. Everhart, North American Fungi No. 2443], ILL 13598.

Gloeosporium tristaniae G. Massee. Kew Bulletin 1912:190. 1912. **Isotype**: On leaves of *Tristania laurina* R. Br. ex Aiton, [Australia], Queensland, Virginia Creek, [in 1911], leg. C.T. White s.n., comm. F.M. Bailey, ILL 13629. **Holotype**: K.

Gloeosporium ulmicola L.E. Miles. Botanical Gazette 71:185–186, pl. 9, figs. 9, 10, 17. 1921 [cited as *ulmicolum,* corrected in accordance with Articles 32.6, 60 & 61 of I.C.B.N.]. **Holotype**: On living leaves of *Ulmus americana* L., Wisconsin, Oconomowoc, 22 AUG 1919, leg. L.E. Miles s.n., ILL 13632.

Gloeosporium vogelianum P.A. Saccardo. Annales Mycologici 6:562–563, tab. 24, fig. 6. 1908. **Isotype**: Hab. in foliis languidis *Coryli colurnae* [= *Corylus colurna* L., Brandenburg: Baumschulen], Tamsel [now Poland, 15 OCT 1907], leg. P. Vogel [ex H. Sydow, Mycotheca Germanica No. 730], ILL 13640.

Gloeosporium vogelii H. & P. Sydow. Annales Mycologici 3:233. 1905. **Isotype**: Hab. in foliis vivis vel languidis *Tiliae ulmifoliae* [*Tilia ulmifolia* Scop. = *T. cordata* P. Miller, oft in Gesellschaft von *Cercospora microsora* P.A. Saccardo, Brandenburg: Baumschulen], Tamsel [now Poland], 24 OCT 1904, leg. P. Vogel, ex H. & P. Sydow, Mycotheca Germanica No. 342, ILL 13641.

Gloeotulasnella opalea D.P. Rogers. Annales Mycologici 31:198–199, fig. 12. 1933. **Holotype**: On the underside of old aspen logs [*Populus grandidentata* Michaux], Iowa, Johnson County, Turkey Creek, 21 MAY 1932, leg. M.C. Fisher & H.L. Dean, comm. D.P. Rogers No. 146, ILL 32838.

Glomerella bromeliae F.L. Stevens & A.G. Weedon in F.L. Stevens. Illinois Biological Monographs 11:201. 1927. **Holotype**: On *Bromelia pinguin* L. [misspelled "*penguin*"], Costa Rica, Escasu, 29 JUN 1923, leg. F.L. Stevens No. 166, ILL 10244.

Gloniella rubra F.L. Stevens. Botanical Gazette 69:254, fig. 21. 1920. **Lectotype** [designated herein]: On *Arthrostylidium multispicatum* Pilger, Porto Rico [Puerto Rico], El Alto de la Bandera, [8 NOV 1913], leg. F.L. Stevens No. 4363, ILL 7605a. **Isolectotypes**: ILL 7605b, c, d, e.

Godronia rugosa J.B. Ellis & B.M. Everhart. Journal of Mycology 8:70. 1902. **Isotype**: On dead limbs of *Oxydendrum arboreum* (L.) DC., Alabama, Tuskegee, [7 SEP 1900], leg. G.W. Garver [ex J.B. Ellis and B.M. Everhart, Fungi Columbiani No. 1534], ILL 8092.

Grallomyces portoricensis F.L. Stevens. Botanical Gazette 65:245–246, text fig. 5. 1918. **Holotype**: On *Clusia minor* L., Porto Rico [Puerto Rico], El Alto de la Bandera, 14 JUL 1915, leg. F.L. Stevens No. 8283, ILL 3543 [also the holotype of *Meliola clusiae* F.L. Stevens]. **Paratypes**: On *Guarea trichilioides* L. [= *G. guidonia* (L.) Sleumer], Las Marias, 10 JUL 1915, leg. F.L. Stevens No. 8166, ILL 5171 and 5173 [also the holotype and an isotype, respectively, of *Meliola guareicola* F.L. Stevens and of *Helminthosporium guareicola* F.L. Stevens]; on ? [as *Amomis caryophyllata* (Jacquin) Krug & Urban on the packet = *Pimenta racemosa* (P. Miller) J.W. Moore], Monte Alegrillo, 14 NOV 1913, leg. F.L. Stevens No. 4521, ILL 16590 and 165905.

Graphiothecium vinosum J.J. Davis. Transactions of the Wisconsin Academy of Sciences, Arts and Letters 18:90. 1915. **Assumed isosyntype**: On leaves of *Ribes americanum* P. Miller, Wisconsin, Madison, 3 SEP 1912, leg. J.J. Davis [ex Fungi Wisconsinenses Exsiccati No. 18], ILL 16376.

Graphium dubautiae F.L. Stevens & A.G. Weedon in F.L. Stevens. Bernice P. Bishop Museum Bulletin 19:159, text fig. 35. 1925. **Holotype**: On *Dubautia laxa* Hooker & Arnott, Hawaii, Oahu, Tantalus, 22 JUN [1921], leg. F.L. Stevens No. 650, ILL 16378. **Isotype**: BISH 499003.

Griggsia cyatheae F.L. Stevens & N.E. Dalbey. Botanical Gazette 68:222–225, plates 15, 16. 1919 [as *cyathea*, corrected in accordance with Articles 32.6 & 61 of I.C.B.N.]. **Syntypes**: On *Cyathea arborea* (L.) J.E. Smith, Porto Rico [Puerto Rico], Maricao, 19 JUL 1915, leg. F.L. Stevens No. 8794, ILL 7458; El Alto de la Bandera, 14 JUL 1915, leg. F.L. Stevens No. 8276, ILL 7459.

Guignardia alyxiae F.L. Stevens. Bernice P. Bishop Museum Bulletin 19:101, text fig. 26b. 1925. **Syntypes** [BISH numbers cited are isosyntypes]: On *Alyxia oliviformis* Gaudich. [cited as *olivaeformis*], Hawaii, Oahu, Wahiawa, 3 JUN [1921], leg. F.L. Stevens No. 199, ILL

9666, BISH 145833 and 499906; Palolo Valley, [and Mt. Olympus], 10 JUN [1921], leg. F.L. Stevens No. 308, ILL 9665, BISH 499002 and 499905.

Guignardia bambusina H. Rehm. Leaflets of Philippine Botany 8:2936. 1916. **Assumed isosyntype:** Ad emortuam *Bambusam* [*vulgarem* = *Bambusa vulgaris* Schrader ex J.C. Wendl.], Philippines, [Luzon, Laguna Prov., Mount Maquiling, near] Los Baños, OCT 1913, leg. S.A. Reyes, comm. C.F. Baker No. 1915a [ex Fungi Malayana No. 139a, (in exsiccati set as 139a,b,c, on the same packet label)], ILL 9668 [also, under No. 139b, an assumed isotype of *Massarinula bambusincola* H. Rehm and, under No. 139c, an assumed isotype of *Didymella eutypoides* H. Rehm].

Guignardia clusiae F.L. Stevens. Transactions of the Illinois State Academy of Science 10:183. 1917. **Holotype:** On *Clusia gundlachii* Stahl, Porto Rico [Puerto Rico], Maricao, [3 APR 1913], leg. F.L. Stevens No. 809, ILL 9727. **Isotype:** ILL 9838 [also a paratype of *Mycosphaerella guttiferae* L.E. Miles]. **Paratype:** [14 NOV 1913], leg. F.L. Stevens No. 4774, ILL 9728.

Guignardia dinochloae H. Rehm. Leaflets of Philippine Botany 8:2936. 1916. **Assumed isotype:** Ad *Dinochloam* sp. [= *Dinochloa* sp.], Philippines, [Luzon], Laguna Prov., [Mount Maquiling, near] Los Baños, DEC 1913, leg. C.F. Baker No. 2189b, [ex Fungi Malayana No. 181b, (in exsiccati set as 181a,b, on the same packet label)], ILL 10128 [also, under No. 181a, an assumed isotype of *Physalospora dinochloae* H. Rehm].

Guignardia fusco-coriacea H. Rehm. Leaflets of Philippine Botany 6:2195. 1914. **Assumed isotype:** Ad folia *Antidesma* sp., Philippines, Luzon, Laguna Prov., Los Baños, [19] SEP 1913, leg. S.A. Reyes, comm. C.F. Baker No. 1841, [ex Fungi Malayana No. 30], ILL 9729.

Guignardia helicteres F.L. Stevens. Transactions of the Illinois State Academy of Science 10:183. 1917. **Isotype:** On *Helicteres jamaicensis* Jacquin, Porto Rico [Puerto Rico], Barceloneta, [10 AUG 1915], leg. F.L. Stevens No. 9260, ILL 9730.

Guignardia heterotrichi F.L. Stevens. Transactions of the Illinois State Academy of Science 10:182–183. 1917. **Holotype:** On *Heterotrichum cymosum* (Wendl. ex Sprengel) Urban, Porto Rico [Puerto Rico], Preston's Ranch, [31 DEC 1913], leg. F.L. Stevens No. 6768, ILL 9732. **Paratypes:** Villa Alba, [3 JAN 1912], leg. F.L. Stevens No. 116, ILL 8853 and 9731 [also the holotype and an isotype, respectively, of *Phyllachora heterotrichi* F.L. Stevens & N.E. Dalbey]; Maricao, [10 JAN 1913], leg. F.L. Stevens No. 199 [erroneously as No. 197 in the protologue], ILL 9733.

Guignardia ingae F.L. Stevens. Annales Mycologici 28:367. 1930. **Holotype:** On *Inga* sp., Costa Rica, Peralta, 13 JUL [as 13 JUN 1923 on the packet], leg. F.L. Stevens No. 451, ILL 9734.

Guignardia jussiaeae F.L. Stevens. Bernice P. Bishop Museum Bulletin 19:101. 1925. **Holotype:** On *Jussiaea villosa* sensu Hillebr. non Lam. [= *Ludwigia octovalvis* (Jacquin) Raven], Hawaii, Oahu, Tantalus, 5 SEP 1909, leg. H.L. Lyon No. 86, ILL 9735. **Isotypes:** BISH 499001.

Guignardia justiciae F.L. Stevens. Botanical Gazette 69:255, figs. 23, 24. 1920. **Holotype:** On *Justicia verticillaris* sensu (Nees) Urban non L.f. [= *J. martinsoniana* Howard], Porto Rico [Puerto Rico], Maricao, [3 APR 1913], leg. F.L. Stevens No. 806, ILL 9738a. **Isotypes:** ILL 9736 and 9738b. **Paratypes:** El Yunque, [28 AUG 1913], leg. F.L. Stevens No. 2839, ILL 9741; El Gigante, [16 JUL 1915], leg. F.L. Stevens No. 8557, ILL 9737.

Guignardia musae F.L. Stevens. Bernice P. Bishop Museum Bulletin 19:101. 1925. **Holotype:** On *Musa* × *paradisiaca* L. [cult. = *M.* cf. *acuminata* Colla × *M. balbisiana* Colla], Hawaii, Oahu, Hakipuu, 19 JUN [1921], leg. F.L. Stevens No. 565, ILL 9742. **Isotypes:** BISH 499904 and 596860.

Guignardia nectandrae F.L. Stevens. Botanical Gazette 69:255–256. 1920. **Holotype:** On *Nectandra coriacea* (Sw.) Griseb.(?) [= *Ocotea coriacea* (Sw.) Britton], Porto Rico [Puerto Rico], Quebradillos, [22 NOV 1913], leg. F.L. Stevens No. 4994, ILL 9743.

Guignardia pipericola F.L. Stevens. Transactions of the Illinois State Academy of Science 10:183–184. 1917. **Holotype:** On *Piper medium* Jacquin [as *Piper* sp. on the packet = *P. amalago* L.], Porto Rico [Puerto Rico], Camuy, 22 NOV 1913, leg. F.L. Stevens No. 4998, ILL 9747a. **Isotypes:** ILL 9747b and 9750. **Paratypes:** Aguada, [22 NOV 1915], leg. F.L. Stevens No. 5078, ILL 9752; Vega Baja, [22 FEB 1913,] leg. F.L. Stevens No. 370a, ILL 9751; Trujillo Alto, [15 AUG 1915], leg. F.L. Stevens No. 9401, ILL 9753; Santa Ana, [1 JUL 1915], leg. F.L. Stevens No. 7615, ILL 9757; Rio Tanamá, [7 JUL 1915], leg. F.L. Stevens No. 7949, ILL 9756; [6 JUL 1915], leg. F.L. Stevens No. 7869, ILL 9760; leg. F.L. Stevens No. 7872, ILL 9755; Manati, [2 JUL 1915], leg. F.L. Stevens No. 7701, ILL 9749; Peñuelas, [JUL 1915], leg. F.L. Stevens No. 9148, ILL 9759; leg. F.L. Stevens No. 3592, ILL 9758; on *Piper marginatum* Jacquin, Lajas, [18 JUN 1915], leg. F.L. Stevens No. 7180, ILL 9745; Cabo Rojo, [15 JUN 1913], leg. F.L. Stevens No. 2244, ILL 9744; [27 DEC 1913], leg. F.L. Stevens No. 6472, ILL 9754; Florida Adentro, [7 JUL 1915], leg. F.L. Stevens No. 7646 [erroneously as 7648 in the protologue], ILL 9748.

Guignardia rhynchosporae F.L. Stevens. Transactions of the Illinois State

Academy of Science 10:184. 1917. **Holotype:** On *Rhynchospora cyperoides* C. Martius [= *R. holoschoenoides* (L.C. Richard) Herter], Porto Rico [Puerto Rico], Martin Peña, [11 AUG 1915], leg. F.L. Stevens No. 9302, ILL 9761.

Guignardia sterculiae H. Rehm. Leaflets of Philippine Botany 6:2194. 1914. **Assumed isotype:** Ad *Sterculiae foetidae* [= *Sterculia foetida* L.], Philippines, Luzon, Laguna Prov., Los Baños, SEP 1913, leg. S.A. Reyes, comm. C.F. Baker No. 1814, [ex Fungi Malayana No. 31], ILL 9762.

Guignardia tetrazygiae F.L. Stevens. Botanical Gazette 69:255. 1920. **Holotype:** On *Tetrazygia* sp., Porto Rico [Puerto Rico], San Germán, [8 DEC 1913], leg. F.L. Stevens 4567, ILL 9764. **Paratype:** On *Tetrazygia* sp. [as *T. elaeagnoides* (Sw.) DC. on the packet], Vega Alta, [NOV 1913], leg. F.L. Stevens No. 4148, ILL 9763.

Gymnopeltis trinidadensis F.L. Stevens. Illinois Biological Monographs 8:192, pl. VIII, figs. 63–65. 1923. **Holotype:** On *Mauritia* sp., Trinidad, Cumuto, 16 AUG 1922, leg. F.L. Stevens No. 979 [erroneously as No. 974 on the packet], ILL 6736.

H

Hadrotrichum virescens P.A. Saccardo & C. Roumeguère var. *poae* P.A. Saccardo in H. & P. Sydow. Annales Mycologici 2:529. 1904. **Isotype:** Hab. in foliis vivis *Poae* spec. [= *Poa* sp., Germany, Brandenburg], Zehlendorf, pr. Berlin, [AUG 1903], leg. P. Sydow, ex H. & P. Sydow, Mycotheca Germanica No. 289, ILL 15995.

Halbaniella portoricensis (C.L. Spegazzini) R.A. Toro in F.J. Seaver, C.E. Chardon and R.A. Toro, 1932. **See basionym:** *Asteridium portoricense* C.L. Spegazzini, 1923 (1924).

Halstedia portoricensis F.L. Stevens. Botanical Gazette 69:253, figs. 18, 19. 1920. **Holotype:** On *Sideroxylon foetidissimum* Jacquin, Porto Rico [Puerto Rico], Quebradillos, [10 AUG 1915], leg. F.L. Stevens No. 9239, ILL 8607.

Haplographium portoricense F.L. Stevens & N.E. Dalbey. Mycologia 11(1):6–7, pl. 3, fig. 9. DEC 1918. **Holotype:** On *Canna* sp., Porto Rico [Puerto Rico], El Gigante, [16 JUL 1915], leg. F.L. Stevens No. 8495, ILL 16000. **Paratype:** On *Canna coccinea* P. Miller, Aibonito, [16 JUL 1915], leg. F.L. Stevens No. 8447, ILL 15999.

Haplolepis polyadelpha H. Sydow. Annales Mycologici 23:411–412. 1925. **Isotype:** Hab. in foliis vivis *Buettneriae carthagensis* [= *Byttneria* *carthagensis* Jacquin (this spelling of generic name conserved)], Costa Rica, La Caja, pr. San José, 6 JAN 1925, leg. H. Sydow, Fungi in Itinere Costaricensi Collecti No. 41, ILL 9380 [also an isotype of *Hypostigme polyadelpha* H. Sydow].

Haplosporella commixta E. Bartholomew in J.B. Ellis and B.M. Everhart. Fungi Columbiana, Century 21, No. 2031. Anno 1905; Bulletin of the Torrey Botanical Club 33:219. 1906. **Isotype:** On fallen limbs of *Ulmus pubescens* Walter [= *Ulmus rubra* Muhl.], Kansas, Stockton, 5 JAN 1905, leg. E. Bartholomew No. 3264, ILL 11282.

Haplosporella hippocastani J.B. Ellis & B.M. Everhart. North American Fungi, Century 26, No. 2563. Anno 1891, nom. nud. **Isotype:** On *Aesculus hippocastanum* L., Canada, [Ontario], London, MAY 1890, leg. J. Dearness, ILL 11283.

Haplosporella thespesiae R. Ciferri. Sydowia Annales Mycologici 10:155–156. (1956) 1957. **Isotype:** Hab. in maculis foliaribus siccis sed in foliis vivis *Thespesiae populneae* [= *Thespesia populnea* (L.) Solander ex Correa] (Malvaceae cult.) [Dominican Republic], Santiago Prov., Valley del Cibao, Hato del Yaque, secus viam, JAN 1930, leg. R. Ciferri, Mycoflora Domingensis Exsiccata No. 376, ILL 33067.

Haplothecium dioscoreae F.L. Stevens. Illinois Biological Monographs 11:197, pl. IX, fig. 72. 1927. **Holotype:** On *Dioscorea* sp., Costa Rica, Peralta, 14 JUL 1923, leg. F.L. Stevens No. 475, ILL 8610.

Haplothecium guianense F.L. Stevens. Illinois Biological Monographs 8:191, pl. VII, figs. 60–62; pl. XVII, fig. 104. 1923. **Holotype:** On unknown lactiferous dicotyledonous leaf (Simaroubaceae?) [as Simarubaceae], British Guiana [Guyana], Demerara-Essequibo R.R., 15 JUL 1922, leg. F.L. Stevens No. 376, ILL 8611.

Haraea mauritiae F.L. Stevens. Illinois Biological Monographs 8:195, pl. IX, figs. 71–76. 1923. **Holotype:** On *Mauritia* sp., Trinidad, Guanapo, 16 AUG 1922, leg. F.L. Stevens No. 908, ILL 3725. **Isotype:** ILL 5548 [also a syntype of *Meliola mauritiae* F.L. Stevens].

Harknessia gunnerae F.L. Stevens & P.A. Young in F.L. Stevens. Bernice P. Bishop Museum Bulletin 19:136. 1925. **Holotype:** On living leaves of *Gunnera petaloidea* Gaudich., Hawaii, Maui, Olinda Pipeline, 5 SEP [1921], leg. F.L. Stevens No. 1143a, ILL 11286. **Isotype:** BISH 499908.

Harknessia hawaiiensis F.L. Stevens & P.A. Young in F.L. Stevens. Bernice P. Bishop Museum Bulletin 19:136. 1925. **Holotype:** On living leaves of *Eucalyptus robusta* J.E. Smith, Hawaii, Oahu, Waipio, 1 JUL 1919, leg. H.L. Lyon No. 124, ILL 11287. **Isotypes:** BISH 499901 and 499902.

Harknessia mauritiae F.L. Stevens. Annales Mycologici 28:369. 1930. **Holotype:** On petioles of *Mauritia* sp., British Guiana [Guyana], Coverden, 5 AUG 1922, leg. F.L. Stevens No. 756a, ILL 11288.

Helicogloea contorta G.E. Baker. Mycologia 38:634, figs. 1–6. 1946. **Holotype:** On *Quercus macrocarpa* Michaux, Iowa, West Okoboji, 19 JUL 1932 [as JUN on the packet], leg. D.P. Rogers No. 829, ILL 32815.

Helicogloea longispora G.E. Baker. Mycologia 38:634–635, figs. 15–20. 1946. **Syntype:** On *Pseudotsuga taxifolia* (Lambert) Britton [= *Pseudotsuga menziesii* (Mirbel) Franco], Oregon, [Douglas County], Comstock, 11 DEC 1937, leg. A.M. & D.P. Rogers No. 478, ILL 32816.

Helicoma chlamydosporum C.A. Shearer. Mycologia 79:468, figs. 1–20. 1987. **Isotypes:** A dried culture isolated from an unidentified twig submerged in Shannon Creek, Panama, Barro Colorado Island, MAR 1981, leg. C.A. Shearer No. CS-648-1, ILL 33503, ILLS 45741. **Holotype:** NY.

Helicomyces torquatus L.C. Lane & C.A. Shearer. Mycotaxon 19:291–294, figs. 1–7. 1984. **Isotypes:** From a dried culture on alfalfa, isolated from an unidentified submerged twig, Panama, Barro Colorado Island, Shannon Creek, 30 JUL 1982, leg. C. Shearer No. CS-688-1, ILL 33504, IMI. **Holotype:** NY.

Helminthosporium brassicae P.C. Hennings. Hedwigia 41:117. 1902. **Isotype:** Auf Blättern von *Brassica oleracea* L., [Brazil], São Paulo, Hort. Botan., 23 APR 1900, leg. A. Puttemans, Fungi S. Paulenses No. 115, ILL 16009.

Helminthosporium caladii F.L. Stevens. Transactions of the British Mycological Society 10:209. 1917. **Holotype:** On *Caladium bicolor* (Aiton) Vent., Porto Rico [Puerto Rico], Mayagüez, [27 OCT 1913], leg. F.L. Stevens No. 3860, ILL 16021. **Paratypes:** [23 JUL 1915], leg. F.L. Stevens No. 75, ILL 16020; [16 JAN 1913], leg. F.L. Stevens No. 252, ILL 16012 and 16019; [9 FEB 1915], leg. F.L. Stevens No. 292, ILL 16022; [24 JUN 1915], leg. F.L. Stevens No. 7587, ILL 16013 and 16018; [24 JUN 1915], leg. F.L. Stevens No. 7401, ILL 16016, ex BPI 70868; Añasco, [10 JUL 1915], leg. F.L. Stevens No. 8691, ILL 16014; [21 SEP 1913], leg. F.L. Stevens No. 3220, ILL 16017; Manati, [5 NOV 1913], leg. F.L. Stevens No. 4327, ILL 16015.

Helminthosporium glabroides F.L. Stevens. Botanical Gazette 65:240–241, pl. V, fig. 8; pl. VI, figs. 9, 10. 1918. **Holotype:** On *Meliola glabroides* F.L. Stevens on *Piper aduncum* L., Porto Rico [Puerto Rico], El Alto de la Bandera, [15 JUL 1915], leg. F.L. Stevens No. 9039, ILL 4039 [also a paratype of *Meliola glabroides* F.L. Stevens]. **Paratypes:** On *Meliola comocladiae* F.L. Stevens on *Comocladia glabra* (J.A. Schultes) Sprengel, Mayagüez, [25 JUN 1915], leg. F.L. Stevens No. 7484, ILL 4332; [15 JUN 1915], leg. F.L. Stevens No. 7056, ILL 4331 [Stevens Nos. 7484 and 7056 are also paratypes of *Meliola comocladiae* F.L. Stevens]; Maricao, [3 APR 1913], leg. F.L. Stevens No. 760, ILL 4330;

on *Meliola compositarum* F.S. Earle var. *portoricensis* F.L. Stevens on *Eupatorium portoricense* Urban [all specimens cited under the latter host are also paratypes of *Meliola compositarum* F.S. Earle var. *portoricensis* F.L. Stevens]: Arecibo-Lares Road, [20 JUN 1915], leg. F.L. Stevens No. 7320, ILL 3805; near Utuado [as Dos Bocas below Utuado on the packets, 16 DEC 1913], leg. F.L. Stevens No. 6031, ILL 3796 [also a paratype of *Perisporium meliolae* F.L. Stevens]; leg. F.L. Stevens No. 6032, ILL 3777 and 11033 [also the holotype and an isotype, respectively, of *Perisporium meliolae* F.L. Stevens, as well as paratypes of *Coniothyrium glabroides* F.L. Stevens]; on *Meliola didymopanacis* P.C. Hennings var. *stevensii* C.G. Hansford on *Dendropanax arboreus* (L.) Decaisne & Planchon, [Mayagüez Mesa, 25 JUN 1915], leg. F.L. Stevens No. 7440, ILL 5065; on *Meliola gaillardiana* F.L. Stevens on *Piper aduncum* L., Rio Arecibo, [8 JUL 1915], leg. F.L. Stevens No. 7796, ILL 5187 [also a paratype of *Meliola gaillardiana* F.L. Stevens]; on *Meliola gesneriae* F.L. Stevens on *Cestrum laurifolium* L'Héritier, Maricao, [3 APR 1913], leg. F.L. Stevens No. 824, ILL 5197 [also a paratype of *Meliola gesneriae* F.L. Stevens]; on *Meliola glabroides* F.L. Stevens on *Piper aduncum* L., Añasco, [12 OCT 1913], leg. F.L. Stevens No. 3582, ILL 4037 and 16041 [also paratypes of *Meliola glabroides* F.L. Stevens]; Maricao, [20 SEP 1913], leg. F.L. Stevens No. 3647, ILL 4035; [18 NOV 1913], leg. F.L. Stevens No. 4802, ILL 4066 [also the holotype of *Coniothyrium glabroides* F.L. Stevens, and a paratype of *Meliola glabroides* F.L. Stevens]; Arecibo-Lares Road, [21 JUN 1915], leg. F.L. Stevens No. 7297, ILL 4038 [also a paratype of *Meliola glabroides* F.L. Stevens]; on *Meliola hessii* F.L. Stevens on *Paullinia pinnata* L., Mayagüez, [4 MAY 1913], leg. F.L. Stevens No. 1207b, ILL 6222b [also an isotype of *Perisporium paulliniae*, and a paratype of *Meliola hessii* F.L. Stevens]; on *Meliola maricaensis* F.L. Stevens on *Ilex nitida* (Vahl) Maxim., Maricao, [20 OCT 1913], leg. F.L. Stevens No. 3679, ILL 4406 [also the holotype of *Meliola maricaensis* F.L. Stevens]; leg. F.L. Stevens No. 3607, ILL 4407 [also a paratype of *Meliola maricaensis* F.L. Stevens]; on *Meliola monensis* F.L. Stevens on *Amyris elemifera* L., Mona Island, [20–21 DEC 1913], leg. F.L. Stevens No. 6150, ILL 5657 [also an isotype of *Stevensula monensis* C.L. Spegazzini, a possible isotype of *Micropeltidium monense* C.L. Spegazzini, and a paratype of *Meliola monensis* F.L. Stevens]; on *Meliola nigra* F.L. Stevens on *Laguncularia racemosa* (L.) C.F. Gaertner, Guanajibo, [19 JUN 1915], leg. F.L. Stevens No. 7197, ILL 5706 [also the holotype of *Meliola nigra* F.L. Stevens]; on *Meliola psychotriae* F.S. Earle on *Chiococca alba* (L.) A.S. Hitchc., Martin Peña, [11 AUG 1915], leg. F.L. Stevens No. 9299, ILL 6113; Rio Tanamá, [6 JUL 1915], leg. F.L.

Stevens No. 7859, ILL 6114; Mayagüez Mesa, [25 JUN 1915], leg. F.L. Stevens No. 7467, ILL 6115; on *Meliola pteridicola* F.L. Stevens on *Anemia adiantifolia* (L.) Sw. [as *Aneimia*], Rio Tanamá, [6 JUL 1915], leg. F.L. Stevens No. 7814, ILL 6138 and 6142 [also an isotype and the holotype, respectively, of *Meliola pteridicola* F.L. Stevens]; Quebradillas, [20 JUN 1915], leg. F.L. Stevens No. 7269, ILL 6139, 6151 and 6152; on *Adiantum latifolium* Lam., Las Marias, [10 JUL 1915], leg. F.L. Stevens No. 8182, ILL 6144 and 6148; Mayagüez, [24 JUL 1915], leg. F.L. Stevens No. 7418, ILL 6143 [Stevens Nos. 7269, 8182 and 7418 are also paratypes of *Meliola pteridicola* F.L. Stevens]; on *Meliola toruloidea* F.L. Stevens on *Cassia quinquangulata* sensu auct. non L.C. Richard [misspelled "*quinquadrangulata*" = *Senna* cf. *nitida* (L.C. Richard) Irwin & Barneby], Jajome Alto, [17 JUL 1915], leg. F.L. Stevens No. 8394, ILL 4491 [also the holotype of *Meliola toruloidea* F.L. Stevens].

Helminthosporium gleicheniae F.L. Stevens & P.A. Glick in F.L. Stevens. Bernice P. Bishop Museum Bulletin 19:152–153, text fig. 33a. 1925. **Syntypes** [BISH numbers cited are isosyntypes]: On *Gleichenia dichotoma* Hooker, Hawaii, Oahu, Ahren's ditch trail, 8 JUN [1921], leg. F.L. Stevens No. 283, ILL 16047, BISH 145838, 499896 and 499897; Wahiawa, 3 JUN [1921], leg. F.L. Stevens No. 223, ILL 16045, BISH 499893; Olympus, 24 JUN [1921], leg. F.L. Stevens No. 673, ILL 16044, BISH 145839 and 499900; Palolo Valley and Mt. Olympus, 10 JUN [1921], leg. F.L. Stevens No. 371, ILL 16046, BISH 145840 and 499899; Kauai, Kalalau trail, 16 JUN [1921], leg. F.L. Stevens No. 509, ILL 16048, BISH 145841 and 498898.

Helminthosporium guareicola F.L. Stevens. Botanical Gazette 65:241, pl. VI, fig. 16. 1918 [cited as *guareicolum*, corrected by S.J. Hughes (1953:25) in accordance with the current I.C.B.N.]. **Holotype**: On *Meliola guareicola* F.L. Stevens on *Guarea trichilioides* L. [= *G. guidonia* (L.) Sleumer], Porto Rico [Puerto Rico], Las Marias, [10 JUL 1915], leg. F.L. Stevens No. 8166, ILL 5171 [also the holotype of *Meliola guareicola* F.L. Stevens and a paratype of *Grallomyces portoricensis* F.L. Stevens]. **Isotype**: ILL 5173. **Paratype**: Utuado [as Dos Bocas on the packet, 8 JUL 1915], leg. F.L. Stevens No. 8096, ILL 5169 [also a paratype of *Meliola guareicola* F.L. Stevens]. ≡ *Spiropes guareicola* (F.L. Stevens) R. Ciferri. Sydowia Annales Mycologici 9:303. 1955.

Helminthosporium guianensis F.L. Stevens & R.I. Dowell. Phytopathology 13:249–250, figs. 3, 4. 1923. **Holotype**: On *Meliola guianensis* F.L. Stevens & R.I. Dowell on *Theobroma cacao* L., British Guiana [Guyana], Coverden, 4 AUG 1922, leg. F.L. Stevens No. 974, ILL 4380 [also the holotype of *Meliola guianensis* F.L. Stevens & R.I. Dowell].

Helminthosporium helleri F.L. Stevens. Botanical Gazette 65:242–243, pl. VI, figs. 14, 15. 1918. **Holotype:** On *Meliola helleri* F.S. Earle on *Myrcia deflexa* (Poiret) DC., Porto Rico [Puerto Rico], El Alto de la Bandera, [14 JUL 1915], leg. F.L. Stevens No. 8268, ILL 5229. **Paratypes:** Leg. F.L. Stevens No. 8296, ILL 5231; on *Eugenia stahlii* (Kiaersk.) Krug & Urban in Urban, Luquillo Forest, [4 DEC 1913], leg. F.L. Stevens No. 5343, ILL 5234; Jajome Alto, [17 JUL 1915], leg. F.L. Stevens No. 8436, ILL 5235; on *Meliola thouiniae* F.S. Earle [as *Meliola canellae* R. Ciferri on the packet] on *Winterana canella* L. [= *Canella winterana* (L.) J. Gaertner], Guayanilla, [14 JUL 1915], leg. F.L. Stevens No. 8548, ILL 6382; 26 JUL 1915], leg. F.L. Stevens No. 9075, ILL 6383; on *Meliola dipholidis* F.L. Stevens on *Dipholis salicifolia* (L.) A. DC. [= *Sideroxylon salicifolium* (L.) Lam.], Guayanilla, [14 JUL 1915], leg. F.L. Stevens No. 8549, ILL 5037 [also the holotype of *Meliola dipholidis* F.L. Stevens and a potential isotype of *Scolecopeltella portoricensis* C.L. Spegazzini].

Helminthosporium mayaguezense L.E. Miles. Transactions of the Illinois State Academy of Science 10:253, figs. 1, 2. 1917. **Holotype:** On culms and leaves of *Paspalum conjugatum* Bergius, Porto Rico [Puerto Rico], Mayagüez District, [11 JUN 1915], leg. F.L. Stevens No. 7124, ILL 16072. **Isotype:** ILL 16029. **Paratypes:** Mayagüez District, [Dos Bocas, 8 JUL 1915], leg. F.L. Stevens No. 1066, ILL 16073; leg. F.L. Stevens No. 970, ILL 16028 and 16031; [El Alto de la Bandera, 14 JUL 1915], leg. F.L. Stevens No. 8279, ILL 16070; San Germán, [8 NOV 1915], leg. F.L. Stevens No. 5803, ILL 16085; Añasco, [18 NOV 1913], leg. F.L. Stevens No. 4904, ILL 16082; Maricao, [19 JUL 1915], leg. F.L. Stevens No. 8776, ILL 16084; [Las Marias, listed under Mayagüez in the protologue, 11 JUL 1915], leg. F.L. Stevens No. 8232, ILL 16027.

Helminthosporium melastomatacearum F.L. Stevens. Botanical Gazette 65:242, pl. VI, fig. 11. 1918 [cited as *melastomacearum*, corrected in accordance with current I.C.B.N. (cf. F.C. Deighton, 1968)]. **Holotype:** On *Meliola melastomatacearum* C.L. Spegazzini on *Miconia racemosa* (Aublet) DC., Porto Rico [Puerto Rico], Mayagüez, [24 JUN 1915], leg. F.L. Stevens No. 7389, ILL 4149.

Helminthosporium ocoteae F.L. Stevens. Botanical Gazette 65:241. 1918. **Holotype:** On *Meliola ocoteae* F.L. Stevens on *Ocotea leucoxylon* (Sw.) de Lanessan, Porto Rico [Puerto Rico], Jajome Alto, [17 JUL 1915], leg. F.L. Stevens No. 8428, ILL 4385 [also the holotype of *Meliola ocoteae* F.L. Stevens and a potential isotype of *Asteridium portoricense* C.L. Spegazzini].

Helminthosporium panici F.L. Stevens. Botanical Gazette 65:242. 1918. **Holotype:** On *Meliola panici* F.S. Earle [var. *olyrae* C.G. Hansford] on

Olyra latifolia L., Porto Rico [Puerto Rico], Mayagüez [El Meradero, 4 AUG 1915], leg. F.L. Stevens No. 9159, ILL 5764. **Paratypes:** [24 JUN 1915], leg. F.L. Stevens No. 7390, ILL 5757 [also the holotype of *Meliola panici* F.S. Earle var. *olyrae* C.G. Hansford]; on *Meliola rectangularis* F.L. Stevens on *Coccoloba laurifolia* sensu auct. non Jacquin [= *C. diversifolia* Jacquin, Arecibo-Lares Road, [21 JUN 1915], leg. F.L. Stevens No. 7292, ILL 4398 [also the holotype of *Meliola rectangularis* F.L. Stevens], and ILL 7371 [also an isotype of *M. rectangularis* and a syntype of *Seynesia coccolobae* R.W. Ryan].

Helminthosporium parasiticum P.A. Saccardo & A.N. Berlese. Boletim da Sociedade Broteriana 7:114. 1889. **Isotypes:** Hab. Parasitans in stromato *Diaporthes* cujusdam [= *Diaporthe* sp.] in caule *Musae* viventis [= *Musa* sp.], Afr. Occid. Isl. São Tomé, (alt. 800 m), [OCT 1887], leg. A. Møller s.n., ILL 16086, and in bound set of C. Roumeguère, Fungi Selecti Gallici Exsiccati, as No. 5074.

Helminthosporium parathesicola F.L. Stevens. Botanical Gazette 65:242, pl. 6, fig. 12. 1918 [as *parathesicolum*, corrected in accordance with Articles 32.6 & 61 of I.C.B.N.]. **Holotype:** On *Meliola parathesicola* F.L. Stevens on *Parathesis serrulata* (Sw.) Mez [= *Parathesis crenulata* (Vent.) J.D. Hooker], Porto Rico [Puerto Rico], Las Marias, [10 JUL 1915], leg. F.L. Stevens No. 8192, ILL 4388 [also the holotype of *Meliola parathesicola* F.L. Stevens]. **Paratypes:** Arecibo-Lares Road, [21 JUN 1915], leg. F.L. Stevens No. 7286, ILL 4387 [also a paratype of *Meliola parathesicola* F.L. Stevens]; on *Meliola bicornis* H.G. Winter on *Dalbergia monetaria* L.f., Arecibo-Lares Road, [21 JUN 1915], leg. F.L. Stevens No. 7243, ILL 4696; on *Meliola retangularis* F.L. Stevens on *Banisteria laurifolia* L. [= *Heteropteris laurifolia* (L.) A. Jussieu], Hormigueros, [23 JUN 1915], leg. F.L. Stevens No. 7358, ILL 4393; Mayagüez, [29 JUN 1915], leg. F.L. Stevens No. 7564, ILL 4396 [also a paratype of *Irenopsis banisteriae* C.G. Hansford]; Utuado, [8 NOV 1913], leg. F.L. Stevens No. 4384, ILL 4397; leg. F.L. Stevens 4392, ILL 4395. [Stevens Nos. 7358, 7564 and 4384, as well as No 4392a, under ILL 4395 p.p., are also paratypes of *Meliola rectangularis* F.L. Stevens]; on *Meliola bicornis* H.G. Winter on *Dalbergia monetaria* L.f., Arecibo-Lares Road, [21 JUN 1915], leg. F.L. Stevens No. 7243, ILL 4696.

Helminthosporium philodendri F.L. Stevens. Botanical Gazette 65:242, pl. VI, fig. 13. 1918. **Holotype:** On *Meliola philodendri* F.L. Stevens on *Philodendron krebsii* Schott [as *Philodendrum* = *Philodendron consanguineum* Schott] Porto Rico [Puerto Rico], Ponce, [8 NOV 1913], leg. F.L. Stevens No. 4346, ILL 5995 [also a paratype of *Meliola philodendri* F.L. Stevens and of *Isthmospora spinosa* F.L. Stevens].

Helminthosporium pinicolum J.B. Ellis & B.M. Everhart, Fungi Columbiani, Century 16, No.1535. Anno 1901, nom. nud. **Isotype:** On dead limbs of *Pinus palustris* P. Miller, Alabama, Tuskegee, 30 APR 1901, leg. G.W. Carver, ILL 16088.

Helminthosporium pseudotsugae W.B. Cooke. Mycologia 44:251–252, fig. 3. 1952. **Holotype:** On bark and resin exudations of *Pseudotsuga taxifolia* (Lambert) Britton var. *glauca* (Beissn.) Sudw. [= *P. menziesii* (Mirbel) Franco var. *glauca* (Beissn.) Franco], Idaho, Nez Perce County, 4.8 miles east of Lenore along the Clearwater River, 14 MAY 1949, leg. W.B. and V.G. Cooke No. 25161, ILL 31553.

Helminthosporium stahlii F.L. Stevens. Transactions of the Illinois State Academy of Science 10:208–209. 1917. **Holotype:** On *Passiflora foetida* L., Porto Rico [Puerto Rico], Luquillo, [21 MAY 1912], leg. F.L. Stevens No. 6, ILL 16130. **Paratypes:** Mayagüez, leg. F.L. Stevens No. 1699, ILL 16129; Preston's Ranch, [24 MAY 1913], leg. F.L. Stevens No. 6670, ILL 16128.

Helminthosporium varroniae F.L. Stevens. Transactions of the Illinois State Academy of Science 10:209. 1917. **Holotype:** On *Varronia* sp. [= *Cordia* sp.], Porto Rico [Puerto Rico], Florida Adentro, [1 JUL 1915], leg. F.L. Stevens No. 7663, ILL 16140.

Helotium conscriptum P.A. Karsten var. *oblongisporum* H. Rehm ex H. Sydow. Annales Mycologici 5:397. 1907. **Isotype:** Hab. ad ramulos *Salicis cinereae* [= *Salix cinerea* L.], Germany, [Mark Brandenburg]: Buch pr. Bernau, [25 NOV 1906], leg. H. Sydow, ex Mycotheca Germanica No. 601, ILL 7811.

Hemidothis palmarum R. Ciferri. Sydowia Annales Mycologici 10:156–157. (1956) 1957. **Isotype:** Hab. in frondibus *Coccothrinacis argenteae* (Palmae) [= *Coccothrinax argentea* (Lodd. ex J.A. & J.H. Schultes) Sargent ex Beccari (Arecaceae), Dominican Republic], Santiago Prov., Cordillera Central, Janico, alt. ca. 250 m, 16 NOV 1930 [as 1927 on the printed label], leg. E.L. Ekman & R. Ciferri, (N.H. 16193, Cif. 4691), ex Mycoflora Domingensis Exsiccata No. 363, ILL 33150.

Hemidothis pellitiforme R. Ciferri. Sydowia Annales Mycologici 10:157–158. (1956) 1957. **Isotype:** Hab. in foliis vivis *Thespesiae populneae* [= *Thespesia populnea* (L.) Solander ex Correa] (Malvaceae, cult.), [Dominican Republic], Santiago Prov., Valle des Cibao, Santiago, Hato del Yaque, pr. viam, JAN 1931, leg. R. Ciferri, Mycoflora Domingensis Exsiccata No. 361, ILL 33151.

Hendersonia bliti F.E. & E.S. Clements. Cryptogamae Formationum Coloradensium, Century 5, No. 488. Anno 1908, nom. nud. **Isotype:** ... in foliis vivis *Bliti capitati* [*Blitum capitatum* L. = *Chenopodium capitatum* (L.) Aschers.], Colorado, Sulphur Springs, alt. 2400 m, 19 JUL 1907, leg. F.E. & E.S. Clements, ILL 11289.

Hendersonia fungicola F.L. Stevens. Annales Mycologici 29:106. 1931. **Holotype:** On *Phyllachora minutissima* (F.M.J. Welwitsch & F. Currey) A.L. Smith on *Pennisetum tristachyum* Sprengel, Peru, Tarma, 3 DEC 1924, leg. F.L. Stevens No. 24a, ILL 11293.

Henningsomyces escharoides H. Sydow. Annales Mycologici 23:331–333. 1925. **Isotype:** Hab. in foliis vivis vel languidis *Mauriae biringo* [= *Mauria biringo* Tulasne], Costa Rica, Grecia, 17 JAN 1925, leg. H. Sydow, Fungi in Itinere Costaricensi Collecti No. 248, ILL 6660 [also an isotype of *Episoma parasiticum* H. Sydow and of *Calolepis congesta* H. Sydow].

Heterochaete crassa M.C. Bodman. Mycologia 41:531–532, figs. 1, 3. 1949. **Isotype:** [On *Nectandra coriacea* (Sw.) Griseb. = *Ocotea coriacea* (Sw.) Britton], Florida, [Dade County], Miami, [Simpson Park, 17 NOV] 1942, leg. R. Singer No. F-1494, ILL 32813. **Holotype:** FH.

Heterochaetella bispora E.R. Luck-Allen. Canadian Journal of Botany 38:563–564, figs. 1–22. 1960. **Isotype:** On *Ulmus* sp., Canada, Ontario, Brant County, south of New Durham, 30 AUG 1937, leg. R.F. Cain, ex TRTC No. 17594, ILL 32825. **Holotype:** TRTC.

Heterosporium crunocallis F.E. & E.S. Clements. Cryptogamae Formationum Coloradensium, Century 6, No. 507. Anno 1908, nom. nud. **Isotype:** Hab. ad folia viva *Crunocallis chamissonis* [*C. chamissoi* (Ledebour ex Sprengel) Rydb. = *Montia chamissoi* (Ledebour ex Sprengel) E. Greene], Colorado, Sierra Blanca, alt. 2800 m, 21 JUN 1907, leg. F.E. & E.S. Clements, ILL 16160.

Heterosporium magnoliae A.G. Weedon. Mycologia 18:222–223, pl. 26. 1926. **Holotype:** On *Magnolia grandiflora* L., Florida, St. Petersburg, 15 FEB 1923, leg. A.G. Weedon No. 1, ILL 16167. ≡ *Stenellopsis magnoliae* (A.G. Weedon) G. Morgan-Jones. Mycotaxon 10:406. 1980. ≡ *Parastenella magnoliae* (A.G. Weedon) J.C. David. Mycological Research 95(1):123–128. 1991.

Hexagonella peleae F.L. Stevens & E.F. Guba in F.L. Stevens. Bernice P. Bishop Museum Bulletin 19:89, text fig. 21, pl. IX(H). 1925. **Holotype:** On *Pelea rotundifolia* A. Gray [= *Melicope rotundifolia* (A. Gray) T.G. Hartley & B.C. Stone], Hawaii, Oahu, Wahiawa, 3 JUN [1921], leg. F.L. Stevens No. 248, ILL 6828.

Himantia stellifera J.R. Johnston in J.R. Johnston & J.A. Stevenson. Journal of the Department of Agriculture of Porto Rico 1:188, pl. 19, fig. 2; pl. 31, figs. 1–4. 1917. **Paratype:** On *Cymbopogon citratus* (DC. ex Nees) Stapf, Porto Rico [Puerto Rico], Rio Piedras, OCT 1915, leg. J.A. Stevenson No. 3212, ILL 16596.

Hirudinaria macrospora V. Cesati in L. Rabenhorst. Klotzschii Herbarium Vivum Mycologicum Sistens Fungorum per Totam Germaniam Crescentium Collectionem Perfectam, Edit. Nova, Century 3, No. 269. 1856; Flora 14:373–370. 1856. **Isolectotypes**: Ad *Crataegi oxyacantham* [= *Crataegus oxyacantha* L., n.w. Italy, Piemonte Prov.] pr. Villafranca d' Asti (Montferrato ...) etc., OCT 1855, leg. V. Cesati, ex Klotzsch herb., ILL 16173 [as a microscopic preparation], IMI 28627. **Lectotype**: K [designated by S.J. Hughes, in Mycological Papers. Commonwealth Mycological Institute No. 39:14, 23. 1951].

Homaromyces epieri R.K. Benjamin. El Aliso 3:183–184, figs. 4–6. 1955. **Paratype**: On all parts of *Epierus pulicarius* Er. [Insecta], Louisiana, Norco, 3 NOV 1944, leg. C.L. Remington, comm. R.K. Benjamin No. 915, ILL 21950.

Hormisciomyces bellus A.C. Batista & Carneiro in A.C. Batista, H. da Silva Maia, J.A. de Lima & E.A.F. da Matta. Atas do Instituto de Micologia 1:262–263, fig. 12. 1960. **Isotypes**: In foliis *Vitecis* sp. [= *Vitex* sp.], soc. *Stellopeltis philodendricolae* [= *S. philodendricola* (A.C. Batista) A.C. Batista & A.F. Vital], Porto Rico [Puerto Rico], 3 APR 1913, leg. W.E. Hess [No. 698], ILL 33200, BPI. **Holotype**: URM 133318.

Hormodendron eupatorii R. Ciferri. Sydowia Annales Mycologici 8:251–252. 1954. **Isotype**: Hab. in foliis vivis *Eupatorii* sp. [= *Eupatorium odoratum* L. on the packet, Dominican Republic], Santiago Prov., Valle de Cibao, El Hoyazo, Santiago, 14 FEB 1932 [3 MAR 1922 on the packet], leg. R. Ciferri No. 3972 [ex Mycoflora Domingensis Exsiccata No. 303], ILL 33201.

Hyalomeliolina costaricensis F.L. Stevens. Illinois Biological Monographs 11:170–171. 1927. **Holotype**: On *Tetracera volubilis* L., Costa Rica, Siquirres, 31 JUL 1923, leg. F.L. Stevens No. 668, Ill 3726. **Isotypes**: ILL 3726a, b, c, d [as microscopic preparations]. ≡ *Meliolinopsis costaricensis* (F.L. Stevens) F. Petrak. Sydowia Annales Mycologici 5:335. 1951.

Hyalomeliolina guianensis F.L. Stevens. Illinois Biological Monographs 8: 194, figs. 68–70 and 105. 1923. **Holotype**: On *Licania* ? sp., British Guiana [Guyana], Rockstone, 17 JUL 1922, leg. F.L. Stevens No. 454, ILL 3729. **Paratype**: Kartabo, 24 JUL 1922, leg. F.L. Stevens No. 665, ILL 3728. ≡ *Meliolinopsis guianensis* (F.L. Stevens) F. Petrak. Sydowia Annales Mycologici 5:335. 1951.

Hyalosphaera miconiae F.L. Stevens. Transactions of the Illinois State Academy of Science 10:172–173. 1917. **Lectotype and isolectotype**: On *Miconia laevigata* (L.) D. Don in Sweet, Porto Rico [Puerto Rico], Maricao, [10 JAN 1913], leg. F.L. Stevens No. 207, ILL 8287 [two packets, the lectotype designated and annotated by A.Y. Rossman, in Mycological Papers. International Mycological Institute 157:55. 1987]. **Paratypes**: On *Miconia laevigata* (L.) D. Don in Sweet, Ponce, [8 NOV 1913], leg. F.L. Stevens No. 4338, ILL 8286; Yabucoa [31 DEC 1913], leg. F.L. Stevens No. 6705, ILL 7299 [also the assumed holotype of *Monogrammia miconiae* F.L. Stevens and of *Paranectria miconiae* F.L. Stevens]; probably also F.L. Stevens No. 6705a, ILL 8288b and 8398 [part of the same collection, also probable isotypes of the above names]; Utuado, [30 DEC 1913], leg. F.L. Stevens No. 6862, ILL 13882 [also a paratype of *Borinquenia miconiae* F.L. Stevens]; leg. F.L. Stevens No. 6871, ILL 13883 [also the holotype of *Borinquenia miconiae* F.L. Stevens. Both Nos. 6862 and 6871 are also cited as syntypes of *Blastotrichum miconiae* F.L. Stevens.].

Hyalotexis pellucida H. Sydow. Annales Mycologici 23:326–329. 1925. **Isotype**: Hab. in foliis vivis Melastomataceae cujusdam indet., Costa Rica, San Pedro de San Ramón, 28 JAN 1925, leg. H. Sydow, Fungi in Itinere Costaricensi Collecti No.148, ILL 8289.

Hymenula copelandii P.A. Saccardo. Annales Mycologici 11:558. 1913 [cited as *copelandi*, corrected in accordance with Articles 32.6 & 61 of I.C.B.N.]. **Assumed isotype**: Hab. in foliis emortuis *Diospyri discoloris* [= *Diospyros discolor* Willd.], Philippines, [Luzon, Laguna Prov.], Los Baños, [Mount Maquiling, JUL 1913], leg. E.B. Copeland No. 1269 [ex C.F. Baker, Fungi Malayana No. 140], ILL 16508.

Hymenula epistroma F. v. Höhnel ex H. Sydow. Mycotheca Germanica, Century 13, No. 648. Anno 1907, nom. nud. **Isosyntypes**: Auf altem Stroma von *Diatripella favacea* (E.M. Fries:E.M. Fries) V. Cesati & G. de Notaris, Brandenburg: Schmidt's Grund by Tamsel [now Poland], 24 FEB 1905, leg. P. Vogel, ILL 16505 and 16506. ≡ *Dendrodochium epistroma* F. v. Höhnel ex F. v. Höhnel. Sitzungsberichte der Kaiserl. Akademie der Wissenschaften in Wien. Mathematisch-Naturwissenschaftliche Klasse. Abt. 1, 118:424. 1909 [the earlier invalid name was cited pro syn. and provided the epithet for this new name].

Hyperus costaricensis F.L. Stevens. Illinois Biological Monographs 11:179–180, pl. V, figs. 35–40; pl. XV, fig. 105. 1927. **Holotype**: On unknown dicotyledonous plant, Costa Rica, Experiencia Farm, 18 JUL 1923, leg. F.L. Stevens No. 525, ILL 8613.

Hypholoma candolleanum E.M. Fries:E.M. Fries var. *annulatum* F.E. & E.S. Clements. Cryptogamae Formationum Coloradensium, Century 2, No. 190. Anno 1906, nom. nud. **Isotype**: Geophilus gregarius ad terram ..., Colorado, Mariposa Dell, alt. 2800 m, 7 AUG 1905, leg. F.E. & E.S. Clements, ILL 29988.

Hypochnus peniophoroides E.A. Burt. Annals of the Missouri Botanical Garden 3:234–235, fig. 25. 1916. **Isotype**: On bark of rotten frondose wood in woods [as "on rotten stick" on the packet], Jamaica, Mooretown, [20 OCT–24 NOV 1902], leg. F.S. Earle, Plants of Jamaica No. 540, ILL 32587, ex NY. **Holotype**: NY.

Hyponectria mohavensis L. Bonar. Mycologia 57:381–382. 1965. **Paratype:** On *Yucca brevifolia* Engelm., California, San Bernardino County, south of Victorville, 17 FEB 1957, leg. L. Bonar s.n., ILL 32645, UC 1272172. **Holotype:** UC 1272176.

Hyponectria phaseoli F.L. Stevens. Botanical Gazette 70:401–402. 1920. **Holotype:** On *Vigna vexillata* (L.) A. Richard, Porto Rico [Puerto Rico], Rosario, [27 OCT 1913], leg. F.L. Stevens No. 3602, ILL 8320. **Isotypes:** ILL 8321 p.p., 8324 and 8334. **Paratypes:** Añasco, [12 OCT 1913], leg. F.L. Stevens No. 3509, ILL 8316 and 8325; Vega Baja, [22 FEB 1913], leg. F.L. Stevens No. 374a, ILL 8317 and 8329; Mayagüez, [30 APR 1913], leg. F.L. Stevens No. 978, ILL 8315, 8321 p.p. and 8331; leg. F.L. Stevens No. 1483, ILL 8318 and 8330; leg. F.L. Stevens No.1098, ILL 8319 and 8328; [2 OCT 1913], leg. F.L. Stevens No. 3149, ILL 8323, 8327 and 8340; on *Phaseolus adenanthus* G. Meyer [= *Vigna adenantha* (G. Meyer) Marechal, Mascherpa & Stanier], Mayagüez, [5 JAN 1914], leg. F.L. Stevens No. 6732, ILL 8333 and 8338; San Germán, (Añasco), [12 OCT 1913], leg. F.L. Stevens No. 4903, ILL 8326 and 8342; on *Phaseolus* sp., Luquillo Forest, [2 DEC 1913], leg. F.L. Stevens No. 5555, ILL 8337.

Hypostigme polyadelpha H. Sydow. Annales Mycologici 23:337–340, fig. 3. 1925. **Isotype:** Hab. in foliis *Buettneriae carthagensis* [= *Byttneria carthagensis* Jacquin (this spelling of the generic name conserved)], Costa Rica, La Caja, pr. San José, 6 JAN 1925, leg. H. Sydow, Fungi in Itinere Costaricensi Collecti No. 41, ILL 9380 [also an isotype of *Haplolepis polyadelpha* H. Sydow].

Hysterium angustatum J.B. v. Albertini & L.D. v. Schweinitz var. *didymosporum* F.E. & E.S. Clements. Cryptogamae Formationum Coloradensium, Century 1, No. 45. Anno 1906, nom. nud. **Isotype:** In ligno putrido *Pseudotsugae mucronatae* [*Pseudotsuga mucronata* (Raf.) Sudw. in Holz. = *P. menziesii* (Mirbel) Franco] Colorado, Larkspur Dell, alt. 2600 m, 28 JUL 1905, leg. F.E. & E.S. Clements, ILL 7612.

Hysterium pulcherrimum L.R. Tehon & P.A. Young. Mycologia 16:31–32, fig. 1. 1924. **Holotype:** On bark of *Platanus occidentalis* L., Illinois, [Piatt County], White Heath, 12 MAY 1923, leg. P.A. Young, ILL 7623.

Hysterographium cocos A.G. Weedon. Mycologia 18:218. 1926. **Holotype:** On *Cocos alphonsei* [probably *C. nucifera* L. (cult.)], Florida, St. Petersburg, FEB 1923, leg. A.G. Weedon No. 4, ILL 7626.

Hysterostomella phoebes H. Sydow. Annales Mycologici 25:32–33. 1927. **Isotype:** Hab. in foliis vivis *Phoebes costaricanae* [= *Phoebe costaricana* Mez & Pittier], Costa Rica, San Pedro de San Ramón, 23 JAN 1925, leg. H. Sydow, Fungi in Itinere Costaricensi Collecti No. 170a, ILL 7636 [also an isotype of *Phragmopeltis phoebes* H. Sydow].

Hysterostomina costaricensis H. Sydow. Annales Mycologici 25:34–35. 1927. **Isotype:** Hab. in foliis vivis *Miconiae longifoliae* [= *Miconia longifolia* (Aublet) DC.], Costa Rica, San Pedro de San Ramón, 28 JAN 1925, leg. H. Sydow, Fungi in Itinere Costaricensi Collecti No. 145, ILL 8614.

Hysterostomina palmae F.L. Stevens. Illinois Biological Monographs 8:176, pl. I, figs. 1–4. 1923. **Holotype:** On palm [Arecaceae sp. indet.], British Guiana [Guyana], Tumatumari, 12 JUL 1922, leg. F.L. Stevens No. 199, ILL 8615.

I

Illosporium commelinae F.L. Stevens. Transactions of the Illinois State Academy of Science 10:215–216, fig. 12. 1917. **Syntypes:** On *Commelina elegans* Kunth in H.B.K. [= *C. erecta* L.], Porto Rico [Puerto Rico], Aguada, [22 NOV 1913], leg. F.L. Stevens No. 5109, ILL 16510; Mt. Gigante, [16 JUL 1915], leg. F.L. Stevens No. 8485, ILL 16509; on *Commelina longicaulis* Jacquin [as *longicaules* = *C. diffusa* N.L. Burman], Aguado, [22 NOV 1913], leg. F.L. Stevens No. 5081, ILL 16511; Caguas, [9 FEB 1913], leg. F.L. Stevens No. 287a, ILL 16516; Hormigueros, [14 JAN 1913], leg. F.L. Stevens No. 224, ILL 16512; Rosario, [14 NOV 1913], leg. F.L. Stevens No. 480, ILL 16515; Guayanilla, 13 NOV 1913, leg. F.L. Stevens No. 5923, ILL 16513; Las Marias, [11 JUL 1915], leg. F.L. Stevens No. 8248, ILL 16514.

Illosporium hibisci F.L. Stevens & A.S. Peirce. Indian Journal of Agricultural Science 3:916. 1933. **Assumed holotype:** On *Hibiscus esculentus* L. [= *Abelmoschus esculentus* (L.) Moench], India, Poona, Bombay, 1932, leg. B.N. Uppal No. 32, ILL 16520. **Isotype:** AMH.

Irene acervata (J.B. Ellis & B.M. Everhart) C.G. Hansford, 1955. **See basionym:** *Meliola acervata* J.B. Ellis & B.M. Everhart, 1897.

Irene adelphica (H. Sydow) F.L. Stevens, 1927. **See basionym:** *Appendiculella adelphica* H. Sydow, 1926.

Irene aibonitensis (F.L. Stevens) R.A. Toro, 1925. **See basionym:** *Meliola aibonitensis* F.L. Stevens, 1916.

Irene alchorneae F.L. Stevens & L.R. Tehon. Mycologia 18:21. pl. 2, fig. 27. 1926 [non *Meliola alchorniae* F.L. Stevens & L.R. Tehon, op. cit., p. 12]. **Holotype:** On *Alchornea cordata* (A. Jussieu) Muell. Arg. in DC., non Bentham in Hooker [= *Aparisthmium cordatum* (A. Jussieu) Baillon, British Guiana [Guyana], Tumatumari, 12 JUL 1922, leg. F.L. Stevens No. 245, ILL 3893. **Paratype:** 10 JUL [1922], leg. F.L. Stevens No. 153a, ILL 3892. ≡ *Irenina alchorneae* (F.L. Stevens & L.R. Tehon) F.L.

Stevens. Annales Mycologici 25:452. 1927. ≡ *Appendiculella alchorneae* (F.L. Stevens & L.R. Tehon) C.G. Hansford. Beihefte zur Sydowia Annales Mycologici 2:205. 1961.

Irene amoena H. Sydow. Annales Mycologici 24:315–316. 1926. **Isotype:** Hab. in foliis *Sloaneae fagineae* [= *Sloanea faginea* Standley], Costa Rica, Piedades de San Ramón, 26 JAN 1925, leg. H. Sydow, Fungi in Itinere Costaricensi Collecti No. 162, ILL 3894. ≡ *Irenina amoena* (H. Sydow) F.L. Stevens, Annales Mycologici 25:451. 1927. ≡ *Meliola amoena* (H. Sydow) F. Petrak in H. Sydow & F. Petrak. Annales Mycologici 27:2. 1929. ≡ *Asteridiella amoena* (H. Sydow) C.G. Hansford ex C.G. Hansford. Beihefte zur Sydowia Annales Mycologici 2:173. 1961.

Irene anastomosans (H.G. Winter) F. Theissen & H. Sydow, (1917) 1918. **See basionym:** *Meliola anastomosans* H.G. Winter, 1886.

Irene andromedae (N.T. Patouillard) H. & P. Sydow, 1917. **See basionym:** *Meliola andromedae* N.T. Patouillard, 1888.

Irene atra (E.M. Doidge) E.M. Doidge, 1920. **See basionym:** *Meliola atra* E.M. Doidge, 1920.

Irene aucubae (P.C. Hennings) C.G. Hansford, 1955. **See basionym:** *Meliola aucubae* P.C. Hennings, 1900 (1901).

Irene bonii (A. Gaillard) H. & P. Sydow, 1917. **See basionym:** *Meliola bonii* A. Gaillard, 1892.

Irene brachycera (H. Sydow) C.G. Hansford, 1955. **See basionym:** *Meliola brachycera* H. Sydow, 1926.

Irene buddlejicola (P.C. Hennings) C.G. Hansford, 1955. **See basionym:** *Meliola buddlejicola* P.C. Hennings, 1905.

Irene caaguazuensis C.G. Hansford. Sydowia Annales Mycologici 9:32–33. 1955. **Isotype:** Hab. in foliis Rutacearum spec. indet. [= Rutaceae sp. indet.], Paraguay, Caáguazú, [JAN 1882], leg. B. Balansa No. 3585, ILL 4210, ex herb. Spegazzini [also an isotype of *Meliola obesula* C.L. Spegazzini]. **Holotype:** K.

Irene calophylli (F.L. Stevens) F.L. Stevens, 1927. **See basionym:** *Meliola calophylli* F.L. Stevens, 1916.

Irene cheirodendri F.L. Stevens. Bernice P. Bishop Museum Bulletin 19:44, text fig. 9d. 1925 [as *cheirodendronis*, corrected by F.C. Deighton (1968) in accordance with current I.C.B.N.]. **Holotype:** On *Cheirodendron gaudichaudii* (DC.) Seem. [= *C. trigynum* (Gaudich.) A.A. Heller], Hawaii, Kauai, Alakai swamp, 22 AUG [1921], leg. O.H. Swezey, [as F.L. Stevens No. 1165], ILL 3953. **Isotypes:** BISH 499894 and 499895. **Paratype:** Oahu, Tantalus, 22 JUN [1921], leg. F.L. Stevens No. 641, ILL 3954. ≡ *Irenina cheirodendri* (F.L. Stevens) F.L. Stevens. Annales Mycologici 25:466. 1927. ≡ *Asteridiella cheirodendri* (F.L. Stevens) C.G. Hansford. Beihefte zur Sydowia Annales Mycologici 2:481. 1961. [Epithets of the synonyms were also cited as *cheirodendronis* and are corrected herein. Hansford (1961) erroneously cited the generic name of the basionym as *Meliola*].

Irene confragosa (H. & P. Sydow) H. & P. Sydow, 1917. **See basionym:** *Meliola confragosa* H. & P. Sydow, 1912.

Irene cornu-caprae (P.C. Hennings) F.L. Stevens, 1927. **See basionym:** *Meliola cornu-caprae* P.C. Hennings, 1904.

Irene crotonis F.L. Stevens & L.R. Tehon. Mycologia 18:20–21, pl. 2, fig. 26. 1926. **Holotype:** On *Croton* sp., Trinidad, St. Augustine, 13 AUG 1922, leg. F.L. Stevens No. 837, ILL 4341. ≡ *Irenopsis crotonis* (F.L. Stevens & L.R. Tehon) F.L. Stevens. Annales Mycologici 25:441. 1927.

Irene crustacea (C.L. Spegazzini) F. Theissen & H. Sydow, (1917) 1918. **See basionym:** *Meliola crustacea* C.L. Spegazzini, 1889.

Irene cubitella F.L. Stevens & L.R. Tehon. Mycologia 18:18–19, pl. 2, fig. 23. 1926. **Holotype:** On *Cassia* sp., British Guiana [Guyana], Tumatumari, 11 JUL 1922, leg. F.L. Stevens No. 119, ILL 3926. **Paratypes:** Leg. F.L. Stevens No. 193, ILL 3927 [also the holotype of *Meliola cubitella* sensu C.G. Hansford, 1961, non (F.L. Stevens & L.R. Tehon) R. Ciferri, 1954]; Trinidad, Cumuto, 16 AUG 1922, leg. F.L. Stevens No. 941, ILL 3928. ≡ *Irenina cubitella* (F.L. Stevens & L.R. Tehon) F.L. Stevens. Annales Mycologici 25:461. 1927. ≡ *Meliola cubitella* (F.L. Stevens & L.R. Tehon) R. Ciferri. Mycopathologia 7:87. 1954, nom. invalid. [in violation of Art. 33.2 of I.C.B.N.], non C.G. Hansford, 1961 [the latter name also not validly published].

Irene cubitorum F.L. Stevens & L.R. Tehon. Mycologia 18:19, pl. 2, fig. 24. 1926. **Holotype:** On *Dimorphandra latifolia* L.R. Tulasne., British Guiana [Guyana], Coverden, 8 AUG 1922, leg. F.L. Stevens No. 810, ILL 3932. **Isotypes:** CUP, FH, and K. **Paratypes:** Demerara-Essequibo R.R., 15 JUL [1922], leg. F.L. Stevens No. 417, ILL 3930; on *Dimorphandra* sp., Demerara-Essequibo R.R., 15 JUL [1922], leg. F.L. Stevens 380, ILL 3931 [also an isotype of *Irenopsis dimorphandrae* C.G. Hansford]. ≡ *Irenina cubitorum* (F.L. Stevens & L.R. Tehon) F.L. Stevens. Annales Mycologici 25:466. 1927. ≡ *Asteridiella cubitorum* (F.L. Stevens & L.R. Tehon) C.G. Hansford ex C.G. Hansford. Beihefte zur Sydowia Annales Mycologici 2:244–245. 1961.

Irene cyclopoda (F.L. Stevens) R.A. Toro, 1925. **See basionym:** *Meliola cyclopoda* F.L. Stevens, 1916.

Irene cyrtandrae F.L. Stevens. Bernice P. Bishop Museum Bulletin 19:44–45, text fig. 9e. 1925. **Holotype:** On *Cyrtandra lessoniana* Gaudich., Hawaii, Kauai, Kalalau trail, 16 JUN [1921], leg. F.L. Stevens No. 481, ILL 3939. **Isotypes:** BISH 499891 and 499892, FH. **Paratype:** On

Cyrtandra cordifolia Gaudich., [Island of] Hawaii, Kilauea, 11 JUL [1921], leg. F.L. Stevens No. 793, ILL 3940. ≡ *Irenina cyrtandrae* (F.L. Stevens) F.L. Stevens. Annales Mycologici 25:465. 1927. ≡ *Asteridiella cyrtandrae* (F.L. Stevens) C.G. Hansford. Beihefte zur Sydowia Annales Mycologici 2:656. 1961 [the basionym was erroneously cited by Hansford as *Meliola cyrtandrae*].

Irene echinus (P.C. Hennings) F.L. Stevens, 1927. **See basionym**: *Meliola echinus* P.C. Hennings, 1904.

Irene escharoides H. Sydow. Annales Mycologici 24:316–317. 1926. **Assumed lectotype**: Hab. in foliis vivis *Tabernaemontanae longipedis* [= *Tabernaemontana longipes* J. Donnell Smith], Costa Rica, San Pedro de San Ramón, 5 FEB 1925, leg. H. Sydow, Fungi in Itinere Costaricensi Collecti No. 393a, ILL 3736 [lectotypification by C.G. Hansford (1961:554) is assumed on the strength of his annotation of ILL 3736 as type and citation of F.L.S. herb. (ILL) as location of the type, although the collection number was erroneously cited as 293a instead of 393a]. ≡ *Irenina escharoides* (H. Sydow) F.L. Stevens. Annales Mycologici 25:460. 1927. ≡ *Meliola escharoides* (H. Sydow) R. Ciferri. Mycopathologia 7:128. 1954. ≡ *Meliola tabernaemontanae* C.L. Spegazzini var. *escharoides* (H. Sydow) C.G. Hansford. Beihefte zur Sydowia Annales Mycologici 2:554. 1961.

Irene exilis (H. & P. Sydow) F.L. Stevens, 1925. **See basionym**: *Meliola exilis* H. & P. Sydow, 1904.

Irene glabroides (F.L. Stevens) R.A. Toro, 1925. **See basionym**: *Meliola glabroides* F.L. Stevens, 1916.

Irene gloriosa (E.M. Doidge) E.M. Doidge, 1920. **See basionym**: *Meliola gloriosa* E.M. Doidge, 1920.

Irene hyptidicola (F.L. Stevens) R.A. Toro, 1925. **See basionym**: *Meliola hyptidicola* F.L. Stevens, 1916.

Irene inermis (K. Kalchbrenner & M.C. Cooke) F. Theissen & H. Sydow var. *minor* C.G. Hansford & F.L. Stevens in C.G. Hansford. Journal of the Linnean Society (Botany) London 51:266–267. 1937. **Holotype**: Hab. in foliis *Labiatarum* [Lamiaceae sp. indet.], Uganda, Kampala, FEB 1930, leg. C.G. Hansford No. 1076, ILL 3750. **Paratypes**: On *Coleus* sp. [= *Plectranthus* sp.], Uganda, Entebbe Road, [mile 13, MAR 1930], leg. C.G. Hansford No. 1134, ILL 3749; on *Hyptis pectinata* (L.) Poiteau, Uganda, Kampala, [AUG 1930], leg. C.G. Hansford No. 1330, ILL 3751. ≡ *Appendiculella labiatarum* C.G. Hansford. Beihefte zur Sydowia Annales Mycologici 2:698–699. 1961, nom. and stat. nov.

Irene ingae F.L. Stevens & L.R. Tehon. Mycologia 18:20, pl. 2, fig. 25. 1926. **Holotype**: On *Inga* sp., British Guiana [Guyana], Kartabo, 22 JUL

1922, leg. F.L. Stevens No. 559, ILL 4384. ≡ *Irenopsis ingae* (F.L. Stevens & L.R. Tehon) F.L. Stevens. Annales Mycologici 25:433. 1927.

Irene irregularis (F.L. Stevens) R.A. Toro, 1925. **See basionym:** *Meliola irregularis* F. L. Stevens, 1916.

Irene larviformis (P.C. Hennings) F.L. Stevens var. *arecibensis* (F.L. Stevens) F.L. Stevens, 1927. **See basionym:** *Meliola arecibensis* F.L. Stevens, 1916.

Irene manca (J.B. Ellis & G.W. Martin) F. Theissen & H. Sydow, (1917) 1918. **See basionym:** *Meliola manca* J.B. Ellis & G.W. Martin, 1883.

Irene marcgraviae (L.R. Tehon) F.L. Stevens & L.R. Tehon, 1926. **See basionym:** *Meliola marcgraviae* L.R. Tehon, 1919.

Irene megalospora (C.L. Spegazzini) F. Theissen & H. Sydow, (1917) 1918. **See basionym:** *Meliola megalospora* C.L. Spegazzini, 1881.

Irene melastomatacearum (C.L. Spegazzini) R.A. Toro, 1925. **See basionym:** *Meliola melastomatacearum* C.L. Spegazzini, 1889.

Irene natalensis (E.M. Doidge) E.M. Doidge, 1920. **See basionym:** *Meliola natalensis* E.M. Doidge, 1917.

Irene nectandrae C.G. Hansford. Sydowia Annales Mycologici 9:34. 1955. **Isotype:** Hab in foliis *Nectandrae* sp. [*Nectandra* sp. = *Ocotea* sp.], British Guiana [Guyana], Wismar, [14 JUL 1922], leg. F.L. Stevens No. 315, ILL 4011. ≡ *Asteridiella nectandrae* (C.G. Hansford) C.G. Hansford ex C.G. Hansford. Beihefte zur Sydowia Annales Mycologici 2:46. 1961. **Holotype:** K.

Irene nuxiae H. Sydow in E.M. Doidge & H. Sydow. Bothalia 2(2):433, 463. 1928. **Assumed isotype** [the single collection cited on p. 463 with the Latin diagnosis]: Hab. in foliis *Nuxiae floribundae* [= *Nuxia floribunda* Bentham in Hooker, sub *Meliola heudeloti* A. Gaillard], South Africa, Transvaal, Woodbush, 3 AUG 1911 [as 5 AUG 1913 on the packet], leg. E.M. Doidge, ex PREM No. 1776, ILL 4090. ≡ *Irenina nuxiae* (H. Sydow in E.M. Doidge & H. Sydow) C.G. Hansford. Proceedings of the Linnean Society, London 157:174. 1946. ≡ *Asteridiella nuxiae* (H. Sydow in E.M. Doidge & H. Sydow) C.G. Hansford ex C.G. Hansford. Beihefte zur Sydowia Annales Mycologici 2:522. 1961.

Irene obesa (C.L. Spegazzini) F. Theissen & H. Sydow, (1917) 1918. **See basionym:** *Meliola obesa* C.L. Spegazzini, 1884.

Irene peglerae (E.M. Doidge) E.M. Doidge, 1920. **See basionym:** *Meliola peglerae* E.M. Doidge, 1917.

Irene perseae (F.L. Stevens) R.A. Toro, 1925. **See basionym:** *Meliola perseae* F.L. Stevens, 1916.

Irene peruviana (H. & P. Sydow) C.G. Hansford, 1955. **See basionym:** *Meliola peruviana* H. & P. Sydow, 1916.

Irene podocarpi (E.M. Doidge) E.M. Doidge, 1920. **See basionym:** *Meliola podocarpi* E.M. Doidge, 1917.

Irene puiggarii (C.L. Spegazzini) E.M. Doidge, 1920. **See basionym**: *Meliola puiggarii* C.L. Spegazzini, 1889.

Irene rinoreae E.M. Doidge. Bothalia 1(2):80–81. 1922. **Isotype**: On leaves of *Rinorea ardesiaeflora* (Welw. in Oliver) O. Kuntze, South Africa, Natal [Prov., Pietermaritzburg], Karkloof, 14 JUL 1921, leg. E.M. Doidge, ex PREM No. 14961, ILL 4225. ≡ *Irenina rinoreae* (E.M. Doidge) F.L. Stevens. Annales Mycologici 25:468. 1927. ≡ *Asteridiella rinoreae* (E.M. Doidge) C.G. Hansford ex C.G. Hansford. Beihefte zur Sydowia Annales Mycologici 2:80. 1961.

Irene scabra (E.M. Doidge) E.M. Doidge, 1920. **See basionym**: *Meliola scabra* E.M. Doidge, 1919.

Irene scaevolicola F.L. Stevens. Bernice P. Bishop Museum Bulletin 19:45–46, text fig. 9g. 1925. **Holotype**: On *Scaevola chamissoniana* Gaudich., Hawaii, Oahu, Wahiawa, 31 MAY [1921], leg. F.L. Stevens No. 160, ILL 4423a. **Isotypes**: ILL 4423b, BISH 145902, 499888, 499889 and 499890. **Paratypes**: On *Scaevola chamissoniana* Gaudich., Oahu, Wahiawa, 3 JUN [1921], leg. F.L. Stevens No. 229, ILL 4419; leg. F.L. Stevens No. 243, ILL 4427; Tantalus, 22 JUN [1921], leg. F.L. Stevens No. 616, ILL 4425; Olympus, 24 JUN [1921], leg. F.L. Stevens No. 698, ILL 4430; [Island of] Hawaii, between Hilo and Kilauea, 10 JUL [1921], leg. F.L. Stevens No. 774, ILL 4416; Kauai, Kalalau trail, 16 JUN [1921], leg. F.L. Stevens No. 492, ILL 4415 and 6879 [also an isotype and the holotype, respectively, of *Trichothallus hawaiiensis* F.L. Stevens]; leg. F.L. Stevens No. 497, ILL 4429; leg. F.L. Stevens No. 486, ILL 4431; leg. F.L. Stevens No. 502, ILL 4433; leg. F.L. Stevens No. 510, ILL 4428; on *Scaevola glabra* Hooker & Arnott, [Island of] Hawaii, between Hilo and Kilauea, 10 JUL [1921], leg. F.L. Stevens No. 778, ILL 4426; Kauai, Kalalau trail, 16 JUN [1921], leg. F.L. Stevens No. 472, ILL 4424; on *Scaevola mollis* Hooker & Arnott, Oahu, Olympus, 24 JUN [1921], leg. F.L. Stevens No. 663, ILL 4414 and 4432; leg. F.L. Stevens No. 696, ILL 4417 and 4421; leg. F.L. Stevens No. 703, ILL 4420; Palolo Valley [and Mt. Olympus], 10 JUN [1921], leg. F.L. Stevens No. 331, ILL 4422; Wahiawa, 3 JUN [1921], leg. F.L. Stevens No. 251, ILL 4418. ≡ *Irenopsis scaevolicola* (F.L. Stevens) F.L. Stevens. Annales Mycologici 25:434. 1927. ≡ *Meliola scaevolicola* (F.L. Stevens) C.G. Hansford. Beihefte zur Sydowia Annales Mycologici 2:627. 1961.

Irene sepulta (N.T. Patouillard ex F.L. Stevens) R.A. Toro, 1925. **See basionym**: *Meliola sepulta* N.T. Patouillard ex F.L. Stevens, 1916.

Irene sidicola F.L. Stevens & L.R. Tehon. Mycologia 18:21, pl. 2, fig. 28. 1926. **Holotype**: On *Sida* sp., British Guiana [Guyana], Rockstone, 17 JUL 1922, leg. F.L. Stevens No. 478, ILL 4365 [No. 478 is omitted from the packet label]. ≡ *Irenopsis molleriana* (H.G. Winter) F.L. Stevens var.

sidicola (F.L. Stevens & L.R. Tehon) F.L. Stevens. Annales Mycologici 25:438. 1927. ≡ *Irenopsis sidicola* (F.L. Stevens & L.R. Tehon) C.G. Hansford. Beihefte zur Sydowia Annales Mycologici 2:185–186. 1961.

Irene solanacearum C.G. Hansford. Sydowia Annales Mycologici 9:35. 1955. **Isotype:** Hab. in foliis *Solanacearum* [= Solanaceae sp. indet.], Ecuador, San Miguel, [4 NOV 1924], leg. F.L. Stevens No. 229, ILL 3766. **Holotype:** FH. ≡ *Asteridiella solanacearum* (C.G. Hansford) C.G. Hansford. Sydowia Annales Mycologici 10:50, 59. (1956) 1957 [treated as sp. nov. (p. 59), rather than comb. nov. Evidence that Hansford intended this name as a new combination can be found in op. cit., pp. 46, 50]. Validation as comb. nov., with full reference to the basionym, dates from citation in Beihefte zur Sydowia Annales Mycologici 2:638. 1961.

Irene solani (F.L. Stevens) R.A. Toro, 1927. **See basionym:** *Meliola solani* F.L. Stevens, 1916.

Irene sororcula (C.L. Spegazzini) F.L. Stevens, 1927. **See basionym:** *Meliola sororcula* C.L. Spegazzini, 1889.

Irene sororcula (C.L. Spegazzini) F.L. Stevens var. *portoricensis* (F.L. Stevens) F.L. Stevens, 1927. **See basionym:** *Meliola compositarum* F.S. Earle var. *portoricensis* F.L. Stevens, 1916.

Irene sororcula (C.L. Spegazzini) F.L. Stevens var. *vernoniae* F.L. Stevens. Annales Mycologici 25:424, fig. 4. 1927. **Holotype:** On Compositae: *Vernonia* sp., Panama, Empire, 8 OCT 1925, leg. F.L. Stevens No. 1132, ILL 3854. **Paratypes:** Panama, Summit, 6 SEP 1924, leg. F.L. Stevens No. 308, ILL 3842; 12 SEP 1924, leg. F.L. Stevens No. 465, ILL 3845; Fort Lorenzo trail, 10 OCT 1924, leg. F.L. Stevens No. 1171, ILL 3844; Tapia, 15 AUG 1923, leg. F.L. Stevens No. 1047, ILL 3843; Mandingo, 15 OCT 1924, leg. F.L. Stevens No. 1319, ILL 3846; Costa Rica, Siquirres, 31 JUL 1923, leg. F.L. Stevens No. 706, ILL 3847. ≡ *Appendiculella vernoniae* (F.L. Stevens) C.G. Hansford. Sydowia Annales Mycologici 9:31–32. 1955, stat. and comb. nov.

Irene speciosa (E.M. Doidge) E.M. Doidge, 1920. **See basionym:** *Meliola speciosa* E.M. Doidge, 1917.

Irene splendens F.L. Stevens. Bernice P. Bishop Museum Bulletin 19:41, 43, text fig. 9b. 1925. **Holotype:** On *Alphitonia excelsa* sensu H. Mann, non (Fenzl) Reissek ex Endl. [= *A. ponderosa* Hillebr.], Hawaii, Kauai, Pipe trail, upper Waimea Canyon, 15 JUN 1921, leg. F.L. Stevens No. 430, ILL 3879. **Isotypes:** BISH 499886 and 499887. ≡ *Appendiculella splendens* (F.L. Stevens) C.G. Hansford. Beihefte zur Sydowia Annales Mycologici 2:363. 1961.

Irene strophanthi (E.M. Doidge) E.M. Doidge, 1920. **See basionym:** *Meliola strophanthi* E.M. Doidge, 1917.

Irene subcrustacea (C.L. Spegazzini) F. Theissen & H. Sydow, (1917) 1918. **See basionym:** *Meliola subcrustacea* C.L. Spegazzini, 1889.

Irene tonkinensis (P.A. Karsten & C. Roumeguère) F.L. Stevens, 1927. **See basionym:** *Meliola tonkinensis* P.A. Karsten & C. Roumeguère, 1890.

Irene tonkinensis (P.A. Karsten & C. Roumeguère) F.L. Stevens var. *cecropiae* F.L. Stevens. Annales Mycologici 25:427, fig. 6. 1927. **Holotype:** On *Cecropia arachnoidea* Pittier [= *C. peltata* L.], Panama, New Limon, 4 AUG 1924, leg. F.L. Stevens No. 1016, ILL 3885. **Paratypes:** Corozal [as Corozol], Trail No. 17, 20 AUG 1924, leg. F.L. Stevens No. 122, ILL 3883; on *Cecropia longifera* [nomen = *C. longipes* Pittier on the packet], France Field, 3 AUG 1924, leg. F.L. Stevens No. 972, ILL 3884. ≡ *Appendiculella tonkinensis* (P.A. Karsten & C. Roumeguère) R.A. Toro var. *cecropiae* (F.L. Stevens) C.G. Hansford. Beihefte zur Sydowia Annales Mycologici 2:321. 1961.

Irene toruloidea (F.L. Stevens) F.L. Stevens & L.R. Tehon, 1926. **See basionym:** *Meliola toruloidea* F.L. Stevens, 1916.

Irene trachylaena H. Sydow. Annales Mycologici 24:318–320. 1926. **Isotype:** Hab. in foliis *Zanthoxyli elephantiasis* [= *Zanthoxylum elephantiasis* Macfad.], Costa Rica, San Ramón, 22 JAN 1925, leg. H. Sydow, Fungi in Itinere Costaricensi Collecti No.114, ILL 4261. ≡ *Irenina trachylaena* (H. Sydow) F.L. Stevens. Annales Mycologici 25:466. 1927. ≡ *Asteridiella trachylaena* (H. Sydow) C.G. Hansford. Beihefte zur Sydowia Annales Mycologici 2:378. 1961.

Irene tuberculata (F.L. Stevens) F.L. Stevens, 1927. **See basionym:** *Meliola tuberculata* F.L. Stevens, 1916.

Irene winteri (C.L. Spegazzini) H. & P. Sydow, 1917. **See basionym:** *Meliola winteri* C.L. Spegazzini, 1888.

Irene zeyheri E.M. Doidge. Bothalia 1(2):75, fig. 5. 1922. **Isosyntypes:** On leaves of *Eugenia zeyheri* (Harvey) Harvey [= *E. capensis* (Ecklon & Zeyher) Sonder], South Africa, Howiesons Poort [near Grahamstown], 12 JUL 1919, leg. E.M. Doidge, ex PREM No. 12388, ILL 4286; Cape Prov., Pirie Forest, 8 JUL 1919, leg. E.M. Doidge, ex PREM No. 12272, ILL 4285. ≡ *Irenina zeyheri* (E.M. Doidge) F.L. Stevens. Annales Mycologici 25:467. 1927. ≡ *Asteridiella zeyheri* (E.M. Doidge) C.G. Hansford ex C.G. Hansford. Beihefte zur Sydowia Annales Mycologici 2:138. 1961.

Irenina aberrans F.L. Stevens. Annales Mycologici 25:462, 1927, nom. and stat. nov. **Based on:** *Meliola tomentosa* H.G. Winter var. *calva* H. Rehm, 1907 [Stevens applied the term "n. sp.", but he probably intended it as a new name; the epithet "*calva*" was pre-occupied at the rank of species under both *Irenina* and *Meliola*]. **Isotype:** On *Styrax* sp., Brazil, [Rio Grande do Sul, São Leopoldo,1906], leg. J. Rick, ex H.

Rehm, Ascomycetes Exsiccati No. 1707, ILL 4259 [only 2 microscopic preparations remain; the specimen used by Stevens was obtained from B (comm. H. Sydow)]. ≡ *Asteridiella aberrans* (F.L. Stevens) C.G. Hansford. Beihefte zur Sydowia Annales Mycologici 2:516. 1961.

Irenina acalyphae F.L. Stevens & E.F. Roldan. Philippine Journal of Science 56:51–52, fig. 1c. 1935, nom. invalid. sine diagn. lat. **Holotype:** On *Acalypha* sp., Philippines, Luzon, Benguet, Acop's, 30 DEC 1930, leg. F.L. Stevens No. 1211, ILL 3889. **Isotypes:** CALP, Phil. Bur. Sci.

Irenina aibonitensis (F.L. Stevens) F.L. Stevens, 1927. **See basionym:** *Meliola aibonitensis* F.L. Stevens, 1916.

Irenina alchorneae (F.L. Stevens & L.R. Tehon) F.L. Stevens, 1927. **See basionym:** *Irene alchorneae* F.L. Stevens & L.R. Tehon, 1926.

Irenina amoena (H. Sydow) F.L. Stevens, 1927. **See basionym:** *Irene amoena* H. Sydow, 1926.

Irenina anastomosans (H.G. Winter) F.L. Stevens, 1927. **See basionym:** *Meliola anastomosans* H.G. Winter, 1886.

Irenina andromedae (N.T. Patouillard) F.L. Stevens, 1927. **See basionym:** *Meliola andromedae* N.T. Patouillard, 1888.

Irenina anguriae F.L. Stevens. Annales Mycologici 25:464, fig. 24. 1927. **Holotype:** On *Anguria* sp. [Cucurbitaceae, probably *Psiguria* sp.], British Guiana [Guyana], Tumatumari, 11 JUL 1922, leg. F.L. Stevens No. 205, ILL 3897. **Paratypes:** On unknown Cucurbitaceae, Panama, Gatun, 26 SEP 1924, leg. F.L. Stevens No. 834, ILL 3898; Empire, 8 OCT 1924, leg. F.L. Stevens No. 1136, ILL 3899; Costa Rica, Port Limon, 10 AUG 1923, leg. F.L. Stevens No. 854, ILL 3900. ≡ *Asteridiella angureae* (F.L. Stevens) C.G. Hansford ex C.G. Hansford. Beihefte 2:123–124. 1961.

Irenina angustispora F.L. Stevens & E.F. Roldan. Philippine Journal of Science 56:53, fig. 1f. 1935, nom. invalid. sine diagn. lat. **Assumed holotype:** On *Neonauclea* sp., [Philippines], Luzon, Benguet, Naguilian Road, 7 JAN 1931, leg. F.L. Stevens No. 1620, ILL 3901 [only a microscopic preparation remains]. **Isotypes:** CALP, Phil. Bur. Sci. ≡ *Asteridiella angustispora* F.L. Stevens & E.F. Roldan ex C.G. Hansford, Sydowia Annales Mycologici 16:312. (1962) 1963.

Irenina angustispora F.L. Stevens & E.F. Roldan (nom. invalid.) var. *laevis* F.L. Stevens & E.F. Roldan. Philippine Journal of Science 56:53, fig. 1g. 1935, nom. invalid. sine diagn. lat. **Holotype:** On Rubiaceae: *Neonauclea* [sp.], Philippines, Luzon, Benguet, Kennon Road, 8 JAN 1931, leg. F.L. Stevens No. 1633, ILL 3902. **Isotypes:** CALP, Phil. Bur. Sci.

Irenina aracearum F.L. Stevens. Annales Mycologici 25:458. 1927. **Holotype:** On Araceae: *Dieffenbachia longispatha* Engler & Krause, Panama, Tapia, 15 AUG 1923, leg. F.L. Stevens No. 1021, ILL 3903.

Irenina atra (E.M. Doidge) F.L. Stevens, 1927. **See basionym:** *Meliola atra* E.M. Doidge, 1920.

Irenina aucubae (P.C. Hennings) F.L. Stevens, 1927. **See basionym:** *Meliola aucubae* P.C. Hennings. 1900 (1901).

Irenina barbaceniae C.G. Hansford. Proceedings of the Linnean Society of London 160:136. (1948) 1949 [as *barbacenniae*, corrected by the author (Hansford, 1961:731), in accordance with current I.C.B.N.]. **Isotypes:** In foliis *Barbaceniae purpureae* [as *Barbacennia = Barbacenia purpurea* Hooker], Brazil, Rio de Janeiro, JUN 1887, leg. E.H.G. Ule, ex Rabenhorst-Winter-Pazschke, Fungi Europaei No. 3849, sub *Meliola glabra* M.J. Berkeley & M.A. Curtis, ILL 4064 and 4069 [as microscopic preparations]. ≡ *Asteridiella barbaceniae* C.G. Hansford ex C.G. Hansford. Beihefte zur Sydowia Annales Mycologici 2:731–732. 1961.

Irenina bonii (A. Gaillard) F.L. Stevens, 1927. **See basionym:** *Meliola bonii* A. Gaillard, 1892.

Irenina buddlejicola (P.C. Hennings) F.L. Stevens, 1927. **See basionym:** *Meliola buddlejicola* P.C. Hennings, 1905.

Irenina callicarpae F.L. Stevens & E.F. Roldan. Philippine Journal of Science 56:53–54. 1935, nom. invalid. sine diagn. lat. **Holotype:** On *Callicarpa magna* J.C. Schauer in DC., [Philippines], Luzon, Benguet, Naguilian Road, 6 JAN 1931, leg. F.L. Stevens No. 1468, ILL 3957. **Isotypes:** CALP, Phil. Bur. Sci. **Paratype:** On *Callicarpa* sp., Luzon, La Union Prov., Santo Tomas, 31 DEC 1930, leg. F.L. Stevens No. 1291, ILL 3956. ≡ *Asteridiella callicarpae* F.L. Stevens & E.F. Roldan ex C.G. Hansford. Sydowia Annales Mycologici 16:312–313. (1962) 1963.

Irenina callista (H. Rehm) C.G. Hansford, 1946. **See basionym:** *Meliola callista* H. Rehm, 1914.

Irenina calva (C.L. Spegazzini) F.L. Stevens, 1927. **See basionym:** *Meliola calva* C.L. Spegazzini, 1889.

Irenina cheirodendri (F.L. Stevens) F.L. Stevens, 1927. **See basionym:** *Irene cheirodendri* F.L. Stevens, 1925.

Irenina clidemiae F.L. Stevens. Annales Mycologici 25:462–463, fig. 22. 1927. **Syntypes:** On *Clidemia* sp., British Guiana [Guyana], Rockstone, 13 JUL 1922, leg. F.L. Stevens No. 254, ILL 3950; Trinidad, Cumuto, 16 AUG 1922, leg. F.L. Stevens No. 912, ILL 3951. ≡ *Meliola clidemiae* (F.L. Stevens) R. Ciferri. Sydowia Annales Mycologici 36:208. 1938. ≡ *Irenopsis clidemiae* (F.L. Stevens) C.G. Hansford. Beihefte zur Sydowia Annales Mycologici 2:153. 1961.

Irenina colubrinae F.L. Stevens. Annales Mycologici 25:451–452, fig. 14. 1927. **Syntypes:** On *Colubrina rufa* (Vell.) Reisseck in Martius, Panama, France Field, 2 SEP 1924, leg. F.L. Stevens No. 173, ILL 3949; Fort Lorenzo trail, 10 OCT 1924, leg. F.L. Stevens No. 1197, ILL 3948. ≡

Asteridiella colubrinae (F.L. Stevens) C.G. Hansford ex C.G. Hansford. Beihefte zur Sydowia Annales Mycologici 2:364. 1961.

Irenina combreti F.L. Stevens. Annales Mycologici 25:465, fig. 25. 1927. **Holotype:** On *Combretum farinosum* Kunth in H.B.K., Panama, Culebra, 2 OCT 1924, leg. F.L. Stevens No. 952, ILL 3946. ≡ *Asteridiella combreti* (F.L. Stevens) C.G. Hansford ex C.G. Hansford. Beihefte zur Sydowia Annales Mycologici 2:159–160. 1961.

Irenina confragosa (H. & P. Sydow) F.L. Stevens, 1927. **See basionym:** *Meliola confragosa* H. & P. Sydow, 1912.

Irenina costi F.L. Stevens. Annales Mycologici 25:458, fig. 19. 1927. **Holotype:** On Zingiberaceae: *Costus* sp., Panama, Brazos Brook Reservoir, 22 SEP 1924, leg. F.L. Stevens No. 728a, ILL 3923. ≡ *Asteridiella costi* (F.L. Stevens) C.G. Hansford. Beihefte zur Sydowia Annales Mycologici 2:704–705. 1961.

Irenina crustacea (C.L. Spegazzini) F.L. Stevens, 1927. **See basionym:** *Meliola crustacea* C.L. Spegazzini, 1889.

Irenina cryptocarpa (J.B. Ellis & G.W. Martin) C.G. Hansford, 1946. **See basionym:** *Meliola cryptocarpa* J.B. Ellis & G.W. Martin, 1883.

Irenina cubitella (F.L. Stevens & L.R. Tehon) F.L. Stevens, 1927. **See basionym:** *Irene cubitella* F.L. Stevens & L.R. Tehon, 1926.

Irenina cubitorum (F.L. Stevens & L.R. Tehon) F.L. Stevens, 1927. **See basionym:** *Irene cubitorum* F.L. Stevens & L.R. Tehon, 1926.

Irenina cyclopoda (F.L. Stevens) F.L. Stevens, 1927. **See basionym:** *Meliola cyclopoda* F.L. Stevens, 1916.

Irenina cyrtandrae (F.L. Stevens) F.L. Stevens, 1927. **See basionym:** *Irene cyrtandrae* F.L. Stevens, 1925.

Irenina dalechampiae F.L. Stevens. Annales Mycologici 25:449, fig. 13. 1927. **Lectotype** [or perhaps holotype]: On Euphorbiaceae: *Dalechampia scandens* L., Ecuador, Terecita, 29 OCT 1924, leg. F.L. Stevens No. 49, ILL 4084 [citation of this collection as "type" by C.G. Hansford (1961:206) is interpreted as lectotypification. Although No. 49 is not marked "type," it is distinguished by larger print from No. 153 in the protologue and perhaps was intended as type by the author.]. **Isolectotype:** ILL 4384 [a microscopic preparation by C.G. Hansford]. **Residual syntype:** 30 OCT 1924, leg. F.L. Stevens No. 153, ILL 4085.

Irenina escharoides (H. Sydow) F.L. Stevens, 1927. **See basionym:** *Irene escharoides* H. Sydow, 1926.

Irenina exilis (H. & P. Sydow) F.L. Stevens, 1927. **See basionym:** *Meliola exilis* H. & P. Sydow, 1904.

Irenina glabroides (F.L. Stevens) F.L. Stevens, 1927. **See basionym:** *Meliola glabroides* F.L. Stevens, 1916.

Irenina glabroides (F.L. Stevens) F.L. Stevens var. *schlegeliae* (F.L. Stevens)

F.L. Stevens, 1927. **See basionym:** *Meliola glabroides* F.L. Stevens var. *schlegeliae* F.L. Stevens, 1916.

Irenina gymnosporiae (H. & P. Sydow) F.L. Stevens, 1927. **See basionym:** *Meliola gymnosporiae* H. & P. Sydow, 1912.

Irenina hyptidicola (F.L. Stevens) F.L. Stevens, 1927. **See basionym:** *Meliola hyptidicola* F.L. Stevens, 1916.

Irenina iquitosensis (P.C. Hennings) C.G. Hansford, 1946. **See basionym:** *Meliola iquitosensis* P.C. Hennings, 1904.

Irenina irregularis (F.L. Stevens) F.L. Stevens, 1927. **See basionym:** *Meliola irregularis* F.L. Stevens, 1916.

Irenina isertiae F.L. Stevens. Annales Mycologici 25:460–461, fig. 20. 1927. **Syntypes:** On *Isertia haenkeana* DC., Panama, France Field, 2 SEP 1924, leg. F.L. Stevens No. 220, ILL 4104; 3 OCT 1924, leg. F.L. Stevens No. 982, ILL 4112; Agua Clara Reservoir, 17 SEP 1924, leg. F.L. Stevens No. 552, ILL 4114; Fort Randolph, 100 feet hill trail, 23 SEP 1924, leg. F.L. Stevens No. 764, ILL 4113; New Limon, 4 OCT 1924, leg. F.L. Stevens No. 1013, ILL 4109; Bella Vista, 7 OCT 1924, leg. F.L. Stevens No. 1112, ILL 4105; Fort Lorenzo trail, 10 OCT 1924, leg. F.L. Stevens No. 1149, ILL 4106; Mandingo, 15 OCT 1924, leg. F.L. Stevens No. 1354, ILL 4108; on *Psychotria* sp., France Field, 2 SEP 1924, leg. F.L. Stevens No. 172, ILL 4111; on unknown host, France Field, 3 OCT 1924, leg. F.L. Stevens No. 1008, ILL 4116; Sweetwater, Fort Sherman, 6 OCT 1924, leg. F.L. Stevens No. 1092, ILL 4107; Peru, Huacapistana, 6 DEC 1924, leg. F.L. Stevens No. 78, ILL 4115. **Isosyntype:** ILL 4110. ≡ *Asteridiella glabra* (M.J. Berkeley & M.A. Curtis) C.G. Hansford ex C.G. Hansford var. *isertiae* (F.L. Stevens) C.G. Hansford. Beihefte zur Sydowia Annales Mycologici 2:579–580. 1961.

Irenina leucosykes (H.S. Yates) C.G. Hansford, 1946. **See basionym:** *Meliola leucosykes* H.S. Yates, (1917) 1918.

Irenina linocierae (H. & P. Sydow) C.G. Hansford, (1948) 1949. **See basionym:** *Meliola linocierae* H. & P. Sydow, 1914.

Irenina longipedicellata F.L. Stevens. Annales Mycologici 25:465–466, fig. 26. 1927. **Holotype:** On Dilleniaceae sp. indet., British Guiana [Guyana], Kartabo, 24 JUL 1922, leg. F.L. Stevens No. 672, ILL 4125. ≡ *Asteridiella longipedicellata* (F.L. Stevens) C.G. Hansford ex C.G. Hansford. Beihefte zur Sydowia Annales Mycologici 2:102–103. 1961.

Irenina manaosensis (P.C. Hennings) C.G. Hansford, 1946. **See basionym:** *Meliola manaosensis* P.C. Hennings, 1904.

Irenina manca (J.B. Ellis & G.W. Martin) F.L. Stevens, 1927. **See basionym:** *Meliola manca* J.B. Ellis & G.W. Martin, 1883.

Irenina mangostana (P.A. Saccardo) F.L. Stevens, 1927. **See basionym:** *Meliola mangostana* P.A. Saccardo, 1921.

Irenina marcgraviae (L.R. Tehon) F.L. Stevens, 1927. **See basionym:** *Meliola marcgraviae* L.R. Tehon, 1919.

Irenina meibomiae F.L. Stevens. Annales Mycologici 25:454, fig. 17. 1927. **Holotype:** On *Meibomia cana* (J.F. Gmelin) S.F. Blake [= *Desmodium incanum* DC.], Panama, France Field, 2 SEP 1924 [as 2 JUL 1924 on the packet], leg. F.L. Stevens No. 1213, ILL 4206. ≡ *Asteridiella meibomiae* (F.L. Stevens) C.G. Hansford ex C.G. Hansford. Beihefte zur Sydowia Annales Mycologici 2:272. 1961.

Irenina melastomatacearum (C.L. Spegazzini) F.L. Stevens, 1927. **See basionym:** *Meliola melastomatacearum* C.L. Spegazzini, 1889.

Irenina monninae F.L. Stevens. Annales Mycologici 25:452, fig. 20. 1927. **Holotype:** On *Monnina rupestris* Kunth in H.B.K., Ecuador, Terecita, 29 OCT 1924, leg. F.L. Stevens No. 50, ILL 4207. ≡ *Asteridiella monninae* (F.L. Stevens) C.G. Hansford ex C.G. Hansford. Beihefte zur Sydowia Annales Mycologici 2:81. 1961.

Irenina morototonii (C.L. Spegazzini) F.L. Stevens, 1927. **See basionym:** *Meliola morototonii* C.L. Spegazzini, 1924.

Irenina nigra F.L. Stevens. Annales Mycologici 25:462, fig. 21. 1927. **Holotype:** On unknown Cucurbitaceae, Costa Rica, Peralta, 11 JUL 1923, leg. F.L. Stevens No. 312, ILL 4209. ≡ *Asteridiella nigra* (F.L. Stevens) C.G. Hansford ex C.G. Hansford. Beihefte zur Sydowia Annales Mycologici 2:124. 1961.

Irenina nuxiae (H. Sydow in E.M. Doidge & H. Sydow) C.G. Hansford, 1946. **See basionym:** *Irene nuxiae* H. Sydow in E.M. Doidge & H. Sydow, 1928.

Irenina obesa (C.L. Spegazzini) F.L. Stevens, 1927. **See basionym:** *Meliola obesa* C.L. Spegazzini, 1884.

Irenina obscura F.L. Stevens. Annales Mycologici 25:454–455, fig. 18. 1927. **Holotype:** On Dilleniaceae sp. indet., Panama, Corozal, Trail No. 17, 30 AUG 1924, leg. F.L. Stevens No. 117, ILL 4214. **Paratypes:** Leg. F.L. Stevens No. 76, ILL 4215; on *Saurauia* sp., Peru, Chosica, 13 DEC 1924, leg. F.L. Stevens No. 228, ILL 4216. ≡ *Asteridiella obscura* (F.L. Stevens) C.G. Hansford ex C.G. Hansford. Beihefte zur Sydowia Annales Mycologici 2:103. 1961.

Irenina parasitica F.L. Stevens. Annales Mycologici 25:454, fig. 16. 1927. **Holotype:** On *Costus* sp., Ecuador, Terecita, 31 OCT 1924, leg. F.L. Stevens No. 194, ILL 4217. ≡ *Asteridiella parasitica* (F.L. Stevens) C.G. Hansford ex C.G. Hansford. Beihefte zur Sydowia Annales Mycologici 2:705. 1961.

Irenina perseae (F.L. Stevens) F.L. Stevens, 1927. **See basionym:** *Meliola perseae* F.L. Stevens, 1916.

Irenina pileae C.G. Hansford. Proceedings of the Linnean Society of Lon-

don 160:125–126. (1948) 1949. **Isotype:** Hab. in foliis *Pileae parietariae* [= *Pilea parietaria* (L.) Blume], Porto Rico [Puerto Rico], Arecibo-Lares Road, 21 JUN 1915, leg. F.L. Stevens No. 7232, ILL 4275. **Holotype:** PREM. ≡ *Asteridiella pileae* (C.G. Hansford) C.G. Hansford ex C.G. Hansford. Beihefte zur Sydowia Annales Mycologici 2:332. 1961.
Irenina pinicola (J. Dearness) F.L. Stevens, 1927. **See basionym:** *Meliola pinicola* J. Dearness, 1926.
Irenina podocarpi (E.M. Doidge) F.L. Stevens, 1927. **See basionym:** *Meliola podocarpi* E.M. Doidge, 1917.
Irenina pseudanastomosans (H. Rehm) F.L. Stevens, 1927. **See basionym:** *Meliola pseudanastomosans* H. Rehm, 1896.
Irenina rinoreae (E.M. Doidge) F.L. Stevens, 1927. **See basionym:** *Irene rinoreae* E.M. Doidge, 1922.
Irenina rubi F.L. Stevens & E.F. Roldan. Philippine Journal of Science 56:52, fig. 1d. 1935, nom. invalid. sine diagn. lat. **Holotype:** On *Rubus rosifolius* J.E. Smith [as *rosaefolius*], Philippines, [Luzon], Benguet, Naguilian Road, 6 JAN 1931, leg. F.L. Stevens No. 1549, ILL 4227. **Paratypes:** Luzon, Tayabas Prov. [= Quezon Prov.], Sariaya, 9 AUG 1930, leg. F.L. Stevens No. 193, ILL 4228; on *Rubus moluccanus* L., Luzon, Mt. Santo Tomas [as Mt. St. Thomas on the packet], Benguet, 31 DEC 1930, leg. F.L. Stevens No. 1361, ILL 4226 [this number, p.p., is also a paratype of *Meliola rubiella* C.G. Hansford. *Irenina rubi* is cited, pro syn., under *Asteridiella rubi* C.G. Hansford, but it is not the basionym of the latter name because Hansford used a different type].
Irenina rubi F.L. Stevens & E.F. Roldan (nom. invalid.) var. *angulata* F.L. Stevens & E.F. Roldan. Philippine Journal of Science 56:52–53, fig. 1e. 1935, nom. invalid. sine diagn. lat. **Holotype:** On *Rubus moluccanus* L., Philippines, Luzon, Benguet, Naguilian Road, 6 JAN 1931, leg. F.L. Stevens No. 1461, ILL 4233 [this number is also the assumed holotype of *Meliola rubiella* C.G. Hansford]. **Isotypes:** CALP, Phil. Bur. Sci. **Paratype:** On *Rubus rosifolius* J.E. Smith [as *rosaefolius*], 6 JAN 1931, leg. F.L. Stevens No. 1472, ILL 4232 [this number is also the assumed holotype of *Asteridiella rubi* C.G. Hansford].
Irenina sanguinea (J.B. Ellis & B.M. Everhart) F.L. Stevens, 1927. **See basionym:** *Meliola sanguinea* J.B. Ellis & B.M. Everhart, 1886.
Irenina scabra (E.M. Doidge) F.L. Stevens, 1927. **See basionym:** *Meliola scabra* E.M. Doidge, 1919.
Irenina schlegeliae (F.L. Stevens) C.G. Hansford, (1948) 1949. **See basionym:** *Meliola glabroides* F.L. Stevens var. *schlegeliae* F.L. Stevens, 1916.
Irenina sepulta (N.T. Patouillard ex F.L. Stevens) F.L. Stevens, 1927. **See basionym:** *Meliola sepulta* N.T. Patouillard ex F.L. Stevens, 1916.

Irenina shropshiriana F.L. Stevens. Annales Mycologici 25:452–453, fig. 15. 1927. **Holotype**: On *Miconia argentea* (Sw.) DC., Panama, Sweetwater, Fort Sherman, 6 OCT 1924, leg. F.L. Stevens No. 1083, ILL 4248. **Paratypes**: France Field, 2 SEP 1924, leg. F.L. Stevens No. 212, ILL 4247; 3 OCT 1924, leg. F.L. Stevens No. 1005, ILL 4250; Fort Lorenzo trail, 10 OCT 1924, leg. F.L. Stevens No. 1150, ILL 4249; Fort Randolph, 100 feet hill trail, 23 SEP 1924, leg. F.L. Stevens No. 747, ILL 4251; leg. F.L. Stevens No. 767, ILL 4253; Barro Colorado [Island], 29 AUG 1924, leg. F.L. Stevens No. 581, ILL 4252; Paitilla Point, 8 SEP 1924, leg. F.L. Stevens No. 341, ILL 4246; Tapia, 15 AUG 1923, leg. F.L. Stevens No. 1043, ILL 4245; leg. F.L. Stevens No. 1008, ILL 4244. ≡ *Irenopsis shropshiriana* (F.L. Stevens) C.G. Hansford. Beihefte zur Sydowia Annales Mycologici 2:152. 1961 [Hansford erroneously indicated F.L. Stevens No. 581 as the type].

Irenina sinuosa F.L. Stevens & E.F. Roldan. Philippine Journal of Science 56:54–55, fig. 1e. 1935, nom. invalid. sine diagn. lat. **Holotype**: On *Glochidion* sp., [Philippines], Luzon, Nueva Ecija Prov., Balete Pass, 9 JAN 1931, leg. F.L. Stevens No. 1744, ILL 4254. **Isotypes**: CALP, Phil. Bur. Sci.

Irenina strophanthi (E.M. Doidge) F.L. Stevens, 1927. **See basionym**: *Meliola strophanthi* E.M. Doidge, 1917.

Irenina subapoda (H. & P. Sydow) F.L. Stevens, 1927. **See basionym**: *Meliola subapoda* H. & P. Sydow, 1914.

Irenina thunbergiae F.L. Stevens & E.F. Roldan. Philippine Journal of Science 56:54, fig. 1h. 1935, nom. invalid. sine diagn. lat. **Holotype**: On *Thunbergia alata* Bojer ex Sims, [Philippines], Luzon, Benguet, Kennon Road, 8 JAN 1931, leg. F.L. Stevens No. 1642, ILL 4258. **Isotypes**: CALP, CUP, Phil. Bur. Sci. ≡ *Asteridiella thunbergiae* F.L. Stevens & E.F. Roldan ex C.G. Hansford. Sydowia Annales Mycologici 16:313 (1962) 1963.

Irenina trachylaena (H. Sydow) F.L. Stevens, 1927. **See basionym**: *Irene trachylaena* H. Sydow, 1926.

Irenina trematis (C. L. Spegazzini) F.L. Stevens, 1927. **See basionym**: *Meliola trematis* C.L. Spegazzini, 1912.

Irenina umirayensis (H.S. Yates) C.G. Hansford, 1946. **See basionym**: *Meliola umirayensis* H.S. Yates, (1918) 1919.

Irenina uncariae (H. Rehm) F.L. Stevens, 1927. **See basionym**: *Meliola uncariae* H. Rehm, 1914.

Irenina viburni (H. & P. Sydow) F.L. Stevens, 1927. **See basionym**: *Meliola viburni* H. & P. Sydow, 1917.

Irenina zeyheri (E.M. Doidge) F.L. Stevens, 1927. **See basionym**: *Irene zeyheri* E.M. Doidge, 1922.

Irenopsis anastomosans (H.G. Winter) R.A. Toro, 1934. **See basionym:** *Meliola anastomosans* H.G. Winter, 1886.

Irenopsis araneosa (H. & P. Sydow) F.L. Stevens, 1927. **See basionym:** *Meliola araneosa* H. & P. Sydow, 1913.

Irenopsis aristolochiella C.G. Hansford. Sydowia Annales Mycologici 11:46. (1957) 1958. **Holotype:** Hab. in foliis *Aristolochiae* [= *Aristolochia* sp.], British Guiana [Guyana], Tumatumari, [10 JUL 1922], leg. F.L. Stevens No. 107, ILL 4640.

Irenopsis armata (C.L. Spegazzini) F.L. Stevens, 1927. **See basionym:** *Meliola armata* C.L. Spegazzini, 1889.

Irenopsis banisteriae C.G. Hansford. Sydowia Annales Mycologici 10:43. (1956) 1957. **Paratype:** Hab. in foliis *Banisteriae laurifoliae* [*Banisteria laurifolia* L. = *Heteropteris laurifolia* (L.) A. Jussieu], Porto Rico [Puerto Rico], Utuado, [Mayagüez Mesa, 26 JUN 1915], leg. F.L. Stevens No. 7564, ILL 4396 [also a paratype of *Meliola rectangularis* F.L. Stevens]. **Holotype:** H.H. Whetzel & C.E. Chardon No. 3286, CUP.

Irenopsis bayamonensis (L.R. Tehon) F.L. Stevens, 1927. **See basionym:** *Meliola bayamonensis* L.R. Tehon, 1919.

Irenopsis benguetensis F.L. Stevens & E.F. Roldan ex C.G. Hansford. Sydowia Annales Mycologici 16:311–312. (1962) 1963; originally published by F.L. Stevens & E.F. Roldan in Philippine Journal of Science 56:49–50, fig. 1b. 1935, as nom. invalid. sine diagn. lat. **Holotype:** On *Ficus variegata* Blume, Philippines, Luzon, Benguet, Naguilian Road, 5 JAN 1931, leg. F.L. Stevens No. 1566, ILL 4290. **Isotypes:** CALP, Phil. Bur. Sci.

Irenopsis bignoniacearum F.L. Stevens. Annales Mycologici 25:442. 1927. **Holotype:** On Bignoniaceae sp. indet., Panama, Corozal, Trail No. 17, 30 AUG 1924, leg. F.L. Stevens No. 81, ILL 4292.

Irenopsis bosciae (E.M. Doidge) F.L. Stevens, 1927. **See basionym:** *Meliola bosciae* E.M. Doidge, 1917.

Irenopsis brasiliensis (C.L. Spegazzini) C.G. Hansford, 1961. **See basionym:** *Meliola brasiliensis* C.L. Spegazzini, 1881.

Irenopsis casearina C.G. Hansford. Beihefte zur Sydowia Annales Mycologici 1:93. 1957. **Holotype:** Hab. in foliis *Caseariae arboreae* [= *Casearia arborea* (L.C. Richard) Urban], Porto Rico [Puerto Rico], Monte de Oro, 13 DEC 1913, leg. F.L. Stevens No. 5709, ILL 5879. **Isotype:** ILL 6029. [No. 5709 is also cited as paratype of *Meliola paulliniae* F.L. Stevens.].

Irenopsis chamaecristicola (F.L. Stevens) F.L. Stevens, 1927. **See basionym:** *Meliola chamaecristicola* F.L. Stevens, 1916.

Irenopsis chiococcae (F.L. Stevens) F.L. Stevens, 1927. **See basionym:** *Meliola chiococcae* F.L. Stevens, 1916.

Irenopsis claviculata (E.M. Doidge) F.L. Stevens, 1927. **See basionym:** *Meliola claviculata* E.M. Doidge, 1920.

Irenopsis clidemiae (F.L. Stevens) C.G. Hansford, 1961. **See basionym:** *Irenina clidemiae* F.L. Stevens, 1927.

Irenopsis comata (E.M. Doidge) F.L. Stevens, 1927. **See basionym:** *Meliola comata* E.M. Doidge, 1920.

Irenopsis comocladiae (F.L. Stevens) F.L. Stevens, 1927. **See basionym:** *Meliola comocladiae* F.L. Stevens, 1916.

Irenopsis conferta (L.R. Tehon) F.L. Stevens, 1927, comb. nov. illegit. [cf. Articles 11.3 & 58.3 of I.C.B.N.]. **See basionym:** *Meliola conferta* L.R. Tehon, 1919 (nom. illegit.), non E.M. Doidge, 1917. **See also:** *Meliola tehoniana* A. Trotter, 1926, nom. nov. [this is the oldest legitimate name to be considered for priority]. ≡ *Irenopsis tehoniana* (A. Trotter) C.G. Hansford, 1961.

Irenopsis conostegiae F.L. Stevens. Annales Mycologici 25:439, fig. 10. 1927. **Holotype:** On *Conostegia xalapensis* (Bonpl.) D. Don, Panama, France Field, 2 SEP 1924, leg. F.L. Stevens No. 216, ILL 4337. **Paratypes:** Leg. F.L. Stevens No. 215, ILL 4336; 3 OCT 1924, leg. F.L. Stevens No. 1000, ILL 4338.

Irenopsis coronata (C.L. Spegazzini) F.L. Stevens, 1927. **See basionym:** *Meliola coronata* C.L. Spegazzini, 1884.

Irenopsis coronata (C.L. Spegazzini) F.L. Stevens var. *hibisci* C.G. Hansford. Journal of the Linnean Society (Botany) London 51:267. 1937. **Isotype:** Hab. in foliis *Hibisci* sp. [= *Hibiscus* sp.], Uganda, Entebbe Road, 1931, leg. C.G. Hansford No. 1544, ILL 4319.

Irenopsis coronata (C.L. Spegazzini) F.L. Stevens var. *philippinensis* F.L. Stevens & E.F. Roldan ex C.G. Hansford. Sydowia Annales Mycologici 16:312. (1962) 1963; originally published by F.L. Stevens & E.F. Roldan in Philippine Journal of Science 56:50, fig. 3j. 1935, as nom. invalid. sine diagn. lat. **Holotype:** On Tiliaceae: *Columbia serratifolia* (Cav.) DC. [= *Colona serratifolia* Cav.], Philippines, Luzon, Laguna Prov., Agricultural College, 10 SEP 1930, leg. F.L. Stevens No. 510, ILL 4322. **Isotypes:** CALP, CUP, Phil. Bur. Sci.

Irenopsis coronata (C.L. Spegazzini) F.L. Stevens var. *triumfettae* (F.L. Stevens) F.L. Stevens, 1927. **See basionym:** *Meliola triumfettae* F.L. Stevens, 1916.

Irenopsis costaricensis F.L. Stevens. Annales Mycologici 25:438, fig. 9. 1927. **Holotype:** On *Quercus oocarpa* Liebm. [= *Q. insignis* Martens & Galeotti ssp. *oocarpa* (Liebm.) Murray], Costa Rica, Cartago, 23 JUN 1923, leg. F.L. Stevens No. 64, ILL 4339.

Irenopsis crotonis (F.L. Stevens & L.R. Tehon) F.L. Stevens, 1927. **See basionym:** *Irene crotonis* F.L. Stevens & L.R. Tehon, 1926.

Irenopsis cryptocarpa (J.B. Ellis & G.W. Martin) C.G. Hansford, 1961. **See basionym:** *Meliola cryptocarpa* J.B. Ellis & G.W. Martin, 1883.
Irenopsis cupaniae (F.L. Stevens) F.L. Stevens, 1927. **See basionym:** *Meliola cupaniae* F.L. Stevens, 1916.
Irenopsis curvata (H.S. Yates) F.L. Stevens, 1927. **See basionym:** *Meliola curvata* H.S. Yates, (1918) 1919.
Irenopsis dimorphandrae C.G. Hansford. Sydowia Annales Mycologici 9:38. 1955. **Isotype:** Hab. in foliis *Dimorphandrae* sp. indet. [= *Dimorphandra* sp.], British Guiana [Guyana], Demerara-Essequibo R.R., [15 JUL 1922], leg. F.L. Stevens No. 380, ILL 3931 [also a paratype of *Irene cubitorum* F.L. Stevens & L.R. Tehon]. **Holotype:** K.
Irenopsis guianensis (F.L. Stevens & R.I. Dowell) F.L. Stevens, 1927. **See basionym:** *Meliola guianensis* F.L. Stevens & R.I. Dowell, 1923.
Irenopsis guignardii (A. Gaillard) F.L. Stevens, 1927. **See basionym:** *Meliola guignardii* A. Gaillard, 1892.
Irenopsis gustaviae C.G. Hansford. Beihefte zur Sydowia Annales Mycologici 1:93–94. 1957. **Holotype:** Hab. in foliis *Gustaviae angustae* [= *Gustavia angusta* J.F. Gmelin], British Guiana [Guyana], Coverden, [4 AUG 1922], leg. F.L. Stevens No. 722, ILL 5277.
Irenopsis ingae (F.L. Stevens & L.R. Tehon) F.L. Stevens, 1927. **See basionym:** *Irene ingae* F.L. Stevens & L.R. Tehon, 1926.
Irenopsis macrochaeta (H. & P. Sydow) F.L. Stevens, 1927. **See basionym:** *Meliola macrochaeta* H. & P. Sydow, 1912.
Irenopsis marcgraviae (L.R. Tehon) C.G. Hansford, 1961. **See basionym:** *Meliola marcgraviae* L.R. Tehon, 1919.
Irenopsis maricaensis (F.L. Stevens) F.L. Stevens, 1927. **See basionym:** *Meliola maricaensis* F.L. Stevens, 1916.
Irenopsis martiniana (A. Gaillard) F.L. Stevens, 1927. **See basionym:** *Meliola martiniana* A. Gaillard, 1892.
Irenopsis miconiae (F.L. Stevens) F.L. Stevens, 1927. **See basionym:** *Meliola miconiae* F.L. Stevens, 1916.
Irenopsis miconiicola (F.L. Stevens) F.L. Stevens, 1927. **See basionym:** *Meliola miconiicola* F.L. Stevens, 1916.
Irenopsis molleriana (H.G. Winter) F.L. Stevens var. *sidicola* (F.L. Stevens & L.R. Tehon) F.L. Stevens, 1927. **See basionym:** *Irene sidicola* F.L. Stevens & L.R. Tehon, 1926.
Irenopsis myrciae C.G. Hansford. Sydowia Annales Mycologici 9:38–39. 1955. **Assumed isotype:** Hab. in foliis *Myrciae deflexae* [= *Myrcia deflexa* (Poiret) DC.], Porto Rico [Puerto Rico], Bandera, [El Alto de la Bandera, 15 JUL 1915], leg. F.L. Stevens No. 8672 p.p., ILL 5230 p.p. **Holotype:** K.
Irenopsis ocoteae (F.L. Stevens) F.L. Stevens, 1927. **See basionym:** *Meliola ocoteae* F.L. Stevens, 1916.

Irenopsis oreocnides C.G. Hansford. Sydowia Annales Mycologici 11:46–47. (1957) 1958 [as *oreocnidae*, corrected in accordance with Articles 32.6 & 61 of I.C.B.N.]. **Holotype:** Hab. in foliis *Oreocnides* [= *Oreocnide* sp.], Philippines, Luzon, Benguet, Acop's, [30 DEC 1930], leg. F.L. Stevens No. 1204, ILL 6529.

Irenopsis parathesicola (F.L. Stevens) F.L. Stevens, 1927. **See basionym:** *Meliola parathesicola* F.L. Stevens, 1916.

Irenopsis portoricensis F.L. Stevens. Annales Mycologici 25:433, fig. 7. 1927. **Holotype:** On *Turpinia paniculata* sensu auct. non Vent. [= *T. occidentalis* (Sw.) G. Don], Porto Rico [Puerto Rico], Maricao, 20 SEP 1915, leg. F.L. Stevens No. 3635, ILL 4381. **Paratype:** 19 JUL 1915, leg. F.L. Stevens No. 8922, ILL 4382. ≡ *Meliola portoricensis* (F.L. Stevens) R. Ciferri. Mycoflora domingensis exsiccata, Century III, No. 209. 1932; Annales Mycologici 36:219. 1938.

Irenopsis pteridicola (F.L. Stevens) C.G. Hansford, 1955. **See basionym:** *Meliola pteridicola* F.L. Stevens, 1916.

Irenopsis rectangularis (F.L. Stevens) F.L. Stevens, 1927. **See basionym:** *Meliola rectangularis* F.L. Stevens, 1916.

Irenopsis rolandrae C.G. Hansford. Beihefte zur Sydowia Annales Mycologici 1:94–95. 1957 [as *rollandiae*, corrected in accordance with Rec. 60H.1 and Articles 60 & 61 of I.C.B.N.]. **Holotype:** Hab. in foliis *Rollandiae argenteae* [an error in the generic name of the host: *Rolandra argentea* Rottb. = *R. fruticosa* (L.) O. Kuntze], British Guiana [Guyana], Tumatumari, [8 JUL 1922], leg. F.L. Stevens No. 55a, ILL 3807. **Paratype:** In foliis *Rollandiae fruticosae* [same error as above = *Rolandra fruticosa* (L.) O. Kuntze], leg. F.L. Stevens No. 55, ILL 3813.

Irenopsis scaevolicola (F.L. Stevens) F.L. Stevens, 1927. **See basionym:** *Irene scaevolicola* F.L. Stevens, 1925.

Irenopsis shropshiriana (F.L. Stevens) C.G. Hansford, 1961. **See basionym:** *Irenina shropshiriana* F.L. Stevens, 1927.

Irenopsis sidicola (F.L. Stevens & L.R. Tehon) C.G. Hansford, 1961. **See basionym:** *Irene sidicola* F.L. Stevens & L.R. Tehon, 1926.

Irenopsis solani (F.L. Stevens) F.L. Stevens, 1927. **See basionym:** *Meliola solani* F.L. Stevens, 1916.

Irenopsis tehoniana (A. Trotter) C.G. Hansford, 1961. **See basionym:** *Meliola tehoniana* A. Trotter, 1926, nom. nov. **See also:** *Meliola conferta* L.R. Tehon, 1919 (nom. illegit.), non E.M. Doidge, 1917.

Irenopsis tenuissima (F.L. Stevens) F.L. Stevens, 1927. **See basionym:** *Meliola tenuissima* F.L. Stevens, 1916.

Irenopsis tortuosa (H.G. Winter in A. Gaillard) F.L. Stevens, 1927. **See basionym:** *Meliola tortuosa* H.G. Winter in A. Gaillard, 1892.

Irenopsis toruloidea (F.L. Stevens) F.L. Stevens, 1927. **See basionym:** *Meliola toruloidea* F.L. Stevens, 1916.

Irenopsis triumfettae (F.L. Stevens) C.G. Hansford & F.C. Deighton, 1948. **See basionym:** *Meliola triumfettae* F.L. Stevens, 1916.

Irenopsis varroniae (F.C. Deighton) C.G. Hansford, 1961. **See basionym:** *Meliola varroniae* F.C. Deighton, 1944.

Irenopsis viticifolii A.C. Batista & L.S. Carneiro in A.C. Batista, M.L. Nascimento & H. da Silva Maia. Atas do Instituto de Micologia 1:25–26. 1960 [as *vitexifolii*, corrected in accordance with Rec. 60H.1 and Articles 60 & 61 of I.C.B.N.]. **Isotypes:** In foliis *Viticis* sp. [cited as *Vitecis* = *Vitex* sp.], socii *Helminthosporii parathesicola* [as *porothesicolum* ≡ *Helminthosporium parathesicola* F.L. Stevens], Puerto Rico, Maricao, 20 SEP 1913, leg. F.L. Stevens & W.E. Hess No. 3732 [erroneously as 3722 in the protologue], ILL 33561, BPI. **Holotype:** URM [I.M.U.R.] 13394.

Isaria palmae F.L. Stevens & C.M. King. Illinois Biological Monographs 11:211. 1927. **Holotype:** On the inflorescence of palm [Arecaceae], Panama, Frijoles, 20 AUG 1923, leg. F.L. Stevens No. 1206, ILL 16384.

Isipinga areolata E.M. Doidge. Bothalia 1(1):15. 1921. **Isosyntypes:** On leaves of *Euclea natalensis* A. DC. in DC., [South Africa], Natal [Prov.], Isipingo Beach, 21 MAY 1917 [as 1915 on the packet], leg. E.M. Doidge, ex PREM No.10153, ILL 6803; Umgeni, 25 MAY 1915 [as 27 MAY 1915 on the packet], leg. E.M. Doidge, ex PREM No. 8986, ILL 6802.

Isthmospora glabra F.L. Stevens. Botanical Gazette 65:244–245, pl. VI, fig. 18. 1918. **Syntypes:** On *Meliola glabroides* F.L. Stevens on *Nectandra patens* (Sw.) Griseb. [= *Ocotea patens* (Sw.) Nees], Porto Rico [Puerto Rico], Maricao, [20 JUL 1915], leg. F.L. Stevens No. 8973, ILL 4003 [also a paratype of *Meliola glabroides* F.L. Stevens]; on *Simarouba tulae* Urban [as *Simaruba*], Mayagüez, [26 JUN 1915], leg. F.L. Stevens No. 7588, ILL 3965 [also the holotype of *Asteridiella simaroubae* C.G. Hansford]; on *Meliola melastomatacearum* C.L. Spegazzini on *Clidemia hirta* (L.) D. Don, Utuado, [AUG 1915], leg. F.L. Stevens No. 9479, ILL 4145; on *Meliola bicornis* H.G. Winter on *Meibomia supina* (Sw.) Britton [= *Desmodium incanum* DC.], Maricao, [20 JUL 1915], leg. F.L. Stevens No. 8975, ILL 4700 and 16176 [the latter an isosyntype]; on *Meliola psychotriae* F.S. Earle [as var. *gonzalagunae* R. Ciferri on the packet], on *Gonzalagunia spicata* (Lam.) M. Gómez, Rio Arecibo, [8 JUL 1915], leg. F.L. Stevens No. 7793, ILL 6098; Mayagüez, [14 JUN 1915], leg. F.L. Stevens No. 7044, ILL 6099; leg. F.L. Stevens No. 7046, ILL 6100.

Isthmospora spinosa F.L. Stevens. Botanical Gazette 65:244, pl. VI, fig. 17. 1918. **Holotype:** On *Meliola psidii* E.M. Fries:E.M. Fries on *Psidium* *guajava* L., Porto Rico [Puerto Rico], Yauco, [3 OCT 1913], leg. F.L. Stevens No. 3120, ILL 6064 [also an isotype of *Trichomerium portoricense* C.L. Spegazzini]. **Paratypes:** On *Meliola byrsonimae* F.L. Stevens on *Byrsonima lucida* (P. Miller) DC., Guayanilla, [14 JUL 1915], leg. F.L. Stevens No. 3541 [as No. 8541 on the packets], ILL 4790 and 4794 [also the holotype and an isotype, respectively, of *Meliola byrsonimae* F.L. Stevens]; on *Meliola chiococcoae* F.L. Stevens on *Chiococca alba* (L.) A.S. Hitchc., Vega Baja, [2 JUL 1915], leg. F.L. Stevens No. 7743, ILL 4325 [also the holotype of *Meliola chiococcae* F.L. Stevens]; on *Meliola helleri* F.S. Earle on *Myrcia splendens* (Sw.) DC., Jajome Alto, [3 DEC 1913], leg. F.L. Stevens No. 5646, ILL 5233; on *Meliola philodendri* F.L. Stevens on *Philodendron krebsii* Schott [= *P. consanguineum* Schott], Ponce, [8 NOV 1913], leg. F.L. Stevens No. 4346, ILL 5995 [also the holotype of *Helminthosporium philodendri* F.L. Stevens and a paratype of *Meliola philodendri* F.L. Stevens]; on *Meliola praetervisa* A. Gaillard on *Coccoloba sintenisii* Urban ex Lindau [cited as *Coccolobis*], Mayagüez, [15 JUN 1915], leg. F.L. Stevens No. 7066, ILL 6014 [also a paratype of *Meliola praetervisa* A. Gaillard var. *stevensii* C.G. Hansford]; on *Coccoloba pyrifolia* Desf., Mayagüez, [15 JUN 1915], leg. F.L. Stevens No. 7065, ILL 6019 and 6020 [also probable isotypes of *Lembosidium portoricense* C.L. Spegazzini]; on *Meliola psidii* E.M. Fries:E.M. Fries on *Psidium guajava* L. [as *Psidium* sp. on the packet], Jajome Alto, [3 DEC 1913], leg. F.L. Stevens No. 5642a, ILL 6065 and 6088]; on *Meliola smilacis* F.L. Stevens on *Smilax coriacea* Sprengel [= *S. havanensis* Jacquin], Manati, [25 NOV 1913], leg. F.L. Stevens No. 5261, ILL 6283 [also the holotype of *Meliola smilacis* F.L. Stevens].

K

Kabatia fragariae W.G. Solheim. Mycologia 41:628–629. 1949. **Isotype:** On living leaves of *Fragaria ovalis* (Lehm.) Rydb. [= *Fragaria virginiana* P. Miller], Wyoming, Albany County, Laramie Mountains, Happy Jack Picnic Area, [alt. 8400 feet], 8 AUG 1942, leg. W.G. & R. Solheim No. 2114, ILL 20888.

Kalmusia pinicola L. Bonar. Mycologia 57:386–388, figs. 2–6. 1965. **Isotype:** On fallen needles of *Pinus sabiniana* Douglas ex Douglas, California, El Dorado County, Coloma, Sutter's Mill Site, [Marshall Gold Discovery State Park], 16 APR 1962, leg. L. Bonar, California Fungi No. 1268, ILL 33553. **Holotype:** UC 1272190.

Kellermannia biseptis F.E. & E.S. Clements. Cryptogamae Formationum Coloradensium, Century 5, No. 487. Anno 1908, nom. nud. **Isotype:** ... in caulibus vetustis *Leptotaeniae multifidae* [*Leptotaenia multifida* Nutt. ex Torrey & Gray = *Lomatium dissectum* (Nutt. ex Torrey & Gray) Mathias & Constance var. *multifidum* (Nutt. ex Torrey & Gray) Mathias & Constance], Colorado, Silverton, alt. 2800 m, 8 JUL 1907, leg. F.E. & E.S. Clements, ILL 11304.

Kusanoopsis guianensis F.L. Stevens & A.G. Weedon. Mycologia 15:200, figs. 3, 4, 10–15. 1923. **Holotype:** On unknown dicotyledonous host, British Guiana [Guyana], Coverden, 4 AUG 1922, leg. F.L. Stevens No. 818, ILL 33535.

L

Lactarius rubrifulvus F.E. & E.S. Clements. Cryptogamae Formationum Coloradensium, Century 2, No. 184. Anno 1906, nom. nud. **Isotype:** Geophilus ..., Colorado, Mount Palsgrove, alt. 2700 m, 2 SEP 1904, leg. F.E. & E.S. Clements, ILL 30237.

Lactarius villosus F.E. & E.S. Clements. Cryptogamae Formationum Coloradensium, Century 4, No. 364. Anno 1907, nom. nud. **Isotype:** Geophilus ..., Colorado, Ruxton Park, alt. 2900 m, 21 AUG 1906, leg. F.E. & E.S. Clements, ILL 30242.

Laestadia rimula F.E. & E.S. Clements. Cryptogamae Formationum Coloradensium, Century 3, No. 207. Anno 1907, nom. nud. **Isotype:** ... in caulibus emortuis stantibusque *Arenariae fendleri* [= *Arenaria fendleri* A. Gray], Colorado, Lake Moraine, alt. 3000 m, 17 JUL 1906, leg. F.E. & E.S. Clements, ILL 9777.

Lageniforma bambusae O.A. Plunkett in F.L. Stevens. Bernice P. Bishop Museum Bulletin 19:99–101, text figs. 25, 26a. 1925. **Assumed holotype:** On *Bambusa* sp., Hawaii, Kauai, Kalalau trail, 21 JUN [1921], leg. F.L. Stevens No. 489 [as No. 87 on the packet], ILL 9780.

Lateropeltis bambusarum L. Shanor. Mycologia 38:337. 1946. **Isotypes:** On *Arthrostylidium racemiflorum* Steudel, El Salvador, [Mt. Cocoquatique, 8 JAN 1941], leg. John Tucker No. 751B, ILL 31384, FH, UC. **Holotype:** BPI. **Paratype:** Panama, Canal Zone, JUL 1923, leg. H. Johnson No. 17, ILL 31383, ex US 1167472 [also a paratype of *Ciliochorella bambusarum* L. Shanor].

Lembosia coccolobae F.S. Earle. Bulletin of the New York Botanical Garden 3:301–302. 1904. **Isotype:** On upper surfaces of leaves of *Cocco-*

loba uvifera L., Porto Rico [Puerto Rico, Santurce (San Antonio Station), 7 JAN 1903], leg. A.A. Heller No. 6375, ILL 7283.

Lembosia eucalypti F.L. Stevens & H.L. Dixon in F.L. Stevens. Bernice P. Bishop Museum Bulletin 19:75. 1925. **Holotype:** On *Eucalyptus* sp., Hawaii, [Island of] Hawaii, Kilauea, 16 JUL 1921, [as JUN on the packet], leg. F.L. Stevens No. 874, ILL 7286. **Isotypes:** BISH 502014 and 502015.

Lembosia fumago (G. Niessl) H.G. Winter, 1884. **See basionym:** *Meliola fumago* G. Niessl, 1881.

Lembosia pandani H. Rehm. Leaflets of Philippine Botany 8:2932. 1915. **Assumed isotype:** Ad *Pandanum* [= *Pandanus* sp.], Philippines, [Luzon, Laguna Prov.], hills back of Paete, APR 1914, leg. C.F. Baker No. 3113b, [ex Fungi Malayana No. 148], ILL 7290.

Lembosia poasensis H. Sydow. Annales Mycologici 23:397–399. 1925. **Isotype:** Hab. in foliis *Chamaedoreae bifurcatae* [= *Chamaedorea bifurcata* Oersted], Costa Rica, in Monte Poas pr. Grecia, 15 JAN 1925, leg. H. Sydow, Fungi in Itinere Costaricensi Collecti No. 196, ILL 8677 [also an isotype of *Phoenicostroma chamaedoreae* H. Sydow].

Lembosia portoricensis R.W. Ryan. Mycologia 16:190. 1924. **Holotype:** On *Coccoloba laurifolia* sensu auct. non Jacquin [= *C. diversifolia* Jacquin], Porto Rico [Puerto Rico], Santa Ana, [1 JUL 1915], leg. F.L. Stevens No. 7611, ILL 7292. **Isotype:** ILL 7293. **Paratype:** Martin Peña, [11 AUG 1915], leg. F.L. Stevens No. 9716, ILL 7282.

Lembosia rapaneae R.W. Ryan. Mycologia 16:191. 1924. **Holotype:** On *Rapanea* sp., Porto Rico [Puerto Rico], Santa Ana, [1 JUL 1915], leg. F.L. Stevens No. 7610, ILL 7294.

Lembosia sapotae R.W. Ryan. Mycologia 16:191. 1924 [as *sepotae*, corrected in accordance with Articles 60 & 61 of I.C.B.N.]. **Holotype:** On *Sapota* sp. [misspelled "Sepota" in the protologue but correctly spelled on the packet], Porto Rico [Puerto Rico], Manati, [25 NOV 1913], leg. F.L. Stevens No. 5320, ILL 7296.

Lembosidium portoricense C.L. Spegazzini. Boletín de la Academia Nacional de Ciencias en Córdoba 26:342–343, illustr. Preprint 1923 [journal part issued in 1924]. **Probable isotypes:** Hab. Sobre las hojas vivas de *Coccoloba pyrifolia* Desf. [cited as *Coccolobis*], Porto Rico [Puerto Rico], en los alrededores de Mayagüez, [15 JUN 1915], leg. F.L. Stevens No. 7065, ILL 6019 and 6020 [also paratypes of *Isthmospora spinosa* F.L. Stevens]. **Holotype:** LPS 1049.

Lembosina cocculi (F.L. Stevens & R.W. Ryan in F.L. Stevens) J.A. v. Arx, 1962. **See basionym:** *Echidnodella cocculi* F.L. Stevens & R.W. Ryan in F.L. Stevens, 1925.

Lepiota cystidiosa A.H. Smith. Papers of the Michigan Academy of Sciences, Arts, and Letters 27:58–60, plates 1, 2. 1942. **Isotype:** Michigan, Ann Arbor, 9 SEP 1940, leg. A.H. Smith No. 15268, ILL 21358. **Holotype:** MICH.

Leptodiscus terrestris J.W. Gerdemann. Mycologia 45:548–553 (554), figs. 1–7. 1953. **Lectotype** [designated herein]: ... single-spore culture isolated from a diseased red clover root [= *Trifolium pratense* L.], Illinois, [Champaign County], Urbana, Agronomy South Farm, Illinois Agricultural Experiment Station, SEP 1951, leg. J.W. Gerdemann s.n., ILL 31238. **Isolectotype:** BPI. ≡ *Mycoleptodiscus terrestris* (J.W. Gerdemann) S.A. Ostazeski, Mycologia 59:970. (1967) 1968. [*Mycoleptodiscus* Ostazeski, 1968, is a substitute name for *Leptodiscus* J.W. Gerdemann, 1953 (nom. illegit.), non R. Hertwig, 1877.].

Leptomeliola cryptocarpa (J.B. Ellis & G.W. Martin) S.J. Hughes. 1993. See **basionym:** *Meliola cryptocarpa* J.B. Ellis & G.W. Martin, 1883.

Leptomeliola hyalospora (J.H. Léveillé) F. v. Höhnel, 1919. See **basionym:** *Meliola hyalospora* J.H. Léveillé, 1846.

Leptomeliola quercina (N.T. Patouillard) F. v. Höhnel, 1919. See **basionym:** *Meliola quercina* N.T. Patouillard, 1890.

Leptomeliola torta (E.M. Doidge) S.J. Hughes, 1993 (nom. illegit.?). See **basionym:** *Meliola torta* E.M. Doidge, 1917, non E.M. Doidge, char. emend., 1919.

Leptosphaeria astericola J.B. Ellis & B.M. Everhart. Journal of Mycology 8:17. 1902. **Assumed isotype:** On dead stems of *Aster multiflorus* Aiton [as *multiflora* = *A. ericoides* L.], Kansas, Rooks County, JUN 1901, leg. E. Bartholomew No. 2885 [ex J.B. Ellis and B.M. Everhart, Fungi Columbiani No. 1537], ILL 9989. **Holotype:** NY.

Leptosphaeria chrysanthemi F.E. & E.S. Clements. Cryptogamae Formationum Coloradensium, Century 1, No. 24. Anno 1906, nom. nud. **Isotype:** ... in caulibus vetustis *Cardui scopulorum* [*Carduus scopulorum* E. Greene = *Cirsium scopulorum* (E. Greene) Cockerell ex Daniels], Colorado, Cabin Canyon, alt. 2700 m, 12 JUL 1905, leg. F.E. & E.S. Clements, ILL 9992.

Leptosphaeria clivensis (M.J. Berkeley & C.E. Broome) P.A. Saccardo var. *constricta* F.E. & E.S. Clements. Cryptogamae Formationum Coloradensium, Century 5, No. 433. Anno 1908, nom. nud. **Isotype:** ... in caulibus vetustis *Senecionis atrati* [= *Senecio atratus* E. Greene], Colorado, Spanish Peaks, alt. 3800 m, 20 JUN 1907, leg. F.E. & E.S. Clements, ILL 9993.

Leptosphaeria erigerontis F.E. & E.S. Clements. Cryptogamae Formationum Coloradensium, Century 1, No. 28. Anno 1906, nom. nud. **Isotype:** ... in caulibus vetustis *Erigerontis viscidui* [*Erigeron viscidus* Rydb. = *E. formosissimus* E. Greene var. *viscidus* (Rydb.) Cronquist],

Colorado, Cabin Canyon, alt. 2700 m, 12 AUG 1905, leg. F.E. and E.S. Clements, ILL 10009.

Leptosphaeria fuscella (M.J. Berkeley & C.E. Broome) V. Cesati & G. de Notaris var. *sydowiana* P.A. Saccardo in H. & P. Sydow. Annales Mycologici 4:484. (1906) JAN 1907. **Isotype:** Hab. in ramis *Hippophaes rhamnoidis* [= *Hippophae rhamnoides* L., Germany], Rüdersdorfer Kalkberge pr. Berolinum, [28 MAY 1905], leg. H. Sydow, ex H. & P. Sydow, Mycotheca Germanica No. 485, ILL 10010.

Leptosphaeria lythri C.H. Peck. Bulletin of the Torrey Botanical Club 33:220–221. 1906. **Isotype:** On dead stems of *Lythrum alatum* Pursh, Kansas, Stockton, [2] OCT [1905], leg. E. Bartholomew, [Fungi Columbiani No. 2229], ILL 10016.

Leptosphaeria modestula F.E. & E.S. Clements. Cryptogamae Formationum Coloradensium, Century 3, No. 237. Anno 1907, nom. nud. **Isotype:** ... in caulibus emortuis *Geranii richardsonii* [= *Geranium richardsonii* Fischer & Trautv.], Colorado, Larkspur Dell, alt. 2500 m, 15 AUG 1906, leg. F.E. & E.S. Clements, ILL 10022.

Leptostroma vestita A.B. Seymour & F.W. Patterson in C.G. Pringle. Mexican Fungi, Decade 1, No. 10. Anno 1896; Botanical Gazette 22:423. 1896. **Isotypes:** On leaves of *Agave vestita* S. Watson, Mexico, barranca, near Guadalajara, MAY 1891, leg. C.G. Pringle, ILL 13137, 13138 and 33204.

Leptothyrium gleicheniae F.L. Stevens & P.A. Young in F.L. Stevens. Bernice P. Bishop Museum Bulletin 19:143. 1925. **Syntypes** [additional ILL number and BISH numbers cited are isosyntypes]: On living leaves of *Gleichenia longissima* Blume [= *G. glauca* (Thunb.) Hooker], Hawaii, Oahu, Wahiawa, 31 MAY [1921], leg. F.L. Stevens No. 153, ILL 7797 and 13144, BISH 146929 and 499884; on *Gleichenia* sp., Maui, Pogue's ditch trail, 6 SEP [1921], leg. F.L. Stevens No. 1158, ILL 7796.

Leptothyrium pothi A.G. Weedon in F.L. Stevens. Bernice P. Bishop Museum Bulletin 19:143. 1925. **Holotype:** On cultivated *Pothos* sp., Hawaii, [Island of] Hawaii, Kapapala Ranch, 18 JUL [1921], leg. F.L. Stevens No. 883, ILL 13157. **Isotypes:** BISH 499882 and 499883. **Paratype:** Oahu, Honolulu, 20 MAY [1921], leg. F.L. Stevens No. 26, ILL 13156.

Leptothyrium sidae F.L. Stevens & P.A. Young in F.L. Stevens. Bernice P. Bishop Museum Bulletin 19:142–143, text figs. 30a, b. 1925. **Syntypes** [BISH numbers cited are isosyntypes]: On living leaves of *Sida spinosa* L., Hawaii, [Island of] Hawaii, Kealakekua [as Bishop Estate Road, Keauhou, Kona on the packet], 21 JUL [1921], leg. F.L. Stevens No. 912, ILL 13159, BISH 499879 and 499880; Maui, Iao Valley, 7 SEP

[1921], leg. F.L. Stevens No. 1152, ILL 13158, BISH 145931 and 499881.

Leveillinopsis palmicola F.L. Stevens. Illinois Biological Monographs 8:179, pl. II, figs. 15–16. 1923. **Holotype:** [On] unknown species of palm [Arecaceae], British Guiana [Guyana], Kartabo, 24 JUL 1922, leg. F.L. Stevens No. 674, ILL 8617.

Limacinia biseptata P.A. Saccardo. Annales Mycologici 13:127. 1915. **Assumed isotype:** Hab. in foliis emorientibus *Macarangae* sp. [as *Macaranga tanarius* (L.) Muell. Arg. on the packet], Philippines, [Luzon, Laguna Prov., Mount Maquiling, near] Los Baños, JAN 1914, leg. C.F. Baker No. 2583 [ex Fungi Malayana No. 150], ILL 6662.

Limaciniella psidii J.M. Mendoza in F.L. Stevens. Bernice P. Bishop Museum Bulletin 19:58, pl. IV(38–40). 1925. **Holotype:** On *Psidium guajava* L. [cited as *guayava*], Hawaii, Kauai, Waimea, 16 JUN [1921], leg. F.L. Stevens No. 542, ILL 6661 [also the holotype of *Deslandesia ficina* (H. Sydow) A.C. Batista var. *microspora* A.C. Batista & A.F. Vital]. **Isotypes:** BISH 499935 and 499936.

Limaciniopsis rollandiae J.M. Mendoza in F.L. Stevens. Bernice P. Bishop Museum Bulletin 19:58, pl. IV(34–37). 1925. **Holotype:** On *Rollandia racemosa* (H. Mann) Hillebr. [= *R. humboldtiana* Gaudich.], Hawaii, Oahu, Waiahole ditch trail, 12 JUN [1921], leg. F.L. Stevens No. 407, ILL 6663 [also the holotype of *Philonectria insignis* (F. Petrak & R. Ciferri) J.A. v. Arx var. *macrospora* A.C. Batista & A.F. Vital]. **Isotypes:** BISH 499970 and 499971.

Linospora trichostigmatis F.L. Stevens. Botanical Gazette 70:399–401, fig. 2. 1920 [as *trichostigmae*, corrected in accordance with Articles 32.6, 60 & 61 of I.C.B.N.]. **Holotype:** On *Trichostigma octandrum* (L.) H. Walter, Porto Rico [Puerto Rico], Guayanilla, [13 NOV 1913], leg. F.L. Stevens No. 5924, ILL 10339.

Lophionema apoclastosporum W.G. Solheim. Mycologia 41:624–625. 1949 [as *apoclastospora*, corrected in accordance with Articles 32.6, 60 & 61 of I.C.B.N.]. **Isotype:** On dead decorticated stems of *Salix* sp., Wyoming, Albany County, Medicine Bow Mountains, near road camp on Libby Creek, 27 JUN 1942, leg. W.G. Solheim No. 2025, ex Mycoflora Saximontanensis Exsiccata No. 430, ILL 20940.

Lophiostoma lophis F.E. & E.S. Clements. Cryptogamae Formationum Coloradensium, Century 5, No. 463. Anno 1908, nom. nud. **Isotype:** ... in ramulis dejectis *Lepargyraeae canadensis* [*Lepargyraea canadensis* (L.) E. Greene = *Shepherdia canadensis* (L.) Nutt.], Colorado, Sulphur Springs, alt. 2400 m, 23 JUL 1907, leg. F.E. & E.S. Clements, ILL 9650.

Lophiotrema vagabundum P.A. Saccardo var. *hydrolapathi* P.A. Saccardo in H. & P. Sydow. Annales Mycologici 3:232. 1905. **Isotype:** Hab. in caulibus emortuis *Rumicis hydrolapathi* [= *Rumex hydrolapathum* Hudson], Germania, [Brandenburg], Wannsee pr. Berlin, [9 OCT 1904], leg. H. Sydow, ex H. & P. Sydow, Mycotheca Germanica No. 322, ILL 9659.

Lophodermellina stevensii L.R. Tehon. Illinois Biological Monographs 13: 311–312. 1935, nom. invalid. sine diagn. lat. **Holotype:** On *Vincentia angustifolia* Gaudich. [= *Machaerina angustifolia* (Gaudich.) T. Koyama], Hawaii, Oahu, Mt. Olympus, [24 JUN 1921], leg. F.L. Stevens No. 727, ILL 7570. **Paratypes:** [Wahiawa, 3 JUN 1921], leg. F.L. Stevens No. 246, ILL 7573; [Palolo Valley, Mt. Olympus, 10 JUN 1921], leg. F.L. Stevens No. 373, ILL 7574; [Tantalus, 22 JUN 1921], leg. F.L. Stevens No. 622, ILL 7572, leg. F.L. Stevens No. 652, ILL 7569. ≡ *Lophodermium stevensii* L.R. Tehon ex C.A. Terrier. Beiträge zur Kryptogamenflora der Schweiz 9(2):32. 1942.

Lophodermium acacicolum L.R. Tehon. Illinois Biological Monographs 13:269–270. 1935, nom. invalid. sine diagn. lat. **Holotype:** On *Acacia koa* A. Gray, Hawaii, Oahu, Wahiawa, [3 JUN 1921], leg. F.L. Stevens No. 234, ILL 7580. **Paratype:** Maui, [Pogue's ditch trail, 6 SEP 1921], leg. F.L. Stevens No. 1156, ILL 7579.

Lophodermium amplum J.J. Davis. Transactions of the Wisconsin Academy of Sciences, Arts and Letters 19(2):695–696. 1919. **Isosyntype:** On leaves of *Pinus banksiana* Lambert, Wisconsin, Millston, [24 JUN 1916], leg. J.J. Davis [Fungi Wisconsinenses Exsiccati No. 70], ILL 7561.

Lophodermium andropogonis L.R. Tehon. Illinois Biological Monographs 13:271. 1935, nom. invalid. sine diagn. lat. **Holotype:** On *Andropogon bicornis* L., Costa Rica, Peralta, [no date], leg. F.L. Stevens No. 463, ILL 7571.

Lophodermium miscanthi L.R. Tehon. Illinois Biological Monographs 13: 284, pl. II, fig. 8; pl. III, fig. 1. 1935, nom. invalid. sine diagn. lat. **Holotype:** Host: *Miscanthus sinensis* N.J. Andersson [the packet labeled *Lophodermium arundinaceum* F.F. Chevallier forma *vulgare* L. Fuckel, on dead *Miscanthus japonicus* sensu auct. non (Thunb.) N.J. Andersson], Philippines, [Luzon], Laguna Prov., Mount Maquiling, near Los Baños, [June 1914], leg. C.F. Baker, ex Fungi Malayana No. 155, ILL 7568.

Lophodermium planchoniae H. Rehm. Leaflets of Philippine Botany 8:2925. 1915. **Assumed isotype:** Ad folium *Planchoniae spectabilis* [= *Planchonia spectabilis* Merrill], Philippines, [Luzon, Laguna Prov.], Los Baños, [Mount Maquiling], APR 1914, leg. C.H. Baker No. 3080 [ex Fungi Malayana No. 156], ILL 7593.

Lophodermium rotundatum H. & P. Sydow. Annales Mycologici 12:201.

1914. **Assumed isotype:** Hab. in foliis emortuis *Dilleniae* spec. [= *Dillenia* sp.], Philippines, [Luzon], Laguna Prov., [Mount Maquiling], near Los Baños, 1 DEC 1913, leg. C.F. Baker No. 2099 [ex Fungi Malayana No. 39], ILL 7595.

Lophodermium stevensii L.R. Tehon ex C.A. Terrier. Beiträge zur Kryptogamenflora der Schweiz 9(2):32. 1942. **Based on:** *Lophodermellina stevensii* L.R. Tehon, 1935, nom. invalid sine diagn. lat., the latter name cited pro syn. **Holotype:** On *Vincentia angustifolia* Gaudich. [= *Machaerina angustifolia* (Gaudich.) T. Koyama], Hawaii, Oahu, Mt. Olympus, [24 JUN 1921], leg. F.L. Stevens No. 727, ILL 7570.

Lophodermium thujae J.J. Davis. Transactions of the Wisconsin Academy of Sciences, Arts and Letters 20:424–425, fig. 2. 1921 [cited as *thuyae*, corrected in accordance with Rec. 60H.1 and Articles 60 & 61 of I.C.B.N.]. **Isotype:** On leaves of *Thuja occidentalis* L. [cited as *Thuya* (orth. var.)], Wisconsin, Saxon, 16 AUG 1919, leg. J.J. Davis [Fungi Wisconsinenses Exsiccati No. 114], ILL 7597.

Lophotrichus ampullus R.K. Benjamin. Mycologia 41:347–349. 1949. **Holotype:** Isolated from goat dung, Illinois, Champaign County, Urbana, NOV 1947, leg. R.K. Benjamin s.n., ILL 1763. **Isotypes:** ILL 1763b, c, d, BPI, FH, ISC.

Lophotrichus martinii R.K. Benjamin. Mycologia 41:349–353, figs. 17–33. 1949. **Holotype:** Isolated from [rabbit?] dung, Peru, Talara, leg. G.W. Martin No. 6290, 5 SEP 1945, ILL 1762. **Isotypes:** ILL 1762b, c, BPI, FH, ISC.

Lyophyllum urbanense A.H. Smith & H. Clémençon. Nova Hedwigia 14:127, figs. 1–9, 15–17. 1968. **Isotypes:** Illinois, Urbana, [Crystal Lake Park], 1 NOV 1966, leg. H. Clémençon No. 661101, ILL 33536, MICH.

M

Macrophoma parca J.B. Ellis & B.M. Everhart. North American Fungi, Series II, Century 22, No. 2159. Anno 1889, nom. nud. **Isotype:** On leaves of *Abies grandis* (Douglas ex D. Don) Lindley, Colorado, Sangre de Christo Range, JUL 1888, leg. C.H. Demetrio, ILL 11325.

Macrophoma subiculis F.E. & E.S. Clements. Cryptogamae Formationum Coloradensium, Century 5, No. 474. Anno 1908, nom. nud. **Isotype:** ... in caulibus et pedunculis emortuis *Gutierreziae sarothrae* [= *Gutierrezia sarothrae* (Pursh) Britton & Rusby], Colorado, Durango, alt. 2000 m, 3 JUL 1907, leg. F.E. & E.S. Clements, ILL 11330.

Macrophoma tiliacea C.H. Peck. Bulletin of the Torrey Botanical Club 34:348. 1907. **Isotype:** On branches of ...*Tilia americana* L., Ohio, Oberlin [College campus, 15] MAR [1907], leg. F.O. Grover [ex Oberlin College Herb. No. 471], ILL 11331.

Macrophomella pandani H. Diedicke. Annales Mycologici 14:63. 1916. **Assumed isotype:** Auf Früchten von *Pandanus luzonensis* Merrill, Philippines, [Luzon, Laguna Prov.], Mount Maquiling, [near Los Baños], 1 APR 1914, leg. C.F. Baker No. 3160 [ex Fungi Malayana No. 157], ILL 11332.

Macrophomopsis dracaenae F.L. Stevens & Baechler in A.G. Weedon. Mycologia 18:222. 1926. **Syntypes:** On *Dracaena* sp., Costa Rica, Guapiles, 18 JUL 1923, leg. F.L. Stevens No. 508, ILL 11334; on [South African] *Iris* sp. (cult.), Florida, St. Petersburg, 30 MAR 1923, leg. A.G. Weedon No. 9, ILL 11333.

Marasmius nucicola W.B. McDougall. Transactions of the Illinois State Academy of Science 17:84. 1925. **Lectotype** [designated by D.E. Desjardin, in litt., 1992, and accepted herein]: Illinois, Urbana, University Woods, 1924, leg. W.B. McDougall s.n., ILL 31106.

Marasmius tritici P.A. Young. Phytopathology 15:118, figs. 1–5. 1925. **Holotype:** Parasitic on culms of *Triticum vulgare* Vill. [= *T. aestivum* L.], Illinois, [Knox County], Abingdon, 12 JUL 1924, leg. P.A. Young, Plant Disease Survey No. 18116, ILL 30170. **Isotype:** ILL 30171.

Maravalia ingae H. Sydow. Mycologia 17:257. 1925. **Holotype:** On *Inga* sp., British Guiana [Guyana], Vreed-en-Hoop, 1 AUG 1922, leg. F.L. Stevens No. 715, ILL 28263. **Paratype:** Guyana [erroneously cited as Trinidad], Coverden, 8 AUG 1922, leg. F.L. Stevens No. 790, ILL 28262. ≡ *Bitzea ingae* (H. Sydow) E.B. Mains. Mycologia 31:38–39. 1939. ≡ *Chaconia ingae* (H. Sydow) G.B. Cummins. Mycologia 48:602. 1956.

Maravalia pura (H. Sydow) E.B. Mains, 1939. **See basionym:** *Argomycetella pura* H. Sydow, 1925.

Maravalia utriculata H. Sydow. Annales Mycologici 23:314. 1925. **Isotype:** Hab. in foliis *Ingae* species ex affinitato *Ingae verae* [= affin. *Inga vera* Willd.], Costa Rica, La Caja pr. San José, 6 JAN 1925, leg. H. Sydow, Fungi in Itinere Costaricensi Collecti No. 279, ILL 28264.

Marssonia chamaenerii F.G.E. Rostrup var. *germanica* H. &. P. Sydow. Annales Mycologici 2:529. 1904. **Isotype:** Hab. in foliis vivis *Epilobii hirsuti* [= *Epilobium hirsutum* L., Germany], Thuringia, pr. Sondershausen, [20 JUN 1903], leg. G. Oertel, ex H. & P. Sydow, Mycotheca Germanica No. 278, ILL 13666.

Marssonia extremorum H. & P. Sydow. Annales Mycologici 2:192. 1904. **Isotype:** Hab. in foliis siccis *Acori calami* [= *Acorus calamus* L., [Germany], Anhalt: Kühnauer See pr. Dessau, [OCT 1903], leg. R. Staritz, ex H. & P. Sydow, Mycotheca Germanica No. 144, ILL 13674.

Masonia crescentiae C.G. Hansford. Sydowia Annales Mycologici 10:97. (1956) 1957. **Isotype**: Hab. in foliis *Crescentiae* sp. [= *Crescentia* sp.], Trinidad, Cumuto, [18 AUG 1922], leg. F.L. Stevens No. 940 p.p., ILL 4992 [also the assumed lectotype of *Meliola crescentiae* F.L. Stevens]. **Holotype**: FH.

Massalongiella canavaliae F.L. Stevens & P.A. Young in F.L. Stevens. Bernice P. Bishop Museum Bulletin 19:98. 1925. **Holotype**: On dead stems of *Canavalia* sp., Hawaii, Oahu, Honolulu, 16 APR 1913, leg. H.L. Lyon No. 312, ILL 9781. **Isotype**: BISH 499036.

Massaria bataanensis H. Rehm. Leaflets of Philippine Botany 8:2951. 1916. **Assumed isotype**: Ad ramum *Eugeniae bataanensis* [= *Eugenia bataanensis* Merrill], Philippines, [Luzon, Laguna Prov.], Mount Maquiling, [near Los Baños], MAY 1914, leg. C.G. Baker No. 3481b [ex Fungi Malayana No. 127b (in exsiccati set as No. 127a, b, on the same packet label)], ILL 9362 [also, under 127a, an assumed isotype of *Clypeosphaeria bakeriana* H. Rehm].

Massaria sieversiae F.E. & E.S. Clements. Cryptogamae Formationum Coloradensium, Century 3, No. 234. Anno 1907, nom. nud. **Isotype**: ... in caulibus emortuis stantibusque *Sieversiae turbinatae* [*Sieversia turbinata* (Rydb.) E. Greene = *Geum rossii* (R. Br.) Ser. var. *turbinatum* (Rydb.) C.L. Hitchc.], Colorado, Bottomless Pit, alt. 3600 m, 13 JUL 1906, leg. F.E. & E.S. Clements, ILL 10236.

Massarinula bambusincola H. Rehm. Leaflets of Philippine Botany 8:2944. 1916. **Assumed isotype**: Ad emortuam *Bambusam vulgarem* [= *Bambusa vulgaris* Schrader ex J.C. Wendl.], Philippines, [Luzon, Laguna Prov., Mount Maquiling], near Los Baños, OCT 1913, leg. S.A. Reyes, comm. C.F. Baker No. 1915b [ex Fungi Malayana No. 139b (in exsiccati set as 139a, b, c, on the same packet label)], ILL 9668 [also, under No. 139a, an assumed isosyntype of *Guignardia bambusina* H. Rehm, and under 139c, an assumed isotype of *Didymella eutypoides* H. Rehm].

Mazzantia arundinellae F.L. Stevens. Illinois Biological Monographs 11:202, pl. X, figs. 78–80; pl. XVII, fig. 120. 1927. **Holotype**: On *Arundinella hispida* (Willd.) O. Kuntze, Costa Rica, Peralta, 12 JUL 1923, leg. F.L. Stevens No. 343, ILL 8345.

Melampsorella blechni H. & P. Sydow. Annales Mycologici 1:537. 1903. **Isotype**: Hab. in frondibus *Blechni spicant* [= *Blechnum spicant* (L.) Roth, Germany], Sachsen, Grosser Winterberg ... pr. Schmilka, [28 AUG 1903], leg. H. & P. Sydow, Mycotheca Germanica No. 61, ILL 18326.

Melampsorella dieteliana H. & P. Sydow. Annales Mycologici 1:537. 1903. **Isotype**: Hab. in frondibus *Polypodii vulgaris* [= *Polypodium vulgare*

L., Germany], Sachsen, Grosser Winterberg ... pr. Schmilka, [26 AUG 1903], leg. H. & P. Sydow, Mycotheca Germanica No. 62, ILL 18335.

Melanconis everhartii J.B. Ellis in J.B. Ellis and B.M. Everhart. Bulletin of the Torrey Botanical Club 10:117. 1883. **Isotype:** On dead maple limbs [= *Acer* sp.], Pennsylvania, West Chester, JUN 1882, [leg. H.H. Haines and B.M. Everhart, North American Fungi No. 1565], ILL 13771.

Melanconium gracile J.B. Ellis & B.M. Everhart. Journal of Mycology 1:44. 1885. **Isotype:** On limbs of dead hickory [= *Carya* sp.], New Jersey, Plainfield, leg. G.F. Meschutt [ex J.B. Ellis and B.M. Everhart, North American Fungi No.2864], ILL 13677. **Holotype:** NY.

Melanomma schizosporum F.E. & E.S. Clements. Cryptogamae Formationum Coloradensium, Century 5, No. 436. Anno 1908, nom. nud. **Isotype:** ... in caulibus emortuis *Chrysothamni graveolentis* [*Chrysothamnus graveolens* (Nutt.) E. Greene = *C. nauseosus* (Pallas ex Pursh) Britton ssp. *graveolens* (Nutt.) Piper], Colorado, Fort Garland, alt. 2400 m, 23 JUN 1907, leg. F.E. & E.S. Clements, ILL 9393.

Melanomma sporadicum J.B. Ellis & B.M. Everhart. The North American Pyrenomycetes. A contribution to Mycologic Botany, p. 186. 1892. **Isotype:** On decorticated wood of *Platanus* sp., Canada, [Ontario, London, JUL 1891], leg. J. Dearness [ex J.B. Ellis and B.M. Everhart, North American Fungi No. 2753], ILL 9394.

Melanopsamma chrysothamni F.E. & E.S. Clements. Cryptogamae Formationum Coloradensium, Century 5, No. 427. Anno 1908, nom. nud. **Isotype:** ... in caulibus emortuis *Chrysothamni graveolentis* [*Chrysothamnus graveolens* (Nutt.) E. Greene = *C. nauseosus* (Pallas ex Pursh) Britton ssp. *graveolens* (Nutt.) Piper], Colorado, Fort Garland, alt. 2400 m, 23 JUN 1907, leg. F.E. & E.S. Clements, ILL 9396.

Melanopsichium pennsylvanicum E. Hirschhorn. Notas del Museo de La Plata, Buenos Aires 6:149–151, fig. 2b. 1941. **Paratype:** On *Polygonum lapathifolium* L., Illinois, Urbana, [8] SEP 1892, [leg. G.P. Clinton No. 11281], C.L. Spegazzini No. 3137, ILL 17094.

Melasmia coccolobae F.L. Stevens. Transactions of the Illinois State Academy of Science 10:197. 1917 [as *coccolobiae*, corrected in accordance with current I.C.B.N. (cf. F.C. Deighton, 1968)]. **Holotype:** On *Coccoloba* sp. [as *Coccolobis*], Porto Rico [Puerto Rico], Maricao, 1915, leg. F.L. Stevens No. 3712, ILL 13168.

Melasmia empetri P. Magnus. Berichte der Deutschen Botanischen Gesellschaft 4:104–107, figs. 1–3. 1886. **Isotype:** Auf *Empetrum nigrum* L., [Poland], Insel Wollin, Pritter Wald, Misdroy, AUG 1884, leg. P. Magnus [No. 44], ILL 13169.

Melasmia ingae F.L. Stevens. Transactions of the Illinois State Academy of

Science 10:197–198. 1917. **Holotype**: On *Inga laurina* (Sw.) Willd., Porto Rico [Puerto Rico], Las Marias, [22 MAR 1913], leg. F.L. Stevens No. 423, ILL 13182.

Melasmia thouiniae H. & P. Sydow. Annales Mycologici 2:171–172. 1904. **Isosyntype**: Hab. in foliis vivis *Thouiniae pringlei* [= *Thouinia pringlei* S. Watson], Mexico, [barranca, Cuernavaca, 11 NOV 1895], leg. C.G. Pringle, ILL 13186.

Meliola abrupta H. & P. Sydow. Annales Mycologici 15:181. 1917. **Isotype**: Hab. in foliis *Derridis* sp. [= *Derris* sp.], Philippines, Luzon, Rizal Prov., DEC 1915, leg. M. Ramos, ex Phil. Bur. Sci. No. 24068, ILL 4503. **Paratype**: On *Derris diadelpha* (Blanco) Merrill [= *D. heptaphylla* (L.) Merrill], NOV 1915, leg. M. Ramos, ex Phil. Bur. Sci. No. 23904, ILL 4504.

Meliola acalyphae H. Rehm. Philippine Journal of Science, Section C (Botany) 8:252. 1913. **Isotype**: Ad folia *Acalyphae stipulaceae* [= *Acalypha stipulacea* Klotzsch], Philippines, Luzon, Laguna Prov., Los Baños, [30 NOV 1912], leg. C.F. Baker No. 483, ILL 3541. ≡ *Amazonia acalyphae* (H. Rehm) F. Theissen. Annales Mycologici 14:407. 1916. ≡ *Asteridiella acalyphae* (H. Rehm) C.G. Hansford ex C.G. Hansford. Beihefte zur Sydowia Annales Mycologici 2:208. 1961.

Meliola acervata J.B. Ellis & B.M. Everhart. Bulletin of the Torrey Botanical Club 24:126. 1897. **Isotypes**: On leaves of *Physalis peruviana* L. [host misidentified as *Alphitonia ponderosa* Hillebr. on one of the packets], Hawaii, Kauai, [Kaholua-Manoa, above Waimea, 30 AUG 1895], leg. A.A. Heller No. 2773, ILL 3764 [ex MO 15350] and probably also ILL 3769 [ex FH, lacking the collection number on the packet]. ≡ *Irene acervata* (J.B. Ellis & B.M.Everhart) C.G. Hansford. Sydowia Annales Mycologici 9:6. 1955. ≡ *Asteridiella acervata* (J.B. Ellis & B.M. Everhart) C.G. Hansford ex C.G. Hansford. Beihefte zur Sydowia Annales Mycologici 2:638–639. 1961.

Meliola acristae C.G. Hansford. Beihefte zur Sydowia Annales Mycologici 1:99. 1957. **Holotype**: Hab. in foliis *Acristae monticolae* [*Acrista monticola* O.F. Cook = *Prestoa montana* (Graham) G. Nicolson], Porto Rico [Puerto Rico, El Alto de la Bandera], leg. F.L. Stevens No. 8303a, ILL 5150. **Paratype**: [Luquillo Forest, 4 DEC 1913], leg. F.L. Stevens No. 5400, ILL 5142.

Meliola acrotricha H. Sydow. Leaflets of Philippine Botany 9:3113–3114. 1925. **Isotype**: On leaves of *Trigonachras membranacea* Radlk., [Philippines], Luzon, Sorsogon Prov., Irosin, JUN 1916, leg. A.D.E. Elmer No. 16426, ILL 4506.

Meliola acutiseta H. & P. Sydow. Leaflets of Philippine Botany 6:1921. 1913 [as *acutisecta*, a typographical error corrected by F.L. Stevens

(1928:285) in accordance with current I.C.B.N.]. **Isotype**: On the under side of the old leaves of *Persea pyriformis* A.D.E. Elmer, Philippines, Mindanao [Island], Cabadbaran, (Mt. Urdaneta), JUL 1912, leg. A.D.E. Elmer No. 13312, ILL 4507.

Meliola adelphica (H. Sydow) F. Petrak in H. Sydow & F. Petrak, 1929. **See basionym**: *Appendiculella adelphica* H. Sydow, 1926.

Meliola aegiphilae F.L. Stevens. Annales Mycologici 26:208, pl. III, fig. 27. 1928. **Holotype**: On *Aegiphila* sp., British Guiana [Guyana], Tumatumari, 12 AUG 1922, leg. F.L. Stevens No. 221, ILL 4508. ≡ *Meliola cookeana* C.L. Spegazzini var. *aegiphilae* (F.L. Stevens) C.G. Hansford. Beihefte zur Sydowia Annales Mycologici 2:691. 1961.

Meliola aethiops P.A. Saccardo. Bullettino dell' Orto Botanico della Universita di Napoli 6:41–42. 1921. **Isotypes**: Hab. in foliis vivis *Cassiae fistulae* [= *Cassia fistula* L., Singapore], AUG 1917, leg. C.F. Baker No. 5165, [ex Fungi Malayana, No. 449], ILL 4534 [as a microscopic preparation], K, S.

Meliola agelaeae F.L. Stevens & E.F. Roldan. Philippine Journal of Science 56:64, fig. 3g. 1935, nom. invalid. sine diagn. lat., non C.G. Hansford, 1938. **Holotype**: On *Agelaea* sp., Philippines, Luzon, Tayabas, Quezon Forest Park, 30 NOV 1930, leg. F.L. Stevens No. 439, ILL 4512. [This name is cited pro syn. under *Meliola agelaeae* C.G. Hansford var. *philippinense* C.G. Hansford, 1963, and ILL 4512 is also the holotype of the varietal name].

Meliola agelaeae C.G. Hansford var. *philippinensis* C.G. Hansford. Sydowia Annales Mycologici 16:316. (1962) 1963. **Holotype**: Hab. in foliis *Agelaeae* spec. indet. [= *Agelaea* sp.], Philippines, [Luzon, Tayabas, Quezon Forest Park, 30 NOV 1930], leg. F.L. Stevens No. 439, ILL 4512 [also the holotype of *Meliola agelaeae* F.L. Stevens & E.F. Roldan, 1935, nom. invalid. sine diagn. lat., non C.G. Hansford, 1938].

Meliola aglaiae H. & P. Sydow. Philippine Journal of Science, Section C (Botany) 9:159–160. 1914. **Isotype**: On leaves of *Aglaia* sp., [Philippines], Palawan [Island], Taytay, MAY 1913, leg. E.D. Merrill No. 8884, ILL 5949.

Meliola aibonitensis F.L. Stevens. Illinois Biological Monographs 2:484–485. 1916. **Holotype**: On unknown dicotyledonous host [identified as *Daphnopsis* sp. in C.G. Hansford, 1961], Porto Rico [Puerto Rico], Aibonito, 16 JUL 1915, leg. F.L. Stevens No. 8470, ILL 3890. ≡ *Irene aibonitensis* (F.L. Stevens) R.A. Toro. Mycologia 17:140. 1925. ≡ *Irenina aibonitensis* (F.L. Stevens) F.L. Stevens. Annales Mycologici 25:451. 1927. ≡ *Asteridiella aibonitensis* (F.L. Stevens) C.G. Hansford ex C.G. Hansford. Beihefte zur Sydowia Annales Mycologici 2:95–96. 1961.

Meliola alangii H. & P. Sydow. Annales Mycologici 14:355. 1916. **Assumed residual isosyntype** [perhaps an isolectotype]: Hab. in foliis *Alangii* *begoniaefolii* [*Alangium begoniifolium* (Roxb.) Baillon = *A. chinense* (Lour.) Rehder], Philippines, [Luzon, Laguna Prov., Mount Maquiling, near] Los Baños, DEC 1915/JAN 1916, leg. C.F. Baker No. ? [ex Fungi Malayana No. 247], ILL 4516 [as a microscopic preparation. Both F.L. Stevens (1928) and C.G. Hansford (1961) cite C.F. Baker No. 4019 as "type" (interpreted as lectotypification), which leaves Baker No. 4052 as residual syntype. Fungi Malayana No. 247 almost certainly represents one of the two collections.].

Meliola alchorneae F.L. Stevens & L.R. Tehon. Mycologia 18:12, pl. 2, fig. 13. 1926. **Holotype**: On *Alchornea* sp., British Guiana [Guyana], Tumatumari, 11 JUL 1922, leg. F.L. Stevens No. 198, ILL 3891.

Meliola alibertiae F.L. Stevens. Annales Mycologici 26:258, pl. V, fig. 67. 1928. **Holotype**: On *Alibertia edulis* (L.C. Richard) A. Richard in DC., Panama, Las Cruces trail, 1 SEP 1924, leg. F.L. Stevens No. 145, ILL 4521.

Meliola aliena H. & P. Sydow. Leaflets of Philippine Botany 5:1535–1536. 1912. **Isosyntypes**: On fallen twigs [the host later identified as *Afzelia rhomboidea* (Blanco) S. Vidal], Philippines, [Island of] Palawan, [Palawan Prov., Brooks Point (Addison Peak)], MAY 1911 [as FEB on the packet], leg. A.D.E. Elmer No. 12586, ILL 4522 and 4523; Puerto Princesa, (Mt. Pulgar), MAR 1911, leg. A.D.E. Elmer No. 12812, ILL 4525. [The inscription on the packet of ILL 4524, a specimen of Phil. Bur. Sci. No. 21786 that is cited as "type" by C.G. Hansford (1961), indicates that the latter collection (Luzon, Bulacan Prov., Angat, SEP 1913, leg. M. Ramos) was made after publication of the protologue and cannot be used for lectotypification.].

Meliola alocasiae H. Sydow. Leaflets of Philippine Botany 9:3114. 1925. **Isotype**: On leaves of *Alocasia vulcanica* A.D.E. Elmer [= *A. maquilingensis* Merrill, Philippines], Luzon, Sorsogon Prov., Irosin, JUN 1916, leg. A.D.E. Elmer No.16333, ILL 4526.

Meliola alyxiae F.L. Stevens. Bernice P. Bishop Museum Bulletin 19:30, 32, text fig. 7d. 1925. **Holotype**: On *Alyxia oliviformis* Gaudich. [cited as *olivaeformis*], Hawaii, Oahu, Wahiawa, 3 JUN [1921], leg. F.L. Stevens No. 217, ILL 4530. **Isotypes**: BISH 499034 and 499969. **Paratypes**: [Island of] Hawaii, Hamakua, Upper ditch trail, 31 JUL [1921], leg. F.L. Stevens No. 1062, ILL 4520; leg. F.L. Stevens No. 1075, ILL 4517; Keauhou, Kona, Bishop Estate Road, 25 JUL [1921]; leg. F.L. Stevens No. 975, ILL 4518; Puna, 9 JUL [1921], leg. F.L. Stevens No. 756, ILL 4519; Kauai, Kalalau trail, 16 JUN [1921]; leg. F.L. Stevens No. 514, ILL 4532; Oahu, Ahren's ditch trail, 8 JUN [1921], leg. F.L. Stevens No. 409, ILL 4531; leg. F.L. Stevens No. 985, ILL 6877 [also a paratype of *Trichothallus hawaiiensis* F.L. Stevens and the holotype of *Trichopeltis rhyacoides* F.L. Stevens]; Oahu, Wahiawa, 3 JUN [1921], leg. F.L.

Stevens No. 210, ILL 4533; on *Vaccinium reticulatum* J.E. Smith, [Island of] Hawaii, Kilauea, 13 JUL [as 14 JUL 1921 on the packet], leg. F.L. Stevens No. 821, ILL 4086 [also a paratype of *Meliola vaccinii* F.L. Stevens and the holotype of *Schiffnerula vaccinii* C.G. Hansford].

Meliola amadelpha H. Sydow. Leaflets of Philippine Botany 9:3114–3115. 1925. **Isotype**: On leaves of a palm [Arecaceae, Philippines], Luzon, Sorsogon Prov., Irosin, JUL 1916, leg. A.D.E. Elmer No. 16689, ILL 4535.

Meliola amaniensis C.G. Hansford. Proceedings of the Linnean Society of London 157:182–183. 1946. **Isotypes**: Hab. in foliis *Pittospori undulati* [= *Pittosporum undulatum* Vent., E. Africa], Tanganyika [Tanzania], Amani, Usambara, 30 SEP 1913, leg. M. Grote [ex H. Sydow, Fungi Exotici Exsiccati No. 249], ILL 5368 [as a microscopic preparation], and in bound exsiccati set [sub *Meliola lanceolato-setosa* H. & P. Sydow].

Meliola amboinensis H. Sydow. Philippine Journal of Science 21:133–134. 1922. **Isotype**: On leaves of *Aganosma* sp., Indonesia, Amboina, Gelala, 19 SEP 1913, leg. C.B. Robinson, Reliquiae Robinsonianae No. 2150, ILL 4550.

Meliola amerimni (F.L. Stevens) C.G. Hansford, 1961. **See basionym**: *Meliola bicornis* H.G. Winter var. *amerimni* F.L. Stevens, 1928.

Meliola amoena (H. Sydow) F. Petrak in H. Sydow & F. Petrak, 1929. **See basionym**: *Irene amoena* H. Sydow, 1926.

Meliola amomicola F.L. Stevens. Illinois Biological Monographs 2:508, fig. 37. 1916. **Holotype**: On *Amomis caryophyllata* (Jacquin) Krug & Urban [= *Pimenta racemosa* (P. Miller) J.W. Moore], Porto Rico [Puerto Rico], Mayagüez Mesa, 15 JUN 1915, leg. F.L. Stevens No. 7054, ILL 4551. **Paratype**: Mayagüez, 25 JUN 1915, leg. F.L. Stevens No. 7483, ILL 4552.

Meliola amomicola F.L. Stevens var. *longispora* A.C. Batista. Atas do Instituto de Micologia 1:26–28., fig. 9a. 1960. **Assumed isotypes**: [On] ... *Amomis caryophyllata* (Jacquin) Krug & Urban [= *Pimenta racemosa* (P. Miller) J.W. Moore] associado a *Phialetea aerospora* A.C. Batista & M.L. Nascimento in A.C. Batista, H. da Silva Maia, J.A. de Lima & E.A.F. da Matta e *Stomiopeltella machadoi* A.C. Batista & J.A. de Lima in A.C. Batista, Porto Rico [Puerto Rico], Monte Alegrillo, 14 NOV 1913, leg. F.L. Stevens No. 4757, ILL 33502, BPI. [also assumed isotypes of *Phialetea aerospora* A.C. Batista & M.L. Nascimento in A.C. Batista, H. da Silva Maia, J.A. de Lima & E.A.F. da Matta]. **Holotype**: URM [I.M.U.R.] 14100.

Meliola amoorae H.S. Yates. Philippine Journal of Science, Section C (Botany) 13:364. (1918) 1919. **Isotype**: On leaves of *Amoora* sp. [= *Aglaia* sp., Philippines], Luzon, Tayabas Prov., Mount Binuang, 20 MAY 1917, leg. M. Ramos & G. Edaño, Phil. Bur. Sci. No. 28908, ILL 5725.

Meliola amphigena F.L. Stevens & L.R. Tehon. Mycologia 18:16–17, pl. 2, fig. 21. 1926. **Holotype**: On Rubiaceae sp. indet., British Guiana [Guyana], Tumatumari, 11 JUL 1922, leg. F.L. Stevens No. 168, ILL 4555. **Paratypes**: Rockstone, 17 JUL 1922, leg. F.L. Stevens No. 450, ILL 4554; Demerara-Essequibo R.R., 15 JUL 1922, leg. F.L. Stevens No. 413, ILL 4553.

Meliola anacardiacearum (F.L. Stevens) C.G. Hansford, 1955. **See basionym**: *Amazonia anacardiacearum* F.L. Stevens, 1927.

Meliola anastomosans H.G. Winter. Hedwigia 25:96–97. 1886. **Isotype**: In foliis vivis Labiatarum [= Lamiaceae, W. Africa, Republ. São Tomé e Principe], São Tomé, [Island, JUN 1885], leg. A. Møller s.n., ILL 3895a [ex herb. H.G. Winter]. ≡ *Irene anastomosans* (H.G. Winter) F. Theissen & H. Sydow. Annales Mycologici 15:461. (1917) 1918. ≡ *Irenina anastomosans* (H.G. Winter) F.L. Stevens. Annales Mycologici 25:456. 1927. ≡ *Irenopsis anastomosans* (H.G. Winter) R.A. Toro. Monographs of the University of Puerto Rico. Series B, Physical and Biological Sciences 2:114. 1934. ≡ *Asteridiella anastomosans* (H.G. Winter) C.G. Hansford ex C.G. Hansford. Beihefte zur Sydowia Annales Mycologici 2:699. 1961.

Meliola anceps H. & P. Sydow var. *mussaendae* (H. & P. Sydow) F.L. Stevens, 1928. **See basionym**: *Meliola mussaendae* H. & P. Sydow, 1917.

Meliola andirae F.S. Earle. Bulletin of the New York Botanical Garden 3:303–304. 1904. **Isotype**: On leaves of *Andira inermis* (W. Wright) Kunth ex DC., Porto Rico [Puerto Rico, Santurce, 22 JAN 1903], leg. A.A. Heller No. 6448, ILL 4611. **Holotype**: NY.

Meliola andromedae N.T. Patouillard. Revue Mycologique 10:137. 1888. **Isotype**: Hab. la face inférieure de feuilles de l' *Andromeda salicifolia* Lam. [= *Agarista salicifolia* (Lam.) G. Don, France], Ile de France, leg. Vincent s.n., ILL 3896 [as a microscopic preparation]. **Holotype**: PC. ≡ *Irene andromedae* (N.T. Patouillard) H. & P. Sydow. Annales Mycologici 15:194–195. 1917. ≡ *Irenina andromedae* (N.T. Patouillard) F.L. Stevens. Annales Mycologici 25:447. 1927. ≡ *Asteridiella andromedae* (N.T. Patouillard) C.G. Hansford ex C.G. Hansford. Beihefte zur Sydowia Annales Mycologici 2:488. 1961.

Meliola andropogonis F.L. Stevens & E.F. Roldan ex W. Yamamoto. Hyogo Noka Daigaku, Sasayama, Japan. Science Reports Series: Plant Protection (Agricultural Biology), [Hyogo University of Agriculture. Science Reports] 3(2):65. 1958; originally published by F.L. Stevens & E.F. Roldan in Philippine Journal of Science 56:61, fig. 31. 1935, as nom. invalid. sine diagn. lat. **Holotype**: On *Andropogon halepensis* (L.) Brot. [= *Sorghum halepense* (L.) Persoon], Philippines, Luzon, Naguilian Road, Benguet, 7 JAN 1931, leg. F.L. Stevens No. 1577, ILL 4573.

Meliola angusta F.L. Stevens & L.R. Tehon. Mycologia 18:6, pl. 1, fig. 3. 1926. **Holotype:** On *Coccoloba* sp. [as *Coccolobis*], British Guiana [Guyana], Kartabo, 22 JUL 1922, leg. F.L. Stevens No. 558, ILL 4575. **Paratypes:** Leg. F.L. Stevens No. 514, ILL 4581; leg. F.L. Stevens No. 539, ILL 4578 [also the holotype of var. *minor* C.G. Hansford ex C.G. Hansford]; 23 JUL 1922, leg. F.L. Stevens No. 576, ILL 4577 and 4580 [also isotypes of *Meliola praetervisa* A. Gaillard var. *stevensii* C.G. Hansford]; Rockstone, 17 JUL 1922, leg. F.L. Stevens No. 478, ILL 4579; leg. F.L. Stevens No. 487, ILL 4576.

Meliola angusta F.L. Stevens & L.R. Tehon var. *minor* C.G. Hansford ex C.G. Hansford. Sydowia Annales Mycologici 16:304. (1962) 1963. **Holotype:** Hab. in foliis *Coccolobae* spec. indet. [= *Coccoloba* sp.], British Guiana [Guyana, Kartabo, 22 JUL 1922], leg. F.L. Stevens No. 539, ILL 4578 [also a paratype of the species name].

Meliola angustispora F.L. Stevens. Annales Mycologici 26:264, pl. VI, fig. 71. 1928. **Holotype:** On *Baccharis rhexioides* Kunth in H.B.K., Panama, Paitilla Point, 9 SEP 1924 [as 24 SEP on the packet], leg. F.L. Stevens No. 344, ILL 4582.

Meliola annonacearum F.L. Stevens. Annales Mycologici 26:245–246, pl. V, fig. 51. 1928 [as *anonacearum*, corrected in accordance with Rec. 60H.1 and Articles 60 & 61 of I.C.B.N.]. **Holotype:** On *Annona* sp. [as *Anona*], Ecuador, Barrn'nital, 17 NOV 1924, leg. F.L. Stevens No. 320, ILL 4585.

Meliola annonae F.L. Stevens. Annales Mycologici 26:240, pl. IV, fig. 41. 1928 [as *anonae*, corrected by C.G. Hansford (1961:39) in accordance with current I.C.B.N.]. **Assumed lectotype:** On *Annona purpurea* Moc. & Sessé ex Dunal [as *Anona*], Panama, Paitilla Point, 8 SEP 1924, leg. F.L. Stevens No. 342, ILL 4584 [citation of specimen in the F.L.S. herb. (ILL) as "type" by C.G. Hansford (1961:39) is interpreted as lectotypification]. **Residual syntype:** Mandingo, 15 OCT 1924, leg. F.L. Stevens No. 1238, ILL 4583 [this collection cited for illustration in fig. 41].

Meliola apayaoensis H.S. Yates. Philippine Journal of Science, Section C (Botany) 13:364. (1918) 1919. **Isotype:** On leaves of *Macaranga tanarius* (L.) Muell. Arg., [Philippines], Luzon, Apayao Subprov., 7 MAY 1917, leg. E. Fenix, Phil. Bur. Sci. No. 28331, ILL 5524. ≡ *Meliola macarangae* H. & P. Sydow var. *apayaoensis* (H.S. Yates) C.G. Hansford. Beihefte zur Sydowia Annales Mycologici 2:228. 1961.

Meliola apiculata C.G. Hansford. Proceedings of the Linnean Society of London 160:137. (1948) 1949. **Isotype:** Hab. in foliis *Scleriae* [= *Scleria* sp.], Porto Rico [Puerto Rico, Manati, 25 NOV 1913], leg. F.L. Stevens No. 5252 p.p., ILL 4625. **Holotype:** PREM.

Meliola araneosa H. & P. Sydow. Leaflets of Philippine Botany 6:1922. 1913. **Isotype:** On older leaves of *Guioa microcarpa* DC., Philippines, Mindanao [Island], Cabadbaran, (Mt. Urdaneta), AUG 1912, leg. H. & P. Sydow, ex Phil. Bur. Sci. No. 13553, ILL 4287. ≡ *Irenopsis araneosa* (H. & P. Sydow) F.L. Stevens. Annales Mycologici 25:434. 1927.

Meliola arcuata E.M. Doidge. Transactions of the Royal Society of South Africa 5:737, pl. LXVI, fig. 37. 1917. **Isosyntype:** On stems of *Viscum* sp., [South Africa, Natal Prov.], Kentani, 4 AUG 1914, leg. A. Pegler No. 1949, ILL 4624, ex PREM 8389. **Syntype:** PREM.

Meliola ardisiae H. Sydow. Leaflets of Philippine Botany 9:3116. 1925. **Assumed lectotype:** On leaves of *Ardisia jagorii* Mez, [Philippines], Luzon, Sorsogon Prov., Irosin, SEP 1916, leg. A.D.E. Elmer No. 17327, ILL 4623 [citation of specimen in F.L.S. herb. (ILL) as "type" by C.G. Hansford (1961:513) is interpreted as lectotypification].

Meliola arecibensis F.L. Stevens. Illinois Biological Monographs 2:491, fig. 18. 1916. **Holotype:** On *Acalypha bisetosa* Sprengel, Porto Rico [Puerto Rico], Vega Baja, 21 FEB 1913, leg. F.L. Stevens No. 365a, ILL 3868. **Paratype:** Dos Bocas, below Utuado, 30 DEC 1913, leg. F.L. Stevens No. 6547, ILL 3867. ≡ *Appendiculella arecibensis* (F.L. Stevens) R.A. Toro. Mycologia 17:144. 1925. ≡ *Irene larviformis* (P.C. Hennings) F.L. Stevens var. *arecibensis* (F.L. Stevens) F.L. Stevens. Annales Mycologici 25:425. 1927.

Meliola argentina C.L. Spegazzini. Anales de la Sociedad Cientifica Argentina 9:177–178. 1880; Fungi Argentini, Pugillus I, pp. 177–178, No. 72. 1880. **Isotype:** Hab. ad folia viva vel languida Cyperaceae cujusdam in paludosis, [Argentina], El Riachuelo, JAN 1880 [as FEB on the packet], leg. C.L. Spegazzini [No. 519 (cf. C.G. Hansford, 1961)], ILL 4636. **Holotype:** LPS.

Meliola argentina C.L. Spegazzini var. *hawaiiensis* C.G. Hansford. Sydowia Annales Mycologici. 11:51–52. (1957) 1958. **Holotype:** Hab. in foliis *Gahniae leptostachyae* [*Gahnia leptostachya* Boeckeler = *G. beecheyi* H. Mann], Hawaii, [Oahu, Olympus, 24 JUN 1921], leg. F.L. Stevens No. 672, ILL 4616. **Paratypes:** [Kauai, Pipe trail, Waimea Canyon, 15 JUN 1921], leg. F.L. Stevens No. 435, ILL 4629; [Oahu, Palolo Valley and Mt. Olympus, 10 JUN 1921], leg. F.L. Stevens No. 361, ILL 4628 and 5006; [Wahiawa, 3 JUN 1921], leg. F.L. Stevens No. 226, ILL 4618.

Meliola aristolochiae F.L. Stevens & L.R. Tehon. Mycologia 18:4–5, pl. 1, fig. 1. 1926. **Holotype:** On *Aristolochia* sp., British Guiana [Guyana], Rockstone, 17 JUL 1922, leg. F.L. Stevens No. 459, ILL 4639. **Paratypes:** Tumatumari, 11 JUL 1922, leg. F.L. Stevens No. 165, ILL 4641; Kartabo, 24 JUL 1922, leg. F.L. Stevens No. 673, ILL 4644; 22 JUL 1922, leg. F.L. Stevens No. 543, ILL 4642.

Meliola aristolochiicola F.L. Stevens. Annales Mycologici 26:278, fig. 76. 1928. **Holotype:** On Aristolochiaceae: *Aristolochia maxima* Jacquin, Panama, Tapia, 15 AUG 1923, leg. F.L. Stevens No. 1005, ILL 4638. **Isotype:** FH.

Meliola armata C.L. Spegazzini. Boletín de la Academia Nacional de Ciencias en Córdoba 11:493–494. 1889; Fungi Puiggariani, Pugillus I, pp. 115–116, No. 231. 1889. **Isolectotype:** Ad folia coriacea viva *Myrsines*? speciei [= *Myrsine*? sp.] cujusdam in sylvis, [Brazil], pr. Apiahy [Apiaí], MAY 1888, leg. J.I. Puiggari No. 2382 p.p., comm. C.L. Spegazzini, ILL 4288 [including a microscopic preparation]. **Assumed lectotype:** LPS [citation of the above collection as "type" by F.L. Stevens (Botanical Gazette 64:421-422. 1917) is interpreted as lectotypification with the specimen in the Spegazzini herb.]. **Residual syntype:** J.I. Puiggari No. 2381, LPS. ≡ *Irenopsis armata* (C.L. Spegazzini) F.L. Stevens. Annales Mycologici 25:437. 1927.

Meliola arrabidaeae C.G. Hansford. Beihefte zur Sydowia Annales Mycologici 1:99. 1957. **Holotype:** Hab. in foliis *Arrabidaeae* sp. indet. [= *Arrabidaea* sp.], Panama, [Agua Clara Reservoir, 17 SEP 1924], leg. F.L. Stevens No. 545, ILL 5984. **Isotype:** ILL 5982.

Meliola arrabidaeae C.G. Hansford var. *irregularis* (F.L. Stevens) C.G. Hansford, 1961. **See basionym:** *Meliola peruviana* H. & P. Sydow var. *irregularis* F.L. Stevens, 1928.

Meliola artocarpi H.S. Yates. Philippine Journal of Science, Section C (Botany) 12:362–363. (1917) 1918 [as *artocarpiae*, corrected in accordance with Articles 32.6 & 61 of I.C.B.N.]. **Isotype:** On leaves of *Artocarpus* sp., Philippines, Samar [Island], Catubig River, 7 FEB 1916, leg. M. Ramos, ex Phil. Bur. Sci. No. 24692, ILL 4645.

Meliola artocarpicola F.L. Stevens ex C.G. Hansford. Sydowia Annales Mycologici 11:52. (1957) 1958. **Isotype:** Hab. in foliis *Artocarpi* sp. [= *Artocarpus* sp.], Philippines, Mindanao [Island], Zamboanga, Malangas, [OCT–NOV 1919], leg. M. Ramos & G. Edaño, ex Phil. Bur. Sci. No. 36433, ILL 4648. **Paratypes:** [Samar Island, Catubig River, FEB–MAR 1916], leg. M. Ramos, ex Phil. Bur. Sci. No. 24617, ILL 4646; [Panay Island, Capiz Prov., Libacao, MAY–JUN 1919], leg. A. Martelino & G. Edaño, ex Phil. Bur. Sci. No. 35861, ILL 4649.

Meliola arundinis N.T. Patouillard var. *angulosa* C.G. Hansford. Proceedings of the Linnean Society of London 160:138–139. (1948) 1949. **Isotype:** Hab. in foliis *Phragmitis* sp. [= *Phragmites* sp.], Philippines, [Luzon, Bataan Prov., Lamao, JUL 1913, leg. E.D. Merrill], ex Phil. Bur. Sci. No. 9101, ILL 5069. **Holotype:** PREM.

Meliola aterrima H. Sydow. Annales Mycologici 24:294–296. 1926. **Isotype:** Hab. in petiolis, rarius in foliis *Xanthoxyli proceri* [= *Zanthoxylum* *procerum* J. Donnell Smith], Costa Rica, San Pedro de San Ramón, 5 FEB 1925, leg. H. Sydow, Fungi in Itinere Costaricensi Collecti No. 113a, ILL 4653.

Meliola atra E.M. Doidge. Transactions of the Royal Society of South Africa 8:137–138. 1920. **Isotype:** In foliis fruticis ignotis [later identified as *Eugenia* sp.], South Africa, Zwartkop, 19 JUL 1918, leg. E.M. Doidge, ex PREM No. 11594, ILL 3918 [as microscopic preparations, the specimen lost from the packet]. ≡ *Irene atra* (E.M. Doidge) E.M. Doidge. South African Journal of Natural History 2:40. 1920. ≡ *Irenina atra* (E.M. Doidge) F.L. Stevens. Annales Mycologici 25:467. 1927. ≡ *Asteridiella atra* (E.M. Doidge) C.G. Hansford ex C.G. Hansford. Beihefte zur Sydowia Annales Mycologici 2:136. 1961.

Meliola aucubae P.C. Hennings. Botanische Jahrbücher für Systematik, Pflanzengeschichte und Pflanzengeographie 29:150. 1900 (1901). **Isotype:** Auf Blättern von *Aucuba japonica* L., Japan, Ise Prov., JUN 1899, leg. M. Shirai s.n., ILL 3916. ≡ *Irenina aucubae* (P.C. Hennings) F.L. Stevens. Annales Mycologici 25:455. 1927. ≡ *Irene aucubae* (P.C. Hennings) C.G. Hansford. Sydowia Annales Mycologici 9:4. 1955. ≡ *Asteridiella aucubae* (P.C. Hennings) C.G. Hansford. Beihefte zur Sydowia Annales Mycologici 2:478. 1961.

Meliola autumnalis H. & P. Sydow. Annales Mycologici 2:169. 1904. **Isotype:** Hab. in foliis vivis *Gei chilensis* [*Geum chilense* Balbis ex Lindley = *G. quellyon* Sweet], Chile, Concepcion, [5 JUN 1895], leg. F.W. Neger s.n., ex Herb. H. Sydow, ILL 3745.

Meliola bakeri H. & P. Sydow. Annales Mycologici 14:355–356. 1916. **Assumed isotype:** Hab. in foliis *Tetrastigmatis* spec. [= *Tetrastigma* sp.], Philippines, [Luzon, Laguna Prov.], Mount Maquiling [near Los Baños], DEC 1915, leg. C.F. Baker No. 3987 [ex. Fungi Malayana No. 249], ILL 4658 [as a microscopic preparation], CUP, FH.

Meliola balansae A. Gaillard. Le Genre *Meliola*, p. 95, pl. 17, fig. 1. 1892. **Isotype:** Ad paginam superiorem foliorum coriaceorum, Paraguay, Carapegua, NOV 1883, leg. B. Balansa No. 4018, ILL 4673. **Holotype:** PC.

Meliola bambusae N.T. Patouillard. Revue Mycologique 10:140. 1888. **Isotype:** Sur les feuilles d'un *Bambusa* sp., Tonkin [Vietnam], leg. B. Balansa [ex C. Roumeguère, Fungi Selecti Exsiccati No. 4433], ILL 5140. **Holotype:** PC.

Meliola banahaensis H.S. Yates. Philippine Journal of Science, Section C (Botany) 13:364–365. (1918) 1919. **Isotype:** On *Dysoxylum* sp.?, [Philippines], Luzon, Laguna Prov., Mount Banahao, 8 MAY 1917, leg. M. Ocampo, ex Phil. Bur. Sci. No. 28011, ILL 4671.

Meliola banarae F.L. Stevens. Annales Mycologici 26:249, pl. V, fig. 57.

1928. **Holotype**: On Flacourtiaceae: *Banara guianensis* Aublet, Panama, [New Limon], Fort Lorenzo trail, 10 OCT 1924 [as 4 OCT on the packet], leg. F.L. Stevens No. 1189, ILL 4670. **Paratype**: New Limon, 4 OCT 1924, leg. F.L. Stevens No. 1017, ILL 4669.

Meliola banguiensis H.S. Yates. Philippine Journal of Science, Section C (Botany) 13:365. (1918) 1919. **Isolectotype**: On leaves of one of the Menispermaceae, Philippines, Luzon, Ilocos Norte Prov., Bangui, FEB–MAR 1917, leg. M. Ramos, ex Phil. Bur. Sci. No. 27697 [erroneously cited as No. 27696 in the protologue], ILL 4654. **Lectotype**: FH. [Both F.L. Stevens (1928:275) and C.G. Hansford (1961:64) cited No. 27697 as the "type." Hansford's decision for location of the lectotype is accepted herein.].

Meliola banosensis H. & P. Sydow. Annales Mycologici 14: 356. 1916. **Assumed isotypes**: Hab. in foliis *Puerariae* spec. [= *Pueraria* sp.], Philippines, [Luzon, Laguna Prov., Mount Maquiling, near] Los Baños, DEC 1915, leg. C.B. Baker No. 4016 [ex Fungi Malayana No. 250], ILL 4666 [as a microscopic preparation], CUP, FH.

Meliola barringtoniicola F.L. Stevens & E.F. Roldan ex C.G. Hansford. Sydowia Annales Mycologici 16:318. (1962) 1963; originally published by F.L. Stevens & E.F. Roldan in Philippine Journal of Science 56:74–75, fig. 3h. 1935, as nom. invalid. sine diagn. lat. **Holotype**: On *Barringtonia* sp., Philippines, Luzon, Tayabas Prov., Quezon Forest Park, 30 NOV 1930, leg. F.L. Stevens No. 440, ILL 4665. **Isotype**: CUP.

Meliola bataanensis H. & P. Sydow. Annales Mycologici 12:551. 1914. **Isotype**: Hab. in foliis *Millettiae* [= *Millettia* sp., Philippines], Luzon, Bataan Prov., Lamao, JUL 1913, leg. E.D. Merrill No. 9106, ILL 4662.

Meliola batangasensis C.G. Hansford. Beihefte zur Sydowia Annales Mycologici 1:100. 1957. **Holotype**: Hab. in foliis *Homalii barandanae* [= *Homalium barandae* S. Vidal ex Fernández-Villar in Blanco], Philippines, Luzon, Batangas [Prov.], APR–MAY 1915, leg. M. Ramos & G. Edaño, ex Phil. Bur. Sci. No. 22682, ILL 5267. **Paratypes**: In foliis *Homalii villariani* [= *Homalium villarianum* S. Vidal], leg. M. Ramos & D. Devoy, ex Phil. Bur. Sci. No. 22681 p.p., ILL 5268; leg. M. Ramos & G. Edaño, ex Phil. Bur. Sci. No. 22681 p.p., ILL 5269 [there are two packets of this number, the labels differing in the second collector].

Meliola bauhiniae H.S. Yates. Philippine Journal of Science, Section C (Botany) 13:365. (1918) 1919. **Isotypes**: On *Bauhinia* sp., [Philippines], Luzon, Ilocos Norte Prov., [Burgos], 14 MAR 1917, leg. M. Ramos, ex Phil. Bur. Sci. No. 27801, ILL 4660 and 4661 [also isotypes of *Meliola burgosensis* C.G. Hansford].

Meliola bayamonensis L.R. Tehon. Botanical Gazette 67:506. 1919. **Holotype**: On *Psychotria pubescens* Sw., Porto Rico [Puerto Rico], Bayamon, 19 FEB 1913, leg. F.L. Stevens No. 392, ILL 4289. ≡ *Irenopsis bayamonensis* (L.R. Tehon) F.L. Stevens. Annales Mycologici 25:437. 1927.

Meliola beebei F.L. Stevens. Annales Mycologici 26:273, pl. VI, fig. 74. 1928. **Holotype**: On *Tabernaemontana* sp., British Guiana [Guyana], Kartabo, 21 JUL 1922, leg. F.L. Stevens No. 506, ILL 4659.

Meliola behniae H. Sydow in E.M. Doidge & H. Sydow. Bothalia 2(2):444, 464. 1928. **Paratype**: Host: *Behnia reticulata* (Thunb.) D.F. Didrichsen, South Africa, [Natal Prov.], Duncairn near Pietermaritzburg, 13 JUL 1921, leg. E.M. Doidge, ex PREM No. 14955, ILL 5024. **Holotype**: PREM [No. 11818, indicated on p. 464 with the Latin diagnosis].

Meliola benguetensis F.L Stevens & E.F. Roldan. Philippine Journal of Science 56:58, fig. 2c. 1935, nom. invalid. sine diagn. lat. **Holotype**: On *Otophora* sp. [= *Lepisanthes* sp.], Philippines, Luzon, Benguet, Kennon Road, 8 JAN 1931, leg. F.L. Stevens No. 1670 [erroneously cited as No. 670 in the protologue], ILL 4655. **Paratype**: Nueva Ecija Prov., Balete Road, San José, 10 JAN 1931 [as 19 JAN 1931 on the packet], leg. F.L. Stevens No. 1794, ILL 4656.

Meliola bicornis H.G. Winter var. *amerimni* F.L Stevens. Annales Mycologici 26:189, pl. II, fig. 12. 1928. **Holotype**: On *Amerimnon brownei* Jacquin [as *brownii* = *Dalbergia brownei* (Jacquin) Schinz], Panama, Paitilla Point, 8 SEP 1924, leg. F.L. Stevens No. 355, ILL 4710. ≡ *Meliola amerimni* (F.L. Stevens) C.G. Hansford. Beihefte zur Sydowia Annales Mycologici 2:273. 1961.

Meliola bicornis H.G. Winter var. *calopogonii* F.L. Stevens. Illinois Biological Monographs 2:532. 1916. **Assumed lectotype**: On *Calopogonium orthocarpum* Urban [= *C. mucunoides* Desv.], Porto Rico [Puerto Rico], Dos Bocas, 8 JUL 1915, leg. F.L. Stevens No. 8060, ILL 4707 [choice of this collection as "type" by C.G. Hansford (1961:277) is interpreted as lectotypification with the specimen at ILL, even though Stevens (1928:189) cited No. 3492 under type locality. The latter number is misquoted (should be 3942), and no choice of type is indicated on either of the two packets at ILL]. **Residual syntypes**: 16 DEC 1913, leg. F.L. Stevens No. 6035, ILL 4708; Mayagüez, 3 OCT 1913, leg. F.L. Stevens No. 3942 [erroneously as No. 3492 in the protologue], ILL 4709 and 6461 [the latter number is an isosyntype]; 10 APR 1913, leg. F.L. Stevens No. 372, ILL 6460; Aguada, 22 NOV 1913, leg. F.L. Stevens No. 5087, ILL 6459. ≡ *Meliola scabriseta* C.G. Hansford & F.C. Deighton var. *calopogonii* (F.L. Stevens) C.G. Hansford. Beihefte zur Sydowia Annales Mycologici 2:277. 1961.

Meliola bicornis H.G. Winter var. *constipata* C.L. Spegazzini. Anales de la Sociedad Cientifica Argentina 26:22. 1888; Fungi Guaranitici, Pugillus II, p. 20, No. 57. 1888. **Isotype**: Ad folia Leguminosae sp. indet. [= Fabaceae], Paraguay, Sierra de Peribebuy, AUG 1883, leg. B. Balansa No. 4022, comm. C.L. Spegazzini, ILL 5421. **Holotype**: LPS. ≡ *Meliola constipata* (C.L. Spegazzini) C.L. Spegazzini. Anales del Museo Nacional de Historia Natural de Buenos Aires 32:370. 1924.

Meliola bicornis H.G. Winter var. *galactiae* F.L. Stevens. Illinois Biological Monographs 2:533. 1916. **Holotype**: On *Galactia dubia* DC., Porto Rico [Puerto Rico], Rio Tanamá, 6 JUL 1915, leg. F.L. Stevens No. 7857 [erroneously as 7856 in the protologue], ILL 4677. **Isotype**: PREM. ≡ *Meliola galactiae* (F.L. Stevens) C.G. Hansford. Beihefte zur Sydowia Annales Mycologici 2:287–288. 1961.

Meliola bicornis H.G. Winter var. *robinsonii* (H. Sydow) F.L. Stevens, 1928. **See basionym**: *Meliola robinsonii* H. Sydow, 1922.

Meliola bicornis H.G. Winter var. *tephrosiae* M. Beeli. Bulletin du Jardin Botanique de L'État à Bruxelles 8:1–2. DEC 1922. **Isotype**: Sur les feuilles de *Tephrosia elegans* Schumacher in Schumacher & Thonning, [Angola, Kwango], Kilebe, 1914, leg. H.J.R. Vanderyst No. 4126, ILL 4726. **Holotype**: BR.

Meliola bidentata M.C. Cooke. Grevillea 11: 37–38. 1882. **Probable isotypes**: On leaves of *Bignonia* sp., Florida, [Gainesville], leg. ?G.W. Martin, ex H.W. Ravenel, Fungi Americani No. 330, ILL 5137 [as a microscopic preparation], and in bound exsiccati set [published by Cooke in 1877 and distributed as *Meliola furcata* J.H. Léveillé. Cooke clearly made a mistake in the protologue by citing No. 128. Label data and packet contents from Ravenel No. 128 in our exsiccati set do not correspond to the other information in the protologue. The packet is part of Century 2, issued in 1877; it is labelled *Corticium punctulatum* M.C. Cooke, ad asseros *Pini*, Aiken, South Carolina. Cooke described the latter species with this specimen as type of the name in Grevillea 6:132. 1878, and it has nothing to do with the name considered here. C.G. Hansford, who annotated the ILL material of *Meliola bidentata* in 1956, cited both Ravenel numbers in his monograph (1961), but he did not identify either as the type collection. He also cited J.B. Ellis's North American Fungi No. 1297(a, b). The packet marked "a" (also distributed under the name *Meliola furcata* J.H. Léveillé) may be part of the type collection, as follows]: **Possible isotypes**: [as *Bignonia capreolata* L. on the packet], Florida, leg. G.W. Martin, [ex J.B. Ellis, North American Fungi No. 1297(a)], ILL 5133 p.p. [fragment of a leaf] and 5138 [a microscopic preparation], also the packet of No. 1297 marked "a" in the bound exsiccati set [leaves in this packet match those of H.W. Ravenel, Fungi Americana No. 330 and the fragment in the packet of ILL 5133. Ellis's North American

Fungi, Century 13, was issued in 1884, and the specimens under consideration possibly are duplicates of the material available to M.C. Cooke. The packets of No. 1297 marked "b" and "c" represent different collections and host plants not under consideration here].

Meliola bignoniacearum F.L Stevens. Annales Mycologici 26:196, pl. III, fig. 19. 1928. **Holotype:** On Bignoniaceae sp. indet., Panama, Culebra, 2 OCT 1924, leg. F.L. Stevens No. 925, ILL 31732. **Paratypes:** On *Phryganocydia corymbosa* Bureau ex K. Schum., Panama, Agua Clara Reservoir, 17 SEP 1924, leg. F.L. Stevens No. 576, ILL 31733; on *Arrabidaea* sp., British Guiana [Guyana], Coverden, 8 AUG 1922, leg. F.L. Stevens No. 789, ILL 31741; on *Tabebuia* sp., Ecuador, Terecita, 29 OCT 1924, leg. F.L. Stevens No. 76, ILL 31742; on *Adenocalymma* sp., Ecuador, Terecita, 29 OCT 1924, leg. F.L. Stevens No. 77, ILL 31740; on Bignoniaceae indet., Panama, Juan Diaz, 18 AUG 1923, leg. F.L. Stevens No. 1160, ILL 31730; 21 AUG 1923, leg. F.L. Stevens No. 1256, ILL 31734; Gamboa, 17 AUG 1923, leg. F.L. Stevens No. 1081, ILL 31737; leg. F.L. Stevens No. 1110, ILL 31725; Corozal, Trail No. 17, 30 AUG 1924, leg. F.L. Stevens No. 74, ILL 31736; Paitilla Point, 8 SEP 1924, leg. F.L. Stevens No. 371, ILL 31739; Darien, 10 SEP 1924, leg. F.L. Stevens No. 407, ILL 31729; Chiva-Chiva trail, 18 SEP 1924, leg. F.L. Stevens No. 608, ILL 31727; Mandingo, 15 OCT 1924, leg. F.L. Stevens No. 1323, ILL 31743; Baille Mona, 20 JUL 1924, leg. F.L. Stevens No. 668, ILL 31735; Fort Sherman, Sweetwater, 6 OCT 1924, leg. F.L. Stevens No. 1059, ILL 31731; Ecuador, Terecita, 29 OCT 1924 [as Panama, Corozal, Trail No. 17, 30 AUG 1924 on the packet], leg. F.L. Stevens No. 82 p.p., ILL 31726a [also an isotype of *Meliola bignoniacearum* F.L. Stevens var. *parasitica* C.G. Hansford]; Ecuador, Ambato, 14 NOV 1924, leg. F.L., Stevens No. 316, ILL 31738.

Meliola bignoniacearum F.L. Stevens var. *irregularis* C.G. Hansford. Beihefte zur Sydowia Annales Mycologici 1:100–101. 1957. **Holotype:** Hab. in Bignoniacearum [= Bignoniaceae sp. indet.], British Guiana [Guyana], Rockstone, [13 JUL 1922], leg. F.L. Stevens No. 249 p.p., ILL 4762 p.p. [erroneously cited as 2762 in the protologue. ILL 4762 is also the holotype of *Meliola bignoniacearum* F.L. Stevens var. *tenuis* C.G. Hansford and of *Meliola rockstonensis* C.G. Hansford.].

Meliola bignoniacearum F.L. Stevens var. *major* C.G. Hansford. Sydowia Annales Mycologici 9:60. 1955. **Isotype:** Hab. in foliis Bignoniacearum [= Bignoniaceae sp. indet.], British Guiana [Guyana], Tumatumari, [8 JUL 1922], leg. F.L. Stevens No. 105, ILL 31724. **Holotype:** FH.

Meliola bignoniacearum F.L. Stevens var. *parasitica* C.G. Hansford. Sydowia Annales Mycologici 9:60. 1955. **Isotype:** Hab. in foliis Bignoniacearum [= Bignoniaceae sp. indet.], Ecuador, Terecita, [29 OCT 1924], leg. F.L. Stevens No. 82, ILL 31726 p.p. [also (p.p.) an isotype of *Meliola stevensiana* C.G. Hansford and, as ILL 31726a, a paratype of the species name]. **Holotype:** FH.

Meliola bignoniacearum F.L Stevens var. *tenuis* C.G. Hansford. Beihefte zur Sydowia Annales Mycologici 1:101. 1957. **Holotype:** Hab. in foliis Bignoniacearum [= Bignoniaceae sp. indet.], British Guiana [Guyana], Rockstone, 13 JUL 1922, leg. F.L. Stevens No. 249 p.p., ILL 4762 p.p. [also (p.p.) the holotype of *Meliola bignoniacearum* F.L. Stevens var. *irregularis* C.G. Hansford and of *Meliola rockstonensis* C.G. Hansford].

Meliola boerlagiodendri H.S. Yates. Philippine Journal of Science, Section C (Botany) 13:365–366. (1918) 1919 [as *boerlagiodendriae*, corrected in accordance with Articles 32.6 & 61 of I.C.B.N.]. **Isotype:** On leaves of *Boerlagiodendron* sp. [= *Osmoxylon* sp., Philippines], Luzon, Tayabas Prov., Mount Binuang, 9 MAY 1917, leg. M. Ramos & G. Edaño, ex Phil. Bur. Sci. No. 28911, ILL 31746.

Meliola bonii A. Gaillard. Le Genre *Meliola*, p. 39, pl. 8, fig. 3. 1892 [as *boni*, corrected in accordance with Articles 32.6 & 61 of I.C.B.N.]. **Isotype:** Ad paginam superiorem foliorum quorundam, Tonkin [Vietnam], leg. M. Bon No. 3319, ILL 3915. ≡ *Irene bonii* (A. Gaillard) H. & P. Sydow. Annales Mycologici 15:194. 1917. ≡ *Irenina bonii* (A. Gaillard) F.L. Stevens. Annales Mycologici 25:449. 1927. ≡ *Asteridiella bonii* (A. Gaillard) C.G. Hansford ex C.G. Hansford. Beihefte zur Sydowia Annales Mycologici 2:756–757. 1961. [Epithets of the three synonyms were also published as *boni* and are corrected herein.].

Meliola borneensis H. Sydow. Annales Mycologici 21:90. 1923. **Assumed lectotype:** Hab. in foliis *Uvariae* [= *Uvaria* sp.], British North Borneo, [Malaysia], Sandakan Prov., Sibuguey, 5 DEC 1920, leg. M. Ramos No. 2138, ILL 4766 [citation of specimen in F.L.S. herb. (ILL) as "type" by C.G. Hansford (1961:35) is interpreted as lectotypification].

Meliola bosciae E.M. Doidge. Transactions of the Royal Society of South Africa 5:731, 745–746, pl. LXI, fig. 26. 1917. **Isotype:** On leaves of *Boscia caffra* Sonder, [South Africa], Natal [Prov.], Winkle Spruit, 6 JUL 1912, leg. E.M. Doidge, ex PREM No. 2510, ILL 4294 [type specified on p. 746 with the Latin diagnosis]. **Paratype:** Durban, Stella Bush, 7 JUN 1915, leg. K. Lansdell No. 9016, ILL 4295. ≡ *Irenopsis bosciae* (E.M. Doidge) F.L. Stevens. Annales Mycologici 25:435. 1927.

Meliola brachycera H. Sydow. Annales Mycologici 24:297–298. 1926. **Isotype:** Hab. in foliis vivis vel languidis *Conostegiae lanceolatae* [= *Conostegia lanceolata* Cogn.], Costa Rica, San Pedro de San Ramón, 5 FEB 1925, leg. H. Sydow, Fungi in Itinere Costaricensi Collecti No. 142, ILL 4773. ≡ *Irene brachycera* (H. Sydow) C.G. Hansford. Sydowia Annales Mycologici 9:4. 1955. ≡ *Asteridiella brachycera* (H. Sydow) C.G. Hansford ex C.G. Hansford. Beihefte zur Sydowia Annales Mycologici 2:153–154. 1961.

Meliola brachypoda H. Sydow. Annales Mycologici 20:67. 1922, nom. nov. **Based on:** *Meliola macarangae* H.S. Yates, (1917) JAN 1918 (nom. illegit.), non H. & P. Sydow, OCT 1917. **Isotype:** On *Macaranga tanarius* (L.) Muell. Arg., Philippines, Luzon, Tyabas Prov., Basiad, 20 DEC 1916, leg. H.S. Yates, ex Phil. Bur. Sci. No. 25621, ILL 4774.

Meliola brasiliensis C.L. Spegazzini. Anales de la Sociedad Cientifica Argentina 12:102. 1881; Fungi Argentini Pugillus IV, p. 102, No. 116. 1881. **Isotype:** Hab. ad folia viva subcoriacea plantae cujusdam, Brasilia meridionali, [C. Brazil], Apiahy [Apiai], leg. J.I. Puiggari No. 1551, ILL 4775. ≡ *Irenopsis brasiliensis* (C.L. Spegazzini) C.G. Hansford. Beihefte zur Sydowia Annales Mycologici 2:660. 1961.

Meliola brasiliensis C.L. Spegazzini var. *sanguineo-maculans* H. Rehm ex J. Rick. Annales Mycologici 5:337. 1907, nom. nud. **Assumed isotype:** In foliis *Schini* [= *Schinus* sp., Brazil], São Leopoldo, leg. E.H.G. Ule, as J. Rick, Fungi Austro-Americani No. 156, ILL 5369.

Meliola brideliae F.L. Stevens & E.F. Roldan ex C.G. Hansford. Sydowia Annales Mycologici 16:319. (1962) 1963; originally published by F.L. Stevens & E.F. Roldan in Philippine Journal of Science 56:69–70, fig. 3e. 1935, as nom. invalid. sine diagn. lat. **Holotype:** On *Bridelia stipularis* (L.) Blume, Philippines, Luzon, Benguet, Naguilian Road, 6 JAN 1931, leg. F.L. Stevens No. 1543, ILL 4777. **Isotype:** CUP.

Meliola bruguierae H. Sydow. Leaflets of Philippine Botany 9:3116–3117. 1925. **Isotype:** On *Bruguiera eriopetala* Wight & Arnott, [Philippines, Luzon], Sorsogon Prov., Irosin, JUL 1916, leg. A.D.E. Elmer No. 16775, ILL 4778.

Meliola buchananiae F.L. Stevens ex C.G. Hansford. Beihefte zur Sydowia Annales Mycologici 1:101. 1957. **Holotype:** Hab. in foliis on *Buchananiae nitidae* [= *Buchanania nitida* Engler], Philippines, Mindanao [Island], Surigao Prov., [JUN 1919, leg. M. Ramos & J. Pascasio], ex Phil. Bur. Sci. No. 35905, ILL 4780. **Paratype:** [APR 1919], Phil. Bur. Sci. No. 34642, ILL 4779.

Meliola buddlejicola P.C. Hennings. Hedwigia 44:61. 1905 [as *buddleyicola*, corrected in accordance with Rec. 60H.1 and Articles 60 & 61 of I.C.B.N.]. **Isotypes:** Auf Blättern von *Buddleja* sp. [as *Buddleya* = *Buddleja americana* L. on the packet, Peru], Tarapoto, Rio Huallaga, NOV 1902, leg. E.H.G. Ule No. 3187, ILL 3908 [ex herb. H. Sydow] and 3909 [as a microscopic preparation, ex Mycotheca Brasiliensis No. 56]. ≡ *Irenina buddlejicola* (P.C. Hennings) F.L. Stevens. Annales Mycologici 25:455. 1927. ≡ *Irene buddlejicola* (P.C. Hennings) C.G. Hansford. Sydowia Annales Mycologici 9:6. 1955. ≡ *Asteridiella buddlejicola* (P.C. Hennings) C.G. Hansford ex C.G. Hansford. Beihefte zur Sydowia Annales Mycologici 2:524. 1961. [Epithets of the three synonyms were also published as *buddleyicola* and are corrected herein.].

Meliola burgosensis C.G. Hansford. Sydowia Annales Mycologici 9:10–11. 1955. **Isotypes**: Hab. in foliis *Bauhiniae* spec. [= *Bauhinia* sp.], Philippines, Luzon [Ilocos Norte Prov., Burgos, 14 MAR 1917], leg. M. Ramos, ex Phil. Bur. Sci. No. 27801 p.p., ILL 4660 p.p. and 4661 p.p. [also paratypes of *Meliola bauhiniae* H.S. Yates].

Meliola burseracearum F.L. Stevens. Annales Mycologici 26:199, pl. III, fig. 23. 1928. **Holotype**: On *Tetragastris panamensis* (Engler) O. Kuntze, Panama, Tapia, 15 AUG 1923, leg. F.L. Stevens No. 1050, ILL 4782 [F.L. Stevens chose No. 1050 as the type in the legend for fig. 23 on p. 381]. **Paratypes**: Leg. F.L. Stevens No. 1029, ILL 4785; leg. F.L. Stevens No. 1052, ILL 4783; Fort Randolph, 100 feet hill trail, 23 SEP 1924, leg. F.L. Stevens No. 773, ILL 4784.

Meliola burseracearum F.L. Stevens var. *major* C.G. Hansford. Sydowia Annales Mycologici 11:53. (1957) 1958. **Paratype**: On *Icica* sp. [= *Protium* sp. (Burseraceae)], Trinidad, [Cumuto, 16 AUG 1922], leg. F.L. Stevens No. 936, ILL 4009.

Meliola byrsonimae F.L. Stevens. Illinois Biological Monographs 2:517–518. 1916. **Holotype**: On *Byrsonima lucida* (P. Miller) DC., Porto Rico [Puerto Rico], Guayanilla, 14 JUL 1915, leg. F.L. Stevens No. 3541 [as 8541 on the packets], ILL 4790. **Isotype**: ILL 4794. [F.L. Stevens No. 3541 is also cited as a paratype of *Isthmospora spinosa* F.L. Stevens.].

Meliola byrsonimae F.L. Stevens var. *minor* C.G. Hansford. Sydowia Annales Mycologici 10:66. (1956) 1957. **Paratypes**: Hab. in foliis *Byrsonimae crassifoliae* [= *Byrsonima crassifolia* (L.) Kunth in H.B.K.], Panama, [Las Cruces trail, 2 SEP 1924], leg. F.L. Stevens No. 159, ILL 4791 and 4793; [28 SEP 1924], leg. F.L. Stevens No. 888, ILL 4792.

Meliola byrsonimicola F.L. Stevens & L.R. Tehon. Mycologia 18:10, pl. 1, fig. 9. 1926. **Holotype**: On *Byrsonima* sp., British Guiana [Guyana], Demerara-Essequibo R.R., 15 JUL 1922, leg. F.L. Stevens No. 333, ILL 4787. **Paratype**: Leg. F.L. Stevens No. 363, ILL 4789.

Meliola byrsonimina F.L. Stevens & L.R. Tehon. Mycologia 18:10–11, pl. 1, fig. 10. 1926. **Holotype**: On *Byrsonima* sp., British Guiana [Guyana], Tumatumari, 10 JUL 1922, leg. F.L. Stevens No. 106, ILL 4786.

Meliola cadigensis H.S. Yates. Philippine Journal of Science, Section C (Botany) 12:363. (1917) 1918. **Isotype**: On leaves of *Glycosmis cochinchinensis* (Lour.) Pierre ex Engler in Engler & Prantl [= *G. pentaphylla* (Retzius) Correa], Philippines, Luzon, Tayabas Prov., Mount Cadig, 16 DEC 1916, leg. H.S. Yates, ex Phil. Bur. Sci. No. 25822, ILL 4849.

Meliola calatheae F.L. Stevens. Annales Mycologici 26:178, pl. II, fig. 4. 1928. **Holotype:** On *Calathea insignis* Petersen, Costa Rica, Columbiana, 19 AUG 1923 [erroneously as JUL in the protologue], leg. F.L. Stevens No. 578, ILL 4851. **Paratypes:** On *Bihai pendula* (Wawra) O. Kuntze [= *Heliconia pendula* Wawra], San Cecilia, 7 AUG 1923, leg. F.L. Stevens No. 749, ILL 4854, leg. F.L. Stevens No. 766, ILL 4852 and 4855; on *Bihai latispatha* (Bentham) Griggs [= *Heliconia latispatha* Bentham], San Cecelia, 7 AUG 1923, leg. F.L. Stevens No. 749a, ILL 4853.

Meliola calatheicola F.L. Stevens. Annales Mycologici 26:265, pl. VI, fig. 72. 1928. **Holotype:** On petioles of *Calathea lutea* (Aublet) G.F.W. Meyer, Costa Rica, Port Limon, 10 AUG 1923, leg. F.L. Stevens No. 882, ILL 4856.

Meliola callicarpae H. & P. Sydow. Annales Mycologici 10:80. 1912. **Isotype:** Hab. in foliis *Callicarpae canae* [= *Callicarpa cana* L.], Philippines, [Luzon], pr. Manila, NOV–DEC 1910, leg. E.D. Merrill No. 7421, ILL 4887.

Meliola callista H. Rehm. Leaflets of Philippine Botany 6:2191–2192. 1914. **Assumed isolectotype:** Ad folia *Premnae odoratae* [= *Premna odorata* Blanco], Philippines, Luzon, Laguna Prov., [Mount Maquiling, near] Los Baños, OCT 1913, leg. S.A. Reyes, comm. C.F. Baker No. 1545 [ex Fungi Malayana No. 41, the latter cited as "type" by C.G. Hansford (1946 and 1961)], ILL 4860. [Since Hansford lists the other two syntypes by the numbers cited in the protologue, this is interpreted as intended lectotypification, although location of the lectotype is not indicated]. ≡ *Irenina callista* (H. Rehm) C.G. Hansford, Proceedings of the Linnean Society of London 157:169. 1946. ≡ *Asteridiella callista* (H. Rehm) C.G. Hansford ex C.G. Hansford. Beihefte zur Sydowia Annales Mycologici 2:687–688. 1961.

Meliola calochaeta H. Sydow. Leaflets of Philippine Botany 9:3117. 1925. **Isotype:** On leaves of *Cryptocarya foxworthyi* A.D.E. Elmer, [Philippines, Luzon], Sorsogon Prov., Irosin, SEP 1916, leg. A.D.E. Elmer No. 17331, ILL 4850.

Meliola calophylli F.L. Stevens. Illinois Biological Monographs 2:490–491, fig. 17. 1916. **Holotype:** On *Calophyllum calaba* sensu Jacquin non L. [= *C. antillanum* Britton in Britton & Wilson], Porto Rico [Puerto Rico], Mayagüez, 15 JUN 1915, leg. F.L. Stevens No. 7059, ILL 3734 [also a probable isotype of *Meliolidium portoricense* C.L. Spegazzini]. **Paratypes:** 25 JUN 1915, leg. F.L. Stevens No. 7489a, ILL 3733; Vega Baja, 5 NOV 1913, leg. F.L. Stevens No. 4310, ILL 6592 [also a paratype of *Perisporium portoricense* F.L. Stevens & R. Higley ex F.L. Stevens]. ≡ *Appendiculella calophylli* (F.L. Stevens) R.A. Toro. Mycologia 17:144. 1925. ≡ *Irene calophylli* (F.L. Stevens) F.L. Stevens. Annales Mycologici 25:428. 1927.

Meliola calopogonii F.L. Stevens. Annales Mycologici 26:255, pl. V, fig. 64. 1928. **Holotype:** On *Calopogonium* sp., Panama, Punta Bruja, 16 JUL 1924, leg. F.L. Stevens No. 525, ILL 4885.

Meliola calva C.L. Spegazzini. Boletín de la Academia Nacional de Ciencias en Córdoba 11:495. 1889; Fungi Puiggariani, Pugillus I, p. 117, No. 233. 1889. **Isotype:** Ad folia viva subcoriacea Laurinearum [= Lauraceae sp. indet., Brazil], in sylvis pr. Apiahy [Apiai], Aest. 1881, leg. J.I. Puiggari [No.. 1483, fide C.G. Hansford (1961)], ex C.L. Spegazzini [sub Puiggari Nos. 1183–1507], ILL 3955 [the packet is unnumbered but stamped "from type collection"]. ≡ *Irenina calva* (C.L. Spegazzini) F.L. Stevens. Annales Mycologici 25:464–465. 1927. ≡ *Asteridiella calva* (C.L. Spegazzini) C.G. Hansford ex C.G. Hansford. Beihefte zur Sydowia Annales Mycologici 2:46–47. 1961.

Meliola campylopoda H. Sydow. Annales Mycologici 24:298–299. 1926. **Isotypes:** Hab. in foliis *Viticis umbrosae* vel speciei affin. [a misidentification], Costa Rica, Piedades de San Ramón, 7 FEB 1925, leg. H. Sydow, Fungi in Itinere Costaricensi Collecti No. 29, ILL 4884, also in bound set of Fungi exotici exsiccati (No. 617). [The host was later identified (cf. C.G. Hansford, 1961:429) and the identity confirmed by us as *Billia columbiana* Planchon & J.J. Linden in Hippocastanaceae.].

Meliola canangae F.L. Stevens ex C.G. Hansford. Beihefte zur Sydowia Annales Mycologici 1:101–102. 1957. **Holotype:** Hab. in foliis *Canangae odoratae* [= *Cananga odorata* (Lam.) J.D. Hooker & T. Thomson], Philippines, Luzon, Tayabas [Prov.], Basiad, DEC 1916, leg. H.S. Yates, ex Phil. Bur. Sci. No. 25685, ILL 4872.

Meliola canariicola F.L. Stevens ex C.G. Hansford. Beihefte zur Sydowia Annales Mycologici 1:102. 1957. **Holotype:** Hab. in foliis *Canarii* sp. [= *Canarium* sp.], Philippines, Luzon, Camarinea Prov., Paracale, [NOV–DEC 1918], leg. M. Ramos & G. Edaño, ex Phil. Bur. Sci. No. 34009, ILL 4871.

Meliola capensis (K. Kalchbrenner & M.C. Cooke) F. Theissen var. *cupaniae* C.G. Hansford. Sydowia Annales Mycologici 10:66–67. (1956) 1957. **Holotype:** Hab. in foliis *Cupaniae americanae* [= *Cupania americana* L.], Porto Rico [Puerto Rico, Mayagüez, 23 JUN 1915], leg. F.L. Stevens No. 7372, ILL 6394.

Meliola capensis (K. Kalchbrenner & M.C. Cooke) F. Theissen var. *euphoriae* C.G. Hansford. Beihefte zur Sydowia Annales Mycologici 1:102–103. 1957. **Holotype:** Hab. in foliis *Euphoriae ?cinereae* [*Euphoria ?cinerea* Radlk. = *Nephelium* sp., Philippines], Sulu Archipelago, Tawitawi Island, [OCT 1919], leg. H.S. Yates, ex Phil. Bur. Sci. No. 36137, ILL 5105

[Hansford marked the ILL specimen as "type"]. **Paratypes:** Leg. H.S. Yates, ex Phil. Bur. Sci. No. 36155, ILL 5103; in foliis *Euphoriae didymae* [*Euphoria didyma* Blanco = *Nephelium glabrum* Noronha], leg. H.S. Yates, ex Phil. Bur. Sci. No. 36149, ILL 5104 [C.G. Hansford marked ILL specimens of the paratypes as "co-types"].

Meliola capensis (K. Kalchbrenner & M.C. Cooke) F. Theissen var. *mataybae* (F.L. Stevens) C.G. Hansford, 1961. **See basionym**: *Meliola mataybae* F.L. Stevens, 1928.

Meliola capsicicola F.L. Stevens. Illinois Biological Monographs 2:509–510, fig. 39. 1916 [as *capsicola*, corrected by F.C. Deighton (1968) in accordance with current I.C.B.N.]. **Holotype:** On *Capsicum baccatum* sensu auct. non L. [= *C. annuum* L.], Porto Rico (Puerto Rico), Dos Bocas, below Utuado, 8 JUL 1915, leg. F.L. Stevens No. 8019, ILL 4867. **Paratype:** Manati, 2 JUL 1915, leg. F.L. Stevens No. 7698, ILL 4866.

Meliola carissae E.M. Doidge. Bothalia 1(2):72, fig. 1. 1922. **Isotype:** On leaves of *Carissa arduina* Lam. [= *C. bispinosa* (L.) Desf., South Africa], Pirie Forest, Kingwilliamstown, 8 AUG 1919, leg. E.M. Doidge, ex PREM No. 12296, ILL 4865.

Meliola carissae E.M. Doidge var. *indica* C.G. Hansford. Sydowia Annales Mycologici 10:67. (1956) 1957. **Isotypes:** Hab. in foliis *Carissae carandatis* [= *Carissa carandas* L.], India, [Poona, Bombay], Bassein, leg. B.N. Uppal No. 22, ILL 4864, AMH. **Holotype:** FH [Hansford erroneously cites Burma as the country of origin].

Meliola catubigensis H.S. Yates. Philippine Journal of Science, Section C (Botany) 12:363–364. (1917) 1918. **Isotype:** On *Loranthus* sp., Philippines, Samar [Island], Catubig River, 17 FEB 1916, leg. M. Ramos, ex Phil. Bur. Sci. No. 24624, ILL 4861.

Meliola cavitensis H.S. Yates. Philippine Journal of Science, Section C (Botany) 13:366. (1918) 1919. **Isotype:** On *Coleus* sp. [= *Plectranthus* sp.], Philippines, Luzon, Cavite Prov., Talisay Ridge, 21 JAN 1917 [as 20 JAN on the packet], leg. E.D. Merrill, ex Phil. Bur. Sci. No. 10634, ILL 5647.

Meliola celtidicola H.S. Yates. Philippine Journal of Science, Section C (Botany) 13:366–367. (1918) 1919 [as *celticola*, corrected in accordance with Articles 32.6, 60 & 61 of I.C.B.N.]. **Isotype:** On *Celtis philippensis* Blanco, [Philippines], Luzon, Ilocos Norte Prov., Bangui, 25 FEB 1917, leg. M. Ramos, ex Phil. Bur. Sci. No. 27746, ILL 4907.

Meliola celtidis H.S. Yates. Philippine Journal of Science, Section C (Botany) 13:367. (1918) 1919 [as *celtidiae*, corrected by C.G. Hansford (1961:317) in accordance with current I.C.B.N.]. **Isotype:** On *Celtis luzonica* Warburg in Perkins [as *luzonensis*, Philippines], Samar [Island], Catubig River, FEB–MAR 1916, leg. M. Ramos, ex Phil. Bur. Sci. No. 24616, ILL 4914.

Meliola cestri L.R. Tehon. Botanical Gazette 67:505–506. 1919. **Holotype:** On *Cestrum* sp., Porto Rico [Puerto Rico], Mayagüez, [Mora], 29 JUN 1915, leg. F.L. Stevens No. 7576, ILL 4916.

Meliola cestricola F.L. Stevens. Annales Mycologici 26:211, pl. IV, fig. 30. 1928. **Holotype:** On *Cestrum* sp., Costa Rica, Peralta, 12 JUL 1923, leg. F.L. Stevens No. 346, ILL 4915.

Meliola cestri-macrophylii C.G. Hansford. Beihefte zur Sydowia Annales Mycologici 1:103. 1957. **Holotype:** Hab. in foliis *Cestri macrophylli* [*Cestrum macrophyllum* Vent. = *C. laurifolium* L'Héritier], Porto Rico [Puerto Rico], El Gigante, [16 JUL 1915], leg. F.L. Stevens No. 8561, ILL 5202.

Meliola chaetochloae F.L. Stevens. Annales Mycologici 26:283, pl. VI, fig. 79. 1928. **Holotype:** On *Chaetochloa sulcata* (Aublet) A.S. Hitchc. [= *Setaria paniculifera* (Steudel) Fournier], Ecuador, Terecita, 30 OCT 1924 [as SEP on the packet], leg. F.L. Stevens No. 138, ILL 4917.

Meliola chagres F.L. Stevens. Annales Mycologici 26:178–179, pl. II, fig. 5. 1928. **Holotype:** On *Inga* sp., Panama, Chagres mouth, 23 AUG 1923 [as 1924 on the packet], leg. F.L. Stevens No. 1288, ILL 4924.

Meliola chamaecristae F.S. Earle. Bulletin of the New York Botanical Garden 3:304. Preprint 1904 (journal issued in 1905). **Isotype:** On leaves and stems of *Chamaecrista glandulosa* (L.) E. Greene, Porto Rico [Puerto Rico], Santurce (San Antonio Station), [7 JAN 1903], leg. A.A. Heller No. 6371, ILL 4918. **Holotype:** NY.

Meliola chamaecristicola F.L. Stevens. Illinois Biological Monographs 2:494–495, fig. 24. 1916. **Holotype:** On *Chamaecrista granulata* (Urban) Britton [= *C. lineata* (Sw.) E. Greene], Porto Rico [Puerto Rico], Mona Island, 20 DEC 1913, leg. F.L. Stevens No. 6113, ILL 4324. **Isotypes:** ILL 4919, BPI 70997. ≡ *Irenopsis chamaecristicola* (F.L. Stevens) F.L. Stevens. Annales Mycologici 25:436. 1927.

Meliola chelonanthi C.G. Hansford. Beihefte zur Sydowia Annales Mycologici 1:104. 1957. **Isotype:** Hab. in foliis *Chelonanthi acutanguli* [*Chelonanthus acutangulus* (Ruiz & Pavon) Gilg in Engler & Prantl = *Irlbachia alata* (Aublet) Maas ssp. *alata*], Costa Rica, El Alto, leg. F.L. Stevens No. 246, ILL 5391. **Holotype:** FH.

Meliola chiococcae F.L. Stevens. Illinois Biological Monographs 2:495–496, fig. 26. 1916. **Holotype:** On *Chiococca alba* (L.) A.S. Hitchc., Porto Rico [Puerto Rico], Vega Baja, 2 JUL 1915, leg. F.L. Stevens No. 7743, ILL 4325 [also a paratype of *Isthmospora spinosa* F.L. Stevens]. **Paratype:** Hormigueros, 23 JUN 1915, leg. F.L. Stevens No. 7325, ILL 4326. ≡ *Irenopsis chiococcae* (F.L. Stevens) F.L. Stevens. Annales Mycologici 25:434. 1927.

Meliola circinans F.S. Earle. Bulletin of the New York Botanical Garden 3:304–305. Preprint 1904 (journal issued in 1905). **Isotype:** On leaves of *Rhynchospora aurea* Vahl [= *R. corymbosa* (L.) Britton], Porto Rico [Puerto Rico, calcarious hills east of Santurce, 8 JAN 1903], leg. A.A. Heller No. 6384, ILL 4932. **Holotype:** NY.

Meliola cissi-rhombifoliae C.G. Hansford. Beihefte zur Sydowia Annales Mycologici 1:104. 1957, nom. and stat. nov. **Based on:** *Meliola rizalensis* H. & P. Sydow var. *panamensis* F.L. Stevens, 1928, non *M. panamensis* F.L. Stevens, 1928, and on the type of the varietal name. **Holotype:** On *Cissus rhombifolia* Vahl, Panama, Fort Randolph, 100 ft. hill trail, 9 SEP 1923, leg. F.L. Stevens No. 761, ILL 6179.

Meliola citricola H. & P. Sydow. Annales Mycologici 15:183–184. 1917. **Isotype:** In foliis *Citri* (verisimiliter, *Citrus nobilis* Lour.), Philippines, Luzon, Laguna Prov., San Antonio, OCT 1915, leg. M. Ramos, ex Phil. Bur. Sci. No. 23747, ILL 4934.

Meliola cladophaga H. Sydow. Annales Mycologici 24:299–301. 1926. **Isotype:** Hab. in ramis vivis *Crotonis gossypiifolii* [= *Croton gossypiifolius* Vahl], Costa Rica, San Pedro de San Ramón, 1 FEB 1925, leg. H. Sydow, Fungi in Itinere Costaricensi Collecti No. 207, ILL 4939. [This collection is cited in the protologue. C.G. Hansford (1961), however, cited Fungi Exotici Exsiccati No. 618 as "type," which also may be part of the original material.].

Meliola? clavatispora C.L. Spegazzini. Boletín de la Academia Nacional de Ciencias en Córdoba 11:500–501. 1889; Fungi Puiggariani, Pugillus I, p. 122, No. 241. 1889. **Isotype:** Ad folia viva *Apocineae* speciei [*Apocynum* sp.] cujusdam in sylvis [Brazil], pr. Apiahy [Apiai], APR 1881, leg. J.I. Puiggari No. 1701, ILL 6536. **Holotype:** LPS 520. ≡ *Meliolinopsis clavatispora* (C.L. Spegazzini) M. Beeli. Bulletin du Jardin Botanique de L'État à Bruxelles 7:119. 1920.

Meliola claviculata E.M. Doidge. Transactions of the Royal Society of South Africa 8:113. 1920. **Isotype:** In foliis *Oncobae* sp. [= *Oncoba* sp.], Portuguese East Africa, [now Mozambique, Zambezia Prov.], Quelimane, 18 SEP 1913 [as 14 SEP on the packet], leg. I.B. Pole Evans, ex PREM No. 7388, ILL 4327. ≡ *Irenopsis claviculata* (E.M. Doidge) F.L. Stevens. Annales Mycologici 25:440. 1927.

Meliola clavispora N.T. Patouillard. Journal de Botanique 4:61, fig. 4. 1890. **Assumed isotype:** Sur feuilles vivantes, [Vietnam], Tonkin, Fu-Phap, [JAN 1889], leg. B. Balansa s.n., ILL 6478. ≡ *Patouillardina clavispora* (N.T. Patouillard) G. Arnaud. Comtes Rendus Hebdomadaires des Séances de l'Académie des Sciences 164:890. 1917. ≡ *Meliolaster clavisporus* (N.T. Patouillard) F. v. Höhnel. Berichte der Deutschen Botanischen Gesellschaft 35:701. (1917) 1918. ≡ *Meliolinopsis clavispora* (N.T. Patouillard) M. Beeli. Bulletin du Jardin Botanique de L'État à Bruxelles 7:119. 1920.

Meliola clavulata H.G. Winter var. *batatae* F.L. Stevens. Annales Mycologici 26:241. 1928. **Holotype**: On *Ipomoea batatas* (L.) Lam., British Guiana [Guyana], Tumatumari, 11 JUL 1922, leg. F.L. Stevens No. 214, ILL 4823. **Paratypes**: 12 JUL 1922, leg. F.L. Stevens No. 229, ILL 4826; Kartabo, 23 JUL 1922, leg. F.L. Stevens No. 632, ILL 4818 and 4820; on *Ipomoea* sp., Costa Rica, Experiencia Farm, 18 JUL 1923, leg. F.L. Stevens No. 516, ILL 4834; Peralta, 11 JUL 1923, leg. F.L. Stevens No. 344, ILL 4830; 12 JUL 1923, leg. F.L. Stevens No. 355, ILL 4827; leg. F.L. Stevens No. 348, ILL 4828; Panama, Summit, 6 SEP 1924, leg. F.L. Stevens No. 329, ILL 4838; Pedro Miguel, 9 SEP 1924, leg. F.L. Stevens No. 392, ILL 4836; Chiva-Chiva trail, 18 SEP 1924, leg. F.L. Stevens No. 600, ILL 4829; leg. F.L. Stevens No. 616, ILL 4831; Baille Mona, 20 SEP 1924, leg. F.L. Stevens No. 683, ILL 4833; Gatun, 26 SEP 1924, leg. F.L. Stevens No. 836, ILL 4819; 11 OCT 1924, leg. F.L. Stevens No. 1210, ILL 4824; Mandingo, 15 OCT 1924, leg. F.L. Stevens No. 1333, ILL 4832; leg. F.L. Stevens No. 1345, ILL 4821; Las Cruces trail, 28 SEP 1924, leg. F.L. Stevens No. 866, ILL 4845.

Meliola clerodendricola P.C. Hennings. Hedwigia 37:288–289. 1898. **Holotype**: On *Clerodendrum capitatum* (Willd.) Schumacher & Thonning [as *Clerodendron*], C. Africa, Bongoland, Boiko bei Sabbi, 8 DEC 1869, leg. G. Schweinfurth No. 2753, ILL 4945.

Meliola clerodendricola P.C. Hennings var. *micromera* (H. & P. Sydow) C.G. Hansford, 1961. **See basionym**: *Meliola micromera* H. & P. Sydow, 1914.

Meliola clidemiae (F.L. Stevens) R. Ciferri, 1938. **See basionym**: *Irenina clidemiae* F.L. Stevens, 1927.

Meliola clusiae F.L. Stevens. Illinois Biological Monographs 2:520. 1916. **Holotype**: On *Clusia minor* L., Porto Rico [Puerto Rico], El Alto de la Bandera, 14 JUL 1915, leg. F.L. Stevens No. 8283, ILL 3543 [also the holotype of *Grallomyces portoricensis* F.L. Stevens]. **Paratype**: 15 JUL 1915, leg. F.L. Stevens No. 8571, ILL 3544. ≡ *Amazonia clusiae* (F.L. Stevens) F.L. Stevens. Annales Mycologici 25:415. 1927.

Meliola coccolobae F.L. Stevens & L.R. Tehon. Mycologia 18:5, pl. 1, fig. 2. 1926 [as *coccolobis*, corrected in accordance with Rec. 60.1 and Articles 32.6 & 61 of I.C.B.N.]. **Holotype**: On *Coccoloba* sp. [as *Coccolobis* (orth. var.)], British Guiana [Guyana], Kartabo, 24 JUL 1922, leg. F.L. Stevens No. 655, ILL 4950. **Paratype**: Trinidad, Cumuto, 16 AUG [1922], leg. F.L. Stevens No. 903, ILL 4949.

Meliola columneae F.L. Stevens. Annales Mycologici 26:247, pl. V, fig. 55. 1928. **Holotype**: On *Columnea heterophylla* Hanstein, Costa Rica, Siquirres, 31 JUL 1923, leg. F.L. Stevens No. 677, ILL 4951.

Meliola comata E.M. Doidge. Transactions of the Royal Society of South

Africa 8:111–112. 1920. **Isotype:** On *Pyrenacantha scandens* Planchon ex Harvey [Icacinaceae, originally misidentified as *Ipomoea* sp. on the packet], South Africa, George, Woodville Forest, 15 FEB 1917 [as 11 FEB on the packet], leg. E.M. Doidge, ex PREM No.11020, ILL 4329. ≡ *Irenopsis comata* (E.M. Doidge) F.L. Stevens. Annales Mycologici 25:437. 1927.

Meliola combinans R. Ciferri. Sydowia Annales Mycologici 10:136–137. (1956) 1957, nom. prov. [invalid.]. **Isotype:** Hab. in foliis vivis vel siccis *Phari* sp. (Graminaceae) [= *Pharus* sp. (Poaceae), Dominican Republic], Puerto Plata Prov., Cordillera Septentrional, pr. viam ad Puerto Plata, 6 AUG 1931, leg. R. Ciferri, Mycoflora Dominigensis Exsiccata No. 339, ILL 33108.

Meliola commixta H. Sydow. Leaflets of Philippine Botany 9:3117–3118. 1925. **Isotype:** On leaves of *Nephelium mutabile* Blume, [Philippines, Luzon], Sorsogon Prov., Irosin, APR 1916, leg. A.D.E. Elmer No. 15686, ILL 4952.

Meliola comocladiae F.L. Stevens. Illinois Biological Monographs 2:493–494, fig. 22. 1916. **Holotype:** On *Comocladia glabra* (J.A. Schultes) Sprengel, Porto Rico [Puerto Rico], Rosario, 4 JUL 1915, leg. F.L. Stevens No. 9015, ILL 4334 [also a potential isotype of *Micropeltidium portoricense* C.L. Spegazzini]. **Paratypes:** Mayagüez Mesa, 25 JUN 1915, leg. F.L. Stevens No. 7484, ILL 4332; 15 JUN 1915, leg. F.L. Stevens No. 7056, ILL 4331; Maricao, 3 APR 1913, leg. F.L. Stevens No. 760, ILL 4330 [F.L. Stevens numbers 760, 7056 and 7484 are also paratypes of *Helminthosporium glabroides* F.L. Stevens]; on *Spondias mombin* L., Maricao, 3 APR 1913, leg. F.L. Stevens No. 749, ILL 4333. ≡ *Irenopsis comocladiae* (F.L. Stevens) F.L. Stevens. Annales Mycologici 25:440. 1927.

Meliola compositarum F.S. Earle. Bulletin of the New York Botanical Garden 3:306. Preprint 1904 (journal issued in 1905). **Paratypes:** On *Eupatorium* sp., Porto Rico [Puerto Rico, 7 miles north of Ponce, 2 DEC 1902], leg. A.A. Heller No. 6185, ILL 3785; [near Rio Piedras, 13 JAN 1899], leg. A.A. Heller No. 141, ILL 3786. **Holotype:** NY. ≡ *Appendiculella compositarum* (F.S. Earle) R.A. Toro. Mycologia 17:144. 1925.

Meliola compositarum F.S. Earle var. *portoricensis* F.L. Stevens. Illinois Biological Monographs 2:490, fig. 16. 1916. **Holotype:** On *Eupatorium portoricense* Urban, Porto Rico [Puerto Rico], Vega Baja, 5 NOV 1913, leg. F.L. Stevens No. 4301, ILL 3840. **Paratypes:** Arecibo-Lares Road, 20 JUN 1915, leg. F.L. Stevens No. 7320, ILL 3805 [also a paratype of *Helminthosporium glabroides* F.L. Stevens]; Dos Bocas, below Utuado, 16 DEC 1913 [also as 30 DEC 1913 in the protologue, i.e., cited twice], leg. F.L. Stevens No. 6031, ILL 3796 [also a paratype of *Perisporium meliolae* F.L. Stevens and of *Helminthosporium glabroides* F.L. Stevens]; leg. F.L. Stevens No. 6032, ILL 3777 and 11033 [also the holotype and an isotype, respectively, of *Perisporium meliolae* F.L. Stevens, as well as paratypes of both *Coniothyrium glabroides* F.L. Stevens and *Helminthosporium glabroides* F.L. Stevens]; 29 DEC 1913, leg. F.L. Stevens No. 6830, ILL 3515 and 8670 [also paratypes of *Phaeodothiopsis eupatorii* F.L. Stevens]; 30 DEC 1913, leg. F.L. Stevens No. 6861, ILL 3795 [also a paratype of *Perisporium meliolae* F.L. Stevens]; leg. F.L. Stevens No. 6866, ILL 3794 and 8673 [also an isotype and the holotype, respectively, of *Phaeodothiopsis eupatorii* F.L. Stevens, as well as paratypes of *Perisporium meliolae* F.L. Stevens]; San Sebastian, 22 NOV 1913, leg. F.L. Stevens No. 5192 [p.p. on the packet], ILL 3773, 4955 and 6590 [also paratypes of *Perisporium meliolae* F.L. Stevens]. ≡ *Irene sororcula* (C.L. Spegazzini) F.L. Stevens var. *portoricensis* (F.L. Stevens) F.L. Stevens. Annales Mycologici 25:425. 1927. ≡ *Appendiculella sororcula* (C.L. Spegazzini) C.G. Hansford var. *portoricensis* (F.L. Stevens) C.G. Hansford. Beihefte zur Sydowia Annales Mycologici 2:616–617. 1961.

Meliola conferta L.R. Tehon. Botanical Gazette 67:502–503, pl. XVIII, figs. 17–19. 1919, nom. illegit., non E.M. Doidge, 1917. **Holotype**: On leaves of *Rhacoma crossopetalum* L. [= *Crossopetalum rhacoma* Crantz], Porto Rico [Puerto Rico], Mona Island, 20 DEC 1913, leg. F.L. Stevens No. 6147, ILL 4335. ≡ *Meliola tehoniana* A. Trotter, Sylloge Fungorum 24:276. 1926, nom. nov. ≡ *Irenopsis conferta* (L.R. Tehon) F.L. Stevens. Annales Mycologici 25:434. 1927, comb. nov. illegit. [cf. Articles 11.3 & 58.3 of I.C.B.N.]. ≡ *Irenopsis tehoniana* (A. Trotter) C.G. Hansford. Beihefte zur Sydowia Annales Myciologici 2:339–340. 1961.

Meliola confragosa H. & P. Sydow. Leaflets of Philippine Botany 5:1536. 1912. **Isotype**: On living leaves of a Cucurbit [Cucurbitaceae indet., later identified as *Trichosanthes quinquangulata* A. Gray (cf. C.G. Hansford, 1961)], Philippines, Palawan [Island], Brooks Point, (Addison Peak), FEB 1911, leg. A.D.E. Elmer No. 12625, ILL 3945. ≡ *Irene confragosa* (H. & P. Sydow) H. & P. Sydow. Annales Mycologici 15:195. 1917. ≡ *Irenina confragosa* (H. & P. Sydow) F.L. Stevens. Annales Mycologici 25:465. 1927. ≡ *Asteridiella confragosa* (H. & P. Sydow) C.G. Hansford ex C.G. Hansford. Beihefte zur Sydowia Annales Myciologici 2:124. 1961.

Meliola congoensis (M. Beeli) C.G. Hansford, 1961. **See basionym**: *Meliola perpusilla* H. Sydow var. *congoensis* M. Beeli, 1920.

Meliola conica F.L. Stevens. Annales Mycologici 26:228, pl. IV, fig. 37. 1928.

Lectotype: On Leguminosae (Mimosaceae) sp. indet. [= Fabaceae subfam. Mimosoideae; as "perhaps *Acacia* sp." on the packet. From comparison with other specimens, however, we place all the host material (including the residual syntypes) in the genus *Pentaclethra*, rather than *Acacia*], Costa Rica, Sabario, 8 AUG 1923, leg. F.L. Stevens No. 787, ILL 4956 [this is the specimen illustrated in fig. 37, and confirmation of No. 787 as "type" by C.G. Hansford (1961:266) is interpreted as lectotypification. Note: This is also a paratype of *Asteridiella pentaclethrae* C.G. Hansford.]. **Residual syntypes:** leg. F.L. Stevens No. 795, ILL 4959; Las Mercedes, 17 JUL 1923, leg. F.L. Stevens No. 493, ILL 4957; Parismina Junction, 20 JUL 1923, leg. F.L. Stevens No. 607, ILL 4958.

Meliola conigera F.L. Stevens & L.R. Tehon. Mycologia 18:9, pl. 1, fig. 7. 1926. **Holotype:** On *Pentaclethra* sp. [identified as *P. macroloba* (Willd.) O. Kuntze in C.G. Hansford, 1961: 266], British Guiana [Guyana], Demerara-Essequibo R.R., 15 JUL 1922, leg. F.L. Stevens No. 387a, ILL 4961 [also an isotype of *Asteridiella pentaclethrae* C.G. Hansford]. **Paratypes:** Wismar, 14 JUL [1922], leg. F.L. Stevens No. 290, ILL 4960; on *Pentaclethra macroloba* (Willd.) O. Kuntze, Kartabo, 22 JUL [1922], leg. F.L. Stevens No. 529, ILL 4962.

Meliola connari H.S. Yates. Philippine Journal of Science, Section C (Botany) 12:364. (1917) 1918 [as *connariae*, corrected by later authors, e.g., A. Trotter (1926) and C.G. Hansford (1957), in accordance with current I.C.B.N.]. **Isotypes:** On *Connarus* sp., Philippines, Luzon, Tayabas Prov., Basiad, 20 DEC 1916, leg. H.S. Yates, ex Phil. Bur. Sci. No. 25622, ILL 4963 and 4964.

Meliola connari H.S. Yates var. *panamensis* C.G. Hansford. Beihefte zur Sydowia Annales Mycologici 1:105. 1957. **Holotype:** Hab. in foliis *Connari panamensis* [= *Connarus panamensis* Griseb.], Panama, [Gatuncillo, 18 AUG 1923], leg. F.L. Stevens No. 1143, ILL 4965.

Meliola constipata (C.L. Spegazzini) C.L. Spegazzini, 1924. **See basionym:** *Meliola bicornis* H.G. Winter var. *constipata* C.L. Spegazzini, 1888.

Meliola contigua P.A. Karsten & C. Roumeguère. Revue Mycologique 12:77. 1890. **Isotypes:** In foliis vivis Palmieri acaulis [Arecaceae sp. indet., Tonkin (Vietnam)], Ououlu, NOV 1888, leg. B. Balansa No. 18, [ex C. Roumeguère, Fungi Selecti Exsiccati No. 5421], ILL 5916 and 5917.

Meliola contorta F.L. Stevens. Illinois Biological Monographs 2:500, fig. 30. 1916. **Holotype:** On *Piper hispidum* Sw., Porto Rico [Puerto Rico], Las Marias, 11 JUL 1915, leg. F.L. Stevens No. 8225, ILL 4981.

Meliola cookeana C.L. Spegazzini. Anales de la Sociedad Cientifica Argentina 12:101. 1881. **Isotype:** [In foliis *Callicarpae americanae* (= *Cal*-

licarpa americana L.), Florida, Gainesville], leg. ?H.W. Ravenel, Fungi Americani Exsiccati No. 84 [distributed as *Meliola amphitricha* E.M. Fries], ILL 4564 [as a microscopic preparation], and in bound exsiccati set.

Meliola cookeana C.L. Spegazzini var. *aegiphilae* (F.L. Stevens) C.G. Hansford, 1961. **See basionym:** *Meliola aegiphilae* F.L. Stevens, 1928.

Meliola cordiicola C.G. Hansford. Beihefte zur Sydowia Annales Mycologici 1:105. 1957. **Holotype:** Hab. in foliis *Cordiae* sp. [= *Cordia* sp.], Porto Rico [Puerto Rico], Mayagüez, [25 JUN 1915], leg. F.L. Stevens No. 7472, ILL 4126.

Meliola cornu-caprae P.C. Hennings. Hedwigia 43:362. 1904. **Isotypes:** An Stämmen einer Euphorbiacee [= Euphorbiaceae sp. indet.], Amazonas, [Brazil], Rio Juruá, Juruá-Miry, JUL 1901, leg. E.H.G. Ule No. 2971, ILL 3735 [as microscopic preparations], S. ≡ *Appendiculella cornu-caprae* (P.C. Hennings) F. v. Höhnel. Sitzungsberichte der Kaiserl. Akademie der Wissenschaften in Wien. Mathematisch-Naturwissenschaftliche Klasse. Abt. 1, 128:556. 1919. ≡ *Irene cornu-caprae* (P.C. Hennings) F.L. Stevens. Annales Mycologici 25:426. 1927.

Meliola coronata C.L. Spegazzini. Anales de la Sociedad Científica Argentina 17:133–134. 1884; Fungi Guaranitici, Pugillus I, pp. 70–71, No. 175. 1884. **Isotypes:** Hab. ad folia viva *Lueheae divaricatae* [as *Lueheae* = *Luehea divaricta* C. Martius], in sylvis subvirgineis, [Paraguay], pr. Guarapi, JUL 1883, leg. B. Balansa No. 3847, comm. C.L. Spegazzini, ILL 4312 and 4313. ≡ *Irenopsis coronata* (C.L. Spegazzini) F.L. Stevens. Annales Mycologici 25:435. 1927.

Meliola crenata H.G. Winter in A. Gaillard var. *bunchosiae* C.G. Hansford. Beihefte zur Sydowia Annales Mycologici 1:105–106. 1957. **Holotype:** Hab. in foliis *Bunchosiae cornifoliae* [= *Bunchosia cornifolia* Kunth in H.B.K.], Panama, [Lama Bracha], 13 SEP 1924, leg. F.L. Stevens No. 496, ILL 4989.

Meliola crescentiae F.L. Stevens. Annales Mycologici 26:240–241, pl. IV, fig. 42. 1928. **Lectotype** : On *Crescentia* sp., Trinidad, Cumuto, 18 AUG 1922, leg. F.L. Stevens No. 940, ILL 4992 [this collection is cited in the caption of fig. 42. C.G. Hansford's (1961:673) confirmation of No. 940 as "type" is accepted as lectotypification with the specimen at ILL; this is also an isotype of *Masonia crescentiae* C.G. Hansford.]. **Isolectotype:** FH. **Residual syntype:** On *Heterophragma roxburghii* DC., India, Bombay, Dharwar, DEC 1918, leg. L.J. Sedgwick [as. F.L. Stevens No. 1993 on the packet], ILL 4991.

Meliola cristata F.L. Stevens. Annales Mycologici 26:193, pl. III, fig. 16. 1928. **Holotype:** On *Calopogonium caeruleum* (Bentham) Sauvalle, Panama, Mandingo, 15 OCT 1924, leg. F.L. Stevens No. 1355, ILL

4998. **Isotype:** ILL 5002. **Paratypes:** On *Calopogonium caeruleum* (Bentham) Sauvalle [as unknown legume on the packet], Panama, Summitt, 28 SEP 1924, leg. F.L. Stevens No. 466, ILL 4994; Las Cruces trail, 28 SEP 1924, leg. F.L. Stevens No. 891, ILL 4993; Pedro Miguel, 9 SEP 1924, leg. F.L. Stevens No. 384, ILL 4999 and 5001; France Field, 3 OCT 1924, leg. F.L. Stevens No. 981, ILL 5003; Ecuador, Terecita, 29 OCT 1924, leg. F.L. Stevens No. 44, ILL 4996; on *Phaseolus* sp., British Guiana [Guyana], Kartabo, 22 JUL 1922, leg. F.L. Stevens No. 614, ILL 4997.

Meliola crotonicola F.L. Stevens. Annales Mycologici 26:184, pl. II, fig. 11. 1928. **Holotype:** On *Croton* sp., Costa Rica, Siquirres, 31 JUL 1923, leg. F.L. Stevens No. 687, ILL 4968.

Meliola crustacea C.L. Spegazzini. Boletín de la Academia Nacional de Ciencias en Córdoba 11:496. 1889; Fungi Puiggariani, Pugillus I, p. 118, No. 235. 1889. **Isotype:** Ad folia viva *Drymidis* speciei [= *Drimys* sp.] cujusdam in dumetis, [Brazil], pr. Apiahy [Apiai], Hiem 1881, leg. J.I. Puiggari s.n., ex herb. C.L. Spegazzini, ILL 3924. ≡ *Irene crustacea* (C.L. Spegazzini) F. Theissen & H. Sydow. Annales Mycologici 15:461. (1917) 1918. ≡ *Irenina crustacea* (C.L. Spegazzini) F.L. Stevens. Annales Mycologici 25:468. 1927. ≡ *Asteridiella crustacea* (C.L. Spegazzini) C.G. Hansford ex C.G. Hansford. Beihefte zur Sydowia Annales Mycologici 2:26–27. 1961.

Meliola cryptocarpa J.B. Ellis & G.W. Martin. American Naturalist 17:1284. 1883. **Isotypes:** On leaves of *Gordonia lasianthus* L., Florida, [Green Cove Springs, JAN 1883], leg. G.W. Martin, [ex J.B. Ellis, North American Fungi No. 1293], ILL 4970 [as a microscopic preparation] and in bound exsiccati set, DAOM. ≡ *Irenina cryptocarpa* (J.B. Ellis & G.W. Martin) C.G. Hansford. Proceedings of the Linnean Society of London 157:171. 1946. ≡ *Irenopsis cryptocarpa* (J.B. Ellis & G.W. Martin) C.G. Hansford. Beihefte zur Sydowia Annales Mycologici 2:127–128. 1961. ≡ *Leptomeliola cryptocarpa* (J.B. Ellis & G.W. Martin) S.J. Hughes. Mycological Papers. International Mycological Institute 166:191. 1993.

Meliola cubitella (F.L. Stevens & L.R. Tehon) R. Ciferri, 1954, nom. invalid. **See basionym:** *Irene cubitella* F.L. Stevens & L.R. Tehon, 1926.

Meliola cubitella sensu C.G. Hansford. Beihefte zur Sydowia Annales Mycologici 2:249–250. 1961, non (F.L. Stevens & L.R. Tehon) R. Ciferri, 1954 (nom. invalid.). **Holotype:** On *Cassia* sp., British Guiana [Guyana, Tumatumari, 11 JUL 1922], leg. F.L. Stevens No. 193, ILL 3927. [Although Hansford cites R. Ciferri as the author and *Irene cubitella* F.L. Stevens & L.R. Tehon as the basionym, his name must be treated as a later homonym attributed solely to him as author and dating from 1961 because he explicitly excludes the type of the basionym and that of Ciferri's combination (Article 48.1 of I.C.B.N.). Hansford's name, however, is not validly published either because it lacks a Latin diagnosis based on the new type (cf. Article 36.1).].

Meliola cucurbitacearum F.L. Stevens. Illinois Biological Monographs 2:526, fig. 51. 1916. **Holotype:** On leaves and stems of an unknown cucurb [Cucurbitaceae], probably *Cayaponia* sp., Porto Rico [Puerto Rico], El Alto de la Bandera, 16 JUL 1915, leg. F.L. Stevens No. 8732, ILL 4972.

Meliola culebrensis C.G. Hansford. Beihefte zur Sydowia Annales Mycologici 1:106. 1957. **Holotype:** Hab. in foliis Acanthacearum spec. indet. [= Acanthaceae sp.], Panama, Culebra, [2 OCT 1924], leg. F.L. Stevens No. 912, ILL 6093.

Meliola cupaniae F.L. Stevens. Illinois Biological Monographs 2:497, fig. 28. 1916. **Holotype:** On *Cupania americana* L., Porto Rico [Puerto Rico], El Miradoro near Mayagüez, 4 AUG 1915, leg. F.L. Stevens No. 9143, ILL 4351. **Paratypes:** 4 AUG 1915, leg. F.L. Stevens No. 9489, ILL 4350; Maricao, 19 JUL 1915, leg. F.L. Stevens No. 8948, ILL 4348; Dos Bocas, near Utuado, 8 JUL 1915, leg. F.L. Stevens No. 8080, ILL 4349; Quebradillas, 22 NOV 1913, leg. F.L. Stevens No. 4979, ILL 4346; 11 AUG 1915, leg. F.L. Stevens No. 9318, ILL 4347. ≡ *Irenopsis cupaniae* (F.L. Stevens) F.L. Stevens. Annales Mycologici 25:434. 1927.

Meliola curvata H.S. Yates. Philippine Journal of Science, Section C (Botany) 13:367–368. (1918) 1919. **Isotypes:** On leaves of an unknown host [as *Melochia umbellata* (Houtt.) Stapf on the packet, Philippines], Samar [Island], Catubig River, FEB–MAR 1916, leg. M. Ramos, ex Phil. Bur. Sci. No. 24642, ILL 4351 and 4973 [also isotypes of *Meliola melochiae* C.G. Hansford]. ≡ *Irenopsis curvata* (H.S. Yates) F.L. Stevens. Annales Mycologici 25:437. 1927.

Meliola cyclopoda F.L. Stevens. Illinois Biological Monographs 2:484, fig. 9. 1916. **Holotype:** On *Pseudelephantopus spicatus* (Jussieu ex Aublet) C.F. Baker, Porto Rico [Puerto Rico], Vega Baja, 2 JUL 1915, leg. F.L. Stevens No. 7733, ILL 3937. **Paratype:** Leg. F.L. Stevens No. 7871, ILL 3938. ≡ *Irene cyclopoda* (F.L. Stevens) R.A. Toro. Mycologia 17:140. 1925. ≡ *Irenina cyclopoda* (F.L. Stevens) F.L. Stevens. Annales Mycologici 25:452. 1927. ≡ *Asteridiella cyclopoda* (F.L. Stevens) C.G. Hansford ex C.G. Hansford. Beihefte zur Sydowia Annales Mycologici 2:619. 1961.

Meliola cydistae F.L. Stevens. Annales Mycologici 26:193–194, pl. III, fig. 17. 1928. **Holotype:** On *Cydista* sp., Panama, Ancon, 1 SEP 1924, leg. F.L. Stevens No. 133, ILL 4974.

Meliola cyperi N.T. Patouillard in A. Gaillard. Le Genre *Meliola*, pp. 70–

71, illustr. 1892. **Isotype:** Ad utramque paginam foliorum *Cyperi cujusdam* [= *Cyperus* sp.] in locis paludosis pr. stationem militarem, Congo Française, Komba [River], 11 JUN 1881, leg. J. Dybowski, ILL 4620. **Holotype:** FH.

Meliola cyperi N.T. Patouillard in A. Gaillard var. *italica* P.A. Saccardo. Annales Mycologici 1:24. 1903. **Isotypes:** In foliis languidis v. emortuis *Cladii marisci* [= *Cladium mariscus* (L.) Pohl] in paludibus Meolo (Venezia) [Italy, Venice], SEP 1902, leg. Antonia Saccardo, ex P.A. Saccardo, Mycotheca Italica No. 1022, ILL 5298 and 5299; also ex H. Rehm, Ascomycetes Exsiccatae No. 1498, ILL 5300 [all as microscopic preparations]. **Holotype:** PAD. ≡ *Meliola italica* (P.A. Saccardo) F.L. Stevens. Annales Mycologici 26:282. 1928.

Meliola dalbergiae C.G. Hansford. Sydowia Annales Mycologici 11:55. (1957) 1958. **Holotype:** Hab. in foliis *Dalbergiae* sp. [= *Dalbergia monetaria* L.f. (on the packet and fide Hansford, 1961)], Porto Rico [Puerto Rico], Mayagüez, [25 JUN 1915], leg. F.L. Stevens No. 7476, ILL 4682. **Paratypes:** [4 JUL 1915], leg. F.L. Stevens No. 9016 p.p., ILL 4721; [Mayagüez Mesa, 29 JUN 1915], leg. F.L. Stevens No. 7577, ILL 4704 and 4742.

Meliola decidua C.L. Spegazzini. Boletín de la Academia Nacional de Ciencias en Córdoba 11:499–500. 1889; Fungi Puiggariani, Pugillus I, pp. 121–122, No. 240. 1889. **Isotype:** Ad folia viva Convolvulaceae? cujusdam in dumetis, [Brazil], pr. Apiahy [Apiai], APR 1888, leg. J.I. Puiggari, comm. C.L. Spegazzini No. 2344, ILL 5008. **Holotype:** LPS.

Meliola delicatula C.L. Spegazzini. Anales de la Sociedad Científica Argentina 26:25. 1888; Fungi Guaranitici, Pugillus II, p. 23, No. 63. 1883. **Isotypes:** Hab. ad folia viva *Myrsinis* [= *Myrsine* sp.] cujusdam in dumetis montanis, [Paraguay], Sierra de Peribebuy, 15 SEP 1883, leg. B. Balansa No. 3985, ILL 5010 and 5011.

Meliola depressula H. & P. Sydow. Annales Mycologici 15:184–185. 1917. **Isotype:** Hab. in foliis *Urceolae imberbis* [= Urceola imberbis (A.D.E. Elmer) Merrill], Philippines, [Luzon], Laguna Prov., [Hills Tack] pr. Paete, APR 1914, leg. C.F. Baker No. 3122 [ex Fungi Malayana No. 548], ILL 5014.

Meliola derridis H.S. Yates. Philippine Journal of Science, Section C (Botany) 13:368. (1918) 1919. **Isotype:** On leaves of *Derris* sp., Philippines, Luzon, Ilocos Norte Prov., Burgos, FEB–MAR 1917 [erroneously as FEB–MAY in the protologue], leg. M. Ramos, ex Phil. Bur. Sci. No. 27788, ILL 4505.

Meliola desmodii P.A. Karsten & C. Roumeguère. Revue Mycologique 12:77. 1890. **Isotypes:** Hab. ad folia *Desmodii* cujusdam [= *Desmodium* sp.], Tonkin [Vietnam], Tu-Phap, DEC 1887 [as DEC 1889 on the packet], leg. B. Balansa No. 5, ex C. Roumeguère, Fungi Selecti Exsiccati No. 5420, ILL 5016 [as a microscopic preparation], and in bound exsiccati set.

Meliola dichotoma M.J. Berkeley & M.A. Curtis. Proceedings of the American Academy of Arts and Sciences 4:130. 1860. **Isotype**: On leaves of some climbing plant [= *Hedera* sp., cf. C.G. Hansford (1961)], Japan, leg. C. Wright [No. 171, fide Hansford (1961)], ILL 33501.

Meliola dicranochaeta H. Sydow. Annales Mycologici 24:301–302. 1926. **Isotype**: Hab. in foliis *Cestri megalophylli* [= *Cestrum megalophyllum* Dunal], Costa Rica, San Pedro de San Ramón, 5 FEB 1925, leg. H. Sydow, Fungi in Itinere Costaricensi Collecti No. 390, ILL 5068.

Meliola didymopanacis P.C. Hennings. Hedwigia 34:106. 1895. **Isotype**:Auf Blättern von *Didymopanax* sp., Brasilia [Brazil], Minas Geraës, leg. A. Glaziou No. 1893, ILL 5067.

Meliola didymopanacis P.C. Hennings var. *stevensii* C.G. Hansford. Sydowia Annales Mycologici 9:15–16. 1955. **Lectotype** [designated herein]: Hab. in foliis *Dendropanacis arborei* [= *Dendropanax arboreus* (L.) Decaisne & Planchon], Porto Rico [Puerto Rico, Florida Adentro, 1 JUL 1915], leg. F.L. Stevens No. 7647, ILL 5066. **Isolectotypes**: ILL 5063, K, PREM.

Meliola dieffenbachiae F.L. Stevens. Illinois Biological Monographs 2:530, fig. 56. 1916. **Holotype**: On *Dieffenbachia seguine* (Jacquin) Schott [cited as *sequine*], Porto Rico [Puerto Rico], Las Marias, 10 JUL 1915, leg. F.L. Stevens No. 8148, ILL 5055. **Paratypes**: Leg. F.L. Stevens No. 8210, ILL 5056; Mariacao, 18 OCT 1913, leg. F.L. Stevens No. 3889, ILL 5042 and 5057; 19 JUL 1915, leg. F.L. Stevens No. 8851, ILL 5043; Cataño, 2 JUL 1915, leg. F.L. Stevens No. 7707, ILL 5044 and 5060; Lajas, 17 JUN 1915, leg. F.L. Stevens No. 7155, ILL 5049 and 5058; Monte de Oro, near Cayey, 3 DEC 1913, leg. F.L. Stevens No. 5666, ILL 5050; leg. F.L. Stevens No. 5731, ILL 5051; Dos Bocas below Utuado, 8 JUL 1915, leg. F.L. Stevens No. 8074, ILL 5052; leg. F.L. Stevens No. 8077, ILL 5041 [also the holotype of *Arthrobotryum dieffenbachiae* F.L. Stevens]; Mayagüez, 24 JUN 1915, leg. F.L. Stevens No. 7420, ILL 5053.

Meliola diospyri H. & P. Sydow in H. & P. Sydow & E.J. Butler var. *yatesiana* (A. Trotter) C.G. Hansford & F.C. Deighton, 1948. **See basionym**: *Meliola yatesiana* A. Trotter, 1926, nom. nov. for *M. diospyriae* H.S. Yates, (1917) 1918 (nom. illegit.).

Meliola diospyriae H.S. Yates. Philippine Journal of Science, Section C (Botany) 12:364–365. (1917) 1918, nom. illegit., non *M. diospyri* H. & P. Sydow in H. & P. Sydow & E.J. Butler, 1911. **Isotypes**: On leaves of *Diospyros discolor* Willd., Philippines, Luzon, Tayabas Prov., Basiad, 8 DEC 1916, leg. H.S. Yates, ex Phil. Bur. Sci. No. 25711, ILL 5040, K.

≡ *Meliola yatesiana* A. Trotter. Sylloge Fungorum 24:284. 1926, nom. nov. ≡ *Meliola diospyri* H. & P. Sydow in H. & P. Sydow & E.J. Butler var. *yatesiana* (A. Trotter) C.G. Hansford & F.C. Deighton. Mycological Papers. Commonwealth Mycological Institute 23:50. 1948.

Meliola dipholidis F.L. Stevens. Illinois Biological Monographs 2:512–513. 1916. **Holotype:** On *Dipholis salicifolia* (L.) A. DC. [= *Sideroxylon salicifolium* (L.) Lam.], Porto Rico [Puerto Rico], Guayanilla, 14 JUL 1915, leg. F.L. Stevens No. 8549, ILL 5037 [also a paratype of *Helminthosporium helleri* F.L. Stevens and a potential isotype of *Scolecopeltella portoricensis* C.L. Spegazzini]. **Paratype:** Quebradillas, 20 JUN 1915, leg. F.L. Stevens No. 7265, ILL 5038.

Meliola diphysae F.L. Stevens. Annales Mycologici 26:195, pl. III, fig. 18. 1928. **Holotype:** On *Diphysa robinioides* Bentham ex Bentham & Oersted, Panama, Bellavista, 7 OCT 1924, leg. F.L. Stevens No. 1124, ILL 5036.

Meliola discocalycis F.L. Stevens ex C.G. Hansford. Beihefte zur Sydowia Annales Mycologici 1:106. 1957. **Holotype:** Hab. in foliis *Discocalycis cybianthoidis* [misspelled "*cymbianthoidis*" = *Discocalyx cybianthoides* (A. DC.) Mez], Philippines, Luzon, Rizal Prov., SEP 1915, leg. H.S. Yates, ex Phil. Bur. Sci. No. 25057, ILL 5032.

Meliola doidgeae H. Sydow in E.M. Doidge & H. Sydow. Bothalia 2(2):457–458 and 465–466. 1928. **Isotype:** Hab. in foliis *Sapindi oblongifolii* [= *Sapindus oblongifolius* Sonder in Harvey & Sonder], South Africa, [Natal Prov.], Stella Bush, Durban, 11 JUL 1911, leg. E.M. Doidge, ex PREM No. 1572, ILL 6463.

Meliola dracaenae F.L. Stevens. Bernice P. Bishop Museum Bulletin 19:40, text fig. 8j. 1925. **Holotype:** On *Dracaena aurea* H. Mann [= *Pleomele aurea* (H. Mann) N.E. Brown], Hawaii, Kauai, Pipe trail, Upper Waimea Canyon, 15 JUN [1921], leg. F.L. Stevens No. 419, ILL 5029. **Isotypes:** ILL 5027, BISH 499032 and 499033. **Paratype:** [Oahu], in 1909, leg. C.N. Forbes-F.L. Stevens No. 1393, ILL 5030.

Meliola dracaenicola N.T. Patouillard & P.A. Hariot. Bulletin de la Société Mycologique de France 24:14–15. (1908) 1909. **Assumed Isotype:** In foliis *Dracaenae* cujusdam [= *Dracaena* sp., C. Africa] Congo, Brazzaville, 27–31 DEC 1903, leg. A. Chevalier No. 11212, ILL 5026.

Meliola drepanochaeta H. Sydow. Annales Mycologici 24:302–303. 1926. **Assumed lectotype:** Hab. in foliis vivis *Perseae cordatae* [*Persea cordata* (Vell.) Mez non Meisner in DC. = *P.* cf. *pyrifolia* Nees], Costa Rica, Piedades de San Ramón, 30 JAN 1925, leg. H. Sydow, Fungi in Itinere Costaricensi Collecti No. 163, ILL 5023 [citation of specimen in the F.L.S. herb. (ILL) as "type" by C.G. Hansford (1961:51) is interpreted as lectotypification; the packet has the original Sydow label, and a microscopic preparation of ILL 5023 by Hansford is marked "type"].

Meliola duggenae F.L. Stevens. Annales Mycologici 26:198, pl. III, fig. 20. 1928. **Holotype:** On *Duggena* sp. [= *Gonzalagunia* sp.], Panama, Fort Lorenzo trail, 10 OCT 1924, leg. F.L. Stevens No. 1159, ILL 5022.

Meliola duggenae F.L. Stevens var. *panamensis* F.L. Stevens. Annales Mycologici 26:198. 1928. **Lectotype:** On *Duggena panamensis* (Cav.) Standley [= *Gonzalagunia panamensis* (Cav.) K. Schum.], Panama, Chagres mouth, 23 AUG 1923, leg. F.L. Stevens No. 1314, ILL 5020 [citation of this collection as "type" by C.G. Hansford (1961:587) is accepted as lectotypification]. **Residual syntype:** On *Duggena rudis* Standley [= *Gonzalagunia rudis* (Standley) Standley], Ancon Hill, 24 SEP 1924, leg. F.L. Stevens No. 701, ILL 5021.

Meliola earlei F.L. Stevens. Illinois Biological Monographs 2:515, fig. 45. 1916 [as *earlii*, corrected in accordance with Articles 60 & 61 of I.C.B.N.]. **Holotype:** On *Pilea* sp., Porto Rico [Puerto Rico], Florida Adentro, 1 JUL 1915, leg. F.L. Stevens No. 7685, ILL 5078. **Paratypes:** On *Pilea parietaria* (L.) Blume, Rio Arecibo, K.64.7, 8 JUL 1915, leg. F.L. Stevens No. 7804, ILL 5079; on *Pilea nummularifolia* (Sw.) Weddell, Jajome Alto, 3 DEC 1913, leg. F.L. Stevens No. 5640, ILL 5080.

Meliola echinus P.C. Hennings. Hedwigia 43:363–364, illustr. 1904. **Isotypes:** Auf Blättern von *Coussapoa* sp., Amazonas [Brazil], Rio Juruá, Juruá-Miry, AUG 1901, leg. E.H.G. Ule No. 3134, [also as Mycotheca brasiliensis No. 57] ILL 3738 [fragment of specimen and several microscopic preparations], and in bound exsiccati set. ≡ *Appendiculella echinus* (P.C. Hennings) F. v. Höhnel. Sitzungsberichte der Kaiserl. Akademie der Wissenschaften in Wien. Mathematisch-Naturwissenschaftliche Klasse. Abt.1., 128:556. 1919. ≡ *Irene echinus* (P.C. Hennings) F.L. Stevens. Annales Mycologici 25:426. 1927.

Meliola edanoana F.L. Stevens ex C.G. Hansford. Beihefte zur Sydowia Annales Mycologici 1:107. 1957. **Holotype:** Hab. in foliis *Zanthoxyli* sp. [= *Zanthoxylum* sp.], Philippines, Mindanao [Island], Zamboanga, Malangas, [OCT-NOV 1919], leg. M. Ramos & C. Edaño, ex Phil. Bur. Sci. No. 36336 [as 36366 on the packet], ILL 5081.

Meliola elaeidis F.L. Stevens. Annales Mycologici 26:181, pl. II, fig. 8. 1928 [as *elaeis*, corrected by F.C. Deighton (1968) in accordance with current I.C.B.N.]. **Holotype:** On Palmae indet. [= Arecaceae indet.], Panama, Culebra, 2 OCT 1925 [as 2 SEP 1924 on the packet], leg. F.L. Stevens No. 943, ILL 5083. **Paratypes:** [On *Desmopsis panamensis* (Robinson) Safford], Panama, Chagres mouth, 22 AUG 1923, leg. F.L. Stevens No. 1266, ILL 5085; on *Elaeis melanococca* sensu auct. non Gaertner [= *E. oleifera* (Kunth) Cortes], Costa Rica, Swamp Mouth [as Port Limon on the packet], 9 AUG 1923, leg. F.L. Stevens No. 823, ILL 5084; Limon, 7 AUG 1923, leg. F.L. Stevens No. 770, ILL 5082.

Meliola elaeocarpi H.S. Yates. Philippine Journal of Science, Section C (Botany) 12:365. (1917) 1918 [as *elaeocarpeae*, corrected in accordance with Articles 32.6, 60 & 61 of I.C.B.N.]. **Isotype:** On *Elaeocarpus* sp., Philippines, Luzon, Benguet Subprov., 3 MAY 1916, leg. H.S. Yates, ex Phil. Bur. Sci. No. 25175, ILL 5086.

Meliola elephantopodis C.G. Hansford. Sydowia Annales Mycologici 11:55–56. (1957) 1958 [as *elephantopi*, corrected by F.C. Deighton (1968) in accordance with current I.C.B.N.]. **Holotype:** Hab. in foliis *Elephantopodis* [as *Elephantopi* = *Elephantopus* sp.], Philippines, Luzon, Nueva Ecija [Prov., Balete Pass, 9 JAN 1930], leg. F.L. Stevens No. 1732, ILL 6527.

Meliola ellisii C. Roumeguère. Revue Mycologique 2:200. 1880; Fungi Gallici Exsiccati, Century 9, No. 894. Anno 1880, nom. nud. **Isotype:** Sur les branches seches du *Vaccinium corymb.* [= *V. corymbosum* L.], Amerique [U.S.A.], comm. J.B. Ellis, ex C. Roumeguère, Fungi Gallici Exsiccati No. 896 [based on J.B. Ellis, North American Fungi No. 192, labeled erroneously as *Sphaeria nidulans* L.D. v. Schweinitz], ILL 6486.

Meliola elmeri H. & P. Sydow. Leaflets of Philippine Botany 5:1537. 1912. **Isotype:** On either upper or lower surface of *Pittosporum pentandrum* (Blanco) Merrill, Philippines, Palawan [Island and] Prov., Brooks Point, (Addison Peak), FEB 1911, leg. A.D.E. Elmer No. 12707, ILL 5089.

Meliola epithemae F.L. Stevens & E.F. Roldan. Philippine Journal of Science 56:65. 1935, nom. invalid. sine diagn. lat. Validly published as *Meliola epithematis* F.L. Stevens & E.F. Roldan ex C.G. Hansford. Sydowia Annales Mycologici 16:316. (1962) 1963. [See next entry].

Meliola epithematis F.L. Stevens & E.F. Roldan ex C.G. Hansford. Sydowia Annales Mycologici 16:316. (1962) 1963; originally published [as *M. epithemae*] by F.L. Stevens & E.F. Roldan in Philippine Journal of Science 56:65. 1935, as nom. invalid. sine diagn. lat. **Holotype:** Hab. in foliis *Epithematis* spec. indet. [= *Epithema* sp.], Philippines, [Luzon, Benguet, Naguilian Road, 7 JAN 1931], F.L. Stevens No. 1394, ILL 5091.

Meliola equadorensis F.L. Stevens. Annales Mycologici 26:259, pl. V, fig. 68. 1928. **Holotype:** On Sapindaceae sp. indet., Ecuador, San Miguel, 4 NOV 1924, leg. F.L. Stevens No. 207, ILL 5092.

Meliola erioglossi C.G. Hansford. Beihefte zur Sydowia Annales Mycologici 1:107. 1957. **Holotype:** Hab. in foliis *Erioglossi rubiginosi* [*Erioglossum rubiginosum* (Roxb.) Blume = *Lepisanthes rubiginosa* (Roxb.) Leenh.],

Philippines, Basilan [Island], Isabela, [NOV–DEC 1919], leg. H.S. Yates, ex Phil. Bur. Sci. No. 36237, ILL 5093.

Meliola eriophora C.L. Spegazzini. Anales de la Sociedad Cientifica Argentina 26:24–25. 1888; Fungi Guaranitici, Pugillus II, pp. 22–23, No. 62. 1888. **Isotype:** Hab. ad folia viva *Fici ibapoy* [*Ficus ibapoy* Parodi (nomen?) = *Ficus* sp.], in sylvis montanis, [Paraguay], pr. Paraguari, JAN 1883, leg. B. Balansa [ex C.L. Spegazzini No. 735 (fide C.G. Hansford, 1961)], ILL 5094 [as a microscopic preparation].

Meliola erythrinae H. & P. Sydow. Annales Mycologici 15:185. 1917. **Isotype:** Hab. in foliis *Erythrinae indicae* [*Erythrina indica* Lam. = *E. variegata* L.], Philippines, Luzon, Laguna Prov., NOV 1915, leg. M. Ramos, ex Phil. Bur. Sci. No. 24052, ILL 5096.

Meliola escharoides (H. Sydow) R. Ciferri, 1954. **See basionym:** *Irene escharoides* H. Sydow, 1926.

Meliola eucalypti F.L. Stevens & E.F. Roldan ex C.G. Hansford. Sydowia Annales Mycologici 16:317–318. (1962) 1963; originally published by F.L. Stevens & E.F. Roldan in Philippine Journal of Science 56:74. 1935, as nom. invalid. sine diagn. lat. **Holotype:** On *Eucalyptus* sp., Philippines, Luzon, Nueva Ecija Prov., San José to Balete Pass, 9 JAN 1931, leg. F.L. Stevens No. 1722, ILL 5097. **Isotype:** CUP.

Meliola eugeniae H. Sydow. Philippine Journal of Science 21:133. 1922. **Isotype:** On *Eugenia caryophyllata* Thunb. [= *Syzygium aromaticum* (L.) Merrill & Perry, Indonesia], Amboina, Katikati, 6 SEP 1913, leg. C.B. Robinson, Reliquiae Robinsonianae No. 2163, ILL 5098.

Meliola eugeniae-monticolae C.G. Hansford. Beihefte zur Sydowia Annales Mycologici 1:107–108. 1957. **Holotype:** Hab. in foliis *Eugeniae monticolae* [= *Eugenia monticola* (Sw.) DC.], Porto Rico [Puerto Rico], Manati, [5 NOV 1913], leg. F.L. Stevens No. 4285, ILL 5228.

Meliola eugeniicola F.L. Stevens. Memoirs of the Department of Agriculture in India, Botanical Series 15(5):107–108, pl. II, figs. 9–11. JAN 1928; Annales Mycologici 26:231–232. May 1928. **Isolectotypes:** On *Eugenia eucalyptoides* F. Mueller, India, Pachanadi, Mangalore, 16 APR 1913, leg. L.S. Subramanian, as F.L. Stevens No. 1989, ILL 5100, HCIO. **Assumed lectotype:** IMI 25367 [citation of IMI for location of "type" by C.G. Hansford (1961:144) is interpreted as lectotypification with this specimen].

Meliola euodiae N.T. Patouillard. Revue Mycologique 10:139–140. 1888 [cited as *evodiae*, corrected in accordance with Articles 60 & 61 of I.C.B.N.]. **Isotype:** Sur les deux faces des feuilles d'un *Evodia* [= *Euodia* sp.], Illes Samoa, leg. E. Guillou No. 1841, ILL 5111. **Holotype:** PC.

Meliola euonymi F.L. Stevens ex C.G. Hansford. Beihefte zur Sydowia Annales Mycologici 1:108. 1957. **Holotype:** Hab. in foliis *Euonymi* sp.,

[= *Euonymus* sp.], Philippines, Panay [Island], Capiz [Prov.], Jamindan, [APR–MAY 1918], leg. M. Ramos & G. Edaño, ex Phil. Bur. Sci. No. 32154, ILL 5101.

Meliola euopla H. Sydow ex F.L. Stevens. Annales Mycologici 26:254. 1928, nom. nov. **Based on:** *Meliola vicina* H. Sydow, 1926 (nom. illegit.), non H. Sydow, 1923. **Lectotype** [designated herein—see p. 240]: Hab in foliis *Rauwolfiae nitidae* [*Rauwolfia* = *Rauvolfia nitida* Jacquin], Costa Rica, Los Angeles de San Ramón, 30 JAN 1925, leg. H. Sydow, Fungi in Itinere Costaricensi Collecti No. 133, ILL 6421.

Meliola euphorbiae F.L. Stevens & L.R. Tehon. Mycologia 18:11–12, pl. 2, fig. 12. 1926. **Holotype:** On an undetermined Euphorbeacea [Euphorbiaceae sp. indet.], British Guiana [Guyana], Kartabo, 24 JUL 1922, leg. F.L. Stevens No. 663, ILL 5102.

Meliola evanida A. Gaillard. Le Genre *Meliola*, pp. 102–103, pl. 18, fig. 12. 1892. **Isotype:** Ad paginam inferiorem *Strychni* cujusdam [= *Strychnos* sp., C. Africa, Congo], Loango, OCT 1888, leg. F. Thollon s.n., ILL 5106. **Holotype:** PC.

Meliola evansii E.M. Doidge. Transactions of the Royal Society of South Africa 8:112–113. 1920. **Isolectotype:** On *Scolopia zeyheri* (Nees) Harvey [as *Celastrineae* in the protologue], South Africa, Cape Prov., Mossel Bay, 22 JUL 1915, leg. I.B. Pole Evans, ex PREM No. 9067, ILL 5107 [also a paratype of *Meliola scolopiae* E.M. Doidge in E.M. Doidge & H. Sydow]. **Lectotype:** PREM [citation of PREM No. 9067 as "type" by F.L. Stevens (1928:168) is accepted as lectotypification with the specimen located at PREM].

Meliola eveae F.L. Stevens. Annales Mycologici 26:247, pl. V, fig. 53. 1928. **Holotype:** On *Cephaelis muscosa* (Jacquin) Sw. [= *Psychotria muscosa* (Jacquin) Steyerm.], Trinidad, Cumuto, 16 AUG 1922, leg. F.L. Stevens No. 945, ILL 5108. **Paratypes:** On *Evea* sp. [= *Psychotria* sp.], British Guiana [Guyana], Tumatumari, 9 JUL 1922, leg. F.L. Stevens No. 93, ILL 5109; on *Evea* sp. [as *Cephaelis rosea* Bentham on the packet = *Psychotria* cf. *rosea* (Bentham) Muell. Arg.], leg. F.L. Stevens No. 87, ILL 5110.

Meliola excoecariae E.M. Doidge. Transactions of the Royal Society of South Africa 8:139. 1920. **Isolectotype:** In foliis *Excoecariae caffrae* [= *Excoecaria caffra* Sim], South Africa, Natal Prov., Buccleuch, 17 JUL 1918, leg. E.M. Doidge, ex PREM No. 11566, ILL 5114. **Lectotype:** PREM [citation of PREM No. 11566 as "type" by C.G. Hansford (1961:229) is accepted as lectotypification with the specimen at PREM].

Meliola excoecariicola F.L. Stevens ex C.G. Hansford. Beihefte zur Sydowia Annales Mycologici 1:108. 1957. **Holotype:** Hab. in foliis *Excoecar-*

iae philippinensis [= *Excoecaria philippinensis* Merrill], Philippines, Luzon, Rizal [Prov.], Mt. Lumutan, [JUL 1917], leg. M. Ramos & G. Edaño, ex Phil. Bur. Sci. No. 29809, ILL 5113.

Meliola exilis H. & P. Sydow. Annales Mycologici 2:170. 1904. **Lectotype:** Hab. in foliis vivis *Gaultheriae* spec. [= *Gaultheria* sp.] in Andibus [border between Chile and Argentina], ad lacum Quillen, leg. F.W. Neger s.n., ILL 4086. [Citation of specimen in the F.L.S. herb. (ILL) as "type" by C.G. Hansford (1961:493) is interpreted as lectotypification. The packet was labeled and marked "type" in what appears to be H. Sydow's handwriting.]. ≡ *Irene exilis* (H. & P. Sydow) F.L. Stevens. Bernice P. Bishop Museum Bulletin 19:41. 1925. ≡ *Irenina exilis* (H. & P. Sydow) F.L. Stevens. Annales Mycologici 25:449. 1927. ≡ *Asteridiella exilis* (H. & P. Sydow) C.G. Hansford ex C.G. Hansford. Beihefte zur Sydowia Annales Mycologici 2:492. 1961.

Meliola exocarpi H.S. Yates. Philippine Journal of Science, Section C (Botany) 13:268. (1918) 1919 [as *exocarpiae*; corrected by C.G.Hansford (1961:361) in accordance with current I.C.B.N.]. **Isotype:** On leaves of *Exocarpos latifolius* R. Br. [cited as *Exocarpus* (orth. var.)], Philippines, Luzon, Ilocos Norte Prov., Burgos, 1 MAR 1917, leg. M. Ramos, ex Phil. Bur. Sci. No. 27846, ILL 5115.

Meliola fagraeae H. & P. Sydow. Annales Mycologici 12:549. 1914. **Isotype:** Hab. in foliis *Fagraeae plumeriaefoliae* [= *Fagraea plumeriaeflora* A. DC. in DC.], Philippines, Luzon, Camarines Prov., Mt. Isarog, NOV–DEC 1913, leg. M. Ramos, ex Phil. Bur. Sci. No. 22222, ILL 5116.

Meliola ficuum H.S. Yates. Philippine Journal of Science, Section C (Botany) 13:368–369. (1918) 1919 [as *ficium*; correction recommended by F.C. Deighton (1968:184), in accordance with current I.C.B.N.]. **Isotype:** On *Ficus* sp., [Philippines], Luzon, Laguna Prov., Mt. Banahao, 8 MAY 1917, leg. M. Ocampo, ex Phil. Bur. Sci. No. 28002, ILL 5121.

Meliola ficuum H.S. Yates var. *ugandensis* C.G. Hansford. Sydowia Annales Mycologici 10:72. (1956) 1957 [the specific epithet as *ficium*; corrected herein— see preceding entry]. **Isotype:** Hab. in foliis *Fici* sp. [= *Ficus* sp.], Uganda, [Butambola, NOV 1930], leg. C.G. Hansford No. 1409, ILL 5123.

Meliola forsteroniae (F.L. Stevens) C.G. Hansford, (1948) 1949. **See basionym:** *Meliola tabernaemontanae* C.L. Spegazzini var. *forsteroniae* F.L. Stevens., 1916.

Meliola franciscana C.G. Hansford. Proceedings of the Linnean Society of London 160:123. (1948) 1949. **Isotype:** Hab. in foliis Leguminosarum sp. indet. [= Fabaceae sp.], Brazil, São Francisco, OCT 1884, leg. E.H.G. Ule, ex G.L. Rabenhorst-H.G. Winter, Fungi Europaei No. 3248, [sub *Meliola ludibunda* C.L. Spegazzini], ILL 5422.

Meliola fumago G. Niessl. Hedwigia 20:99. 1881; and in L. Rabenhorst. Fungi Europaei, Klotzschii Herbarii Vivi Mycologici Continuatio. Ausgabe III. Century 26, No. 2513. Anno 1881. **Isotypes:** Auf Blättern einer *Celastrus* Art aus Calcutta [India], leg. W. Kurz, ILL 6480 [as a microscopic preparation], and in bound exsiccati set. ≡ *Dimerosporium fumago* (G. Niessl) P.A. Saccardo. Sylloge Fungorum 1:53. 1882. ≡ *Lembosia fumago* (G. Niessl) H.G. Winter. Flora 67:266. 1884. ≡ *Dimerium fumago* (G. Niessl) P.A. & D. Saccardo & P. Sydow in P.A. & D. Saccardo. Sylloge Fungorum 17:537. 1905. ≡ *Asterina fumago* (G. Niessl) F. v. Höhnel. Sitzungsberichte der Kaiserl. Akademie der Wissenschaften in Wien. Mathematisch-Naturwissenschaftliche Klasse, Abt. 1. 119:435. 1910.

Meliola funerea D. McAlpine. Proceedings of the Linnean Society of New South Wales 21:104, pl. 10, figs. 1–6. 1896. **Isotype:** On leaves of *Grevillea robusta* A. Cunningham ex R. Br., Australia, New South Wales, Lismore, MAR [1896], leg. J.H. Maiden [No. 1751+3], ILL 5376. **Holotype:** VPRI. ≡ *Meliola lanosa* N.T. Patouillard var. *funerea* (D. McAlpine) C.G. Hansford. Beihefte zur Sydowia Annales Mycologici 2:99. 1961.

Meliola furcata J.H. Léveillé. Annales des Sciences Naturelles Botanique, Series 3, 5:266. 1846. **Isotype:** Hab. paramaribo in Guyana Batava [Surinam], ad folia [the host identified as *Cissus* sp. (cf. C.G. Hansford, 1961)], leg. H.A.H. Kegel No. 595, ILL 5135. **Holotype:** PC.

Meliola furcillata E.M. Doidge. Transactions of the Royal Society of South Africa 5:738, 747, pl. LXV, fig. 39. 1917. **Isotype:** On *Allophylus monophyllus* (C. Presl) Radlk. ex Taubert [fide C.G. Hansford (1961), the host cited as *Maesa rufescens* A. DC. in DC. in the protologue, a probable misidentification], South Africa, Natal [Prov.], Amanzimtoti, 10 JUL 1911, leg. I.B. Pole Evans, ex PREM No. 1573, ILL 5157. **Holotype:** PREM.

Meliola gaillardiana F.L. Stevens. Illinois Biological Monographs 2:529–530, fig. 55. 1916. **Holotype:** On *Piper aduncum* L. [misspelled "*adunctum*"], Porto Rico [Puerto Rico], Rio Arecibo, 8 JUL 1915, leg. F.L. Stevens No. 7794, ILL 5185. **Isotype:** ILL 5186. **Paratypes:** Leg. F.L. Stevens No. 7796, ILL 5187 [also a paratype of *Helminthosporium glabroides* F.L. Stevens]; Dos Bocas, below Utuado, 8 JUL 1915; leg. F.L. Stevens No. 8044, ILL 5159; 30 DEC 1913, leg. F.L. Stevens No. 6802, ILL 5158; Las Marias, 11 JUL 1915, leg. F.L. Stevens No. 8223, ILL 5160.

Meliola galipeae H. & P. Sydow. Annales Mycologici 14:77. 1916. **Isotype:** In foliis *Galipeae longiflorae* [= *Galipea longiflora* Krause], Brazil, Seringal São Francisco, Rio Acre, MAR 1911, leg. E.H.G. Ule No. 3433, ILL 5188.

Meliola ganophylli F.L. Stevens & E.F. Roldan ex C.G. Hansford. Sydowia Annales Mycologici 16:314. (1962) 1963; originally published by F.L. Stevens & E.F. Roldan in Philippine Journal of Science 56:56–57, fig. 3k. 1935, as nom. invalid. sine diagn. lat. **Holotype:** On *Ganophyllum falcatum* Blume [the identification later corrected to *G. obliquum* Merrill by S.F. Blake (cf. C.G. Hansford, 1961)], Philippines, Luzon, Benguet, Kennon Road, 8 JAN 1931, leg. F.L. Stevens No. 1671, ILL 5192. **Isotypes:** CALP, Phil. Bur. Sci.

Meliola garciniae H.S. Yates. Philippine Journal of Science, Section C (Botany) 13:369. (1918) 1919. **Isotype:** On leaves of *Garcinia* sp., [Philippines], Luzon, Ilocos Norte Prov., Burgos, 14 MAR 1917, leg. M. Ramos, ex Phil. Bur. Sci. No. 27795, ILL 5193.

Meliola garciniae H.S. Yates var. *mangostana* (P.A. Saccardo) C.G. Hansford, (1948) 1949. **See basionym:** *Meliola mangostana* P.A. Saccardo, 1921.

Meliola garugae F.L. Stevens & E.F. Roldan ex C.G. Hansford. Sydowia Annales Mycologici 16:319–320. (1962) 1963; originally published by F.L. Stevens & E.F. Roldan in Philippine Journal of Science 56:68. 1935, as nom. invalid. sine diagn. lat. **Holotype:** On *Garuga abilo* (Blanco) Merrill, Philippines, Luzon, Nueva Ecija Prov., Muñoz, 3 OCT 1930, leg. F.L. Stevens No. 781, ILL 5194.

Meliola geniculata H. & P. Sydow & E.J. Butler. Annales Mycologici 9:381–382, fig. 5. 1911. **Isotype:** Hab. in foliis *Odinae wodier* [*Odina wodier* Roxb. = *Lannea coromandelia* (Houtt.) Merrill, India], Travancore, Pulliyanur, 8 OCT 1907, leg. E.J. Butler No. 1366, ILL 5195.

Meliola gesneriae F.L. Stevens. Illinois Biological Monographs 2:515–516. 1916. **Holotype:** On *Gesneria albiflora* (Decaisne) O. Kuntze [= *G. pedunculosa* (DC.) Fritsch], Porto Rico [Puerto Rico], Mayagüez Mesa, 25 JUN 1915, leg. F.L. Stevens No. 7431, ILL 5198. **Paratypes:** Leg. F.L. Stevens No. 7465, ILL 5199; Dos Bocas, below Utuado, 8 JUL 1915, leg. F.L. Stevens No. 8018, ILL 5200; 30 NOV 1913, leg. F.L. Stevens No. 6590, ILL 5201; on *Cestrum laurifolium* L'Héritier, Maricao, 3 APR 1913, leg. F.L.Stevens No. 824, ILL 5197 [also a paratype of *Helminthosporium glabroides* F.L. Stevens]; on *Cestrum macrophyllum* Vent. [= *C. laurifolium* L'Héritier], El Alto de la Bandera, 14 JUL 1915, leg. F.L. Stevens No. 8301, ILL 5203.

Meliola glabra M.J. Berkeley & M.A. Curtis var. *psychotriae* F.L. Stevens. Illinois Biological Monographs 2:482. 1916. **Holotype:** On *Palicourea domingensis* (Jacquin) DC. [= *P. crocea* (Sw.) J.A. Schultes in Roemer & J.A. Schultes], Porto Rico [Puerto Rico], Florida Adentro, 1 JUL 1915, leg. F.L. Stevens No. 7649, Ill 4060. **Isotype:** ILL 3972. **Paratypes:** On *Coccocypselum repens* Sw., Maricao, 20 JUL 1915, leg.

F.L. Stevens No. 8961, ILL 3974; on *Palicouria* sp.?, Vega Baja, 20 FEB 1913, leg. F.L. Stevens No. 468, ILL 3967; Ponce, 8 NOV 1913, leg. F.L. Stevens No. 4367, ILL 3973; Dos Bocas, below Utuado, 30 DEC 1913, leg. F.L. Stevens No. 6650, ILL 3968; on *Palicourea* sp. [as *Psychotria* sp. on the packet], El Gigante, 15 DEC 1913, leg. F.L. Stevens No. 5944, ILL 4077; on *Psychotria berteriana* DC. [misspelled "*bertiana*"], El Alto de la Bandera, 15 JUL 1915, leg. F.L. Stevens No. 8278, ILL 3971 and 3978; leg. F.L. Stevens No. 8654, ILL 3970; leg. F.L. Stevens No. 8646, ILL 4070; leg. F.L. Stevens No. 8566, ILL 4071; leg. F.L. Stevens No. 8710, ILL 4079; leg. F.L. Stevens No. 8673, ILL 3979; El Gigante, 17 JUL 1915 [as 16 JUL on the packet], leg. F.L. Stevens No. 8528, ILL 4072; on *Psychotria grandis* Sw., Mayagüez, 25 JUN 1915, leg. F.L. Stevens No. 7487, ILL 33559; on *Psychotria pubescens* Sw., Arecibo-Lares Road, 21 JUN 1915, leg. F.L. Stevens No. 7281, ILL 4073; Vega Baja, 2 JUL 1915, leg. F.L. Stevens No. 7732, ILL 3975; leg. F.L. Stevens No. 7741, ILL 3976; Dos Bocas, below Utuado, 8 JUL 1915, leg. F.L. Stevens No. 8032, ILL 3977; on *Psychotria* sp., Quebradillas, 22 NOV 1913, leg. F.L. Stevens No. 5032, ILL 4061.

Meliola glabriuscula C.L. Spegazzini. Revista del Museo de La Plata 15 (Series 2,2):15–16. 1908. Fungi Aliquot Paulistani, pp. 15–16, No. 35. 1908. **Isotypes**: Hab. ad folia valde coriacea nitidissima (*Photiniae*?) [= *Photinia*? sp.] viva, aqua branca, [Brazil], Isolamento pr. São Paulo, leg. A. Usteri, comm. C.L. Spegazzini [No. 572 (fide C.G. Hansford, 1961), but the number omitted from the packets], ILL 5205 and 5206 [leaf fragment (ILL 5205) and two microscopic preparations]. **Holotype**: LPS. ≡ *Asteridiella glabriuscula* (C.L. Spegazzini) C.G. Hansford ex C.G. Hansford. Beihefte zur Sydowia Annales Mycologici 2:757. 1961.

Meliola glabroides F.L. Stevens. Illinois Biological Monographs 2:486–487, fig. 13. 1916. **Holotype**: On *Piper aduncum* L. [misspelled "*adunctum*"], Porto Rico [Puerto Rico], Maricao, Indiera Fria, 8 OCT 1913, leg. F.L. Stevens No. 3371, ILL 4029. **Paratypes**: On *Piper aduncum* L. [misspelled "*adunctum*"], El Alto de la Bandera, 15 JUL 1915, leg. F.L. Stevens No. 9039, ILL 4039 [also the holotype of *Helminthosporium glabroides* F.L. Stevens]; leg. F.L. Stevens No. 8633, ILL 4036; Las Marias, 10 JUL 1915, leg. F.L. Stevens No. 9603, ILL 4027; leg. F.L. Stevens No. 8133, ILL 4028; Dos Bocas, below Utuado, 8 JUL 1915, leg. F.L. Stevens No. 8064, ILL 4032; Vega Baja, 2 JUL 1915; leg. F.L. Stevens No. 7724, ILL 4034; Mayagüez Mesa, 29 JUN 1915, leg. F.L. Stevens No. 7563, ILL 4030 and 4033; Añasco, 12 OCT 1913, leg. F.L. Stevens No. 3582, ILL 4037 and 16041 [also paratypes of *Helminthosporium glabroides* F.L. Stevens]; 8 NOV 1913 [as 18 NOV on the packet], leg. F.L. Stevens No. 4802, ILL 4066 [also the holotype of *Coniothyrium glabroides* F.L. Stevens and a paratype of *Helminthosporium glabroides* F.L. Stevens]; Maricao, 20 SEP 1913, leg. F.L. Stevens No. 3647, ILL 4035 [also a paratype of *Helminthosporium glabroides* F.L. Stevens]; Arecibo-Lares Road, 21 JUN 1915; leg. F.L. Stevens No. 7297, ILL 4038 [also a paratype of *Helminthosporium glabroides* F.L. Stevens]; Aibonito, 16 JUL 1915, leg. F.L. Stevens No. 8471, ILL 4040; Martin Peña, 11 AUG 1915, leg. F.L. Stevens No. 9334, ILL 4026; Trujillo Alto, 16 AUG 1915, leg. F.L. Stevens No. 9472, ILL 3981; Utuado, 8 NOV 1913, leg. F.L. Stevens No. 4393, ILL 3983 and 4022; on *Nectandra patens* (Sw.) Griseb. [= *Ocotea patens* (Sw.) Nees], Mayagüez Mesa, 25 JUN 1915, leg. F.L. Stevens No. 7466, ILL 3980; 15 JUN 1915, leg. F.L. Stevens No. 7081, ILL 3984; 29 JUN 1915, leg. F.L. Stevens No. 7595, ILL 4007 and 4025 [also the holotype and an isotype, respectively, of *Arthrobotryum glabroides* F.L. Stevens]; Maricao, 20 SEP 1913, leg. F.L. Stevens No. 4852, ILL 4004 and 4006; 20 JUL 1915, leg. F.L. Stevens No. 8867, ILL 4005 [also a paratype of *Arthrobotryum glabroides* F.L. Stevens]; leg. F.L. Stevens No. 8874, ILL 4021; leg. F.L. Stevens No. 8973, ILL 4003 [also a syntype of *Isthmospora glabra* F.L. Stevens]; 19 JUL 1915, leg. F.L. Stevens No. 8750, ILL 3985; on *Solanum persicifolium* Dunal, Quebradillas, 22 NOV 1913, ILL 4056; on *Solanum rugosum* Dunal, Las Marias, 10 JUL 1915, leg. F.L. Stevens No. 8121, ILL 4055; on *Sauvagesia erecta* L., Las Marias, 10 JUL 1915, leg. F.L. Stevens No. 8129, ILL 3988; El Alto de la Bandera, 15 JUL 1915, leg. F.L. Stevens No. 8641, ILL 3990; Maricao, 19 JUL 1915, leg. F.L. Stevens No. 8777, ILL 4054; leg. F.L. Stevens No. 8944, ILL 3989; on *Stachytarpheta cayennensis* (L.C. Richard) Vahl, Trujillo Alto, 15 AUG 1915, leg. F.L. Stevens No. 9405, ILL 3995; Sabana Llana, 13 AUG 1915, leg. F.L. Stevens No. 9380, ILL 3996; [as *Stachytarpheta strigosa* Vahl on the packet], near Puebla Vieja, 13 JAN 1903, leg. A.A. Heller No. 6402, ILL 4984. ≡ *Irene glabroides* (F.L. Stevens) R.A. Toro. Mycologia 17:142. 1925. ≡ *Irenina glabroides* (F.L. Stevens) F.L. Stevens. Annales Mycologici 25:463. 1927. ≡ *Asteridiella glabroides* (F.L. Stevens) C.G. Hansford. Beihefte zur Sydowia Annales Mycologici 2:71. 1961.

Meliola glabroides F.L. Stevens var. *schlegeliae* F.L. Stevens. Illinois Biological Monographs 2:488. 1916. **Holotype:** On *Schlegelia* sp., Porto Rico [Puerto Rico], El Alto de la Bandera, 14 JUL 1915, leg. F.L. Stevens No. 8289, ILL 3999. **Paratype:** Leg. F.L. Stevens No. 8274, ILL 4000. ≡ *Irenina glabroides* (F.L. Stevens) F.L. Stevens var. *schlegeliae* (F.L. Stevens) F.L. Stevens. Annales Mycologici 25:464. 1927. ≡ *Irenina schlegeliae* (F.L. Stevens) C.G. Hansford. Proceedings of the Linnean

Society of London 160:132. (1948) 1949. ≡ *Asteridiella schlegeliae* (F.L. Stevens) C.G. Hansford ex C.G. Hansford. Beihefte zur Sydowia Annales Mycologici 2:661. 1961.

Meliola gleditsiae C.L. Spegazzini. Anales del Museo Nacional de Historia Natural de Buenos Aires 23:41–42. 1912; Mycetes Argentinenses, Series 6, No. 1337. 1912 [as *gleditschiae*, corrected according to Rec. 60H.1 and Articles 60 & 61 of I.C.B.N.]. **Isotypes:** Hab. ad folia viva *Gleditschiae amorphoidis* [= *Gleditsia amorphoides* (Griseb.) Taubert], in silvis, [Argentina], Misiones, pr. Puerto León, JUL 1909, leg. A.L. Venturi, ex C.L. Spegazzini [No. 515 (cf C.G. Hansford, 1961)], ILL 5207 and 5208.

Meliola gliricidiae H. & P. Sydow. Annales Mycologici 12:550–551. 1914. **Isotype:** Hab. in foliis *Gliricidiae sepium* [= *Gliricidia sepium* (Jacquin) Kunth ex Walpers], Philippines, [Luzon], Rizal Prov., [Antipolo, 17] AUG 1913, leg. M. Ramos, ex Phil. Bur. Sci. No. 21929, ILL 5209.

Meliola glochidii F.L. Stevens & E.F. Roldan ex C.G. Hansford. Sydowia Annales Mycologici 16:319. (1962) 1963; originally published by F.L. Stevens & E.F. Roldan in Philippine Journal of Science 56:73–74. 1935, as nom. invalid. sine diagn. lat. **Holotype:** On *Glochidion* sp., Philippines, Luzon, Benguet, Naguilian Road, 6 JAN 1931, leg. F.L. Stevens No. 1561, ILL 5210. **Isotype:** CUP.

Meliola gloriosa E.M. Doidge. Transactions of the Royal Society of South Africa 8:139–140. 1920. **Isotype:** In foliis *Celastrus cordatus* E. Meyer ex Sonder & Harvey? [as *Gymnosporia cordata*? on the packet = *Maytenus* (?) *cordata* (E. Meyer ex Sonder & Harvey) Loes.], South Africa, Natal Prov., Buccleuch, 17 JUL 1918, leg. E.M. Doidge, ex PREM No. 11565, ILL 3939. ≡ *Irene gloriosa* (E.M. Doidge) E.M. Doidge. South African Journal of Natural History 2:40. 1920. ≡ *Appendiculella gloriosa* (E.M. Doidge) C.G. Hansford. Beihefte zur Sydowia Annales Mycologici 2:338–339. 1961.

Meliola gnathonella F.L. Stevens & L.R. Tehon. Mycologia 18:16, pl. 2, fig. 20. 1926. **Holotype:** On *Jacaranda* sp. (?), British Guiana [Guyana], Tumatumari, 12 JUL 1922, leg. F.L. Stevens No. 231, ILL 5211.

Meliola gregoriana F.L. Stevens. Bernice P. Bishop Museum Bulletin 19:39, text fig. 8f. 1925. **Holotype:** On *Dianella odorata* sensu Hillebr. non Blume [= *D. sandwicensis* Hooker & Arnott], Hawaii, Oahu, Kalihi Valley, MAR 1916, leg. C.N. Forbes-F.L. Stevens No. 2306, ILL 5125. **Isotype:** BISH 499031.

Meliola grewiicola C.G. Hansford. Sydowia Annales Mycologici 11:56–57. (1957) 1958. **Holotype:** Hab. in foliis *Grewiae* sp. [= *Grewia* sp.], Philippines, Luzon, Laguna [Prov., Agricultural College, 10 SEP 1930], leg. F.L. Stevens No. 496a, ILL 6531.

Meliola groteana H. & P. Sydow. Annales Mycologici 11:402–403. 1913. **Isotypes**: Hab. in foliis *Maesae lanceolatae* [= *Maesa lanceolata* Forssk.], Deutsch-Ostafrika [E. Africa, Tanzania, Tanga Prov.], Amani, [Usambara], 1 JUL 1913, leg. M. Grote [ex H. Sydow, Fungi Exotici Exsiccati No. 247], ILL 5182 [as a microscopic preparation], and in bound exsiccati set.

Meliola groteana H. & P. Sydow var. *ardisiicola* C.G. Hansford. Sydowia Annales Mycologici 9:18. 1955. **Isotype**: Hab. in foliis *Ardisiae perrottetianae* [*Ardisia perrottetiana* A. DC.], Philippines, [Luzon, Laguna Prov.], Los Baños, Mount Maquiling, [SEP 1914], leg. C.F. Baker, Fungi Malayana No. 254, ILL 6021. **Holotype**: K.

Meliola guamensis H. Sydow. Annales Mycologici 19:304. 1921. **Isotype**: Hab. in foliis *Ochrosiae* spec. [= *Ochrosia* sp.], Island of Guam, OCT 1911, leg. R.C. McGregor No. 586, ILL 5178.

Meliola guaranitica C.L. Spegazzini. Anales de la Sociedad Cientifica Argentina 17:134. 1883; Fungi Guaranitici, Pugillus I, p. 71, No. 177. 1883. **Isotype**: Hab. ad folia viva arborum in sylvis subvirgineis, [Paraguay], pr. Guarapi, JAN 1883, leg. B. Balansa, comm. C.L. Spegazzini No. 3781, ILL 5177. **Holotype**: LPS.

Meliola guareae C.L. Spegazzini. Anales del Museo Nacional de Historia Natural de Buenos Aires 23:42. 1912; Mycetes Argentinenses, Series 6, p. 42, No. 1338. 1912. **Isotype**: Ad folia viva *Guareae balansae* [= *Guarea balansae* C. DC.], in silvis, [Argentina], Misiones, pr. Puerto León, AUG 1909, leg. C.L. Spegazzini [No. 849 (cf. C.G. Hansford, 1961)], ILL 5172 [in association with and also the holotype of *Meliola guareella* C.G. Hansford]. **Holotype**: LPS.

Meliola guareella C.G. Hansford. Beihefte zur Sydowia Annales Mycologici 1:109. 1957 [as *guareiella*, corrected by F.C. Deighton (1968) in accordance with current I.C.B.N.]. **Holotype**: Hab. in foliis *Guareae* sp. [= *Guarea* sp.], socio *Meliola guareae* C.L. Spegazzini, Argentina, Misiones, [pr. Puerto León, AUG 1909], leg. C.L. Spegazzini, ex. ILL 5172 [in association with and also an isotype of *Meliola guareae* C.L. Spegazzini. Under the latter name, Spegazzini's original collection is cited as No. 849 by Hansford (1961).]. **Assumed isotype**: LPS.

Meliola guareicola F.L. Stevens. Illinois Biological Monographs 2:521. 1916. **Holotype**: On *Guarea trichilioides* L. [= *Guarea guidonia* (L.) Sleumer], Porto Rico [Puerto Rico], Las Marias, 10 JUL 1915, leg. F.L. Stevens No. 8166, ILL 5171 [also the holotype of *Helminthosporium guareicola* F.L. Stevens and a paratype of *Grallomyces portoricensis* F.L. Stevens]. **Isotype**: ILL 5173. **Paratypes**: 11 JUL 1915, leg. F.L. Stevens No. 8245, ILL 5167; Mayagüez Mesa, 25 JUN 1915, leg. F.L. Stevens No. 7464, ILL 5166; Adjuntas, 22 NOV 1913, leg. F.L. Stevens No.

4971, ILL 5168; Dos Bocas, below Utuado, 8 AUG 1915, leg. F.L. Stevens No. 8096, ILL 5169 [also a paratype of *Helminthosporium guareicola* F.L. Stevens].

Meliola guianensis F.L. Stevens & R.I. Dowell. Phytopathology 13:248–249, figs. 1, 2, 3A–D. 1923. **Holotype:** On *Theobroma cacao* L., British Guiana [Guyana], Coverden, 4 AUG 1922, leg. F.L. Stevens No. 974, ILL 4380 [also the holotype of *Helminthosporium guianensis* F.L. Stevens & R.I. Dowell]. ≡ *Irenopsis guianensis* (F.L. Stevens & R.I. Dowell) F.L. Stevens. Annales Mycologici 25:441. 1927.

Meliola guignardii A. Gaillard. Bulletin de la Société Mycologique de France 8:176–177, pl. 16, fig. 1. 1892. **Isotype:** Ad paginam superiorem foliorum coriaceorum arboris cujusdam, Ecuador, Canzacoto, JUL 1892, leg. G. v. Lagerheim s.n., ILL 4383. ≡ *Irenopsis guignardii* (A. Gaillard) F.L. Stevens. Annales Mycologici 25:433. 1927.

Meliola gymnanthicola F.L. Stevens. Illinois Biological Monographs 2:517. 1916. **Holotype:** On *Gymnanthes lucida* Sw., Porto Rico [Puerto Rico], Guayanilla, 14 JUL 1915, leg. F.L. Stevens No. 8596, ILL 5161.

Meliola gymnanthicola F.L. Stevens var. *manihot* (F.L. Stevens & L.R. Tehon) F.L. Stevens, 1928. **See basionym:** *Meliola manihot* F.L. Stevens & L.R. Tehon, 1926.

Meliola gymnosporiae H. & P. Sydow. Annales Mycologici 10:79. 1912. **Isotype:** Hab. in foliis *Gymnosporiae spinosae* [*Gymnosporia spinosa* (Blanco) Merrill & Rolfe = *Maytenus* sp., Philippines, Luzon], Manila and vicinity, NOV–DEC 1910, leg. E.D. Merrill No. 7422, ILL 4088. ≡ *Irenina gymnosporiae* (H. & P. Sydow) F.L. Stevens. Annales Mycologici 25:467. 1927. ≡ *Asteridiella gymnosporiae* (H. & P. Sydow) C.G. Hansford ex C.G. Hansford. Beihefte zur Sydowia Annales Mycologici 2:341. 1961.

Meliola hariotii C.L. Spegazzini. Revista Argentina de Historia Natural 1:404–405. 1891 [as *harioti*]; Fungi Guaranitici Nonnulli Novi vel Critici, pp. 404–405, No. 78. 1891. **Isotype:** Hab. ad folia viva Bignoniaceae? (an Leguminosae?) [= Fabaceae (?)] cujusdam in dumetis [Paraguay], pr. Asunción, JAN 1874, leg. B. Balansa, comm. C.L. Spegazzini No. 1291, ILL 5216 [consisting of a leaf fraction and a microscopic preparation; the host treated as a member of Bignoniaceae in Hansford (1961)]. **Holotype:** LPS. Superfluous obligate synonym: *Meliola hariotula* C.G. Orejuela. Mycologia 36:436. 1944, nom. illegit.

Meliola hariotula C.G. Orejuela. Mycologia 36:436. 1944 (nom. illegit.). **Superfluous renaming of:** *Meliola hariotii* C.L. Spegazzini, 1891 [see preceding entry].

Meliola hawaiiensis F.L. Stevens. Bernice P. Bishop Museum Bulletin 19:37–38, text fig. 8c. 1925. **Holotype:** On *Eugenia sandwicensis* A.

Gray, Hawaii, Oahu, Olympus, 24 JUN [1921], leg. F.L. Stevens No. 667, ILL 5220. **Isotypes:** BISH 499030, 499968, 502016 and 502017. **Paratypes:** Kauai, Kalalau trail, 16 JUN [1921], leg. F.L. Stevens No. 490, ILL 5219; Maui, Kaluaaha, AUG 1912, leg. C.N. Forbes-F.L. Stevens No. 315, ILL 5217; 1913, leg. H.L. Lyon No. 275, ILL 5218; [21 SEP 1909], leg. H.L. Lyon No. 60, ILL 5221.

Meliola heliciicola C.G. Hansford. Sydowia Annales Mycologici 11:57. (1957) 1958. **Holotype:** Hab. in foliis *Heliciae* sp. [= *Helicia* sp.], Philippines, Luzon, [Trinidad Valley, 30 DEC 1930], leg. F.L. Stevens No. 1196, ILL 6533.

Meliola heliconiae F.L. Stevens. Annales Mycologici 26:210, pl. IV, fig. 29. 1928. **Holotype:** On *Heliconia* sp., Panama, Barro Colorado [Island], 26 JUL 1924, leg. F.L. Stevens No. 20, ILL 5223. **Paratypes:** 29 AUG 1924, leg. F.L. Stevens No. 44, ILL 5222; 19 SEP 1924, leg. F.L. Stevens No. 633, ILL 5224.

Meliola helleri F.S. Earle. Bulletin of the New York Botanical Garden 3:307. Preprint 1904 (journal issued in 1905). **Isotype:** On leaves of unknown woody plant, perhaps Myrtaceae, Porto Rico [Puerto Rico, three miles west of Ponce, limestone hills along coast, 9 DEC 1902], leg. A.A. Heller No. 6251, ILL 5226. **Holotype:** NY.

Meliola hessii F.L. Stevens. Illinois Biological Monographs 2:527, fig. 52. 1916. **Holotype:** On *Paullinia pinnata* L., Porto Rico [Puerto Rico], Sabana Llana, 13 AUG 1915, leg. F.L. Stevens No. 9367, ILL 6221. **Paratype:** Mayagüez, 4 JUN 1913, leg. F.L. Stevens No. 1207b, ILL 6222b [a microscopic preparation of ILL 6222. ILL 6222b is also an isotype of *Perisporium paulliniae* F.L. Stevens and a paratype of *Helminthosporium glabroides* F.L. Stevens].

Meliola heterocephala H. & P. Sydow. Annales Mycologici 14:356. 1916. **Assumed isotype:** Hab in foliis *Desmodii* spec. [= *Desmodium* sp.], Philippines, [Luzon, Laguna Prov.], Mount Maquiling [as Makiling], Dec. 1915, leg. C.F. Baker No. 3986 [ex Fungi Malayana No. 251], ILL 5243 [as a microscopic preparation].

Meliola heterodonta H. & P. Sydow. Annales Mycologici 14:357. 1916. **Possible isotype:** Hab. in foliis arboris ignotae [as *Dracontomelum* = *Dracontomelon* sp. on the packet, the host cited in C.G. Hansford (1961), as *D. dao* (Blanco) Merrill & Rolfe], Philippines, [Luzon, Laguna Prov.], Mount Maquiling [as Makiling], DEC 1915, leg. C.F. Baker No. 4031 [ex Fungi Malayana No. 252], ILL 5259.

Meliola heterotricha H. & P. Sydow. Leaflets of Philippine Botany 6:1923–1924. 1913. **Isotype:** Hab. upon the lower side of the leaves of *Donax cannaeformis* (G. Forster) K. Schum., Philippines, Mindanao [Island, Agusan], Cabadbaran, (Mt. Urdaneta), AUG 1912, leg. unknown, ex Phil. Bur. Sci. No. 13541, ILL 5260.

Meliola hippomanes F.L. Stevens. Annales Mycologici 26:284, pl. VI, fig. 80. 1928 [as *hippomaneae*, corrected by F.C. Deighton (1968) in accordance with current I.C.B.N.]. **Holotype:** On *Hippomane mancinella* L., Panama, Paitilla Point, 8 SEP 1924 [as 8 JUL on the packet], leg. F.L. Stevens No. 375, ILL 5262.

Meliola hispida F.L. Stevens. Annales Mycologici 26:241–242, pl. IV, fig. 44. 1928. **Holotype:** On *Calathea macrosepala* K. Schum., Costa Rica, Sabario, 8 AUG 1923, leg. F.L. Stevens No. 797, ILL 5264 [also considered "type" by C.G. Hansford (1961)]. **Paratypes**(?) [cited as a form that differed in certain characteristics]: On *Calathea insignis* Petersen, Costa Rica, Columbiana, 19 AUG 1923, leg. F.L. Stevens No. 569, ILL 5263 and 5538 [also paratypes of *Meliola marantacearum* F.L. Stevens].

Meliola holigarnae F.L. Stevens. Memoirs of the Indian Department of Agriculture. Botanical Series 15:108, pl. I, figs. 5–8. JAN 1928; Annales Mycologici 26:260–261. MAY 1928. **Lectotype:** On *Holigarna grahamii* J.D. Hooker, India, Anmod, N. Kanara, 25 DEC 1917, leg. L.J. Sedgwick, as F.L. Stevens No. 1981, ILL 5265 [lectotypification by F.L. Stevens in the second reference]. **Isolectotype:** HC 10. **Paratype:** India, Ekambi, N. Kanara, OCT 1919, leg. L.J. Sedgwick, as F.L. Stevens No. 1986a, ILL 5266.

Meliola hopeae H.S. Yates. Philippine Journal of Science, Section C (Botany) 13:369. (1918) 1919. **Isotype:** On leaves of *Hopea* sp., [Philippines], Luzon, Tayabas Prov., Mount Cadig, DEC 1916, leg. H.S. Yates, ex Phil. Bur. Sci. No. 25774, ILL 5270.

Meliola horrida H. Rehm. Philippine Journal of Science, Section C (Botany) 8:393. 1913, nom. illegit., non J.B. Ellis & B.M. Everhart, 1893. **Isotype:** Ad folia coriacea, Philippines, Luzon, Laguna Prov., Los Baños, APR 1913, leg. C.F. Baker No. 976, ILL 6174 [as a microscopic preparation]. ≡ *Meliola rehmii* F.L. Stevens. Annales Mycologici 26:222. 1928, nom. nov.

Meliola hoyae P.A. Saccardo. Atti dell' Accademia Scientifica Veneto-Trentino-Istriana, Series 3, 10:60–61. Preprint 1917 (journal issued in 1919). **Isotype:** On *Hoya luzonensis* [probably: *H. luzonica* Schlechter in Perkins = *H.* cf. *melliflua* (Blanco) Merrill] with *Gloeosporium hoyae* H. & P. Sydow, Philippines, [Luzon, Laguna Prov.], Los Baños, SEP 1913, leg. C.F. Baker No. 1842, ILL 5272 [as a microscopic preparation].

Meliola hyalospora J.H. Léveillé. Annales des Sciences Naturellas (Botanique) Series 3, 5:266. 1846. **Isotype:** Ad foliis *Desmonchi* [*Desmonchus* = *Desmoncus* sp.], Guyana Batava [Surinam], leg. M. Kegel No. 594, ILL 6481. **Holotype:** PC. ≡ *Leptomeliola hyalospora* (J.H. Léveillé)

F. v. Höhnel. Sitzungsberichte der Kaiserl. Akademie der Wissenschaften in Wien. Mathematisch-Naturwissenschaftliche Klasse, Abt. 1. 128:558. 1919. ≡ *Meliolinopsis hyalospora* (J.H. Léveillé) M. Beeli. Bulletin du Jardin Botanique de L'État à Bruxelles 7:119. 1920.

Meliola hyptidicola F.L. Stevens. Illinois Biological Monographs 2:484, fig. 8. 1916. **Holotype:** On *Hyptis lantanifolia* Poiteau, Porto Rico [Puerto Rico], Las Marias, 10 JUL 1915, leg. F.L. Stevens No. 8130, ILL 31705. **Paratypes:** On *Hyptis capitata* Jacquin, El Gigante, 16 JUL 1915, leg. F.L. Stevens No. 8526 [fig. 8], ILL 31704; on *Hyptis pectinata* (L.) Poiteau, Dos Bocas, below Utuado, 7 JUL 1915, leg. F.L. Stevens No. 7981, ILL 4100; Maricao, 19 JUL 1915, leg. F.L. Stevens No. 8791, ILL 4098; on *Hyptis* sp., Monte de Oro, 13 DEC 1913, leg. F.L. Stevens No. 5760, ILL 4099 [also the holotype of *Naemosphaera hyptidicola* F.L. Stevens]. ≡ *Irene hyptidicola* (F.L. Stevens) R.A. Toro. Mycologia 17:139–140. 1925. ≡ *Irenina hyptidicola* (F.L. Stevens) F.L. Stevens. Annales Mycologici 25:455. 1927.

Meliola ichnocarpi F.L. Stevens & E.F. Roldan. Philippine Journal of Science 56:72–73. 1935, nom. invalid. sine diagn. lat., non C.G. Hansford & M.J. Thirumalachar, 1948. **Holotype:** On *Ichnocarpus volubilis* (Lour.) Merrill, Philippines, Luzon, Batangas Prov., Benguet, Cuenca, Naguilian Road, 28 SEP 1930, leg. F.L. Stevens No. 722a, ILL 5273. **Isotypes:** CALP, Phil. Bur. Sci. **Paratype:** Luzon, Benguet, Naguilian Road, 5 JAN 1931, leg. F.L. Stevens No. 1465, ILL 5272 [this number is also the holotype of *Meliola ichnocarpi-volubili* C.G. Hansford, 1963, and the above invalid name is cited, pro syn., under the latter, validly published name. See next entry.].

Meliola ichnocarpi-volubili C.G. Hansford. Sydowia Annales Mycologici 16:320–321. (1962) 1963. **Holotype:** In foliis *Ichnocarpi volubilis* [*Ichnocarpus volubilis* (Lour.) Merrill], Philippines, [Luzon, Benguet, Naguilian Road, 5 JAN 1931], leg. F.L. Stevens No. 1465, ILL 5272.

Meliola illigerae F.L. Stevens & E.F. Roldan ex C.G. Hansford. Sydowia Annales Mycologici 16:316–317. (1962) 1963; originally published by F.L. Stevens & E.F. Roldan in Philippine Journal of Science 56:66, fig. 3a. 1935, as nom. invalid. sine diagn. lat. **Holotype:** On *Illigera luzonensis* (C. Presl) Merrill, Philippines, Luzon, Benguet, Naguilian Road, 5 JAN 1931, leg. F.L. Stevens No. 1524, ILL 5274.

Meliola imperatae H. & P. Sydow. Annales Mycologici 15:186. 1917. **Isotype:** Hab. in foliis *Imperatae cylindricae* [= *Imperata cylindrica* (L.) Beauvois], Philippines, Luzon, Rizal Prov., DEC 1915, leg. M. Ramos, ex Phil. Bur. Sci. No. 24069, ILL 5276.

Meliola inconspicua C.G. Hansford. Beihefte zur Sydowia Annales Mycologici 1:110. 1957. **Holotype:** Hab. in foliis Melastomatacearum spec.

[= Melastomataceae sp. indet.], Panama, Culebra, [2 OCT 1924], leg. F.L. Stevens No. 926, ILL 4172.

Meliola indica H. & P. Sydow in H. & P. Sydow & E.J. Butler var. *careyae* F.L. Stevens. Memoirs of the Indian Department of Agriculture. Botanical Series 15(5):109. JAN 1928 [originally cited as *caryae*; corrected by the author in Annales Mycologici 26:223. MAY 1928]. **Assumed holotype:** On *Careya arborea* Roxb., India, N. Kanara, Gairsoppa Falls, OCT 1919, leg. L.J. Sedgwick, as F.L. Stevens No. 1985, ILL 5278. **Isotypes:** HCIO, IMI.

Meliola indigoferae H. Sydow in E.M. Doidge & H. Sydow. Bothalia 2(2):451, 456. 1928. **Paratype:** [Sub *Meliola malacotricha* C.L. Spegazzini; hab. in foliis *Indigoferae natalensis* [= *Indigofera natalensis* Bolus], South Africa, Natal Prov., Buccleuch, 23 MAR 1916, leg. E.M. Doidge, ex PREM No. 9703, ILL 5450. [cited in English language description on p. 451].

Meliola inocarpi F.L. Stevens. Annales Mycologici 26:232, pl. IV, fig. 40. 1928. **Lectotype** [tentatively designated herein]: On *Inocarpus* sp. [as *I. edulis* J.R. & G. Forster on the packet, Republic of Singapore], Singapore Island, Straits Settlements, leg. C.F. Baker, Fungi Malayana No. 459, ILL 5281 [as two microscopic preparations].

Meliola integriseta (C.L. Spegazzini) C. L. Spegazzini var. *lepisanthea* (P.A. Saccardo) F.L. Stevens, 1928. **See basionym:** *Meliola lepisanthea* P.A. Saccardo, 1917.

Meliola integriseta (C.L. Spegazzini) C. L. Spegazzini var. *stevensii* (M. Beeli) F.L. Stevens, 1928. **See basionym:** *Meliola stevensii* M. Beeli, 1920.

Meliola intermedia A. Gaillard. Le Genre *Meliola*, p. 94–95, pl. 17, fig. 2. 1892. **Residual isosyntype:** Ad ... foliorum Rubiaceae, [C. Africa], Congo Française, Oubanghi, leg. F. Thollon No. 40, ILL 5287 [the name was lectotypified by C.G. Hansford (1961:546), with citation of F. Thollon No. 31 (no material of the latter collection found at ILL or seen by Hansford)].

Meliola intricata H. & P. Sydow. var. *major* M. Beeli. Bulletin du Jardin Botanique de L'État à Bruxelles 7:96. 1920. **Assumed isotype:** Épiphylle sur les feuilles d'une monocotylée [*Panicum* sp., C. Africa], Congo [Zaire], Kwilu River, H.J.R. Vanderyst No. 2689, ILL 5291. **Holotype:** BR.

Meliola ipomeae H. Rehm. Annales Mycologici 12:171. 1914, nom. nud., non *M. ipomoeae* F.S. Earle, 1901. **Assumed isotype:** Ad folia *Ipomeae* [*Ipomea* (orth. var.) = *Ipomoea* sp.], Philippines, Luzon, Laguna Prov., Los Baños, leg. C.F. Baker, ex H. Rehm, Ascomycetes No. 2104, ILL 5457.

Meliola ipomoeae F.S. Earle. Muhlenbergia 1:10–11. 1901. **Isotype**: On living leaves of *Ipomoea* sp., Porto Rico [Puerto Rico], near Mayagüez, alt. 400 ft., JAN 1900, leg. A.A. Heller No. 4358, ILL 5457 [as microscopic preparations].

Meliola ipomoeicola M. Beeli. Bulletin de Jardin Botanique de L'État à Bruxelles 7:96. 1920. **Isotype**: Épiphylle sur feuil. d' *Ipomoea* sp. (Convolvulacée), [C. Africa], Congo [Zaire], Wombali, 1913, leg. H.J.R. Vanderyst No. 2061, ILL 5292. **Holotype**: BR.

Meliola iquitosensis P.C. Hennings. Hedwigia 43:361. 1904. **Isotype**: Auf Blättern einer Zwergpalme [Arecaceae], Peru, Iquitos, Rio Amazonas, JUL 1902, leg. E.H.G. Ule No. 3211, ILL 5294. ≡ *Meliolinopsis iquitosensis* (P.C. Hennings) M. Beeli. Bulletin du Jardin Botanique de L'État à Bruxelles 7:119. 1920. ≡ *Meliolina iquitosensis* (P.C. Hennings) F.L. Stevens. Annales Mycologici 25:419–420. 1927. ≡ *Irenina iquitosensis* (P.C. Hennings) C.G. Hansford. Proceedings of the Linnean Society of London 157:169. 1946. ≡ *Asteridiella iquitosensis* (P.C. Hennings) C.G. Hansford ex C.G. Hansford. Beihefte zur Sydowia Annales Mycologici 2:722. 1961.

Meliola irosinensis H. Sydow. Leaflets of Philippine Botany 9:3118–3119. 1925. **Isotype**: On leaves of *Boerlagiodendron*, probably *B. mindanaense* Merrill [= *Osmoxylon* cf. *eminens* (Bull) W.R. Philipson, Philippines], Luzon, Sorsogon Prov., Irosin, DEC 1915, leg. A.D.E. Elmer No. 14526, ILL 5295.

Meliola irregularis F.L. Stevens. Illinois Biological Monographs 2:483, fig. 6. 1916. **Holotype**: On *Hygrophila brasiliensis* (Sprengel) Lindau, Porto Rico [Puerto Rico], Rio Piedras, 11 AUG 1915, leg. F.L. Stevens No. 9283, ILL 4103. ≡ *Irene irregularis* (F.L. Stevens) R.A. Toro. Mycologia 17:139. 1925. ≡ *Irenina irregularis* (F.L. Stevens) F.L. Stevens. Annales Mycologici 25:455. 1927. ≡ *Asteridiella irregularis* (F.L. Stevens) C.G. Hansford ex C.G. Hansford. Beihefte zur Sydowia Annales Mycologici 2:677. 1961.

Meliola isothea H. Sydow. Annales Mycologici 24:303–304. 1926. **Isolectotype**: Hab. in foliis vivis *Tabernaemontanae citrifoliae* [= *Tabernaemontana citrifolia* L.], Costa Rica, Piedades de San Ramón, 7 FEB 1925, leg. H. Sydow, Fungi in Itinere Costaricensi Collecti No. 131, ILL 5296 [citation of this collection as "type" by C.G. Hansford (1961:547) is interpreted as lectotypification, although no location for the lectotype is indicated]. **Residual isosyntype**: Hab. in fol. *Tabernaemontanae oppositifoliae* [*T. oppositifolia* (Sprengel) Urban = *T. citrifolia* L.], 26 JAN 1925, leg. H. Sydow, Fungi in Itinere Costaricensi Collecti No. 132, ILL 5297.

Meliola italica (P.A. Saccardo) F.L. Stevens, 1928. **See basionym**: *Meliola cyperi* N.T. Patouillard in A. Gaillard var. *italica* P.A. Saccardo, 1903.

Meliola ixorae H.S. Yates. Philippine Journal of Science, Section C (Botany) 12:365. (1917) 1918 [as *ixoriae*, corrected by both F.L. Stevens (1928:280) and C.G. Hansford (1961:606) in accordance with current I.C.B.N.]. **Isotype:** On *Ixora philippinensis* Merrill, [Philippines], Luzon, Manila and vicinity, FEB 1917, leg. H.S. Yates, ex Phil. Bur. Sci. No. 25841, ILL 5302.

Meliola janeirensis C.G. Hansford. Proceedings of the Linnean Society of London 160:121. (1948) 1949. **Isotype:** Hab. in foliis *Crotonis* sp. [= *Croton* sp.], Brazil, Rio de Janeiro, JUL 1887, leg. E.H.G. Ule, ex G.L. Rabenhorst, H.G. Winter, and F.O. Pazschke, Fungi Europeae No. 3848, sub *M. biocornis* H.G. Winter var. *constipata* C.L. Spegazzini, ILL 4967. **Holotype:** PREM.

Meliola jasminicola P.C. Hennings. Hedwigia 34:11. 1895. **Isotype:** In Gärten auf Blättern von *Jasminum* sp., Tonkin [Vietnam], Hanoi, AUG 1890, leg. B. Balansa No. 4542, ILL 5315.

Meliola jatrophae F.L. Stevens. Illinois Biological Monographs 2:516. 1916. **Holotype:** On *Jatropha hernandiifolia* Vent., Porto Rico [Puerto Rico], Rio Tanamá, near Arecibo, 6 JUL 1915, leg. F.L. Stevens No. 7873, ILL 5318. **Paratype:** Dos Bocas, near Utuado, 8 AUG 1915, leg. F.L. Stevens No. 7930, ILL 5319.

Meliola juddiana F.L. Stevens. Bernice P. Bishop Museum Bulletin 19:32–33, text fig. 7e, pl. II(F). 1925. **Holotype:** On *Pelea* sp. [= *Melicope* sp.], Hawaii, [Island of] Hawaii, Kona, Keauhou, Bishop Estate Road, 25 JUL [1921], leg. F.L. Stevens No. 986, ILL 5337. **Isotypes:** BISH 499029 and 499967. **Paratypes:** On *Pelea* sp. [= *Melicope* sp., Island of] Hawaii, Kona, Keauhou, 25 JUL [1921], leg. F.L. Stevens No. 974, ILL 5332 and 5334; Waimea, 30 JUL [1921], leg. F.L. Stevens No. 1048, ILL 5342; Hamakua, upper ditch trail, 28 JUL [1921], leg. F.L. Stevens No. 1034, ILL 5338; Maui, Olinda pipeline, 5 SEP [1921], leg. F.L. Stevens No. 1148, ILL 5341; Molokai, Halawa, [AUG 1912], leg. C.N. Forbes-F.L. Stevens No. 483, ILL 5320; Oahu, Palolo Valley [& Mt. Olympus], 10 JUN [1921], leg. F.L. Stevens No.297, ILL 5340; Olympus, 24 JUN [1921], leg. F.L. Stevens No. 712, ILL 5330 and 5339; leg. F.L. Stevens No. 704, ILL 5343; Tantalus, 27 MAY 1913, leg. H.L. Lyon No. 346, ILL 5321 and 5329; 7 SEP 1913, leg. H.L. Lyon s.n., ILL 5328; on *Pelea elliptica* (A. Gray) Hillebr. [= *Melicope elliptica* A. Gray], Kauai, Kalalau trail, 16 JUN [1921], leg. F.L. Stevens No. 526, ILL 5333; on *Pelea rotundifolia* A. Gray [= *Melicope rotundifolia* (A. Gray) T.G. Hartley & B.C. Stone], Oahu, leg. C.N. Forbes-F.L. Stevens No. 1328, ILL 5327 [also a paratype of *Meliola peleae* F.L. Stevens]; on *Pelea clusiaefolia* A. Gray [= *Melicope clusiifolia* (A.Gray) T.G. Hartley & B.C. Stone], Lanai, 1915, leg. Munro-F.L. Stevens s.n., ILL 5325;

Maui, 1910, leg. C.N. Forbes-F.L. Stevens s.n., ILL 5322; on *Pelea sandwicensis* (Hooker & Arnott) A. Gray [= *Melicope sandwicensis* (Hooker & Arnott) T.G. Hartley & B.C. Stone], 1920, leg. C.N. Forbes-F.L. Stevens No. 235, ILL 5335; on *Pelea parvifolia* Hillebr. [= *Melicope lydgatei* (Hillebr.) T.G. Hartley & B.C. Stone], Molokai, [Pukoo Ridge, AUG 1912], leg. C.N. Forbes-F.L. Stevens No. 411, ILL 5324 and 5968 [also paratypes of *Meliola peleae* F.L. Stevens]; on *Pelea cinerea* (A. Gray) Hillebr. [= *Melicope cinerea* A. Gray], Oahu, 1912, leg. C.N. Forbes-F.L. Stevens No. 1816, ILL 5323.

Meliola juruana P.C. Hennings. Hedwigia 43:365–366, illustr. 1904. **Paratype:** Auf Blättern von *Lonchocarpus ulei* H.A.T. Harms, nom. nud. [the host identified as *Swartzia* sp. in C.G. Hansford (1961)], Brazil, Rio Juruá, leg. E.H.G. Ule No. 2935, ILL 5344. [characterized as "unreif" (immature) in the protologue].

Meliola kaduae F.L. Stevens. Bernice P. Bishop Museum Bulletin 19:30, text fig. 7c. 1925. **Holotype:** On *Gouldia* sp. [= *Hedyotis* sp.], Hawaii, Oahu, Tantalus, 22 JUN [1921], leg. F.L. Stevens No. 601, ILL 5346. **Isotypes:** ILL 6261, BISH 145986 and 499966. **Paratypes:** On *Kadua* sp. [= *Hedyotis* sp.], Oahu, Tantalus, 22 JUN [1921], leg. F.L. Stevens No. 601a, ILL 5345; on *Straussia kaduana* (Cham. & Schlecht.) A. Gray [= *Psychotria kaduana* (Cham. & Schlecht.) Fosb.], Oahu, Olympus [Palolo Valley and Mt. Olympus on the packet], 10 JUN [1921], leg. F.L. Stevens No. 335, ILL 3601 and 3607; on *Straussia* sp. [= *Psychotria* sp.], Kauai, Kalalau trail, 16 JUN [1921], leg, F.L. Stevens No. 511, ILL 3606; leg. F.L. Stevens No. 512, ILL 5348 and 5351; Oahu, Tantalus, 22 JUN [1921], leg. F.L. Stevens No. 617, ILL 5347; on *Gouldia terminalis* (Hooker & Arnott) Hillebr. [= *Hedyotis terminalis* (Hooker & Arnott) W.L. Wagner & Herbst], 22 JUN [1921], Oahu, Tantalus, 22 JUN [1921], leg. F.L. Stevens No. 604, ILL 6255; on *Gouldia* sp. [= *Hedyotis* sp.], Oahu, Tantalus, 22 JUN [1921], leg. F.L. Stevens No. 597, ILL 5349; on *Gouldia lanceolata* (Wawra) A.A. Heller [= *Hedyotis terminalis* (Hooker & Arnott) W.L. Wagner & Herbst, [Island of] Hawaii, Waimea, 30 JUL [1921], leg. F.L. Stevens No. 1049, ILL 5350.

Meliola kartaboensis F.L. Stevens. Annales Mycologici 26:249, pl. V, fig. 56. 1928. **Holotype:** On *Solanum* sp., British Guiana [Guyana], Kartabo, 24 JUL 1922, leg. F.L. Stevens No. 635, ILL 5356.

Meliola kauaiensis F.L. Stevens. Bernice P. Bishop Museum Bulletin 19:39–40, text fig. 8h. 1925. **Holotype:** On *Kadua* sp. [= *Hedyotis* sp.], Hawaii, Kauai, Kalalau trail, 16 JUN [1921], leg. F.L. Stevens No. 531, ILL 5355. **Isotypes:** ILL 5352, BISH 499028 and 499965. **Paratypes:** On *Kadua knudsenii* Hillebr. [= *Hedyotis knudsenii* (Hillebr.) Fosb., Kauai,

Pipe trail, Waimea Canyon, 15 JUN [1921], leg. F.L. Stevens No. 436, ILL 5353; leg. F.L. Stevens No. 437, ILL 5354.

Meliola koae F.L. Stevens. Bernice P. Bishop Museum Bulletin 19:34, text fig. 7h, pl. II(K). 1925. **Holotype:** On *Acacia koa* A. Gray, Hawaii, Oahu, Wahiawa, 31 MAY [1921], leg. F.L. Stevens No. 163, ILL 5363. **Isotypes:** ILL 5360 and 5364, BISH 145987, 145988, 499027 and 499964. **Paratypes:** [Island of] Hawaii, OCT 1913, leg. R.S. Hosmer, as H.L. Lyon No. 415, ILL 5361; Kauai, Kalalau trail, 31 MAY [1921, as 16 JUN on the packet], leg. F.L. Stevens No. 521, ILL 5362.

Meliola kydia P.A. Saccardo. Bulletino dell' Orto Botanico della Regia Universita di Napoli 6:43. 1921, nom. invalid. [as "nomen ad interim," cf. Article 34.1b of I.C.B.N.]. **Assumed isotypes:** Hab. in foliis *Garciniae kydiae* [*Garcinia kydia* Roxb. = *G. cowa* Roxb., Republic of Singapore, as "Island of Singapore, Straits Settlements" on the original exsiccati label, AUG 1917], leg. C.F. Baker No. 4986 [ex Fungi Malayana No. 450], ILL 6482 [as a microscopic preparation], CUP, FH.

Meliola laevigata H. & P. Sydow. Leaflets of Philippine Botany 5:1537. 1912. **Isotype:** On *Paralstonia clusiacea* Baillon [= *Alyxia clusiacea* (Baillon) Pichon], Philippines, Palawan [Island], Puerto Princesa, (Mt. Pulgar), MAR 1911, leg. A.D.E. Elmer No. 12784, ILL 5365, ex Phil. Bur. Sci.

Meliola laevipoda C.L. Spegazzini. Revista Argentina de Historia Natural 1:403–404. 1891; Fungi Guaranitici Nonnulli novi vel Critici, pp. 403–404, No. 77. 1891. **Isotypes:** Ad folia viva *Aspidospermae quebrachii* [= *Aspidosperma* cf. *quebracho-blanco* Schlecht.], in sylvis, [Paraguay], pr. Yaguaron, NOV 1882, leg. B. Balansa, comm. C.L. Spegazzini No. 3589, ILL 5373 and 5374. **Holotype:** LPS.

Meliola lagunensis C.G. Hansford ex C.G. Hansford. Sydowia Annales Mycologici 16:302. (1962) 1963; originally published in Beihefte zur Sydowia Annales Mycologici 2:36. 1961, as nom. invalid. sine diagn. lat. **Holotype:** In foliis *Uvariae* spec. indet. [= *Uvaria* sp.], Philippines, [Luzon Island, Laguna Agricultural College, 10 SEP 1930], leg. F.L. Stevens No. 493 p.p., ILL 4767.

Meliola lanceolato-setosa H. & P. Sydow. Annales Mycologici 12:197–198. 1914. **Isotypes:** Hab. in foliis vivis *Markhamiae* sp. [= *Markhamia* sp., identified as *M. platycalyx* (Baker) Sprague in Hansford (1961)], Deutsch-Ostafrika [E. Africa, Tanzania], pr. Tengeni, 11 OCT 1913, leg. M. Grote (sub Botanical Institute Amani No. 5602), ex H. Sydow, Fungi Exotici Exsiccati No. 248, ILL 5367 [as a microscopic preparation], and in bound exsiccati set.

Meliola lanosa N.T. Patouillard var. *funerea* (D. McAlpine) C.G. Hansford, 1961. **See basionym:** *Meliola funerea* D. McAlpine, 1896.

Meliola lepisanthea P.A. Saccardo. Atti dell' Accademia Scientifica Veneto-Trentino-Istriana, Series 3, 10:61. 1917. **Isotype:** In foliis *Lepisanthis* sp. [= *Lepisanthes* sp.], Philippines, [Luzon, Laguna Prov.], Los Baños, JUN 1913, leg. C.F. Baker No. 1279, ILL 5286. ≡ *Meliola integriseta* C.L. Spegazzini var. *lepisanthea* (P.A. Saccardo) F.L. Stevens. Annales Mycologici 26:254. 1928.

Meliola leptochaeta H. & P. Sydow. Annales Mycologici 15:187. 1917. **Isotypes:** Hab. in foliis *Vavaeae* sp. [= *Vavaea* sp.], Philippines, Luzon, Rizal Prov., SEP 1915, leg. H.S. Yates, ex Phil. Bur. Sci. No. 25009, ILL 5370 and 5371.

Meliola leptoclada H. Sydow. Annales Mycologici 20:62. 1922. **Isotype:** Hab. in foliis *Schefflerae octophyllae* [= *Schefflera octophylla* (Lour.) H.A.T. Harms, China, Kwangtung Prov.], Loh Fau Mountain, (Lofaushan), 20 AUG 1917, leg. E.D. Merrill, ex Phil. Bur. Sci. No. 11133, ILL 5380. **Holotype:** FH.

Meliola leptospora A. Gaillard. Le Genre *Meliola*, pp. 87–88, pl. 15, fig. 5. 1892. **Isotype:** Ad paginam inferiorem foliorum quorundam in silva, [C. Africa], Congo [Zaire], pr. Mayumbe, NOV 1888, leg. F. Thollon No. 1207, ILL 5381. **Holotype:** PC.

Meliola leucosykes H.S. Yates. Philippine Journal of Science, Section C (Botany) 12:366. (1917) 1918 [as *leucosykeae*; corrected by F.C. Deighton (1968) in accordance with current I.C.B.N.]. **Isotype:** On *Leucosyke capitellata* (Poiret) Weddell, Philippines, Samar [Island], Catubig River, 17 FEB 1916, leg. M. Ramos, ex Phil. Bur. Sci. No. 24621, ILL 5382. ≡ *Irenina leucosykes* (H.S. Yates) C.G. Hansford. Proceedings of the Linnean Society of London 157:170. 1946. ≡ *Asteridiella leucosykes* (H.S. Yates) C.G. Hansford ex C.G. Hansford. Beihefte zur Sydowia Annales Mycologici 2:333. 1961. [Epithets of the synonyms were also published as *leucosykeae* and are corrected herein.].

Meliola linocierae H. & P. Sydow. Annales Mycologici 12:550. 1914. **Isotypes:** Hab. in foliis *Linocierae cumingianae* [*Linociera cumingiana* S. Vidal = *Chionanthus ramiflora* Roxb.], Philippines, [Luzon], Rizal Prov., Antipolo, 14 AUG 1918, leg. M. Ramos, ex Phil. Bur. Sci. No. 254 and ex H. Sydow, Fungi Exotici Exsiccati No. 375, ILL 5383 and 33539, as well as packet in bound exsiccati set [associated with and also (p.p.) isotypes of *Meliola linociericola* C.G. Hansford — only ILL 5383 annotated and confirmed by Hansford]. ≡ *Irenina linocierae* (H. & P. Sydow) C.G. Hansford. Proceedings of the Linnean Society of London 160:128. (1948) 1949 [erroneously as *linocieriae*]. ≡ *Asteridiella linocierae* (H. & P. Sydow) C.G. Hansford ex C.G. Hansford. Beihefte zur Sydowia Annales Mycologici 2:529–530. 1961 [erroneously as *linocieriae*].

Meliola linociericola C.G. Hansford. Proceedings of the Linnean Society of London 160:129. (1948) 1949 [as *linocieriicola*, corrected by F.C. Deighton (1968) in accordance with current I.C.B.N.]. **Isotype:** Hab. in foliis *Linocierae cumingianae* [as *Linocieriae* = *Linociera cumingiana* S. Vidal = *Chionanthus ramiflora* Roxb.], Philippines, Luzon, Rizal Prov., [Antipolo], 14 AUG 1915, leg. M. Ramos, in consortio *Ireninae linocierae* [erroneously as *linocieriae*] sub *Meliola* in H. & P. Sydow, Fungi Exotici Exsiccati No. 375 p.p., ILL 5383 [this specimen annotated and confirmed by C.G. Hansford]. **Possible isotypes:** ILL 33539 and packet in bound exsiccati set at ILL. **Holotype:** PREM. [All these specimens are also isotypes of *Meliola linocierae* H. & P. Sydow.].

Meliola linocierina C.G. Hansford. Beihefte zur Sydowia Annales Mycologici 1:110–111. 1957. **Holotype:** Hab. in foliis *Linocierae* spec. [*Linociera* = *Chionanthus* sp.], Philippines, Luzon, Ilocos Norte Prov., JUL–AUG 1918, leg. M. Ramos, ex Phil. Bur. Sci. No. 33432, ILL 5384.

Meliola lisianthii F.L. Stevens & L.R. Tehon. Mycologia 18:15–16, pl. 2, fig. 19. 1926 [as *lisianthi*, corrected in accordance with Articles 32.6 & 61 of I.C.B.N.]. **Holotype:** On *Lisianthius grandiflorus* Aublet [as *Chelonanthus chelonoides* (L.) Gilg in Engler & Prantl (a synonym) on the packet = *Irlbachia alata* (Aublet) Maas ssp. *alata*], British Guiana [Guyana], Wismar, 14 JUL 1922, leg. F.L. Stevens No. 316, ILL 5388 [also an isotype of *Schiffnerula lisianthii* C.G. Hansford]. **Isotype:** ILL 5398.

Meliola litseae H. & P. Sydow. Annales Mycologici 15:187–188. 1917. **Isotype:** Hab. in foliis *Litseae perrottetii* [= *Litsea perrottetii* (Blume) Fernandez-Villar in Blanco], Philippines, Luzon, Laguna Prov., Los Baños, 20 NOV 1912, leg. C.F. Baker No. 480, ILL 4761.

Meliola litseae H. & P. Sydow var. *rotundipoda* C.G. Hansford. Reinwardtia 3:88–89. 1954. **Isotype:** Hab. in foliis *Litseae glutinosae* [= *Litsea glutinosa* C.B. Robinson], Philippines, Luzon, Los Baños, ex C.F. Baker, Fungi Malayana No. 362, ILL 6204. **Holotype:** BO 16006. **Paratype:** [Luzon, Rizal Prov., Antipolo], JUN 1913, leg. M. Ramos, ex Phil. Bur. Sci. No. 20994, ILL 6201.

Meliola livistonae H.S. Yates. Philippine Journal of Science, Section C (Botany) 12:366–367. (1917) 1918 [as *livistoniae*, corrected by F.C. Deighton (1968) in accordance with current I.C.B.N.]. **Isotype:** On *Livistona* sp., Philippines, Luzon, Tayabas Prov., Basiad, 19 DEC 1916, leg. H.S. Yates, ex Phil. Bur. Sci. No. 25632, ILL 5405.

Meliola lobeliae F.L. Stevens. Bernice P. Bishop Museum Bulletin 19:29, text fig. 7a. 1925. **Holotype:** On *Clermontia* sp., Hawaii, Maui, Iao Valley, 7 SEP [1921], leg. F.L. Stevens No. 1154, ILL 5408 [also an isolectotype of *Asterina clermontiae* F.L. Stevens & R.W. Ryan in F.L. Stevens and a paratype of *Trichothallus hawaiiensis* F.L. Stevens]. **Isotypes:** BISH 499026 and 499963, as well as ILL 6881 [marked No. 1154a on the packet, the latter also the lectotype of *Asterina clermontiae*] and No. 1154b, ILL 6934, BISH 499858 and 499862 [these are also the holotype and isotypes, respectively, of *Clypeolella clermontiae* F.L. Stevens & R.W. Ryan in F.L. Stevens, as well as isolectotypes of *Asterina clermontiae* and paratypes of *Trichothallus hawaiiensis*]. **Paratypes:** Molokai, leg. C.N. Forbes-F.L. Stevens No. 32, ILL 5406; [Island of] Hawaii, Keauhou, Kona, Bishop Estate Road, 25 JUL [1921], leg. F.L. Stevens No. 979, ILL 5407 and 7227 [also paratypes of *Calothyriopeltis metrosideri* F.L. Stevens & R.W. Ryan in F.L. Stevens (the locality for the latter is cited as Kealakekua in the protologue but is as cited here on the packet)].

Meliola lonchocarpicola F.L. Stevens. Annales Mycologici 26:207, pl. III, fig. 26. 1928. **Syntypes:** On *Lonchocarpus* sp., Panama, Fort Lorenzo trail, 10 OCT 1924, leg. F.L. Stevens No. 1180, ILL 5410; leg. F.L. Stevens No. 1172, ILL 5412; Brazos Brook Reservoir, 22 SEP 1924 [as JUL on the packet], leg. F.L. Stevens No. 721, ILL 5409.

Meliola longistipitata F.L. Stevens. Annales Mycologici 26:191, pl. II, fig. 13; pl. III, fig. 13a. 1928. **Holotype:** On *Dimerocostus uniflorus* (Poeppig ex Petersen) K. Schum. Panama, Fort Lorenzo trail, 10 OCT 1924, leg. F.L. Stevens No. 1186, ILL 5414. **Isotype:** ILL 5416.

Meliola lophopetali F.L. Stevens ex C.G. Hansford. Beihefte zur Sydowia Annales Mycologici 1:111. 1957. **Holotype:** Hab. in foliis *Lophopetali toxici* [= *Lophopetalum toxicum* Loher], Philippines, Luzon, Rizal Prov., [SEP 1915], leg. H.S. Yates, ex Phil. Bur. Sci. No. 25076, ILL 5418.

Meliola loxostylidis E.M. Doidge. Transactions of the Royal Society of South Africa 8:114. 1920. **Isosyntype:** In foliis fruticis incognitis [later identified by Doidge as *Loxostylis alata* Sprengel ex Reichenb., which is also the host cited in the protologue for the other syntype], South Africa, Natal [Prov.], Mayville, 22 JUL 1915, leg. J. Medley Wood, ex PREM No. 9026, ILL 5419. **Syntype:** PREM.

Meliola lucumae F.L. Stevens. Illinois Biological Monographs 2:517, fig. 47. 1916. **Holotype:** On *Lucuma multiflora* A. DC. [= *Pouteria multiflora* (A. DC.) Eyma], Porto Rico [Puerto Rico], Las Marias, 10 JUL 1915, leg. F.L. Stevens No. 8164, ILL 5420. [As a result of a printer's error, the last line in the protologue duplicates part of the type citation for the species preceding in the text: *M. gymnanthicola* F.L. Stevens. C.G. Hansford (1961:505) failed to recognize this error, and he cited the wrong type (on *Gymnanthes lucida* Sw., Guayanilla, F.L. Stevens No. 8596) for *M. lucumae*.].

Meliola ludibunda C.L. Spegazzini. Anales de la Sociedad Cientifica Argentina 18:263. 1884; Fungi Guaranitici, Pugillus I, pp. 71–72, No. 178. 1884. **Isolectotype**: Hab. ad folia viva arborum herbarumque, [Paraguay], pr. Guarapi, Paraguari, Villa Rica, Caáguazú, per ann. 1881–83, leg. B. Balansa No. 3489, comm. C.L. Spegazzini No. 546, ILL 5423. **Lectotype**: LPS [citation of Balansa No. 3489 in the Spegazzini herb. as "type" by C.G. Hansford (1961:389) is accepted as lectotypification].

Meliola lundiae F.L. Stevens. Annales Mycologici 26:179, pl. II, fig. 7. 1928. **Holotype**: On *Lundia umbrosa* Bureau ex Baillon, Ecuador, Barrn'nital, 17 NOV 1924 [as 17 DEC on the packet], leg. F.L. Stevens No. 321, ILL 5424.

Meliola luzonensis H. & P. Sydow. Annales Mycologici 15:188. 1917. **Isotypes**: Hab. in foliis *Antidesmae* sp. [= *Antidesma* sp.], Philippines, Luzon, Bataan Prov., DEC 1915, leg. M. Ramos, ex Phil. Bur. Sci. No. 23976, ILL 5425, FH, S.

Meliola lyoni F.L. Stevens. Bernice P. Bishop Museum Bulletin 19:37, text fig. 8b. 1925. **Holotype**: On *Dodonaea viscosa* (L.) Jacquin, Hawaii, [Island of] Hawaii, Kilauea, 14 JUL [1921], leg. F.L. Stevens No. 843, ILL 5430 [R.D. Goos & D.P. Gowing (1992) cite the host as *Eugenia sandwicensis* A. Gray = *Syzygium sandwicensis* (A. Gray) Niedenzu]. **Isotypes**: BISH 499025 and 499962. **Paratypes**: [Island of] Hawaii, Kilauea, 16 JUL [1921], leg. F.L. Stevens No. 865, ILL 5427; Hualalai, 19 JUL [1921], leg. C. Judd No. 901, ILL 5428; [lava] flow of 1855 below Hale Aloha, 7 JUN 1915, leg. C.N. Forbes-F.L. Stevens No. 754, ILL 5426; Kauai, Kalalau trail, 16 JUN [1921], leg. F.L. Stevens No. 508, ILL 5431; leg. C.N. Forbes-F.L. Stevens No. 87, ILL 5429.

Meliola macarangae H. & P. Sydow. Annales Mycologici 15:188. OCT 1917. **Paratype**: Hab. in caulibus *Macarangae* sp. [= *Macaranga* sp.], Philippines, [Luzon], Laguna Prov., NOV 1915, leg. M. Ramos, ex Phil. Bur. Sci. No. 24045, ILL 5523.

Meliola macarangae H. & P. Sydow var. *apayaosensis* (H.S. Yates) C.G. Hansford, 1961. **See basionym**: *Meliola apayaosensis* H.S. Yates, (1918) 1919.

Meliola macarangae H.S. Yates. Philippine Journal of Science, Section C (Botany) 12:367. (1917) JAN 1918, nom. illegit., non H. & P. Sydow, OCT 1917. **Isotype**: On *Macaranga tanarius* (L.) Muell. Arg., Philippines, Luzon, Tyabas Prov., Basiad, 20 DEC 1916, leg. H.S. Yates, ex Phil. Bur. Sci. No. 25621, ILL 4774. ≡ *Meliola brachypoda* H. Sydow, Annales Mycologici 20:67. 1922, nom. nov.

Meliola macowaniana F. v. Thümen. Mycotheca Universalis, Century 6, No. 568. Anno 1876; Flora 59:569. 1876. **Isotype**: Ad folia viva *Celastri* *buxifolii* [*Celastrus buxifolius* L. = *Maytenus* sp., South Africa], Cape of Good Hope [as Promont. Bonae Spei]: Somerset-East, in dumetis montis "Boschberg" AUG 1876, leg. P. MacOwan No. 1250, ILL 6484. ≡ *Asterina macowaniana* (F. v. Thümen) K. Kalchbrenner & M.C. Cooke. Grevillea 9:33. 1880. ≡ *Dimerosporium macowanianum* (F. v. Thümen) P.A. Saccardo. Sylloge Fungorum 1:53. 1882. ≡ *Englerula macowaniana* (F. v. Thümen) F. v. Höhnel. Sitzungsberichte der Kaiserl. Akademie der Wissenschaften in Wien. Mathematisch-Naturwissenschaftliche Klasse. Abt. 1, 119:420. 1910, nom. illegit. [cf. Article 52 of I.C.B.N. This is an incorrectly cited name; the generic name adopted by the author (in op. cit.) is *Parenglerula*]. ≡ *Parenglerula macowaniana* (F. v. Thümen) F. v. Höhnel, op. cit. 119:456, 465. 1910. ≡ *Dimerium macowanianum* (F. v. Thümen) E.M. Doidge. Transactions of the Royal Society of South Africa 5:718. 1917. ≡ *Englerulaster macowaniana* (F. v. Thümen) G. Arnaud. Thèses 1. Les Astérinées, p. 183. 1918.

Meliola macrochaeta H. & P. Sydow. Leaflets of Philippine Botany 5:1538. 1912. **Isotype**: On the older leaves of *Alsodeia formicaria* A.D.E. Elmer [= *Rinorea formicaria* (A.D.E. Elmer) Merrill], Philippines, Palawan [Island], Puerto Princesa, (Mt. Pulgar), MAR 1911, leg. A.D.E. Elmer No. 12887, ex Phil. Bur. Sci., ILL 4405. ≡ *Irenopsis macrochaeta* (H. & P. Sydow) F.L. Stevens. Annales Mycologici 25:438. 1927.

Meliola macropoda H. Sydow. Annales Mycologici 24:296–297. 1926. **Isotype**: Hab. in foliis *Xanthoxyli proceri* [= *Zanthoxylum procerum* J. Donnell Smith], Costa Rica, San Pedro de San Ramón, 5 FEB 1925, leg. H. Sydow, Fungi in Itinere Costaricensi Collecti No. 113b, ILL 5526.

Meliola maesae H. Rehm. Philippine Journal of Science, Section C (Botany) 8:392. 1913. **Residual isosyntype**: Ad folia et ramos vivos *Maesae laxae* [= *Maesa laxa* Mez, Philippines], Luzon, Laguna Prov., Los Baños, [12] JAN 1913, leg. C.F. Baker No. 699, ILL 5527. **Assumed lectotype**: C.F. Baker No. 718, S [citation of this specimen by C.G. Hansford (1961:511) as "type" is interpreted as lectotypification].

Meliola magna F.L Stevens. Annales Mycologici 26:252, pl. V, fig. 63. 1928. **Holotype**: On *Nectandra* sp. [as *N. sanguinea?* Rolander ex Rottb. on the packet = *Ocotea* sp.], Costa Rica, Peralta, 12 JUL 1923, leg. F.L. Stevens No. 373, ILL 5528.

Meliola magnoliae F.L. Stevens. Illinois Biological Monographs 2:523–524, fig. 50. 1916. **Holotype**: On *Magnolia portoricensis* Bello, Porto Rico [Puerto Rico], Monte Alegrillo, near Maricao, 14 NOV 1913, leg. F.L. Stevens No. 4738, ILL 5530. **Isotype**: ILL 5529.

Meliola makilingiana H. & P. Sydow. Annales Mycologici 15:188–189.

1917. **Isotype:** Hab. in foliis *Psychotriae* sp. [= *Psychotria* sp.], Philippines, [Luzon], Laguna Prov., Mount Maquiling, DEC 1913, leg. C.F. Baker No. 2146, [ex Phil. Bur. Sci. No. 550], ILL 5531.

Meliola malacensis P.A. Saccardo. Bulletino Dell' Orto Botanico Della Universita di Napoli 6:43. Preprint 1918 (journal issued in 1921). **Isolectotype:** Hab. in foliis adhuc subvivis *Wormiae suffruticosae* [*Wormia suffruticosa* W. Griffith = *Dillenia suffruticosa* (W. Griffith) Martelli], Singapore [as "Island of Singapore, Straits Settlements" on the original exsiccati label], JUN 1917. leg. C.F. Baker [ex Fungi Malayana No. 451], ILL 6500 [as a microscopic preparation]. **Lectotype:** K [designated by S.J. Hughes, Mycological Papers. International Mycological Institute 166:239. 1993. Hughes changed spelling of the epithet to "*malaccensis.*"]. ≡ *Meliolina malacensis* (P.A. Saccardo) A. Trotter. Sylloge Fungorum 24:360–361. 1926. ≡ *Meliolina malacensis* (P.A. Saccardo) F.L. Stevens. Annales Mycologici 25:418. 1927 [superfluous (illegit.) later homonym].

Meliola malacotricha C.L. Spegazzini var. *major* M. Beeli. Bulletin du Jardin Botanique de L'État à Bruxelles 7:96–97. 1920. **Isotype:** In foliis Cucurbitacearum [= Cucurbitaceae, C. Africa], Congo [Zaire], Wombali, 1913, leg. H.J.R. Vanderyst No. 2064, ILL 5463. **Holotype:** BR.

Meliola malaneae F.L. Stevens & L.R. Tehon. Mycologia 18:17–18, pl. 2, fig. 22. 1926. **Holotype:** On *Malanea* sp., Trinidad, Cumuto, 16 AUG 1922, leg. F.L. Stevens No. 911, ILL 5586. **Paratypes:** On *Malanea macrophylla* Bartl. in Schomb., Cumuto, 16 AUG [1922], leg. F.L. Stevens No. 949, ILL 5584; on *Psychotria* sp., leg. F.L. Stevens No. 944, ILL 5585.

Meliola malangasensis C.G. Hansford. Beihefte zur Sydowia Annales Mycologici 1:111. 1957. **Holotype:** Hab. in foliis *Eugeniae* sp. [= *Eugenia* sp.], Philippines, Mindanao [Island], Zamboanga District, Malangas, [OCT–NOV 1919], leg. M. Ramos & G. Edaño, ex Phil. Bur. Sci. No. 36335, ILL 5099.

Meliola manaosensis P.C. Hennings. Hedwigia 43:366, illustr. 1904. **Assumed isotypes:** Auf Blättern von *Mauritia martiana* Spruce [as *M. aculeata* Kunth in H.B.K. on the packet], Brazil, Manaos, Rio Negro, [MAY] 1902, [leg. E.H.G. Ule] No. 3145 [ex Mycotheca Brasiliensis No. 59], ILL 6537 and 6538. ≡ *Meliolinopsis manaosensis* (P.C. Hennings) M. Beeli. Bulletin du Jardin Botanique de L'État à Bruxelles 7:120. 1920. ≡ *Irenina manaosensis* (P.C. Hennings) C.G. Hansford. Proceedings of the Linnean Society of London 157:170. 1946. ≡ *Asteridiella manaosensis* (P.C. Hennings) C.G. Hansford ex C.G. Hansford. Beihefte zur Sydowia Annales Mycologici 2:723. 1961.

Meliola manca J.B. Ellis & G.W. Martin. American Naturalist 17:1284. 1883. **Isotypes:** On living leaves of *Myrica cerifera* L., Florida, Green Cove Springs, FEB 1883, leg. G.W. Martin, ex J.B. Ellis, North American Fungi No. 1292, ILL 4193 and 4196 [as microscopic preparations], and in bound exsiccati set. **Holotype:** NY. ≡ *Irene manca* (J.B. Ellis & G.W. Martin) F. Theissen & H. Sydow. Annales Mycologici 15:461. (1917) 1918. ≡ *Irenina manca* (J.B. Ellis & G.W. Martin) F.L. Stevens. Annales Mycologici 25:448. 1927. ≡ *Asteridiella manca* (J.B. Ellis & G.W. Martin) C.G. Hansford ex C.G. Hansford. Beihefte zur Sydowia Annales Mycologici 2:304–305. 1961.

Meliola mandevillae F.L. Stevens. Annales Mycologici 26:245, pl. V, fig. 50. 1928. **Holotype:** On *Mandevilla* sp., British Guiana [Guyana], Kartabo, 23 JUL 1922, leg. F.L. Stevens No. 626, ILL 5529. **Paratypes:** Panama, Corozal, Trail No. 17, 30 AUG 1924, leg. F.L. Stevens No. 102, ILL 5534; Agua Clara Reservoir, 17 SEP 1924, leg. F.L. Stevens No. 553, ILL 5535.

Meliola mandingensis C.G. Hansford. Beihefte zur Sydowia Annales Mycologici 1:112. 1957. **Holotype:** Hab. in foliis Bignoniacearum spec. [= Bignoniaceae sp. indet.], Panama, Mandingo, [15 OCT 1924], leg. F.L. Stevens No. 1339, ILL 31728.

Meliola mangiferae F.S. Earle. Bulletin of the New York Botanical Garden 3:307–308. Preprint 1904 (journal issued in 1905). **Residual isosyntype:** On leaves of the mango, *Mangifera indica* L., Jamaica, Castleton Gardens, 9 JAN 1903, leg. A.A. Heller No. 6393, ILL 5510. **Assumed lectotype:** K [citation of F.S. Earle No. 272 as "type" by C.G. Hansford (1961:464) is interpreted as lectotypification with the specimen at K].

Meliola mangostana P.A. Saccardo. Bulletino dell' Orto Botanico della Regia Universita di Napoli 6:42. 1921. **Assumed Isotype:** Hab. in foliis *Garciniae mangostanae* [= *Garcinia mangostana* L., Republic of Singapore, as "Island of Singapore, Straits Settlements" on the original exsiccati label], AUG 1917, leg. C.F. Baker No. 5084 [ex Fungi Malayana No. 453], ILL 4201 [leaf fragment of host and microscopic preparations], CUP, FH. ≡ *Irenina mangostana* (P.A. Saccardo) F.L. Stevens. Annales Mycologici 25:457. 1927. ≡ *Meliola garciniae* H.S. Yates var. *mangostana* (P.A. Saccardo) C.G. Hansford. Proceedings of the Linnean Society of London 160:120. (1948) 1949.

Meliola manihot F.L. Stevens & L.R. Tehon. Mycologia 18:11, pl. 1, fig. 11. 1926. **Holotype:** On *Manihot palmata* Muell. Arg., British Guiana [Guyana], Tumatumari, 12 JUL 1922, leg. F.L. Stevens No. 217, ILL 5162. ≡ *Meliola gymnanthicola* F.L. Stevens var. *manihot* (F.L. Stevens & L.R. Tehon) F.L. Stevens. Annales Mycologici 26:251. 1928.

Meliola manihoticola P.C. Hennings. Hedwigia 43:364–365 (illustr.). 1904. **Assumed residual isosyntypes**: Auf Blättern von *Manihot utilissima* Pohl [= *M. esculenta* Crantz], Brazil, [Amazonas], Rio Juruá, 1901, leg. E.H.G. Ule, ex Mycotheca Brasiliensis No. 60, ILL 5536 [a microscopic preparation], and in bound exsiccati set. **Assumed lectotype**: S. [citation of Ule No. 2969 as "type" by C.G. Hansford (1961:219) is interpreted as lectotypification with the specimen at S].

Meliola mapaniae H.S. Yates. Philippine Journal of Science, Section C (Botany) 12:367–368. (1917) 1918. **Isotype**: On *Mapania* sp., Philippines, Samar [Island], Catubig River, 20 FEB 1916, leg. M. Ramos, ex Phil. Bur. Sci. No. 24640, ILL 5537.

Meliola marantacearum F.L. Stevens. Annales Mycologici 26:208, pl. III, fig. 28. 1928. **Holotype**: On *Calathea insignis* Petersen, Costa Rica, Siquirres, 31 AUG 1924, leg. F.L. Stevens No. 693, ILL 5541. **Isotype**: ILL 5540. **Paratypes**: Columbiana, 19 AUG 1923, leg. F.L. Stevens No. 569, ILL 5263 and 5538 [also paratypes(?) of *Meliola hispida* F.L. Stevens]; leg. F.L. Stevens No. 587, ILL 5539.

Meliola marantae F.L. Stevens. Annales Mycologici 26:250–251, pl. V, fig. 60. 1928. **Holotype**: On Marantaceae sp., British Guiana [Guyana], Rockstone, 17 JUL 1922, leg. F.L. Stevens No. 465, ILL 5542. **Paratype**: On *Maranta arundinacea* L., Tumatumari, 11 JUL 1922 [as 12 JUL on the packet], leg. F.L. Stevens No. 202, ILL 5544.

Meliola marcgraviae L.R. Tehon. Botanical Gazette 67:506–507. 1919. **Holotype**: On leaves of *Marcgravia rectiflora* Triana & Planchon [= *M. trinitatis* C. Presl], Porto Rico [Puerto Rico, "Alto"], 16 JUL 1915, leg. F.L. Stevens No. 8722, ILL 4202. **Isotype**: ILL 4203. ≡ *Irene marcgraviae* (L.R. Tehon) F.L. Stevens & L.R. Tehon. Mycologia 18:22. 1926. ≡ *Irenina marcgraviae* (L.R. Tehon) F.L. Stevens. Annales Mycologici 25:452. 1927. ≡ *Irenopsis marcgraviae* (L.R. Tehon) C.G. Hansford. Beihefte zur Sydowia Annales Mycologici 2:129–130. 1961.

Meliola maricaensis F.L. Stevens. Illinois Biological Monographs 2:499. 1916. **Holotype**: On *Ilex nitida* (Vahl) Maxim., Porto Rico [Puerto Rico, Rio] Maricao, 20 OCT 1913, leg. F.L. Stevens No. 3679, ILL 4406 [also a paratype of *Helminthosporium glabroides* F.L. Stevens]. **Isotype**: Leg. F.L. Stevens No. 3679-2, ILL 4408. **Paratypes**: Leg. F.L. Stevens No. 3607, ILL 4407 [also a paratype of *Helminthosporium glabroides* F.L. Stevens]; 4 APR 1913 [as 3 APR on the packet], leg. F.L. Stevens No. 824, ILL 4409. ≡ *Irenopsis maricaensis* (F.L. Stevens) F.L. Stevens. Annales Mycologici 25:434. 1927.

Meliola martiniana A. Gaillard. Le Genre *Meliola*, pp. 68, 69, pl. 22, fig. 2. 1892. **Isotype**: Hab. ad ... foliorum *Perseae palustris* [= *Persea palustris* (Raf.) Sargent], Florida, leg. G.W. Martin [ex G.L. Rabenhorst,

Fungi Europaei No. 3852], ILL 4205. ≡ *Irenopsis martiniana* (A. Gaillard) F.L. Stevens. Annales Mycologici 25:437. 1927.

Meliola mataybae F.L. Stevens. Annales Mycologici 26:228, pl. IV, fig. 38. 1928. **Holotype:** On *Matayba scrobiculata* Radlk., Costa Rica, El Alto, 26 JUL 1923, leg. F.L. Stevens No. 245, ILL 5545. **Isotype:** FH. ≡ *Meliola capensis* (K. Kalchbrenner & M.C. Cooke) F. Theissen var. *mataybae* (F.L. Stevens) C.G. Hansford. Beihefte zur Sydowia Annales Mycologici 2:447. 1961.

Meliola mauritiae F.L. Stevens. Annales Mycologici 26:215–216, pl. IV, fig. 33. 1928. **Syntypes plus one isosyntype:** On *Mauritia* sp., Trinidad, Guanapo, 16 AUG 1922, leg. F.L. Stevens No. 925, ILL 5546; leg. F.L. Stevens No. 908, ILL 3725 and 5548 [also the holotype and an isotype, respectively, of *Haraea mauritiae* F.L. Stevens].

Meliola mayaguesiana F.L. Stevens. Illinois Biological Monographs 2:500–501, fig. 31. 1916. **Holotype:** On *Palicourea crocea* (Sw.) J.A. Schultes in Roemer & J.A. Schultes, Porto Rico [Puerto Rico], Lajas, 17 JUN 1915, leg. F.L. Stevens No. 7157, ILL 5549. **Paratypes:** Leg. F.L. Stevens No. 7196, ILL 5563; Las Marias, 10 JUL 1915, leg. F.L. Stevens No. 8162, ILL 5562 and 5566; leg. F.L. Stevens No. 8138, ILL 33560; on *Palicourea domingensis* (Jacquin) DC. [= *P. crocea* (Sw.) J.A. Schultes in Roemer & J.A. Schultes], Piedras, 12 AUG 1915, leg. F.L. Stevens No. 9320, ILL 5553; on *Palicourea riparia* Bentham [= *Palicourea crocea* (Sw.) Roemer & J.A. Schultes], Mayagüez, 25 JUN 1915, leg. F.L. Stevens No. 7403, ILL 5551; leg. F.L. Stevens No. 7019, ILL 5552; on *Palicourea* sp., Mayagüez, 30 APR 1913, leg. F.L. Stevens No. 979, ILL 5550; 3 MAY 1913, leg. F.L. Stevens No. 1131, ILL 5554.

Meliola mayepeae F.L. Stevens. Illinois Biological Monographs 2:516–517, fig. 46. 1916. **Holotype:** On *Mayepea domingensis* (Lam.) Krug & Urban [= *Chionanthus domingensis* Lam.], Porto Rico [Puerto Rico], Mayagüez Mesa, 25 JUN 1915, leg. F.L. Stevens No. 7468, ILL 5559. **Paratype:** El Alto de la Bandera, 16 AUG 1915, leg. F.L. Stevens No. 8703, ILL 5557.

Meliola mayepeicola F.L. Stevens. Illinois Biological Monographs 2:519–520. 1916. **Holotype:** On *Mayepea domingensis* (Lam.) Krug & Urban [= *Chionanthus domingensis* Lam.], Porto Rico [Puerto Rico], Mayagüez Mesa, 29 JUN 1915, leg. F.L. Stevens No. 7556, ILL 5560. **Paratype:** Maricao, 3 APR 1913, leg. F.L. Stevens No. 822, ILL 5561.

Meliola megalocarpa H. Sydow. Annales Mycologici 21:94–95. 1923. **Isotype:** Hab. in foliis *Mabae buxifoliae* [*Maba buxifolia* (Rottb.) Jussieu = *Diospyros ferrea* (Willd.) Bakh., not *D. buxifolia* (Blume) Hiern], Philippines, Luzon, Tayabas Prov., Baler, 26 JUN 1913, leg. L. Escritor, ex Phil. Bur. Sci. No. 21213, ILL 5568.

Meliola megalochaeta H. Sydow. Philippine Journal of Science 21:135. 1922. **Isolectotype**: On *Kibara moluccana* Boerlage ex Perkins, Indonesia [Moluccas], Amboina, Hitoe Messen, 13 OCT 1913, leg. C.B. Robinson, Reliquiae Robinsonianae No. 2078, ILL 5575. **Assumed lectotype**: S [citation of this institution for location of "type" by C.G. Hansford (1961:41) is interpreted as lectotypification].

Meliola megalopoda H. & P. Sydow. Annales Mycologici 15:189. 1917. **Assumed isotype**: Hab. in foliis coriaceis ignotis humi jacentes in silva [host later identified as *Eugenia* sp. (cf. F.L. Stevens & E.F. Roldan, 1935:60)], Philippines, [Luzon], Laguna Prov., Mount Maquiling [as Mt. Makiling], APR 1914, leg. C.F. Baker No. 3070 [ex Fungi Malayana No. 551], ILL 5576 [as a microscopic preparation].

Meliola megalospora C.L. Spegazzini. Anales de la Sociedad Científica Argentina 12 (Series 2,2):100–101. 1881; Fungi Argentini Additis Nonnullis Brasiliensibus Montevideensibusque. Pugillus IV, No. 115. 1881. **Lectotype**: Hab. ad folia viva *Jodinae rhombifoliae* [= *Jodina rhombifolia* (Hooker & Arnott) Reissek in Martius], in sylva australiore, "Tordillo" dicta, ac in dumetis marittimis, [Argentina], "Cabo S. Antonio" [Cape San Antonio], DEC & JAN 1881, leg. ?, comm. C.L. Spegazzini, ILL 6501 [citation of specimen in F.L.S. herb. (ILL) as "type" by C.G. Hansford (1961:360) was interpreted as lectotypification by S.J. Hughes (1993:217) and is accepted herein]. ≡ *Irene megalospora* (C.L. Spegazzini) F. Theissen & H. Sydow. Annales Mycologici 15:461. (1917) 1918. ≡ *Meliolina megalospora* (C.L. Spegazzini) F.L. Stevens. Annales Mycologici 25:416. 1927. ≡ *Asteridiella megalospora* (C.L. Spegazzini) C.G. Hansford. Beihefte zur Sydowia Annales Mycologici 2:360. 1961.

Meliola meibomiae F.L. Stevens & L.R. Tehon. Mycologia 18:7–8, pl. 1, fig. 5. 1926. **Holotype**: On *Meibomia* sp. [= *Desmodium* sp.], British Guiana [Guyana], Rockstone, 16 JUL 1922, leg. F.L. Stevens No. 434, ILL 5579. **Paratypes**: 13 JUL [1922], leg. F.L. Stevens No. 257, ILL 5578; Kartabo, 24 JUL [1922], leg. F.L. Stevens No. 650, ILL 5577.

Meliola melanococcae F.L. Stevens. Annales Mycologici 26:183, pl. II, fig. 10. 1928. **Holotype**: On *Elaeis melanococca* sensu auct. non Gaertner [= *E. oleifera* (Kunth) Cortes], Peru, Huacapistana, 6 DEC 1924, leg. F.L. Stevens No. 79, ILL 5587. **Paratypes**: Panama, Darien, 10 SEP 1924, leg. F.L. Stevens No. 403, ILL 5589; Mandingo, 15 OCT 1924, leg. F.L. Stevens No. 1316, ILL 5588.

Meliola melastomatacearum C.L. Spegazzini. Boletín de la Academia Nacional de Ciencias en Córdoba 11:494–495. 1889; Fungi Puiggariani, Pugillus I, pp. 116–117, No. 232. 1889 [cited as *melastomacearum* in both references, corrected by F.C. Deighton (1968) in accordance with current I.C.B.N.]. **Isotype:** Hab. ad folia viva Melastomacearum [= Melastomataceae] in dumetis, [Brazil], pr. Apiahy [Apiai], MAY 1888, leg. J.I. Puiggari, comm. C.L. Spegazzini No. 2485, ILL 4179. ≡ *Irene melastomatacearum* (C.L. Spegazzini) R.A. Toro. Mycologia 17:141–142. 1925. ≡ *Irenina melastomatacearum* (C.L. Spegazzini) F.L. Stevens. Annales Mycologici 25:459–460. 1927. ≡ *Asteridiella melastomatacearum* (C.L. Spegazzini) C.G. Hansford ex C.G. Hansford. Beihefte zur Sydowia Annales Mycologici 2:154. 1961. [Epithets of the three synonyms were also published as *melastomacearum* and are corrected herein.].

Meliola meliacearum F.L. Stevens & E.F. Roldan ex C.G. Hansford. Sydowia Annales Mycologici 16:320. (1962) 1963; originally published by F.L. Stevens & E.F. Roldan in Philippine Journal of Science 56:66–67, fig. 3b. 1935, as nom. invalid. sine diagn. lat. **Holotype:** On *Dysoxylum cumingianum* C. DC., Philippines, Luzon, Laguna [Prov.], Mount Maquiling, 7 OCT 1930, leg. F.L. Stevens No. 824, ILL 5590.

Meliola melochiae C.G. Hansford. Beihefte zur Sydowia Annales Mycologici 1:112–113. 1957. **Isotypes:** Hab. in foliis *Melochiae umbellatae* [= *Melochia umbellata* (Houtt.) Stapf] socio *Irenopsis curvata* (H.S. Yates) C.G. Hansford., Philippines, Samar [Island], Catubig River, [FEB–MAR 1916], leg. M. Ramos, ex Phil. Bur. Sci. No. 24642 p.p., ILL 4351 and 4973 [also isotypes of *Meliola curvata* H.S. Yates ≡ *Irenopsis curvata* (H.S. Yates) F.L. Stevens].

Meliola memecyli H. & P. Sydow var. *microspora* C.G. Hansford. Sydowia Annales Mycologici 10:78. (1956) 1957. **Isotype:** Hab. in foliis *Memecyli* sp. [= *Memecylon* sp.] socio *Meliola affinis* Syd., Philippines, Luzon, Nueva Ecija [Prov.], San José to Balete Road, [9 JAN 1931], leg. F.L. Stevens No. 1777, ILL 4510. **Holotype:** CUP.

Meliola merrillii H. & P. Sydow. Philippine Journal of Science, Section C (Botany) 8:479. 1913. **Isotypes:** On living leaves of *Cissus* (?*adnata*) [= *C.* cf. *adnata* Roxb.], Philippines, Luzon, Laguna Prov., Mount Maquiling, MAR 1913, leg. E.D. Merrill No. 8672, ILL 5603 and 5623.

Meliola miconiae F.L. Stevens. Illinois Biological Monographs 2:498, fig. 29. 1916. **Holotype:** On *Miconia prasina* (Sw.) DC., Porto Rico [Puerto Rico], Las Piedras, 12 AUG 1915, leg. F.L. Stevens No. 9366, ILL 4410. **Paratype:** Las Marias, 10 JUL 1915, leg. F.L. Stevens No. 8160, ILL 4411 [also an isolectotype of *Echidnodella melastomacearum* R.W. Ryan and a paratype of *Morenoella miconiae* R.W. Ryan]. **Assumed paratypes:** ILL 6930, 7326 and 7345 [also the lectotype and isolectotypes, respectively, of *Echidnodella melastomacearum* R.W. Ryan, as well as paratypes of *Morenoella miconiae* R.W. Ryan]. ≡ *Irenopsis miconiae* (F.L. Stevens) F.L. Stevens. Annales Mycologici 25:436. 1927.

Meliola miconiicola F.L. Stevens. Illinois Biological Monographs 2:491–492, fig. 19. 1916 [as *miconieicola*, corrected in accordance with Articles 60 & 61 of I.C.B.N.]. **Holotype:** On *Miconia sintenisii* Cogn., Porto Rico [Puerto Rico], El Alto de la Bandera, 15 JUL 1915, leg. F.L. Stevens No. 8639, ILL 4412. **Isotype:** ILL 7295. ≡ *Irenopsis miconiicola* (F.L. Stevens) F.L. Stevens. Annales Mycologici 25:436. 1927 [the epithet also published as *miconieicola* and corrected herein].

Meliola micromeli F.L. Stevens & E.F. Roldan. Philippine Journal of Science 56:57–58, fig. 2b. 1935, nom. invalid. sine diagn. lat. **Holotype:** On *Micromelum minutum* (G. Forster) Wight & Arnott [the host possibly misidentified (cf. C.G. Hansford, 1961)], Philippines, Luzon, Nueva Ecija Prov., San José to Balete Pass, 6 JAN 1931, leg. F.L. Stevens No. 1726, ILL 5638.

Meliola micromera H. & P. Sydow. Annales Mycologici 12:552–553. 1914. **Isotype:** Hab. in foliis *Gmelinae philippensis* [= *Gmelina philippensis* Cham.], Philippines, [Luzon], Bulacan Prov., Angat, SEP 1913, leg. M. Ramos, ex Phil. Bur. Sci. No. 21807, ILL 5640. ≡ *Meliola clerodendricola* P.C. Hennings var. *micromera* (H. & P. Sydow) C.G. Hansford. Beihefte zur Sydowia Annales Mycologici 2:694–695. 1961.

Meliola microspora N.T. Patouillard & A. Gaillard. Bulletin du Société Mycologique de France 4:104. 1888. **Isolectotype:** Sur les feuilles d'une plante herbacée [subsequently identified as *Vandellia diffusa* L. in C.G. Hansford (1961) = *Lindernia diffusa* (L.) Wettst., Venezuela], San Fernando, SEP, leg. A. Gaillard No. 262, ILL 5648 [as a microscopic preparation]. **Assumed lectotype:** S [citation of this institution for location of the "type" by C.G. Hansford (1961:655) is interpreted as lectotypification].

Meliola microthecia F. v. Thümen. Flora 1876:569. 1876; Mycotheca Universalis, Century 9, No. 851. Anno 1877. **Isotypes:** In foliis vivis *Barosmae scopariae* [*Barosma scoparia* Ecklon & Zeyher = *Agathosma ovata* (Thunb.) Pillans], South Africa, Cape Prov. [as Promont. bonae spei], Grahamstown, JUL 1876, leg. P. MacOwan No. 1260, ex Mycotheca Universalis No. 851, ILL 6471 [as a microscopic preparation], and in bound exsiccati set [also an isotype of *Meliola thuemeniana* F.L. Stevens, nom. nov., based on the description of *M. microthecia* sensu A. Gaillard, 1892, non F. v. Thümen, 1876 — see next entry].

Meliola microthecia sensu A. Gaillard, Le Genre *Meliola*, p. 73. 1892, non F. v. Thümen, 1876. **Based on** the same type collection as *Meliola microthecia* F. v Thümen — see preceding entry. [According to F.L. Stevens' (1928) interpretation, Gaillard's description differs significantly from the diagnosis published by F. v. Thümen, and characteristics of the spores observed by Gaillard suggest that he had a different species on the same host.]. ≡ *Meliola thuemeniana* F.L. Stevens. Annales Mycologici 26:259. 1928, nom. nov.

Meliola modesta H. Sydow. Annales Mycologici 24:304–306. 1926. **Isotypes**: Hab. in foliis vivis *Thevetiae neriifoliae* [= *Thevetia neriifolia* A. Jussieu ex Steudel = *T. peruviana* (Persoon) K. Schum.], Costa Rica, San Pedro de San Ramón, 23 JAN 1925, leg. H. Sydow, Fungi in Itinere Costaricensi Collecti No. 231, ILL 5653 and H. Sydow, Fungi Exotici Exsiccati No. 619 in bound set at ILL [cf. C.G. Hansford (1961)].

Meliola monensis F.L. Stevens. Illinois Biological Monographs 2:506, fig. 35. 1916. **Holotype**: On *Amyris elemifera* L., Porto Rico [Puerto Rico], Mona Island, 20–21 DEC 1913, leg. F.L. Stevens No. 6158, ILL 5658. **Isotype**: ILL 5654. **Paratypes**: Leg. F.L. Stevens No. 6146, ILL 5655 and 5659; leg. F.L. Stevens No. 6150, ILL 5657 and 5660 [also isotypes of *Stevensula monensis* C.L. Spegazzini and possible isotypes of *Micropeltidium monense* C.L. Spegazzini, as well as paratypes of *Helminthosporium glabroides* F.L. Stevens.

Meliola monilispora A. Gaillard. Le Genre *Meliola*, p. 101, pl. 18, fig. 2. 1892. **Isotype**: Ad paginam inferiorem foliorum quorundam, [Apocynaceae sp. indet. (? *Strophanthus* sp. fide C.G. Hansford, 1961), C. Africa], Congo Française, Niari, JUN 1880, leg. F. Thollon s.n., ILL 5661. **Holotype**: PC.

Meliola monnieriae F.L. Stevens. Annales Mycologici 26:244–245, pl. IV, fig. 49. 1928. **Holotype**: On *Monnieria trifolia* L., British Guiana [Guyana], Kartabo, 21 JUL [1922], leg. F.L. Stevens No. 532, ILL 5662.

Meliola monochroma R. Ciferri. Sydowia Annales Mycologici 10:138–139. (1956) 1957. **Isotype**: Hab. in foliis vivis *Mimosae ceratoniae* (Leguminosae) [= *Mimosa ceratonia* L. (Fabaceae subfam. Mimosoideae), Dominican Republic], Llano Costero, Santo Domingo Prov., Ciudad Trujillo, Los Alcarrizos, 11 FEB 1930 [erroneously as 1830 in the protologue], leg. R. Ciferri, Mycoflora Domingensis Exsiccata No. 344, ILL 33059.

Meliola morbosa F.L. Stevens. Bernice P. Bishop Museum Bulletin 19:38, text fig. 8d. 1925. **Holotype**: On *Claoxylon sandwicense* Müll. Arg., Hawaii, Kauai, Upper Waimea Canyon, 15 JUN [1921], leg. F.L. Stevens No. 452, ILL 5675. Isotypes: BISH 499024 and 499961.

Meliola morototonii C.L. Spegazzini. Anales del Museo Nacional de Historia Natural de Buenos Aires 32:360–361. 1924 [as *morototoni*, corrected in accordance with Articles 32.6 & 61 of I.C.B.N.]. **Isolectotype**: Sobre las hojas vivas de *Didymopanax morototonii* (Aublet) Decaisne & Planchon, [Argentina], cerca de Puerto León, JUL 1909, y de Puerto Aguirre, AUG 1923, leg. C.L. Spegazzini [No. 1812; ILL

4208 [the packet lacks a number but is labeled "part of type collection"]. **Lectotype**: LPS [citation of specimen in the Spegazzini herb. as "type" by C.G. Hansford (1961:482) is accepted as lectotypification]. ≡ *Irenina morototonii* (C.L. Spegazzini) F.L. Stevens. Annales Mycologici 25:468. 1927. ≡ *Asteridiella morototonii* (C.L. Spegazzini) C.G. Hansford ex C.G. Hansford. Beihefte zur Sydowia Annales Mycologici 2:481–482. 1961. [Epithets in the two synonyms were also published as *morototoni* and are corrected herein.].

Meliola morrowii F.L. Stevens. Annales Mycologici 26:183, pl. II, fig. 9. 1928. **Holotype**: On Palmae [= Arecaceae sp. indet.], Panama, Baillemona, 20 SEP 1924, leg. F.L. Stevens No. 680, ILL 5676.

Meliola mucronata (J.P.F.C. Montagne) P.A. Saccardo, 1882. **See basionym**: *Capnodium mucronatum* J.P.F.C. Montagne, 1860.

Meliola multiseta M. Beeli. Bulletin de la Société Royale de Botanique de Belgique 60:84–85, fig. 37. 1927. **Isotype**: Sur une feuille des dicotylée (Anacardiaceae?), [C. Africa], Congo [Zaire], Kikwit, Ipamu, leg. H.J.R. Vanderyst No. 9872, ILL 5677. Holotype: BR.

Meliola mussaendae H. & P. Sydow. Annales Mycologici 15:190. 1917. **Isotypes**: Hab. in foliis *Mussaendae philippicae* [= *Mussaenda philippica* A. Richard], Philippines, Luzon, Laguna Prov., NOV 1915, leg. M. Ramos, ex Phil. Bur. Sci. No. 24057, ILL 4567 and 4568. ≡ *Meliola anceps* H. & P. Sydow var. *mussaendae* (H. & P. Sydow) F.L. Stevens. Annales Mycologici 26:205-206. 1928.

Meliola mussaendae-arcuatae C.G. Hansford. Sydowia Annales Mycologici 11:57–58. (1957) 1958. **Assumed isotype**: Hab. in foliis *Mussaendae arcuatae* [= *Mussaenda arcuata* Poiret in Lam., as *Craterispermum* sp. on the packet], Uganda, [Entebbe Wood, FEB 1931], leg. C.G. Hansford No. 1441, ILL 6423 [this collection is cited in the protologue; later, however, C.G. Hansford (1961:600), cited his No. 3668 as "type"].

Meliola mycetiae F.L. Stevens ex F.C. Deighton. Sydowia Annales Mycologici 21:186. (1967) 1968; originally published by F.L. Stevens in F.L. Stevens & E.F. Roldan, Philippine Journal of Science 56:70–71, fig. 3f. 1935, as nom. invalid. sine diagn. lat. **Holotype**: On *Mycetia javanica* (Blume) Reinw. ex Korth., Philippines, Luzon, Benguet, Kennon Road, below Camp 3, 8 JAN 1931, leg. F.L. Stevens No. 1669, ILL 5655.

Meliola myrsinacearum F.L. Stevens. Illinois Biological Monographs 2:508–509. 1916. **Holotype**: On *Ardisia guadalupensis* Duchass. ex Griseb. [= *Ardisia obovata* Desv. ex Hamilton], Porto Rico [Puerto Rico], Mayagüez Mesa, 29 JUN 1915, leg. F.L. Stevens No. 7576, ILL 5687. **Paratypes**: 15 JUN 1915, leg. F.L. Stevens No. 7057, ILL 5686; on Myrsinaceae sp. indet., Maricao, 19 JUL 1915, leg. F.L. Stevens No.

8905, ILL 5689; [as Rio Maricao, above Maricao on the packet], 20 OCT 1913, leg. F.L. Stevens No. 3681, ILL 5688 and 5691.

Meliola myrtacearum F.L. Stevens & E.F. Roldan ex C.G. Hansford. Sydowia Annales Mycologici 16:317. (1962) 1963; originally published by F.L. Stevens & E.F. Roldan in Philippine Journal of Science 56:73. 1935, as nom. invalid. sine diagn. lat. **Holotype**: On *Eugenia* sp., Philippines, Luzon, Laguna Prov., Mount Maquiling, 18 JAN 1931, leg. F.L. Stevens No. 1946, ILL 5692.

Meliola natalensis E.M. Doidge. Transactions of the Royal Society of South Africa 5:724, 743, pl. LIX, fig. 12. 1917. **Isotypes**: On leaves of unknown shrub [later identified as *Dovyalis rhamnoides* (Burchell ex DC.) Harvey on the packet of ILL 3870], South Africa, Natal [Prov.], Umgeni, near Durban, 27 MAY 1915, leg. E.M. Doidge, ex PREM No. 8980, ILL 3869 and 3870. **Holotype**: PREM. ≡ *Irene natalensis* (E.M. Doidge) E.M. Doidge. South African Journal of Natural History 2:40. 1920. ≡ *Appendiculella natalensis* (E.M. Doidge) C.G. Hansford. Beihefte zur Sydowia Annales Mycologici 2:106–107. 1961.

Meliola neotorta S.J. Hughes, 1993, nom. nov. (illegit.?). **See basionym**: *Meliola torta* E.M. Doidge, char. emend., 1919, non E.M. Doidge, 1917. **Isotype**: On *Trichocladus crinitus* (Thunb.) Persoon, [South Africa, Cape Prov.], Kingwilliamstown District, Izelini Forest, 8 JUN 1915 [as 14 JUN on the typed packet label], leg. Forester Emmett, ex PREM No. 9064, ILL 6400. **Holotype**: PREM.

Meliola nephelii P.A. Saccardo. Bulletino dell' Orto Botanica della Regia Universita di Napoli 6:42. 1921. **Assumed isotype**: Hab. in foliis *Nephelii cappacei* (sic) [probably = *Nephelium lappaceum* L., Republic of Singapore, as "Island of Singapore, Straits Settlements" on the original exsiccati label], SEP 1917, leg. C.F. Baker No. 5172 [ex Fungi Malayana No. 454], ILL 5693 [as a microscopic preparation].

Meliola nephelii P.A. Saccardo var. *major* C.G. Hansford & F.L. Stevens in C.G. Hansford. Journal of the Linnean Society (Botany) London 51:278. 1937. **Assumed holotype**: In foliis *Allophyli* sp. [= *Allophylus* sp., as Sapindaceae sp. indet. on the packet], Uganda, Entebbe Road, [JAN 1931], leg. C.G. Hansford No. 1443, ILL 5155 [the packet annotated and marked as "type" by Hansford].

Meliola nepheliicola F.L. Stevens & E.F. Roldan ex W. Yamamoto. Hyogo Noka Daigaku, Sasyama, Japan. Science Reports Series: Plant Protection (Agricultural Biology), [Hyogo University of Agriculture. Science Reports] 3(2):62. 1958; originally published by F.L. Stevens & E.F. Roldan in Philippine Journal of Science 56:60, fig. 2f. 1935, as nom. invalid. sine diagn. lat. **Holotype**: On *Nephelium intermedium* Radlk. in Perkins, Philippines, Luzon, Laguna Prov., Mount Maquiling, 22

JUL 1930, leg. F.L. Stevens No. 77, ILL 5694. **Isotypes:** CALP, Phil. Bur. Sci.

Meliola niessleana H.G. Winter. Hedwigia 24:260–261. 1885 and in L. Rabenhorstii, Fungi Europaei Exsiccati, Series II, Century 34, No. 3339. Anno 1885. **Isotypes:** In foliis vivis *Rhododendri chamaecisti* [*Rhododendron chamaecistus* L. = *Rhodothamnus chamaecistus* (L.) Reichenb.], Austria, Salzburg, AUG 1884, leg. G. Mayendorf v. Niessl, ex Rabenhorst, Fungi Europaei No. 3339, ILL 5701 [as a microscopic preparation], and in bound exsiccati set.

Meliola nigra F.L. Stevens. Illinois Biological Monographs 2:505–506. 1916. **Holotype:** On *Laguncularia racemosa* (L.) C.F. Gaertner, Porto Rico [Puerto Rico], Guanajibo, 19 JUN 1915, leg. F.L. Stevens No. 7197, ILL 5706 [also a paratype of *Helminthosporium glabroides* F.L. Stevens]. **Isotype:** ILL 5704. **Paratype:** Jayuda [erroneously as Joyuda in the protologue], 31 MAY 1915 [as 1913 on the packet], leg. F.L. Stevens No. 363, ILL 5705.

Meliola nigro-rufescens P.A. Saccardo. Atti della Accademia Scientifica Veneto-Trentino-Istriano. Padova. Series 3, 10:60. Preprint 1917 (journal issued in 1919). **Assumed isotype:** Hab. in foliis *Canarii* sp. [= *Canarium* sp.], Philippines, [Luzon, Laguna Prov.], Los Baños, [Mount Maquiling], JUL 1913, leg. C.F. Baker No. 1420 [ex Fungi Malayana No. 363], ILL 4874.

Meliola nigro-rufescens P.A. Saccardo var. *teramni* P.A. Saccardo. Atti della Accademia Scientifica Veneto-Trentino-Istriano. Padova. Series 3, 10:60. Preprint 1917 (journal issued in 1919). **Assumed isotype:** Hab. in foliis *Teramni uncinati* [= *Teramnus uncinatus* (L.) Sw.], Philippines, Luzon, Laguna Prov., Los Baños, Mount Maquiling, SEP 1913, leg. C.F. Baker No. 1657 [ex Fungi Malayana No. 364], ILL 4873 [as a microscopic preparation]. ≡ *Meliola teramni* (P.A. Saccardo) H. & P. Sydow. Annales Mycologici 15:193. 1917.

Meliola obesa C.L. Spegazzini. Anales de la Sociedad Científica Argentina 18:264. 1884. **Isotype:** Ad folia viva Rutaceae sp. [misspelled "Ruthaceae," Paraguay], pr. Pirayú, JUL 1883, leg. B. Balansa No. 3834, comm. C.L. Spegazzini, ILL 4211. ≡ *Irene obesa* (C.L. Spegazzini) F. Theissen & H. Sydow. Annales Mycologici 15:461. (1917) 1918. ≡ *Irenina obesa* (C.L. Spegazzini) F.L. Stevens. Annales Mycologici 25:450. 1927. ≡ *Asteridiella obesa* (C.L. Spegazzini) C.G. Hansford ex C.G. Hansford. Beihefte zur Sydowia Annales Mycologici 2:378–379. 1961.

Meliola obesula C.L. Spegazzini. Revista Argentina de Historia Natural 1:402–403. 1891; Fungi Guaranitici Nonnulli Novi vel Critici, p. 27, No. 75. 1891. **Isotype:** Hab. ad folia viva Rutaceae cujusdam in sylva, [Paraguay], Caáguazú, JAN 1882, leg. B. Balansa No. 3585, comm. C.L.

Spegazzini, ILL 4210 [also an isotype of *Irene caaguazuensis* C.G. Hansford]. ≡ *Asteridiella obesa* (C.L. Spegazzini) C.G. Hansford ex C.G. Hansford var. *obesula* C.G. Hansford. Beihefte zur Sydowia Annales Mycologici 2:379. 1961.

Meliola obvallata H. Sydow. Annales Mycologici 21:90. 1923. **Isolectotype:** Hab. in foliis *Aglaiae palembanicae* [= *Aglaia palembanica* Miq.], British North Borneo [Malaysia, Sabah], Sandakan Prov., Sibuguey, 22 NOV 1920, leg. M. Ramos No. 2206, ILL 5707. **Assumed lectotype:** S, ex Sydow herb. [citation of this institution for location of "type" by C.G. Hansford (1961:410) is interpreted as lectotypification].

Meliola ochnae E.M. Doidge. Transactions of the Royal Society of South Africa 8:141. 1920. **Isosyntype:** In foliis *Ochnae atropurpureae* var. *natalitiae* [= *Ochna atropurpurea* DC. var. *natalitia* (Meisner) Harvey], South Africa, Natal [Prov.], Buccleuch, 17 JUL 1918, leg. E.M. Doidge, ex PREM No. 11567, ILL 5708. **Syntype:** PREM.

Meliola ocoteae F.L. Stevens. Illinois Biological Monographs 2:497–498. 1916. **Holotype:** On *Ocotea leucoxylon* (Sw.) de Lanessan., Porto Rico [Puerto Rico], Jajome Alto, 17 JUL 1915, leg. F.L. Stevens No. 8428, ILL 4385 [also the holotype of *Helminthosporium ocoteae* F.L. Stevens and a potential isotype of *Asteridium portoricense* C.L. Spegazzini]. ≡ *Irenopsis ocoteae* (F.L. Stevens) F.L. Stevens. Annales Mycologici 25:436. 1927.

Meliola ocoteicola F.L. Stevens. Illinois Biological Monographs 2:513, fig. 43. 1916. **Lectotype:** On *Ocotea leucoxylon* (Sw.) de Lanessan., Porto Rico [Puerto Rico], Mayagüez Mesa, 29 JUN 1915, leg. F.L. Stevens No. 7560, ILL 5709 [the author inadvertently indicated both No. 7560 and No. 4731 as "type" in the protologue; however, in his monograph (F.L. Stevens, 1928:279) he cited only No. 7560 under the type locality. This is interpreted as lectotypification, especially since the decision was upheld by C.G. Hansford (1961:57)]. **Assumed isolectotypes:** ILL 7335, 7337, and 7340 [these packets were filed under an unpublished name, later identified by R.W. Ryan (1924:193) as belonging in *Morenoella portoricensis* C.L. Spegazzini]. **Residual syntype and isosyntype:** On *Chrysophyllum* sp., Monte Alegrillo, 14 NOV 1913, leg. F.L. Stevens No. 4731, ILL 5711 and 7159 [also an isotype and the holotype, respectively, of *Asterina sydowiana* R.W. Ryan]. **Paratypes:** On *Ocotea leucoxylon* (Sw.) de Lanessan, Monte Alegrillo, near Maricao, 14 NOV 1913, leg. F.L. Stevens No. 4762, ILL 5710; on *Chrysophyllum* sp., Monte Alegrillo, 18 NOV 1913, leg. F.L. Stevens No. 4579 [erroneously as No. 4519 in the protologue], ILL 5713.

Meliola odontocephala H. Sydow. Leaflets of Philippine Botany 9:3119. 1925. **Isotype:** On leaves of *Harpullia arborea* (Blanco) Radlk. [misspelled "*Harpulia*"], Philippines, Luzon, Sorsogon Prov., Irosin, AUG 1916, leg. A.D.E. Elmer No. 17012, ILL 5714, ex Sydow herb.

Meliola olecranoni F.L. Stevens & L.R. Tehon. Mycologia 18:15, pl. 2, fig. 18. 1926 [as *olecranonis*, corrected by F.C. Deighton (1968) in accordance with current I.C.B.N.]. **Holotype:** On *Guajava* sp. [= *Psidium guajava* L. (cf. C.G. Hansford, 1961)], British Guiana [Guyana], Tumatumari, 8 AUG 1922, leg. F.L. Stevens No. 64, ILL 5715. ≡ *Meliola trichostroma* (G. Kunze) R.A. Toro var. *olecranoni* (F.L. Stevens & L.R. Tehon) C.G. Hansford. Beihefte zur Sydowia Annales Mycologici 2:147. 1961 [the epithet also published as *olecranonis* and corrected herein].

Meliola oleicola E.M. Doidge, Bothalia 1(2):73–74, fig. 4. 1922. **Isotype:** On leaves of *Olea laurifolia* Lam., South Africa, Natal Prov., Buccleuch, 17 JUL 1918, leg. E.M. Doidge, ex PREM No. 11557, ILL 5990 [also an isotype of *Meliola petiolaris* E.M. Doidge, ex PREM No. 11558; both PREM numbers are on the same packet].

Meliola oligomera H. & P. Sydow. Annales Mycologici 15:190–191. 1917. **Isotype:** Hab. in foliis *Turpiniae* spec. [a misidentification in the protologue, the host correctly named *Hippocratea* sp. on the packet at ILL; probably *H. indica* Willd.], Philippines, Luzon, Rizal Prov., NOV 1915, leg. M. Ramos, ex Phil. Bur. Sci. No. 23882, ILL 5722.

Meliola oligopoda H. Sydow. Annales Mycologici 21:89–90. 1923. **Isolectotype:** Hab. in foliis Melastomataceae, British North Borneo, [Malaysia], Sandakan Prov., Batolima, 14 OCT 1920, leg. M. Ramos No. 2079, ILL 5723. **Assumed lectotype:** S [citation of this institution for location of "type" by C.G. Hansford (1961:157) is interpreted as lectotypification].

Meliola oncinotidis E.M. Doidge. Bothalia 4(4):851–853, text fig. 5. 1948. **Paratype:** [On leaves of] *Oncinotis inandensis* J.M. Wood & Evans, South Africa, Natal Prov., Buccleuch near Cramond, [23 MAR 1916], leg. E.M. Doidge, ex PREM No. 9722, ILL 4749. **Assumed holotype:** J. Gerstner, ex PREM No. 33509, PREM [the only collection cited with the Latin diagnosis and confirmed as "type" in C.G. Hansford (1961:543)].

Meliola opaca H. & P. Sydow. Leaflets of Philippine Botany 6:1924. 1913. **Isotype:** On *Dracontomelon dao* (Blanco) Merrill & Rolfe [as *Dracontomelum*], Philippines, Mindanao [Island, Agusan del Norte Prov.], Cabadbaran, (Mt. Urdaneta), AUG 1912, A.D.E. Elmer No. 13457, ex Phil. Bur. Sci., ILL 5724.

Meliola opposita H. & P. Sydow. Leaflets of Philippine Botany 6:1924–1925. 1913. **Isolectotype:** On leaves of Meliaceae sp. indet., Philippines, Mindanao [Island, Agusan del Norte Prov.], Cabadbaran, (Mt. Urdaneta), AUG 1912, leg. A.D.E. Elmer No. 13657, ex Phil. Bur. Sci. No.

13659, ILL 5726. **Assumed lectotype**: S, ex herb. Sydow [citation of this institution for location of "type" by C.G. Hansford (1961:411) is interpreted as lectotypification].

Meliola osmanthi H. & P. Sydow var. *hawaiiensis* C.G. Hansford. Sydowia Annales Mycologici 9:44–45. 1955. **Isotypes**: Hab. in foliis *Osmanthi sandwicensis* [*Osmanthus sandwicensis* (A. Gray) Knobloch = *Nestegis sandwicensis* (A.Gray) O. & I. Degener & L. Johnson], Hawaii, Oahu, Kuliouou [erroneously as Kulwuon in the protologue, 29 MAY 1921], leg. E.L. Caum, as F.L. Stevens No. 146, ILL 5730, BISH 145996. **Holotype**: K.

Meliola osmanthicola C.G. Hansford. Sydowia Annales Mycologici 9:70. 1955. **Isotypes**: Hab. in foliis *Osmanthi sandwicensis* [*Osmanthus sandwicensis* (A. Gray) Knobloch = *Nestegis sandwicensis* (A. Gray) O. & I. Degener & L. Johnson], Hawaii, [Kauai, Kalalau trail (erroneously as Oahu in the protologue), 16 JUN 1921], leg. F.L. Stevens No. 513 p.p., ILL 5732, 5733 and 5734, BISH 499023, BPI 71000 p.p. [also isotypes of *Meliola osmanthina* C.G. Hansford and, as ILL 7223, a paratype of *Calothyrium osmanthi* F.L. Stevens & R.W. Ryan in F.L Stevens]. **Holotype**: S, ex BPI 71000 p.p.

Meliola osmanthina C.G. Hansford. Sydowia Annales Mycologici 9:70–71. 1955. **Isotypes**: Hab. in foliis *Osmanthi sandwicensis* [*Osmanthus sandwicensis* (A. Gray) Knobloch = *Nestegis sandwicensis* (A. Gray) O. & I. Degener & L. Johnson], Hawaii, [Kauai, Kalalau trail (erroneously as Oahu in the protologue), 16 JUN 1921], leg. F.L. Stevens No. 513, ILL 5732, 5733 and 5734, BISH 499960, BPI 71000 p.p. [also isotypes of *Meliola osmanthicola* C.G. Hansford and, as ILL 7223, a paratype of *Calothyrium osmanthi* F.L. Stevens & R.W. Ryan in F.L Stevens]. **Holotype**: S., ex BPI 71000 p.p.

Meliola ouroupariae F.L. Stevens. Annales Mycologici 26:246, pl. V, fig. 52. 1928. **Holotype**: On *Ourouparia tomentosa* (DC.) K. Schum. (*Uncaria*) [= *Uncaria tomentosa* DC.], Costa Rica, Sabario, 8 AUG 1923 [as 8 JUL on the packet], leg. F.L. Stevens No. 800, ILL 5735.

Meliola palaquii F.L. Stevens & E.F. Roldan ex C.G. Hansford. Sydowia Annales Mycologici 16:313–314. (1962) 1963; originally published by F.L. Stevens & E.F. Roldan in Philippine Journal of Science 56:55, fig. 2a. 1935, as nom. invalid. sine diagn. lat. **Holotype**: On *Palaquium* sp., Philippines, Luzon, Laguna Prov., Mount Maquiling, 18 JAN 1931, leg. F.L. Stevens No. 1900, ILL 5904. **Isotypes**: CALP, Phil. Bur. Sci.

Meliola palaquiicola C.G. Hansford. Sydowia Annales Mycologici 11:58. (1957) 1958. **Holotype**: Hab. in foliis *Palaquii* sp. [= *Palaquium* sp.], Philippines, Luzon, Laguna Prov., [Mount Maquiling, 18 JAN 1931], leg. F.L. Stevens No. 1884, ILL 6530.

Meliola palawanensis H. & P. Sydow. Leaflets of Philippine Botany 5:1539. 1912. **Isotype:** Upon both sides of older leaves of *Morinda bartlingii* A.D.E. Elmer, [Philippines], Palawan [Island], Puerto Princesa, (Mt. Pulgar), APR 1911, leg. A.D.E. Elmer No. 13040, ex Phil. Bur. Sci., ILL 5903.

Meliola pallida F.L. Stevens. Annales Mycologici 26:177, pl. II, fig. 3. 1928. **Holotype:** On *Ipomoea* sp., British Guiana [Guyana], Tumatumari, 12 JUL 1922, leg. F.L. Stevens No. 228, ILL 5905.

Meliola palmicola H.G. Winter in L. Rabenhorst. Fungi Europaei Exsiccati, Centuri 36, No. 3547. Anno 1886, and in Hedwigia 26:31–32. (1886) 1887. **Isotypes:** Ad *Sabalidis serrulatae* [*Sabal serrulata* Michaux) J.H. Schultes in Roemer & J.A. Schultes = *Serenoa repens* (Bartram) Small], Florida, pr. Green Cove Springs, MAR 1886, leg. G. Martin, ex L. Rabenhorst, Fungi Europaei Exsiccati No. 3547, ILL 5912 [as a microscopic preparation], and in the bound exsiccati set.

Meliola panamensis F.L. Stevens. Annales Mycologici 26:212, pl. IV, fig. 31. 1928. **Holotype:** On *Coccoloba* sp. [cited as *Coccolobis*], Panama, Frijoles, 14 OCT 1924, leg. F.L. Stevens No. 1256, ILL 5921.

Meliola pandani H. Sydow. Annales Mycologici 26:89–90. 1928. **Isosyntype:** Hab. in foliis *Pandani affinis* [= *Pandanus affinis* Kurz], British North Borneo, [Malaysia, Myburgh Prov.], Sandakan, [OCT–DEC 1921], leg. A.D.E. Elmer No. 20075, ILL 5922, ex Sydow herb.

Meliola panici F.S. Earle, Muhlenbergia 1:12. 1901. **Isotype:** On *Panicum latifolium* L. [= *Dichanthelium latifolium* (L.) Gould & Clark], Porto Rico [Puerto Rico], calcareous hills east of Santurce, alt. 25 ft., [19] JAN 1900, leg. A.A. Heller No. 4343, ILL 5799. **Holotype:** NY.

Meliola panici F.S. Earle var. *olyrae* C.G. Hansford. Sydowia Annales Mycologici 10:82. (1956) 1957. **Holotype:** Hab. in foliis *Olyrae latifoliae* [= *Olyra latifolia* L.], Porto Rico [Puerto Rico, Mayagüez, 24 JUN 1915], leg. F.L. Stevens No. 7390, ILL 5757 [also a paratype of *Helminthosporium panici* F.L. Stevens].

Meliola panici F.S. Earle var. *panicicola* (H. & P. Sydow) C.G. Hansford, 1961. **See basionym:** *Meliola panicicola* H. & P. Sydow, 1914.

Meliola panicicola H. & P. Sydow. Annales Mycologici 12:552. 1914. **Isotype:** Hab. in foliis *Panici palmaefolii* [*Panicum palmifolium* Koenig = *Setaria palmifolia* (Koenig) Stapf], Philippines, [Luzon], Camarines [Camarines Sur Prov.], Mt. Isarog, NOV–DEC 1913, leg. M. Ramos, ex Phil. Bur. Sci. No. 22121, ILL 5928. ≡ *Meliola panici* F.S. Earle var. *panicicola* (H. & P. Sydow) C.G. Hansford. Beihefte zur Sydowia Annales Mycologici 2:748. 1961.

Meliola paraensis P.C. Hennings. Hedwigia 39:(77). 1900. **Isotype:** Auf Blättern von *Vitex* sp., [Brazil], Para, Botanischer Garten, JUN 1899, leg. J. Huber No. 13, ILL 5939 [microscopic preparations of the original material of J. Huber No. 13 obtained from B. C.G. Hansford (1961:696), however, cited as "type" Huber No. 4, a collection not mentioned in the protologue.].

Meliola parasitica F.L. Stevens. Annales Mycologici 26:282, pl. VI, fig. 78. 1928. **Holotype**: On Guarea sp., British Guiana [Guyana], Kartabo, 23 JUL 1922, leg. F.L. Stevens No. 625, ILL 5940.

Meliola parathesicola F.L. Stevens. Illinois Biological Monographs 2:492–493, fig. 20. 1916. **Holotype**: On *Parathesis serrulata* (Sw.) Mez [= *P. crenulata* (Vent.) J.D. Hooker in Hemsley], Porto Rico [Puerto Rico], Las Marias, 10 JUL 1915, leg. F.L. Stevens No. 8192, ILL 4388 [also the holotype of *Helminthosporium parathesicola* F.L. Stevens]. **Paratypes**: Las Marias, 10 JUL 1915, leg. F.L. Stevens No. 8204, ILL 4391; Maricao, 19 JUL 1915, leg. F.L. Stevens No. 8947, ILL 4390; Arecibo-Lares Road, 21 JUN 1915, leg. F.L. Stevens No. 7286, ILL 4387 [also a paratype of *Helminthosporium parathesicola* F.L. Stevens]. ≡ *Irenopsis parathesicola* (F.L. Stevens) F.L. Stevens. Annales Mycologici 25:435. 1927.

Meliola paropsiae (M. Beeli) C.G. Hansford, 1961. **See basionym**: *Meliola polytricha* K. Kalchbrenner & M.C. Cooke var. *paropsiae* M. Beeli, 1927.

Meliola parvula H. & P. Sydow. Leaflets of Philippine Botany 6:1925–1926. 1913. **Isotype**: On ... leaves of Meliaceae sp. indet., [Philippines], Mindanao [Island, Agusan del Norte Prov.], Cabadbaran, (Mt. Urdaneta), AUG 1912, leg. A.D.E. Elmer No. 13450, ex Phil Bur. Sci. No. 13452, ILL 5950.

Meliola patens H. & P. Sydow. Leaflets of Philippine Botany 5:1538–1539. 1912. **Isotype**: On *Lunasia amara* Blanco, [Philippines], Palawan [Island], Puerto Princesa, (Mt. Pulgar), APR 1911, leg. A.D.E. Elmer No. 13023, ex Phil. Bur. Sci., ILL 5951.

Meliola paucipes F.L. Stevens. Illinois Biological Monographs 2:510, fig. 40. 1916. **Holotype**: On *Piper blattarum* Sprengel, Porto Rico [Puerto Rico], Mayagüez Mesa, 25 JUN 1915, leg. F.L. Stevens No. 7463, ILL 5954.

Meliola paulliniae F.L. Stevens. Illinois Biological Monographs 2:513–514. 1916. **Holotype**: On *Paullinia pinnata* L., Porto Rico [Puerto Rico], Mayagüez, 3 MAY 1913, leg. F.L. Stevens No. 1149, ILL 5891. **Paratypes**: On *Paullinia pinnata* L. Mayagüez, 31 OCT 1913, leg. F.L. Stevens No. 3956, ILL 5899; leg. F.L. Stevens No. 3914, ILL 5898; leg. F.L. Stevens No. 3967a, ILL 5900; Vega Baja, 22 FEB 1913, leg. F.L. Stevens No. 376, ILL 14494 [also the holotype of *Acremonium meliolae* F.L. Stevens]; El Alto de la Bandera, 16 JUL 1915, leg. F.L. Stevens

No. 8722, ILL 5895; Rio Arecibo, K. 64.7, 8 JUL 1915, leg. F.L. Stevens No. 7787, ILL 5901; Barros, 2 JAN 1913, leg. F.L. Stevens No. 55, ILL 5902; on *Casearia aculeata* Jacquin, Lajas, 17 JUN 1915, leg. F.L. Stevens No. 7151, ILL 6033; on *Casearia arborea* (L.C. Richard) Urban, Monte de Oro, 13 DEC 1913, leg. F.L. Stevens No. 5709, ILL 5879 and 6029 [also the holotype and an isotype, respectively, of *Irenopsis casearina* C.G. Hansford]; on *Casearia ramiflora* Vahl [= *C. guianensis* (Aublet) Urban], Barceloneta, 10 AUG 1915, leg. F.L. Stevens No. 9256, ILL 6032; Manati, 2 JUL 1915, leg. F.L. Stevens No. 7688, ILL 5671; Martin Peña, 11 AUG 1915; leg. F.L. Stevens No. 9306, ILL 5864 and 6024; leg. F.L. Stevens No. 9328, ILL 5897; Santa Ana, 31 DEC 1913, leg. F.L. Stevens No. 6683, ILL 5886 and 5887; San Germán, 12 DEC 1913, leg. F.L. Stevens No. 5844, ILL 5885 and 5888; Vega Baja, 2 MAR 1913, leg. F.L. Stevens No. 510, ILL 5672; 1 MAR 1913, leg. F.L. Stevens No. 512, ILL 5862; 5 NOV 1913, leg. F.L. Stevens No. 4262, ILL 5674; 2 JUL 1915, leg. F.L. Stevens No. 7745, ILL 5861; on *Casearia sylvestris* Sw., Arecibo-Lares Road, 21 JUN 1915, leg. F.L. Stevens No. 7285, ILL 6025; Mayagüez Mesa, 1 MAY 1913, leg. F.L. Stevens No. 1051, ILL 5884 [also the holotype of *Fusarium meliolicola* F.L. Stevens and of *Nectria meliolicola* F.L. Stevens]; 14 JUN 1915, F.L. Stevens No. 7017, ILL 5882; 4 MAY 1913, leg. F.L. Stevens No. 1200, ILL 5894; Miradero, 4 AUG 1915, leg. F.L. Stevens No. 9136, ILL 5896; San Germán, 12 DEC 1913, leg. F.L. Stevens No. 5837, ILL 5668 and 5880; leg. F.L. Stevens No. 5864, ILL 5673; on *Casearia* sp., [River junction], Dos Bocas, below Utuado, 17 DEC 1915, leg. F.L. Stevens No. 6071, ILL 6028; Mayagüez, 31 OCT 1913, leg. F.L. Stevens No. 3935, ILL 5669 and 5670; leg. F.L. Stevens No. 3920, ILL 5893; 15 JUN 1915, leg. F.L. Stevens No. 7074, ILL 6030; on *Mammea americana* L., [Rio Maricao above] Maricao, 20 SEP 1913, leg. F.L. Stevens No. 3641, ILL 5881; Las Marias, 10 JUL 1915, F.L. Stevens No. 8207, ILL 6031.

Meliola paulliniae F.L. Stevens var. *dentata* F.L. Stevens. Annales Mycologici 26:197. 1928, non *M. serjaniae* F.L. Stevens var. *dentata* F.L. Stevens, op. cit. p. 280. **Lectotype** [designated herein]: On *Paullinia* sp., Panama, Corozal, Trail No. 17, 30 AUG 1924, leg. F.L. Stevens No. 97, ILL 5868 [we decided to lectotypify because C.G. Hansford (1961:432) erroneously cited the holotype of *Meliola serjaniae* F.L. Stevens var. *dentata* F.L. Stevens as the "type" of this variety, which is represented in the protologue by 14 syntypes]. **Residual syntypes**: Las Cruces trail, 2 SEP 1924, leg. F.L. Stevens No. 168, ILL 5867; 28 SEP 1924, leg. F.L. Stevens No. 878, ILL 5869; leg. F.L. Stevens No. 894, ILL 5875; Summit, 6 SEP 1924, leg. F.L. Stevens No. 346, ILL 5865; Tumba Muerta, 27 SEP 1924, leg. F.L. Stevens No. 852, ILL 5876; Culebra, 2

SEP 1924, leg. F.L. Stevens No. 953, ILL 5870; Gamboa, 16 AUG 1923, leg. F.L. Stevens No. 1080, ILL 5872; on *Paullinia cururu* L., Panama, Las Cruces trail, 2 SEP 1924, leg. F.L. Stevens No. 148, ILL 5878; leg. F.L. Stevens No. 158, ILL 5871; on *Serjania* sp., Panama, Brazos Brook Reservoir, 9 SEP 1924, leg. F.L. Stevens No. 724, ILL 5866; Costa Rica, Siquirres, 31 JUL 1923, leg. F.L. Stevens No. 703, ILL 5874; on unknown host, Costa Rica, [Columbiana], leg. F.L. Stevens No. 593C, ILL 5877; Panama, Tumba Muerta, 12 OCT 1924, leg. F.L. Stevens No. 1224, ILL 5873.

Meliola pazschkeana A. Gaillard. Le Genre *Meliola*, pp. 95–96, pl. XXI, figs. 4, 4a & 4b. 1892. **Isotypes**: Hab. ad paginam superiorem foliorum *Bauhiniae* cujusdam [= *Bauhinia* sp., Brazil], Rio de Janeiro, [JUN] 1887, leg. E.H.G. Ule, sub No. 1002, ex Fungi Europaei Exsiccati No. 3854, ILL 5956 [as a microscopic preparation], and in bound exsiccati set.

Meliola peglerae E.M. Doidge. Transactions of the Royal Society of South Africa 5:730, pl. LXIII, fig. 24. 1917; The Annals of the Bolus Herbarium 2: 109–110. 1918. **Isolectotypes**: On leaves of *Anastrabe integerrima* E. Meyer ex Bentham, [South Africa], Kentani, 4 JUN 1912, leg. A. Pegler No. 1883, ILL 3875 and 3876 [the latter ex PREM 2363]. **Lectotype**: PREM 2363 [this number attributed to I.B. Pole Evans by E.M. Doidge, 1918. It is the only collection cited by Doidge (1920) in her transfer of the species to *Irene*. We interpret this as lectotypification. Annotations on the packets at ILL by C.G. Hansford indicate that he regarded both Pegler No. 1883 and PREM 2363 as part of the type collection. See also F.L. Stevens (1927) and C.G. Hansford (1961).]. **Residual isosyntype**: On leaves of an unknown shrub [identified as *Anastrabe integerrima* E. Meyer ex Bentham on the packet], Natal Prov., Umgeni, near Durban, 16 JUL 1915, leg. J. Medley Wood, ex PREM No. 9036, ILL 3877. ≡ *Irene peglerae* (E.M. Doidge) E.M. Doidge. South African Journal of Natural History 2:40. 1920. ≡ *Asteridiella peglerae* (E.M. Doidge) C.G. Hansford ex C.G. Hansford. Beihefte zur Sydowia Annales Mycologici 2:653. 1961.

Meliola peleae F.L. Stevens. Bernice P. Bishop Museum Bulletin 19:34–35, text fig. 7g, pl. II(G). 1925. **Holotype**: On *Pelea* sp. [= *Melicope* sp.], Hawaii, [Island of] Hawaii, Kilauea, 14 JUL [1921], leg. F.L. Stevens No. 840, ILL 5958. **Isotypes**: ILL 5966, BISH 499022 and 499959. **Paratypes**: On *Pelea* sp. [= *Melicope* sp.], Oahu, Olympus, 24 JUN [1921], leg. F.L. Stevens No. 669, ILL 5960; leg. F.L. Stevens No. 726, ILL 5967; Kauai, [Pipe trail], Waimea Canyon, 15 JUN [1921], leg. F.L. Stevens No. 434, ILL 5969; [Island of] Hawaii, Hamakua, upper ditch trail, 31 JUL [1921], leg. F.L. Stevens No. 1073, ILL 5973; Kona, Keauhou, [Bishop Estate Road], 25 JUL [1921], leg. F.L. Stevens No. 988, ILL 5957; Molokai, Pukoo Ridge, AUG 1912, leg. C.N. Forbes-F.L. Stevens No. 411, ILL 5324 and 5968 [also paratypes of *Meliola juddiana* F.L. Stevens]; on *Pelea rotundifolia* A. Gray [= *Melicope rotundifolia* (A. Gray) T.G. Hartley & B.C. Stone], Oahu, Wahiawa, 30 JUN [1921], leg. F.L. Stevens No. 200, ILL 5961; on *Pelea rotundifolia* [erroneously as *P. cinerea* (A. Gray) Hillebr. in the protologue = *Melicope* sp.], Oahu, in 1912 (?), leg. C.N. Forbes-F.L. Stevens No. 1328, ILL 5327 [also a paratype of *Meliola juddiana* F.L. Stevens]; on *Pelea elliptica* (A. Gray) Hillebr. [= *Melicope elliptica* A. Gray], Lanai, in 1915 and 1916, Munro s.n., ILL 5971 and 5965; Oahu, Wahiawa, 3 JUN [1921], leg. F.L. Stevens No. 203, ILL 5962; on *Pelea barbigera* (A. Gray) Hillebr. [= *Melicope barbigera* A. Gray], Kauai, Waimea Canyon, 15 JUN [1921], leg. F.L. Stevens No. 440, ILL 5964; on *Pelea sandwicensis* (Hooker & Arnott) A. Gray [= *Melicope sandwicensis* (Hooker & Arnott) T.G. Hartley & B.C. Stone], Kauai, [Pipe trail], Waimea Canyon, 15 JUN [1921], leg. F.L. Stevens No. 449, ILL 5959; on *Pelea cinerea* (A. Gray) Hillebr. [= *Melicope cinerea* A. Gray], Lanai, in 1913, leg. C.N. Forbes-F.L. Stevens No. 251, ILL 5972; Oahu, in 1912, leg. C.N. Forbes-F.L. Stevens No. 1776, ILL 5963; on *Cryptocarya mannii* Hillebr., Kauai, Kalalau trail, 16 JUN [1921], leg. F.L. Stevens No. 506, ILL 5970.

Meliola pelliculosa H. & P. Sydow. Philippine Journal of Science, Section C (Botany) 8:480. 1913. **Isotype:** On living or languishing leaves of *Lumnitzera racemosa* Willd., Philippines, Luzon, Manila, 22 FEB 1913, leg. E.D. Merrill S-115, ex Fungi Exotici Exsiccati No. 252, ILL 5974.

Meliola peltata E.M. Doidge. Transactions of the Royal Society of South Africa 5:727, 744–745, pl. LXI, fig. 18. 1917. **Isotype:** On leaves of *Podocarpus thunbergii* Hooker, South Africa, Cape Prov., Knysna Forest, 3 JUN 1912, leg. P.J. Pienaar, ex PREM No. 2436, ILL 5976. **Holotype:** PREM.

Meliola peregrina H. & P. Sydow. Philippine Journal of Science, Section C (Botany) 8:479. 1913. **Isotype:** On living leaves of *Maesa laxa* Mez, Philippines, Luzon, Nueva Vizcaya Prov., JAN 1913, leg. R.C. McGregor, ex Phil. Bur. Sci. No. 20255, ILL 3564. ≡ *Amazonia peregrina* (H. & P. Sydow) H. & P. Sydow. Annales Mycologici 15:238. 1917.

Meliola perexigua A. Gaillard. Le Genre *Meliola*, p. 98. 1892. **Isotype:** Hab. ad paginam inferiorem foliorum coriaceorum, [C. Africa], Congo, leg. F.R. Thollon s.n., ILL 5977. **Holotype:** PC.

Meliola permixta H. Sydow. Annales Mycologici 21:90–91. 1923. **Isolectotype:** Hab. in foliis *Ipomoeae* sp. [= *Ipomoea* sp.], British North Borneo [Malaysia], Sandakan, 16 DEC 1920, leg. M. Ramos No. 2146, ILL 5979.

Assumed lectotype: S [citation of this institution for location of "type" by C.G. Hansford (1961:646) is interpreted as lectotypification].

Meliola perpusilla H. Sydow var. *congoensis* M. Beeli, Bulletin du Jardin Botanique de L'État à Bruxelles 7:97–98. 1920. **Isotype:** In foliis Asclepiadacearum [= Asclepiadaceae], C. Africa, Congo [Zaire], Kikwit, 1914, leg. H.J.R. Vanderyst No. 2744, ILL 5981. **Holotype:** BR. ≡ *Meliola congoensis* (M. Beeli) C.G. Hansford. Beihefte zur Sydowia Annales Mycologici 2:567. 1961.

Meliola perseae F.L Stevens. Illinois Biological Monographs 2:485, fig. 10. 1916. **Holotype:** On *Persea gratissima* C.F. Gaertner [= *P. americana* P. Miller], Porto Rico [Puerto Rico], Las Marias, 10 JUL 1915, leg. F.L. Stevens No. 8212, ILL 4218. ≡ *Irene perseae* (F.L. Stevens) R.A. Toro. Mycologia 17:140. 1925. ≡ *Irenina perseae* (F.L. Stevens) F.L. Stevens. Annales Mycologici 25:465. 1927. ≡ *Asteridiella perseae* (F.L. Stevens) C.G. Hansford ex C.G. Hansford. Beihefte zur Sydowia Annales Mycologici 2:47. 1961.

Meliola perseae F.L. Stevens forma *setulifera* C.L. Spegazzini. Boletín de la Academia Nacional de Ciencias en Córdoba 26:380. Preprint 1923 [journal part issued in 1924]. **Isotypes:** Hab. Sobre las hojas vivas de *Persea gratissima* C.F. Gaertner [= *P. americana* P. Miller], Florida, cerca de Gainesville, leg. ?, ex H.W. Ravenel, Fungi Americani Exsiccati No. 82, ILL 4562 [as a microscopic preparation], and in bound exsiccati set [sub *Meliola amphitricha* E.M. Fries]. ≡ *Meliola setulifera* (C.L. Spegazzini) F.L. Stevens. Annales Mycologici 26:285. 1928.

Meliola peruviana H. & P. Sydow. Annales Mycologici 14:75–76. 1916. **Isolectotype:** In foliis Bignoniaceae sp. indet., Peru, Seringal Auristella, Rio Acre, AUG 1911, leg. E.H.G. Ule No. 3452, ILL 5989. **Assumed lectotype:** S [citation of this institution for location of "type" by C.G. Hansford (1961:662) is interpreted as lectotypification]. ≡ *Irene peruviana* (H. & P. Sydow) C.G. Hansford. Sydowia Annales Mycologici 9:56–57. 1955. ≡ *Asteridiella peruviana* (H. & P. Sydow) C.G. Hansford ex C.G. Hansford. Beihefte zur Sydowia Annales Mycologici 2:662. 1961.

Meliola peruviana H. & P. Sydow var. *irregularis* F.L. Stevens. Annales Mycologici 26:256, pl. VI, fig. 70. 1928. **Syntypes:** On Bignoniaceae sp. indet., Panama, Barro Colorado [Island], 10 SEP 1924, leg. F.L. Stevens No. 420, ILL 5988; leg. F.L. Stevens No. 421, ILL 5986; Agua Clara Reservoir, 17 SEP 1924, leg. F.L. Stevens No. 547, ILL 5987. ≡ *Meliola arrabidaeae* C.G. Hansford var. *irregularis* (F.L. Stevens) C.G. Hansford. Beihefte zur Sydowia Annales Mycologici 2:674. 1961.

Meliola petiolaris E.M. Doidge. Transactions of the Royal Society of South Africa 8:142. 1920. **Isotype:** In foliis et petiolis *Oleae laurifoliae* [=

Olea laurifolia Lam.], South Africa, Natal Prov., Buccleuch, 17 JUL 1918, leg. E.M. Doidge, ex PREM No. 11558, ILL 5990. [also an isotype of *Meliola oleicola* E.M. Doidge, as PREM No. 11557; both PREM numbers on the same packet].

Meliola philippinensis (F.L. Stevens) C.G. Hansford, 1961. **See basionym**: *Meliolina philippinensis* F.L. Stevens, 1927.

Meliola philodendri F.L. Stevens. Illinois Biological Monographs 2:528, fig. 54. 1916. **Holotype**: On *Philodendron krebsii* Schott [= *P. consanguineum* Schott], Porto Rico [Puerto Rico], Arecibo-Lares Road, 21 JUN 1915, leg. F.L. Stevens No. 7225, ILL 5991 [also a potential isotype of *Scolecopeltella microcarpa* C.L. Spegazzini]. **Paratypes**: Jayuya, 2 MAR 1917, leg. F.L. Stevens No. 377, ILL 5992; 1 MAR 1913, leg. F.L. Stevens No. 437, ILL 5998; Ponce, 8 NOV 1913, leg. F.L. Stevens No. 4346, ILL 5995 [also the holotype of *Helminthosporium philodendri* F.L. Stevens and a paratype of *Isthmospora spinosa* F.L. Stevens], ILL 5997; Jajome Alto, 17 JUL 1915, leg. F.L. Stevens No. 8424, ILL 5999; Maricao, 20 JUL 1915, leg. F.L. Stevens No. 8994, ILL 5994 and 5996; El Alto de la Bandera, 16 JUL 1915, leg. F.L. Stevens No. 8712, ILL 5993.

Meliola picramniae C.G. Hansford. Beihefte zur Sydowia Annales Mycologici 1:114. 1957. **Holotype**: Hab. in foliis *Picramniae antidesmae* [= *Picramnia antidesma* Sw.], Costa Rica, El Alto, [6 JUL 1923], leg. F.L. Stevens No. 242a, ILL 5118.

Meliola pilocarpi F.L Stevens. Illinois Biological Monographs 2:509. 1916. **Holotype**: On *Pilocarpus racemosus* Vahl, Porto Rico [Puerto Rico], Mayagüez, 15 JUN 1915, leg. F.L. Stevens No. 7080, ILL 6001.

Meliola pinicola J. Dearness. Mycologia 18:244. 1926. **Assumed lectotype**: On living needles of *Pinus echinata* P. Miller, North Carolina, Pisgah National Forest, Pisgah, 6 JUL 1925, leg. G.G. Hedgcock No. 24394, comm. J. Dearness No. 5878, ILL 4219 [citation of specimen in the F.L.S. herb. (ILL) as "type" by C.G. Hansford (1961:751) is interpreted as lectotypification]. ≡ *Irenina pinicola* (J. Dearness) F.L. Stevens. Annales Mycologici 25:449, fig. 12. 1927. ≡ *Asteridiella pinicola* (J. Dearness) C.G. Hansford. Beihefte zur Sydowia Annales Mycologici 2:751. 1961.

Meliola piperina H. & P. Sydow. Annales Mycologici 14:358. 1916. **Assumed isotypes**: Hab. in foliis *Piperis* spec. [= *Piper* sp.], Philippines, [Luzon, Laguna Prov.], Los Baños, JAN 1916, leg. C.F. Baker No. 4046 [ex Fungi Malayana No. 367], ILL 6002 [as a microscopic preparation], CUP, FH.

Meliola pisoniae F.L. Stevens & E.F. Roldan ex W. Yamamoto. Hyogo Noka Daigaku, Sasayama, Japan. Science Reports Series: Plant Protection (Agricultural Biology), [Hyogo University of Agriculture. Science

Reports] 3(2):68. 1958; originally published by F.L. Stevens & E.F. Roldan in Philippine Journal of Science 56:62–63, fig. 2h. 1935, as nom. invalid. sine diagn. lat. **Holotype**: On *Pisonia umbellifera* (J.R. & G. Forster) Seem., Philippines, Luzon, Laguna [Prov.], Mount Maquiling, Los Baños, 8 AUG 1930 [as 26 AUG on the packet], leg. F.L. Stevens No. 373, ILL 6056.

Meliola pisoniicola F.L. Stevens & E.F. Roldan ex C.G. Hansford. Sydowia Annales Mycologici 16:317. (1962) 1963; originally published by F.L. Stevens & E.F. Roldan in Philippine Journal of Science 56:69, fig. 3d. 1935, as nom. invalid. sine diagn. lat. **Holotype**: On *Pisonia* sp., Philippines, Luzon, Nueva Ecija Prov., San José-Balete Pass, 10 JAN 1931, leg. F.L. Stevens No. 1813, ILL 6059.

Meliola pistaciae F.L. Stevens & E.F. Roldan. Philippine Journal of Science 56:58–59, fig. 2d. 1935, nom. invalid. sine diagn. lat. **Holotype**: On *Pistacia* sp. [the host possibly misidentified (cf. C.G. Hansford, 1961)], Philippines, Luzon, Nueva Ecija Prov., San José to Balete Pass, 9 JAN 1931, leg. F.L. Stevens No. 1712, ILL 6060. **Isotypes**: CALP, Phil. Bur. Sci.

Meliola pithecellobii F.L. Stevens & L.R. Tehon. Mycologia 18:9–10, pl. 1, fig. 8. 1926 [as *pithecolobii*, corrected in accordance with current I.C.B.N. (cf. F.C. Deighton, 1968)]. **Holotype**: On *Pithecellobium jujunba* (Willd.) Urban [as *Pithecolobium*], Trinidad, Cumuto, 16 AUG 1922, leg. F.L. Stevens No. 966, ILL 6035.

Meliola podocarpi E.M. Doidge. Transactions of the Royal Society of South Africa 5:725–726, 743, pl. LIX, fig. 14. 1917. **Residual isosyntypes**: On leaves of *Podocarpus elongata* L'Héritier ex Persoon, [South Africa, Cape Prov.], Fort Cunningham, Toise River, 20 MAR 1915, leg. Forest Officer, ex PREM No. 8897, ILL 4221 and 4222. **Residual syntype**: PREM [in her transfer to *Irene*, Doidge (1920) cited only her own collection: PREM No. 1748 (not at ILL); this is interpreted as lectotypification; see also F.L. Stevens (1927:447)]. ≡ *Irene podocarpi* (E.M. Doidge) E.M. Doidge. South African Journal of Natural History 2:40. 1920. ≡ *Irenina podocarpi* (E.M. Doidge) F.L. Stevens. Annales Mycologici 25:447–448. 1927. ≡ *Asteridiella podocarpi* (E.M. Doidge) C.G. Hansford ex C.G. Hansford. Beihefte zur Sydowia Annales Mycologici 2:750. 1961.

Meliola polyodonta H. Sydow. Annales Mycologici 24:306. 1926. **Isotypes**: Hab. in foliis Leguminosae cujusdam adhuc omnino in incertae [= Fabaceae sp. indet.], Costa Rica, San Pedro de San Ramón, 5 FEB 1925, leg. H. Sydow, Fungi in Itinere Costaricensi Collecti No. 385, ILL 6034; also in bound set of Fungi Exotici Exsiccati under No. 620.

Meliola polytricha K. Kalchbrenner & M.C. Cooke var. *abyssinica* P.C. Hennings. Bulletin de l'Herbier Boissier, Series 1, 1:117. 1893. **Isotype**:

In foliis vivis *Osyridis abyssinicae* [= *Osyris abyssinica* Hochst. ex A. Richard], Abyssinia [Ethiopia], Col. Eritrea, pr. Saganeiti, alt. 2200 m, MAY 1892, leg. G. Schweinfurth s.n., ILL 6026, ex herb. H. Sydow.

Meliola polytricha K. Kalchbrenner & M.C. Cooke var. *paropsiae* M. Beeli. Bulletin de la Société Royale de Botanique de Belgique 60:85. 1927. **Isotype:** Sur les feuilles de *Paropsia* sp. [Passifloraceae, C. Africa], Congo [Zaire], Ipamu, Kikwit, 1921, leg. H.J.R. Vanderyst No. 9777, ILL 6027. **Holotype:** BR. ≡ *Meliola paropsiae* (M. Beeli) C.G. Hansford. Beihefte zur Sydowia Annales Mycologici 2:122. 1961.

Meliola portoricensis (F.L. Stevens) R. Ciferri, 1932, 1938. **See basionym:** *Irenopsis portoricensis* F.L. Stevens, 1927.

Meliola praetervisa A. Gaillard var. *stevensii* C.G. Hansford. Sydowia Annales Mycologici 9:73–74. 1955. **Isotypes:** Hab in foliis *Coccolobae* spec. [= *Coccoloba* sp.], British Guiana [Guyana], Kartabo, [23 JUL 1922], leg. F.L. Stevens No. 576, ILL 4577 and 4580 [also paratypes of *Meliola angusta* F.L. Stevens & L.R. Tehon]. **Holotype:** FH. **Paratype:** in foliis *Coccolobae sintenisii* [misspelled "*sintensii*" = *Coccoloba sintenisii* Urban ex Lindau], Porto Rico [Puerto Rico], Mayagüez [15 JUN 1915], leg. F.L. Stevens No. 7066, ILL 6014 [also a paratype of *Isthmospora spinosa* F.L. Stevens].

Meliola premnae C.G. Hansford. Proceedings of the Linnean Society of London 160:135. (1948) 1949. **Isotypes:** Hab. in caulibus *Premnae odoratae* [= *Premna odorata* Blanco], Philippines, [Luzon], Laguna Prov., Mount Maquiling near Los Baños, SEP 1913, leg. C.F. Baker, Fungi Malayana Supplement No. 41, sub *Meliola callista* H. Rehm, ILL 4857 and 4859. **Holotype:** PREM.

Meliola protii F.L. Stevens. Annales Mycologici 26:199, pl. III, fig. 22. 1928. **Holotype:** On *Protium panamense* (Rose) I.M. Johnston, Panama, Agua Clara Reservoir, 17 SEP 1924, leg. F.L. Stevens No. 583, ILL 6013.

Meliola pseudanastomosans H. Rehm. Hedwigia 35:(150). 1896. **Isotype:** Ad folia *Psoraleae* [= *Psoralea* sp. (sensu lato)], Ecuador, Pichincha Prov., San Jorge, JUL 1892, leg. N.G. v. Lagerheim s.n., ILL 4223 [as a microscopic preparation]. ≡ *Irenina pseudanastomosans* (H. Rehm) F.L. Stevens. Annales Mycologici 25: 469. 1927. ≡ *Asteridiella pseudanastomosans* (H. Rehm) C.G. Hansford ex C.G. Hansford. Beihefte zur Sydowia Annales Mycologici 2:271–272. 1961.

Meliola psilostomatis F. v. Thümen. Mycotheca Universalis, Century 8, No. 775. Anno 1877; Flora 60:408. 1877 [as *psilostomae*, corrected in P.A. Saccardo (1882:54) in accordance with the current I.C.B.N.]. **Isotype:** In foliis vivis *Psilostomae ciliatae* [*Psilostoma ciliata* Klotzsch in Ecklon & Zeyher = *Canthium ciliatum* (Klotzsch in Ecklon & Zeyher) O. Kuntze, South Africa], Cape of Good Hope [as Promont. Bonae Spei],

Somerset East, ad pedem montis "Boschberg," in sylvis, DEC 1876, leg. P. MacOwan No. 1291, ILL 6488 [as a microscopic preparation]. ≡ *Dimerosporium psilostomatis* (F. v. Thümen) P.A. Saccardo. Sylloge Fungorum 1:54. 1882. ≡ *Dimerium psilostomatis* (F. v. Thümen) P.A. & D. Saccardo & P. Sydow in P.A. & D. Saccardo. Sylloge Fungorum 17:537. 1905; emend., as [*D. psilostomae*] by F. v. Höhnel. Sitzungsberichte der Kaiserl. Akademie der Wissenschaften in Wien. Mathematisch-Naturwissenschaftliche Klasse, Abt. 1. 119:465-466. 1910.

Meliola psychotriae F.S. Earle. Bulletin of the New York Botanical Garden 3:308. 1905. **Isolectotypes**: On leaves of *Psychotria* sp., Porto Rico [Puerto Rico, limestone hills along coast, three miles west of Ponce, 9 DEC 1902], leg. A.A. Heller No. 6252, ILL 6103 and 6119, NY. **Assumed lectotype**: K [F.L. Stevens (1928:246) cited the above collection under "type locality" but did not indicate a lectotype; citation of K for location of "type" by C.G. Hansford (1961:597) is interpreted as lectotypification]. **Residual isosyntypes**: On *Erithalis fruticosa* L. [misspelled "*Erithelis*"], Porto Rico [Puerto Rico], in sands near sea at Santurce, 19 JAN 1903, leg. A.A. Heller No. 6430, ILL 6101, NY.

Meliola pteridicola F.L. Stevens. Illinois Biological Monographs 2:496, fig. 27. 1916. **Holotype**: On *Anemia adiantifolia* (L.) Sw. [as *Aneimia*], Porto Rico [Puerto Rico], Rio Tanamá, near Arecibo, 6 JUL [1915], leg. F.L. Stevens No. 7814, ILL 6142 [also a paratype of *Helminthosporium glabroides* F.L. Stevens]. **Isotypes**: ILL 6138, BPI 70894. **Paratypes**: Quebradillas, 23 JUN [1915], leg. F.L. Stevens No. 7269, ILL 6139, 6151 and 6152; Dos Bocas, below Utuado, 8 JUL [1915], leg. F.L. Stevens No. 8015, ILL 6149 and 6150; on *Anemia* sp. [as *Aneimia*], Dos Bocas, below Utuado, 30 DEC 1913, leg. F.L. Stevens No. 6594, ILL 6145; on *Adiantum latifolium* Lam., Las Marias, 10 JUL [1915], leg. F.L. Stevens No. 8182, ILL 6144 and 6148; Mayagüez, 24 JUN [1915], leg. F.L. Stevens No. 7418, ILL 6143; on *Adiantum* sp. aff. *A. cristatum* L. [= *A.* cf. *pyramidale* (L.) Willd.], Mayagüez, 19 JUL [1915], leg. F.L. Stevens No. 8795, ILL 6147; on *Adiantum* sp., Mayagüez, 1 MAY 1913, leg. F.L. Stevens No. 1063, ILL 6146 [F.L. Stevens numbers 7269, 8182 and 7418 are also cited as paratypes of *Helminthosporium glabroides* F.L. Stevens]. ≡ *Irenopsis pteridicola* (F.L. Stevens) C.G. Hansford. Sydowia Annales Mycologici 9:39. 1955.

Meliola pterocarpi H.S. Yates. Philippine Journal of Science, Section C (Botany) 13:235. (1918) 1919 [as *pterocarpiae*, corrected in accordance with Articles 32.6 & 61 of I.C.B.N.]. **Isotype**: On leaves of *Pterocarpus indicus* Willd., British North Borneo [Malaysia], Tenom, 17 OCT 1917, leg. H.S. Yates No. 102, ILL 6010.

Meliola pterospermi F.L. Stevens. Memoirs of the Indian Department of

Agriculture. Botanical Series 15(5):108–109, pl. III, figs. 12–14. JAN 1928; Annales Mycologici 26:260. MAY 1928. **Isolectotypes**: On *Pterospermum* sp., Burma, Bassein, 31 NOV 1912, leg. E.J. Butler, as F.L. Stevens No. 1987, ILL 6009, HCIO. **Assumed lectotype**: IMI 25710 [citation of this specimen as "type" by C.G. Hansford (1961:181) is interpreted as lectotypification. F.L. Stevens did not indicate a lectotype.].

Meliola pterospermicola F.L. Stevens & E.F. Roldan ex C.G. Hansford. Sydowia Annales Mycologici 16:318–319. (1962) 1963; originally published by F.L. Stevens & E.F. Roldan in Philippine Journal of Science 56:68–69, fig. 3e. 1935, as a nom. invalid. sine diagn. lat. **Holotype**: On *Pterospermum obliquum* Blanco, Philippines, Luzon, Laguna Prov., Agricultural College, 10 SEP 1930, leg. F.L. Stevens No. 498, ILL 6008.

Meliola puiggarii C.L. Spegazzini. Boletín de la Academia Nacional de Ciencias en Córdoba 11:492. 1889; Fungi Puiggariani, Pugillus I, p. 114, No. 228. 1889. **Isotype**: Hab. ad folia viva *Rubi* speciei cujusdam [= *Rubus* sp.] in dumetis, [Brazil], pr. Apiahy [Apiai], May 1888, leg. J.I. Puiggari No. 2722, ex C.L. Spegazinni herb., ILL 3752. ≡ *Irene puiggarii* (C.L. Spegazzini) E.M. Doidge. South African Journal of Natural History 2:39. 1920.

Meliola pulchella C.L. Spegazzini. Boletín de la Academia Nacional de Ciencias en Córdoba 11:491–492. 1889; Fungi Puiggariani, Pugillus I, pp. 113–114, No. 227. 1889. **Isotype**: Hab. ad folia viva Myrtaceae cujusdam in dumetis, [Brazil], pr. Apiahy [Apiai], Hiem 1881, leg. J.I. Puiggari No. 1699, ex C.L. Spegazzini herb., ILL 6007. **Holotype**: LPS.

Meliola pulcherrima H. & P. Sydow. Annales Mycologici 11:254. 1913. **Isotypes**: In foliis vivis *Fici benjamine* [*Ficus benjamina* L., a misidentification, the host later identified as *Syzygium cumini* (L.) Skeels in the Myrtaceae (cf. S.J. Hughes, 1993:123)], Philippines, Luzon, Rizal Prov., Antipolo, OCT 1912, leg. M. Ramos, ex H. Sydow, Fungi Exotici Exsiccati No. 124 [erroneously as No. 121 in the protologue], ILL 6508 [as a microscopic preparation], DAOM 163861, and in bound exsiccati set. ≡ *Meliolina pulcherrima* (H. & P. Sydow) H. &. P. Sydow. Annales Mycologici 12:553. 1914.

Meliola pululahuensis A. Gaillard. Bulletin de la Société Mycologique de France 8:183, pl. 15, fig. 2. 1892. **Isotype**: Ad paginam superiorem foliorum *Piperis* [= *Piper* sp.] cujusdam, circa craterium "Pululahua," Ecuador, Pichincha Prov., missit N.G. v. Lagerheim s.n., FEB 1892, ILL 6006. **Holotype**: PC [cf. C.G. Hansford (1961)].

Meliola quercina N.T. Patouillard. Journal de Botanique 4:61–62, fig. 5. 1890. **Isotype**: Sur feuilles vivantes du Chêne [= *Quercus* sp.] Bois près de la rive gauche de la rivière noire, en face de Fu-Phap, Tonkin [Vietnam,

JAN 1889], leg. B. Balansa, [ex C. Roumeguère, Fungi Selecti Exsiccati No. 5945], ILL 6540. **Assumed holotype:** FH. ≡ *Leptomeliola quercina* (N.T. Patouillard) F. v. Höhnel. Sitzungsberichte der Kaiserl. Akademie der Wissenschaften in Wien. Mathematisch-Naturwissenschaftliche Klasse, Abt. 1. 128:558. 1919. ≡ *Meliolinopsis quercina* (N.T. Patouilllard) M. Beeli. Bulletin du Jardin Botanique de L'État à Bruxelles 7:119. 1920.

Meliola quinquespora F. v. Thümen. Flora 59:568–569. 1876; Mycotheca Universalis, Century 7, No. 657. 1877. **Isotypes:** In foliis languescentibus *Buddleiae auriculatae* [= *Buddleja auriculata* Bentham in Hooker], South Africa, [Cape Prov.] Promont. Bonae spei in dumetis montis "Boschberg," Somerset East, AUG 1876, leg. P. MacOwan No. 1251, ILL 6489 [as a microscopic preparation], and in bound exsiccati set.

Meliola radians (H. & P. Sydow) M. Beeli, 1920. **See basionym:** *Meliolina radians* H. & P. Sydow, 1914.

Meliola ramosii H. & P. Sydow. Annales Mycologici 12:552. 1914. **Isosyntype:** Hab in foliis *Homonoiae ripariae* [= *Homonoia riparia* Lour.], Philippines, Luzon, Rizal Prov., Antipolo, 13 AUG 1913, leg. M. Ramos, ex H. Sydow, Fungi Exotici Exsiccati No. 378, ILL 6171.

Meliola rectangularis F.L. Stevens. Illinois Biological Monographs 2:495, fig. 25. 1916. **Holotype:** On *Coccoloba laurifolia* sensu auct. non Jacquin [as *Coccolobis* = *Coccoloba diversifolia* Jacquin], Porto Rico [Puerto Rico], Arecibo-Lares Road, 21 JUN 1915, leg. F.L. Stevens No. 7292, ILL 4398. **Isotype:** ILL 7371. [Stevens No. 7292 is also cited as syntype of *Seynesia coccolobae* R.W. Ryan and a paratype of *Helminthosporium panici* F.L. Stevens. The two specimens at ILL are also potential isotypes of *Scolecopeltis pachyasca* C.L. Spegazzini.]. **Paratypes:** On *Banisteria laurifolia* L. [= *Heteropteris laurifolia* (L.) A. Jussieu], Utuado, 8 NOV 1913, leg. F.L. Stevens No. 4384, ILL 4397; leg. F.L. Stevens 4392a, ILL 4395 p.p.; Mayagüez Mesa, 29 JUN 1915, leg. F.L. Stevens No. 7564, ILL 4396 [also a paratype of *Irenopsis banisteriae* C.G. Hansford and of *Helminthosporium parathesicola* F.L. Stevens]; Maricao, 20 SEP 1913, leg. F.L. Stevens No. 4852, ILL 4394; Hormigueros, K. 7, 23 JUN 1915, leg. F.L. Stevens No. 7358, ILL 4393 [also a paratype of *Helminthosporium parathesicola* F.L. Stevens]; Martin Peña, 11 OCT 1915 [as AUG on the packet], leg. F.L. Stevens No. 9298, ILL 4392. ≡ *Irenopsis rectangularis* (F.L. Stevens) F.L. Stevens. Annales Mycologici 25:436. 1927.

Meliola recurvipoda C.G. Hansford. Beihefte zur Sydowia Annales Mycologici 1:114–115. 1957. **Holotype:** Hab. in foliis *Peleae hawaiiensis* [*Pelea hawaiensis* Wawra = *Melicope hawaiensis* (Wawra) T.G. Hartley & B.C. Stone], Hawaii, Kauai, Waimea Canyon, [Pipe trail], 15 JUL 1921, leg. F.L. Stevens No. 411, ILL 5336.

Meliola rehmii F.L. Stevens. Annales Mycologici 26:222. 1928, nom. nov. **Based on:** *Meliola horrida* H. Rehm, 1913, nom. illegit., non J.B. Ellis & B.M. Everhart, 1893. **Isotype:** Ad folia coriacea, Philippines, Luzon, Laguna Prov., Los Baños, APR 1913, leg. C.F. Baker No. 976, ILL 6174 [as a microscopic preparation].

Meliola rhamnicola F.L. Stevens & L.R. Tehon. Mycologia 18:14–15, pl. 2, fig. 17. 1926. **Holotype:** On *Gouania* sp. (?), British Guiana [Guyana], Tumatumari, 11 JUL 1922, leg. F.L. Stevens No. 203, ILL 6188.

Meliola rhoina E.M. Doidge in E.M. Doidge & H. Sydow. Bothalia 2(2):454, 466–467. 1928. **Isotype:** On *Rhus longispina* Ecklon & Zeyher, South Africa, Cape Prov., Despatch, near Uitenhage, 24 MAR 1911, leg. E.M. Doidge, ex PREM No. 1239, ILL 6184. **Holotype:** PREM. **Paratype:** on *Rhus crenata* Thunb., Natal [Prov.], Verulam, 3 SEP 1913, leg. I.B. Pole Evans, ex PREM No. 6804, ILL 6185.

Meliola rigida E.M. Doidge. Transactions of the Royal Society of South Africa 5:736, 747, pl. LXIV, fig. 35. 1917. **Isotype:** On leaves of *Xymalos monospora* (Harvey) Baillon ex Warburg, [South Africa, Transvaal Prov.], Zoutpansberg District, Woodbush, 3 AUG 1911, leg. E.M. Doidge, ex PREM No. 1775, ILL 6182. **Holotype:** PREM [this is the single collection cited with the Latin diagnosis on p. 747]. **Paratype:** Natal Prov., Buccleuch, 18 MAR 1915, leg. I.B. Pole Evans, ex PREM No. 8894, ILL 6183.

Meliola rizalensis H. & P. Sydow. Annales Mycologici 12:551–552. 1914. **Isotypes:** Hab. in foliis *Viticis parviflorae* [= *Vitex parviflora* A.L. Jussieu], Philippines, [Luzon], Rizal Prov., Antipolo, 6 JAN 1914, leg. M. Ramos, ex Phil. Bur. Sci. No. 294, ILL 6178; also ex H. Sydow, Fungi Exotici Exsiccati No. 379. ILL 6176 [as microscopic preparations], and in bound exsiccati set.

Meliola rizalensis H. & P. Sydow var. *panamensis* F.L. Stevens. Annales Mycologici 26:250, pl. V, fig. 58. 1928. **Holotype:** On *Cissus rhombifolia* Vahl, Panama, Fort Randolph, 100 ft. hill trail, 9 SEP 1923, leg. F.L. Stevens No. 761, ILL 6179. ≡ *Meliola cissi-rhombifoliae* C.G. Hansford. Beihefte zur Sydowia Annales Mycologici 1:104. 1957, nom. and stat. nov.

Meliola robinsonii H. Sydow. Philippine Journal of Science 21:135. 1922. **Isolectotype:** On leaves of *Entada phaseoloides* (L.) Merrill, Indonesia, [Moluccas], Amboina, Soja, 2 AUG 1913, leg. C.G. Robinson, Reliquiae Robinsonianae No. 2119, ILL 4727. **Assumed lectotype:** S [citation of this institution for location of the "type" by C.G. Hansford (1961:260-261) is interpreted as lectotypification]. ≡ *Meliola bicornis* H.G. Winter var. *robinsonii* (H. Sydow) F.L. Stevens. Annales Mycologici 26:189. 1928.

Meliola rockstonensis C.G. Hansford. Beihefte zur Sydowia Annales Mycologici 1:115. 1957. **Holotype:** Hab. in foliis Bignoniacearum spec. indet. [= Bignoniaceae sp.], British Guiana [Guyana], Rockstone, 13 JUL 1922, leg. F.L. Stevens No. 249 p.p., ILL 4762 [also the holotype of *Meliola bignoniacearum* F.L. Stevens var. *irregularis* C.G. Hansford and of *Meliola bignoniacearum* F.L. Stevens var. *tenuis* C.G. Hansford].

Meliola roureae H. & P. Sydow. Annales Mycologici 15:191–192. 1917. **Isotype:** Hab. in foliis *Roureae erectae* [= *Rourea erecta* Merrill], Philippines, Luzon, Rizal Prov., [24] NOV 1915, leg. M. Ramos, ex Phil. Bur. Sci. No. 23926, ILL 6191. **Paratype:** Bataan Prov., DEC 1915, leg. M. Ramos, ex Phil. Bur. Sci. No. 23994, ILL 6189.

Meliola roureae H.S. Yates. Philippine Journal of Science, Section C (Botany) 13:370. (1918) 1919, nom. illegit., non H. & P. Sydow, 1917. **Isotype:** On leaves of *Rourea erecta* Merrill, [Philippines], Luzon, Ilocos Norte Prov., Bangui, FEB–MAR 1917, leg. M. Ramos, ex Phil. Bur. Sci. No. 27724, ILL 6192.

Meliola rubi F.L. Stevens & E.F. Roldan ex C.G. Hansford. Sydowia Annales Mycologici 16:315. (1962) 1963; originally published by F.L. Stevens & E.F. Roldan in Philippine Journal of Science 56:63–64, fig. 2i. 1935, as nom. invalid. sine diagn. lat. **Holotype:** On *Rubus moluccanus* L., Philippines, Luzon, Benguet, Naguilian Road, 6 JAN 1931, leg. F.L. Stevens No. 1469, ILL 6193. **Isotypes:** CALP, CUP, Phil. Bur. Sci.

Meliola rubiella C.G. Hansford. Beihefte zur Sydowia Annales Mycologici 1:115–116. 1957. **Assumed holotype:** Hab. in foliis *Rubi moluccani* [= *Rubus moluccanus* L.], Philippines, [Luzon, Benguet, Naguilian Road, 6 JAN 1931], leg. F.L. Stevens No. 1461, ILL 4233 p.p. [also the holotype of *Irenina rubi* F.L. Stevens & E.F. Roldan (nom. invalid.) var. *angulata* F.L. Stevens & E.F. Roldan, nom. invalid. sine diagn. lat.]. **Paratype:** [Luzon, Benguet, Mt. St. Thomas, 31 DEC 1930], leg. F.L. Stevens No. 1361, ILL 4226 p.p. [also a paratype of *Irenina rubi* F.L. Stevens & E.F. Roldan, nom. invalid. sine diagn. lat.].

Meliola rudolphiae F.L. Stevens. Illinois Biological Monographs 2:511–512, fig. 41. 1916. **Holotype:** On *Rudolphia volubilis* Willd. [= *Neorudolphia volubilis* (Willd.) Britton], Porto Rico [Puerto Rico], Maricao, Monte Alegrillo, 14 NOV 1913, leg. F.L. Stevens No. 4791, ILL 6196. **Paratypes:** Maricao, 18 NOV 1913, leg. F.L. Stevens No. 4835, ILL 6194; Luquillo Forest, 2 DEC 1913, leg. F.L. Stevens No. 5439, ILL 6200; El Alto de la Bandera, 10 JUL 1915, leg. F.L. Stevens No. 8698, ILL 6198; Aibonito, 16 JUL 1915, leg. F.L. Stevens No. 8467, ILL 6197.

Meliola sacchari H. & P. Sydow. Annales Mycologici 12:548–549. 1914. **Isotypes:** Hab. in foliis *Sacchari spontanei* [= *Saccharum spontaneum*

L.], Philippines, Luzon, Ifugao Subprov., FEB 1913, leg. R.C. MacGregor, ex Phil. Bur. Sci. No. 20051, ILL 6205 and 6206.

Meliola sakawensis P.C. Hennings. Hedwigia 43:141. 1904. **Isotype**: Auf Blättern von *Clerodendrum trichotomum* Thunb. [as *Clerodendron*], Japan, Tosa, Sakawa-machi, AUG 1901, leg. T. Yoshinago No. 76, ILL 6213.

Meliola sakawensis P.C. Hennings var. *longispora* M. Beeli. Bulletin du Jardin Botanique de L'État à Bruxelles 7:98. 1920. **Isotype**: Épiphylle sur les feuilles de *Clerodendrum* [as *Clerodendron*] (Verbenaceae), Africa, Congo [Zaire], Wombali, 1913, leg. H.J.R. Vanderyst No. 2065, ILL 6211. **Holotype**: BR.

Meliola samarensis H.S. Yates. Philippine Journal of Science, Section C (Botany) 12:368. (1917) 1918. **Isotype**: On the petioles of an unknown host [identified as *Lepisanthes* sp. (cf. C.G. Hansford, 1961)], Philippines, Samar [Island], Catubig River, MAR 1916, leg. M. Ramos, ex Phil. Bur. Sci. No. 24919, ILL 6219.

Meliola sandwicensis J.B. Ellis & B.M. Everhart var. *major* C.G. Hansford ex C.G. Hansford. Sydowia Annales Mycologici 16:308–309. (1962) 1963; originally published in Beihefte zur Sydowia Annales Mycologici 2:589. 1961, as nom. invalid. sine diagn. lat. **Holotype**: On *Gouldia coriacea* (Hooker & Arnott) Hillebr. [= *Hedyotis terminalis* (Hooker & Arnott) W.L. Wagner & Herbst], Hawaii, [Kauai, Pipe trail, Waimea Canyon, 15 JUN 1921], leg. F.L. Stevens No. 446, ILL 6248. **Isotypes**: BISH 146031 and 486218.

Meliola sanguinea J.B. Ellis & B.M. Everhart. Journal of Mycology 2:42. 1886. **Isotypes**: On leaves, stems, and petioles of *Rubus trivialis* Michaux, Louisiana, Plaquemines Parish, Pointe a la Hache, [5] JAN 1886, leg. A.B. Langlois No. 74, ILL 4239, K. **Holotype**: NY. ≡ *Irenina sanguinea* (J.B. Ellis & B.M. Everhart) F.L. Stevens. Annales Mycologici 25:448. 1927.

Meliola sapindacearum C.L. Spegazzini. Revista Argentina de Historia Natural 1:405–406. 1891; Fungi Guaranitici Nonnulli Novi vel Critici, p. 29, No. 79. 1891. **Isotypes**: Hab. ad folia viva Sapindaceae cujusdam in sylva, [Argentina], Caáguazú, JAN 1882, leg. B. Balansa No. 3600 [comm. C.L. Spegazzini No. 559], ILL 6220 and 6223. **Holotype**: LPS.

Meliola sapindi F.L. Stevens. Annales Mycologici 26:199, pl. III, fig. 21. 1928. **Holotype**: On leaves and petioles of *Sapindus saponaria* L., Panama, Culebra, 2 OCT 1924, leg. F.L. Stevens No. 932, ILL 6224.

Meliola sauropicola H.S. Yates. Philippine Journal of Science, Section C (Botany) 12:368–369. (1917) 1918. **Isotype**: On leaves of *Sauropus* sp., Philippines, Samar, Catubig River, 15 FEB 1916, leg. M. Ramos, ex Phil. Bur. Sci. No. 24705, ILL 6225.

Meliola scabra E.M. Doidge. Transactions of the Royal Society of South Africa 7:194-195, fig. 2. 1919. **Isolectotype:** Hab. in foliis *Trichocladi criniti* [= *Trichocladus crinitus* (Thunb.) Persoon, South Africa, Cape Prov.], Izelini [Forest], Kingwilliamstown District, 8 JUN 1915 [as 14 JUN on the typed packet label], leg. Emmett, ex PREM No. 9064, ILL 6400 [also an isotype of *Meliola torta* E.M. Doidge, 1917, char. emend., 1919, of *Perisporina meliolicola* E.M. Doidge, and of *Meliola neotorta* S.J. Hughes, nom. nov. (illegit.?)]. **Lectotype:** PREM [No. 9064 is the only collection cited by Doidge (1920) for her comb. nov. under *Irene*. We accept this as lectotypification.]. ≡ *Irene scabra* (E.M. Doidge) E.M. Doidge. South African Journal of Natural History 2:40. 1920. ≡ *Irenina scabra* (E.M. Doidge) F.L. Stevens. Annales Mycologici 25:464. 1927. ≡ *Asteridiella scabra* (E.M. Doidge) C.G. Hansford. Beihefte zur Sydowia Annales Mycologici 2:302. 1961.

Meliola scabriseta C.G. Hansford & F.C. Deighton var. *calopogonii* (F.L. Stevens) C.G. Hansford, 1961. **See basionym:** *Meliola bicornis* H.G. Winter var. *calopogonii* F.L. Stevens, 1916.

Meliola scaevolae H. & P. Sydow. Annales Mycologici 12:551. 1914. **Isotype:** Hab. in foliis *Scaevolae frutescentis* [misspelled "*fructescentis*" = *Scaevola frutescens* (P. Miller) Krause], Philippines, Luzon, Tayabas Prov., Bales, 26 JUN 1913, leg. L. Escritor, ex Phil. Bur. Sci. No. 21212a, ILL 6226.

Meliola scaevolicola (F.L. Stevens) C.G. Hansford, 1961. **See basionym:** *Irene scaevolicola* F.L. Stevens, 1925.

Meliola schizolobii H. & P. Sydow. Annales Mycologici 14:76. 1916. **Lectotype** [designated herein]: In foliis *Schizolobii excelsi* [= *Schizolobium excelsum* Vogel], Brasilia [Brazil], Seringal São Francisco, Rio Acre, JUN 1911, leg. E.H.G. Ule No. 3495, ILL 6228. **Isolectotype:** S.

Meliola scolopiae E.M. Doidge in E.M. Doidge & H. Sydow. Bothalia 2(2):437–438, 467. 1928. **Paratype:** On *Scolopia zeyheri* (Nees) Harvey, South Africa, Cape Prov., Mossel Bay, 22 JUL 1915, leg. I.B. Pole Evans, ex PREM No. 9067 [cited in the English language description on p. 438], ILL 5107 [also an isolectotype of *Meliola evansii* E.M. Doidge].

Meliola semecarpi H. Sydow. Annales Mycologici 21:95. 1923. **Isolectotype:** Hab. in foliis *Semecarpi* sp. [= *Semecarpus* sp.], Philippines, Palawan [Island], Taytay, 7 APR 1913, leg. E.D. Merrill No. 8753, ILL 6233. **Assumed lectotype:** S [citation of this institution for location of the "type" by C.G. Hansford (1961: 472) interpreted as lectotypification].

Meliola semecarpicola C.G. Hansford. Sydowia Annales Mycologici 11:58–59. (1957) 1958. **Holotype:** Hab. in foliis *Semecarpi* sp. [= *Semecarpus* sp.], Philippines, Luzon, [Laguna Prov.], Mount Maquiling, [18 JAN 1931], leg. F.L. Stevens No. 1921, ILL 6523.

Meliola sepulta N.T. Patouillard ex F.L. Stevens. Illinois Biological Monographs 2:482–483. 1916. **Isotype**: On *Avicennia nitida* Jacquin [= *A. germinans* (L.) L.], Porto Rico [Puerto Rico], 31 JAN 1899 [erroneously cited as 1889], leg. A.A. Heller No. 390, ILL 4243. **Holotype**: NY. **Paratypes**: 17 JAN 1903, leg. A.A. Heller No. 6416, ILL 4242, NY, K. ≡ *Irene sepulta* (N.T. Patouillard ex F.L. Stevens) R.A. Toro. Mycologia 17:139. 1925. ≡ *Irenina sepulta* (N.T. Patouillard ex F.L. Stevens) F.L. Stevens. Annales Mycologici 25:450. 1927. ≡ *Asteridiella sepulta* (N.T. Patouillard ex F.L. Stevens) C.G. Hansford ex C.G. Hansford. Beihefte zur Sydowia Annales Mycologici 2:687. 1961.

Meliola serjaniae F.L. Stevens. Illinois Biological Monographs 2:512, fig. 42. 1916. **Holotype**: On *Serjania polyphylla* (L.) Radlk., Porto Rico [Puerto Rico], Vega Baja, 22 FEB 1913, leg. F.L. Stevens No. 425, ILL 6237. **Paratypes**: Florida Adentro, 1 AUG 1915, leg. F.L. Stevens No. 7654, ILL 6236; Arecibo-Lares Road, 21 JUN 1915, leg. F.L. Stevens No. 7219, ILL 6238; Cataño, 6 NOV 1913, leg. F.L. Stevens No. 4181, ILL 6239.

Meliola serjaniae F.L. Stevens var. *dentata* F.L. Stevens. Annales Mycologici 26:280. 1928, non *M. paulliniae* F.L. Stevens var. *dentata* F.L. Stevens, op. cit. p. 197. **Holotype**: On *Serjania triquetra* Radlk., Panama, Juan Diaz, 12 AUG 1923, leg. F.L. Stevens No. 1243, ILL 6235 [C.G. Hansford (1961) confused the types of the two varieties].

Meliola serjaniicola F.L. Stevens & L.R. Tehon. Mycologia 18:14, pl. 2, fig. 16. 1926. **Holotype**: On *Serjania paucidentata* DC., British Guiana [Guyana], Coverden, 8 AUG 1922, leg. F.L. Stevens No. 798, ILL 6240.

Meliola setulifera (C.L. Spegazzini) F.L. Stevens, 1928. **See basionym**: *Meliola perseae* F.L. Stevens forma *setulifera* C.L. Spegazzini, 1923.

Meliola shropshiriana F.L. Stevens. Annales Mycologici 26:243, pl. IV, fig. 45. 1928. **Holotype**: On Bignoniaceae sp. indet., Panama, Corozal, Trail No. 17, 30 AUG 1924, leg. F.L. Stevens No. 115, ILL 6270. **Paratypes**: Leg. F.L. Stevens No. 110, ILL 6268; Las Cruces trail, 2 SEP 1924, leg. F.L. Stevens No. 153, ILL 6267; leg. F.L. Stevens No. 169, ILL 6269.

Meliola sideroxyli F.L. Stevens. Bernice P. Bishop Museum Bulletin 19:35, text fig. 8a. 1925. **Holotype**: On *Sideroxylon sandwicense* (A. Gray) Bentham & J.D. Hooker ex Hillebr. [= *Pouteria sandwicensis* (A. Gray) Baehni & O. Degener], Hawaii, Kauai, Kokee, 28 AUG [1921], leg. O.H. Swezey, as F.L. Stevens No. 1160, ILL 6277 [cited as isolectotype by R.D. Goos & D.P. Gowing (1992), but no lectotype has been designated, as far as we could determine. **Isotypes**: BISH 499021 and 499958.

Meliola singaporensis C.G. Hansford. Proceedings of the Linnean Society of London 157:179. 1946. **Isotype:** Hab. in foliis *Eugeniae grandis* [*Eugenia grandis* Wight = *Syzygium grande* (Wight) Wight ex Walpers], Singapore [as "Island of Singapore, Straits Settlements" on the original exsiccati label, OCT 1917], leg. C.F. Baker, Fungi Malayana No. 457, ILL 3917 [as a microscopic preparation]. **Holotype:** PREM.

Meliola sinuosa E.M. Doidge. Transactions of the Royal Society of South Africa 5:735, 746, pl. LXIV, fig. 33. 1917. **Isotype:** On leaves of *Trichilia emetica* Vahl, [South Africa, Transvaal Prov.], Zoutpansberg District, Spelonken, Lemana, 14 AUG 1911, leg. E.M. Doidge, ex PREM No. 1783, ILL 6280. **Holotype:** PREM [this is the single collection cited with the Latin diagnosis on p. 746].

Meliola smilacis F.L. Stevens. Illinois Biological Monographs 2:524. 1916. **Holotype:** On *Smilax coriacea* Sprengel [= *S. hawanensis* Jacquin], Porto Rico [Puerto Rico], Manati, 25 NOV 1913, leg. F.L. Stevens No. 5261, ILL 6283 [also a paratype of *Isthmospora spinosa* F.L. Stevens]. **Paratype:** On *Smilax* sp. [as *S. coriacea* Sprengel on the packet = *S. hawanensis* Jacquin], Jajome Alto, 17 JUL 1915, leg. F.L. Stevens No. 8429, ILL 6282.

Meliola solani F.L. Stevens. Illinois Biological Monographs 2:483, fig. 7. 1916. **Holotype:** On *Solanum jamaicense* P. Miller, Porto Rico [Puerto Rico], Monte de Oro, 3 DEC 1913, leg. F.L. Stevens No. 5750, ILL 4403. ≡ *Irene solani* (F.L. Stevens) R.A. Toro. Mycologia 19:73. 1927. ≡ *Irenopsis solani* (F.L. Stevens) F.L. Stevens. Annales Mycologici 25:439. 1927.

Meliola sororcula C.L. Spegazzini. Boletín de la Academia Nacional de Ciencias en Córdoba 11:493. 1889; Fungi Puiggariani, Pugillus I, p. 115, No. 230. Anno 1889. **Isotype:** Hab. ad folia viva *Baccharidis pingraeae* [= *Baccharis pingraea* DC.] in dumetis, [Brazil], pr. Apiahy [Apiai], MAY 1888 [as May 1886 on the retyped packet label], leg. J.I. Puiggari No. 2774, ex C.L. Spegazzini herb., ILL 3778. ≡ *Irene sororcula* (C.L. Spegazzini) F.L. Stevens. Annales Mycologici 25:423. 1927. ≡ *Appendiculella sororcula* (C.L. Spegazzini) C.G. Hansford. Beihefte zur Sydowia Annales Mycologici 2:615–616. 1961.

Meliola speciosa E.M. Doidge. Transactions of the Royal Society of South Africa 5:726, 744, pl. LX, fig. 15. 1917. **Isotype:** On leaves of *Gymnosporia* sp. [= *Maytenus* sp., South Africa, Transvaal Prov.], Zoutpansberg District, Woodbush, 2 AUG 1911 [as 2 MAY on the packet], leg. E.M. Doidge, ex PREM No. 1740, ILL 3878. **Holotype:** PREM. ≡ *Irene speciosa* (E.M. Doidge) E.M. Doidge. South African Journal of Natural History 2:40. 1920. ≡ *Appendiculella speciosa* (E.M. Doidge) C.G. Hansford. Beihefte zur Sydowia Annales Mycologici 2:339. 1961.

Meliola spegazziniana H.G. Winter in C.L. Spegazzini. Anales de la Sociedad Cientifica Argentina 26:25–26. 1888; Fungi Guaranitici, Pugillus II, pp. 23–24, No. 64. 1888. **Isotypes**: Hab. ad folia Compositarum arborescentium [Asteraceae sp. indet.], in sylvis, [Paraguay], pr. Paraguari, 5 MAR 1883, leg. B. Balansa, comm. C.L. Spegazzini No. 3751, ILL 6284 and 6285 [the latter a microscopic preparation, ex C. Roumeguère, Fungi Selecti Exsiccati No. 5238]; also in bound exsiccati set at ILL.

Meliola stemonae H. Sydow. Philippine Journal of Science 21:134–135. 1922. **Isotype**: On leaves of *Stemona tuberosa* Lour., Indonesia, [Moluccas] Amboina, Hitoe, 8 OCT 1913, leg. C.B. Robinson, Reliquiae Robinsonianae No. 2230, ILL 6286.

Meliola stenotaphri F.L. Stevens. Illinois Biological Monographs 2:509, fig. 38. 1916. **Holotype**: On *Stenotaphrum secundatum* (Walter) O. Kuntze, Porto Rico [Puerto Rico], Manati, 5 NOV 1913, leg. F.L. Stevens No. 4304, ILL 6289. **Paratypes**: Rio Tanamá, near Arecibo, 7 JUL 1915, leg. F.L. Stevens No. 7940, ILL 6291; leg. F.L. Stevens No. 7852, ILL 6292; Dos Bocas, below Utuado, 8 JUL 1915, leg. F.L. Stevens No. 8023, ILL 6290; Arecibo, K 64.7, 8 JUL 1915, leg. F.L. Stevens No. 7810, ILL 6293.

Meliola stevensiana C.G. Hansford. Sydowia Annales Mycologici 9:76. 1955, nom. illegit., non R. Ciferri, 1954. **Isotype**: Hab. in foliis Bignoniacearum sp. indet. [= Bignoniaceae sp. indet.], Ecuador, Terecita, 29 OCT 1924, leg. F.L. Stevens No. 82 p.p., ILL 31726 p.p. [also, p.p., an isotype of *Meliola bignoniacearum* F.L. Stevens var. *parasitica* C.G. Hansford]. **Holotype**: FH. ≡ *Meliola thaxteri* C.G. Hansford, Beihefte zur Sydowia Annales Mycologici 2:666-667. 1961, nom. nov.

Meliola stevensii M. Beeli. Bulletin du Jardin Botanique de L'État à Bruxelles 7:98–99. 1920. **Isolectotype**: In foliis Sapindacearum [= Sapindaceae sp. indet.], C. Africa, Congo [Zaire], Wombali, 1913, leg. H.J.R. Vanderyst No. 2031, ILL 5283. **Assumed lectotype**: BR [citation of this institution for location of the "type" by C.G. Hansford (1961:451) is interpreted as lectotypification]. ≡ *Meliola integriseta* (C.L. Spegazzini) C.L. Spegazzini var. *stevensii* (M. Beeli) F.L. Stevens. Annales Mycologici 26:254. 1928.

Meliola straussiae C.G. Hansford. Beihefte zur Sydowia Annales Mycologici 1:117. 1957. **Holotype**: Hab. in foliis *Straussiae* sp. [*Straussia* = *Psychotria* sp.], Hawaii, [Kauai, Kalalau trail, 16 JUN 1921], leg. F.L. Stevens No. 483 p.p., ILL 3583. **Isotype**: ILL 3592.

Meliola strophanthi E.M. Doidge. Transactions of the Royal Society of South Africa 5:729, 745, pl. LXII, fig. 23. 1917. **Isotype**: On leaves of *Strophanthus speciosus* (Ward & Haworth) Reber, [South Africa,

Transvaal Prov.], Zoutpansberg District, Woodbush, [Helpmakaar], 3 AUG 1911, leg. E.M. Doidge, ex PREM No. 1781, ILL 4255. **Holotype**: PREM. ≡ *Irene strophanthi* (E.M. Doidge) E.M. Doidge. South African Journal of Natural History 2:41. 1920. ≡ *Irenina strophanthi* (E.M. Doidge) F.L. Stevens. Annales Mycologici 25:460. 1927. ≡ *Asteridiella strophanthi* (E.M. Doidge) C.G. Hansford ex C.G. Hansford. Beihefte zur Sydowia Annales Mycologici 2:541–542. 1961.

Meliola strychnicola A. Gaillard. Le Genre *Meliola*, p. 72, pl. 12, fig. 4a. 1892. **Isotype**: Ad paginam superiorem foliorum *Strychni* cujusdam [= *Strychnos* sp., C. Africa], Congo, Osika, in 1883, leg. J. de Brazza No. 137, ILL 6298. **Holotype**: PC.

Meliola strychni-multiflorae C.G. Hansford. Sydowia Annales Mycologici 11:59. (1957) 1958 [misspelled "*stychni-multiflorae*," corrected by Hansford (1961:526)]. **Holotype**: Hab. in foliis *Strychni multiflorae* [= *Strychnos multiflora* Bentham], Philippines, Luzon, Nueva Ecija Prov., San José, [Balete Road, 10 JAN 1931], leg. F.L. Stevens No. 1806, ILL 6300.

Meliola styracearum F.L. Stevens. Annales Mycologici 26:229, pl. IV, fig. 39. 1928. **Holotype**: On Styracaceae: *Styrax argenteus* C. Presl, Costa Rica, Cartago, 23 JUN 1923, leg. F.L. Stevens No. 105, ILL 6295. **Paratype**: Leg. F.L. Stevens No. 73, ILL 6294.

Meliola styracicola C.L. Spegazzini. Anales del Museo Nacional de Historia Natural de Buenos Aires 23:44–45. 1912. **Isotypes**: Ad folia viva *Styracis leprosae* [= *Styrax leprosa* Hooker & Arnott, Argentina], in sylvis, Misiones, pr. Puerto León, JUL 1909, leg. (?), comm. C.L. Spegazzini, ILL 6296 [as a microscopic preparation] and 6297 [the packet marked "part of the type collection"]. **Assumed holotype**: LPS 513. ≡ *Asteridiella styracicola* (C.L. Spegazzini) C.G. Hansford ex C.G. Hansford. Beihefte zur Sydowia Annales Mycologici 2:515–516. 1961 [Hansford erroneously cited what seems to be an unpublished name (*Styrax lanosa*) for the host; he cited the type collection as leg. "Venturi, in SPEG No. 513"].

Meliola subapoda H. & P. Sydow. Annales Mycologici 12:547–548. 1914. **Isotype**: Hab. in foliis *Malloti philippensis* [= *Mallotus philippensis* (Lam.) Muell. Arg.], Philippines, Luzon, Bulacan Prov., Angat, SEP 1913, leg. M. Ramos, ex Phil. Bur. Sci. No. 21824, ILL 4256. ≡ *Irenina subapoda* (H. & P. Sydow) F.L. Stevens. Annales Mycologici 25:466–467. 1927. ≡ *Asteridiella subapoda* (H. & P. Sydow) C.G. Hansford ex C.G. Hansford. Beihefte zur Sydowia Annales Mycologici 2:211. 1961.

Meliola subcrustacea C.L. Spegazzini. Boletín de la Academia Nacional de Ciencias en Córdoba 11:496–497. 1889; Fungi Puiggariani, Pugillus I, pp. 118–119, No. 236. 1889. **Isotype**: Hab. ad folia viva plantae ignotae cujusdam in sylvis, [Brazil], pr. Apiahy [Apiai], Hiem. 1888, leg.

J.I. Puiggari No. 2703 [comm. C.L. Spegazzini No. 527 (cf. C.G. Hansford, 1961)], ILL 3925. ≡ *Irene subcrustacea* (C.L. Spegazzini) F. Theissen & H. Sydow. Annales Mycologici 15:461. (1917) 1918. ≡ *Asteridiella subcrustacea* (C.L. Spegazzini) C.G. Hansford ex C.G. Hansford. Beihefte zur Sydowia Annales Mycologici 2:757–758. 1961.

Meliola substenospora F. v. Höhnel forma *rottboelliae* H. Rehm. Leaflets of Philippine Botany 6:2193. 1914, nom. nud. **Assumed isosyntypes:** Ad folia *Rottboelliae exaltatae* [*Rottboellia exaltata* (L.) L.f. = *R. cochinchinensis* (Lour.) Clayton], Philippines, Luzon, Laguna Prov., Los Baños, [Mount Maquiling], SEP 1913, leg. M.B. Raimundo, comm. C.F. Baker No. 1839 [ex Fungi Malayana No. 45], ILL 6306 and 6303 [the latter as a microscopic preparation].

Meliola sydowiana F.L. Stevens & R.H. Larson in F.L. Stevens. Annales Mycologici 26:281, pl. VI, fig. 77. 1928. **Isotype:** Distributed by Philippine Bureau of Science as *Meliola amphitricha* E.M. Fries, on *Sapindus saponaria* L., Philippines, Luzon, Laguna Prov., [NOV–DEC 1910], leg. R.C. McGregor, cited as Phil. Bur. Sci. No. 12499, ILL 6310.

Meliola symphoremae F.L. Stevens & E.F. Roldan. Philippine Journal of Science 56:61, fig. 2g. 1935, nom. invalid. sine diagn. lat. **Holotype:** On *Symphorema luzonicum* (Blanco) Fernandez-Villar in Blanco, Philippines, Luzon, Laguna Prov., Mount Maquiling, 20 SEP 1930, leg. F.L. Stevens No. 655, ILL 6311. **Isotype:** CUP. ≡ *Meliola symphorematis* F. Petrak var. *major* C.G. Hansford. Sydowia Annales Mycologici 16:314–315. (1962) 1963 [see next entry].

Meliola symphorematis F. Petrak var. *major* C.G. Hansford. Sydowia Annales Mycologici 16:314–315. (1962) 1963. **Based on** the type of *M. symphoremae* F.L. Stevens & E.F. Roldan, 1935, nom. invalid. [the latter name cited pro syn.]. **Holotype:** On *Symphorema luzonicum* (Blanco) Fernandez-Villar in Blanco, Philippines, Luzon, Laguna Prov., Mount Maquiling, 20 SEP 1930, leg. F.L. Stevens No. 655, ILL 6311. **Isotype:** CUP. [Note: *Meliola symphorematis* F. Petrak var. *symphorematis* is based on a different type.].

Meliola tabernaemontanae C.L. Spegazzini. Anales del Museo Nacional de Historia Natural de Buenos Aires 23:45. 1912; Mycetes Argentinenses, Series VI, p. 45, No. 1345. 1912 [as *tabernemontanae*, but correctly spelled on packet labels]. **Isotypes:** Hab. ad folia viva *Tabernaemontanae histricis* [=*Tabernaemontana hystrix* Steudel], in silvis, [Argentina], Misiones, pr. Bonpland, NOV 1909, leg. C.L. Spegazzini No. 536 [cf. C.G. Hansford, 1961; no number on the packets], ILL 6315 and 6320. **Holotype:** LPS.

Meliola tabernaemontanae C.L. Spegazzini var. *escharoides* (H. Sydow) C.G. Hansford, 1961. **See basionym:** *Irene escharoides* H. Sydow, 1926.

Meliola tabernaemontanae C.L. Spegazzini var. *forsteroniae* F.L. Stevens. Illinois Biological Monographs 2:518. 1916. **Holotype**: On *Forsteronia corymbosa* sensu auct. non (Jacquin) G. Meyer [= *F. portoricensis* Woodson], Porto Rico [Puerto Rico], Utuado, leg. F.L. Stevens No. 4682, ILL 6316]. ≡ *Meliola forsteroniae* (F.L. Stevens) C.G. Hansford. Proceedings of the Linnean Society of London 160:129. (1948) 1949.

Meliola tabernaemontanicola C.G. Hansford & M.J. Thirumalachar var. *luzonensis* C.G. Hansford. Sydowia Annales Mycologici 11:59. (1957) 1958 [the specific epithet incorrectly cited as *tabernaemonticola* but later corrected by the author (C.G. Hansford, 1961:561). **Holotype**: Hab. in foliis *Tabernaemontanae* sp. [= *Tabernaemontana* sp.], Philippines, Luzon, Nueva Ecija, [Prov., Muñoz, 3 OCT 1930], leg. F.L. Stevens No. 788, ILL 6312. **Paratype**: [San José to Balete Road, 9 JAN 1931], leg. F.L. Stevens No. 1689, ILL 6313.

Meliola tamarindi H. & P. Sydow. Annales Mycologici 10:79–80. 1912. **Isotype**: Hab in foliis *Tamarindi indici* [= *Tamarindus indica* L.], Philippines, [Luzon], Manila and vicinity, NOV–DEC 1910, leg. E.D. Merrill No. 7416, ILL 6335.

Meliola tapirirae F.L. Stevens & L.R. Tehon. Mycologia 18:13, pl. 2, fig. 14. 1926. **Lectotype**: On *Tapirira* sp., [British Guiana (Guyana), R.R., 15 JUL 1922], leg. F.L. Stevens No. 330, ILL 6337 [no type is indicated in the protologue but both F.L. Stevens (1928:182) and C.G. Hansford (1961:458) selected No. 330. We accept this as lectotypification.]. **Residual syntype**: ... leg. F.L. Stevens No. 338, ILL 6336.

Meliola tapiriricola F.L. Stevens & L.R. Tehon. Mycologia 18:13–14, pl. 2, fig. 15. 1926. **Holotype**: On *Tapirira guianensis* Aublet, British Guiana [Guyana], Wismar, 14 JUL 1922, leg. F.L. Stevens No. 283, ILL 6338.

Meliola tayabensis H.S. Yates. Philippine Journal of Science, Section C (Botany) 12:369. (1917) 1918. **Isotype**: On leaves of *Linociera* sp. [= *Chionanthus* sp.], Philippines, Luzon, Tayabas Prov., Basiad, DEC 1916, leg. H.S. Yates, ex Phil. Bur. Sci. No. 25649, ILL 6339.

Meliola tecomae F.L. Stevens. Illinois Biological Monographs 2:521–522, fig. 48. 1916. **Holotype**: On *Tecoma pentaphylla* (L.) Jussieu [= *Tabebuia heterophylla* (DC.) Britton], Porto Rico [Puerto Rico], Martin Peña, 11 AUG 1915, leg. F.L. Stevens No. 9332, ILL 6345. **Isotype**: ILL 6344. **Paratypes**: Las Marias, 10 JUL 1915, leg. F.L. Stevens No. 8177, ILL 6353; Maricao, 20 JUL 1915, leg. F.L. Stevens No. 8960, ILL 6357; Mayagüez, 24 JUN 1915, leg. F.L. Stevens No. 7396, ILL 6342; 15 JUN 1915 [erroneously as 24 JUN in the protologue], leg. F.L. Stevens No. 7078, ILL 6340; on *Tecoma* sp. [= *Tabebuia* sp.], El Miradoro, 4 AUG 1915, leg. F.L. Stevens No. 9163, ILL 6356; Mayagüez, 31 OCT 1913, leg. F.L. Stevens No. 3950, ILL 6351 and 6359; Maricao, 18 NOV 1913, leg. F.L. Stevens No. 4804, ILL 6352 and 6361; Quebradillas, 22 NOV 1913, leg. F.L. Stevens No. 4978, ILL 6350; leg. F.L. Stevens No. 4981, ILL 6346 and 6360; Vega Baja, 5 NOV 1913, leg. F.L. Stevens No. 4310a, ILL 6347; Arecibo-Lares Road, 21 JAN 1914, leg. F.L. Stevens No. 6790, ILL 6355 and 6358; Las Marias, 22 MAR 1913, leg. F.L. Stevens No. 3593, ILL 6354.

Meliola tehoniana A. Trotter. Sylloge Fungorum 24:276. 1926, nom. nov. **Based on**: *Meliola conferta* L.R. Tehon, 1919 (nom. illegit.), non E.M. Doidge, 1917. **Holotype**: On leaves of *Rhacoma crossopetalum* L. [= *Crossopetalum rhacoma* Crantz], Porto Rico [Puerto Rico], Mona Island, 20 DEC 1913, leg. F.L. Stevens No. 6147, ILL 4335. ≡ *Irenopsis tehoniana* (A. Trotter) C.G. Hansford. Beihefte zur Sydowia Annales Mycologici 2:339–340. 1961.

Meliola tenuis M.J. Berkeley & M.C. Cooke ex P.A. Saccardo. Sylloge Fungorum 1:762. 1882; originally published sine diagnosis by M.J. Berkeley & M.C. Cooke in Grevillea 7:49. 1878. **Isotypes**: Hab. in foliis *Arundinariae* [= *Arundinaria* sp., Georgia, Darien], leg. H.W. Ravenel No. 2482, ex Fungi Americani Exsiccati No. 331, ILL 6367 [as a microscopic preparation], and in bound exsiccati set.

Meliola tenuissima F.L. Stevens. Illinois Biological Monographs 2:492. 1916. **Holotype**: On *Gouania lupuloides* (L.) Urban, Porto Rico [Puerto Rico], Yauco, 3 OCT 1913, leg. F.L. Stevens No. 3142, ILL 4434. **Paratype**: Villa Alba, 3 JAN 1913, leg. F.L. Stevens No. 96, ILL 4435. ≡ *Irenopsis tenuissima* (F.L. Stevens) F.L. Stevens. Annales Mycologici 25:439. 1927.

Meliola teramni (P.A. Saccardo) H. & P. Sydow. Annales Mycologici 15:193. 1917. **See basionym**: *Meliola nigro-rufescens* P.A. Saccardo var. *teramni* P.A. Saccardo, 1917.

Meliola teramniae H.S. Yates. Philippine Journal of Science, Section C (Botany) 12:369. (1917) 1918, nom. illegit., non *Meliola teramni* (P.A. Saccardo) H. & P. Sydow, 1917. **Isotype**: On leaves of *Teramnus labialis* (L.f.) Sprengel, Philippines, Luzon, Kalinga Subprov., 27 MAR 1916, leg. H.S. Yates, ex Phil. Bur. Sci. No. 25344, ILL 6370.

Meliola terecitensis C.G. Hansford. Beihefte zur Sydowia Annales Mycologici 1:117–118. 1957. **Holotype**: Hab. in foliis Sapindacearum spec. [= Sapindaceae sp. indet.], Ecuador, Terecita, 29 OCT 1924, leg. F.L. Stevens No. 81 p.p., ILL 4078.

Meliola thaxteri C.G. Hansford. Beihefte zur Sydowia Annales Mycologici 2:666–667. 1961, nom. nov. **Based on**: *Meliola stevensiana* C.G. Hansford, 1955 (nom. illegit.), non R. Ciferri, 1954. **Isotype**: Hab. in foliis Bignoniacearum sp. indet. [= Bignoniaceae sp. indet.], Ecuador,

Terecita, 29 OCT 1924, leg. F.L. Stevens No. 82 p.p., ILL 31726 p.p. **Holotype:** FH.

Meliola theacearum F.L. Stevens. Memoirs of the Department of Agriculture in India. Botanical Series 15(5):107, pl. I, figs. 1–4. JAN 1928; Annales Mycologici 26:207. MAY 1928. **Isolectotypes:** On Theaceae: *Schima* sp., [Malay Peninsula, erroneously as India in the protologue]: Penang, Government Hill, JUL 1918, leg. E.J. Butler, as F.L. Stevens No. 1982, ILL 6373, HCIO. **Assumed lectotype:** IMI 25720 [citation of this specimen as "type" by C.G. Hansford (1961:128) is interpreted as lectotypification].

Meliola themedae F.L. Stevens & E.F. Roldan ex C.G. Hansford. Sydowia Annales Mycologici 16:314. (1962) 1963; originally published by F.L. Stevens & E.F. Roldan in Philippine Journal of Science 56:59, fig. 2e. 1935, as nom. invalid. sine diagn. lat. **Holotype:** On *Themeda gigantea* (Cav.) E. Hackel in A. DC., Philippines, Luzon, Nueva Ecija Prov., Muñoz, 3 OCT 1930, leg. F.L. Stevens No. 794, ILL 6374.

Meliola thomasiana P.A. Saccardo. Boletim da Sociedade Broteriana 21:212–213. (1905) 1906. **Isotype:** In foliis caulibusque vivus *Elatostematis angolensis* [nomen? = *Elatostema* sp., W. Africa], São Tomé, alt. 135 m, 1885, leg. A. Møller s.n., ILL 6375.

Meliola thouiniae F.S. Earle. Bulletin of the New York Botanical Garden 3:308–309. Preprint 1904 (the journal issued in 1905). **Isotype:** On leaves of *Thouinia striata* Radlk. [misspelled "*stiata*"], Porto Rico [Puerto Rico, limestone hills near Bayamon, 21 JAN 1903], leg. A.A. Heller No. 6435, ILL 6404.

Meliola thuemeniana F.L. Stevens. Annales Mycologici 26:259. 1928, nom. nov. **Based on:** *Meliola microthecia* sensu A. Gaillard, 1892, non F. v. Thümen, 1876. **Isotypes:** In foliis vivis *Barosmae scopariae* [*Barosma scoparia* Ecklon & Zeyher = *Agathosma ovata* (Thunb.) Pillans], South Africa, Cape of Good Hope, [Grahamstown, JUL 1876], leg. P. MacOwan No. 1260, ex F. v. Thümen, Mycotheca Universalis No. 851, ILL 6471 [as a microscopic preparation], and in bound exsiccati set [distributed as *Meliola microthecia* F. v. Thümen but published by A. Gaillard, in Le Genre *Meliola*, p. 73, 1892, with a description significantly different from the original diagnosis of F. v. Thümen in Flora 1876:569. 1876. Stevens (1928:295) regarded the spores observed by Gaillard as those of a different species on the same host and collection. Hansford (1961), however, placed this name in synonymy under *Meliola microthecia* F. v. Thümen].

Meliola toddaliae E.M. Doidge. Transactions of the Royal Society of South Africa 5:732, 746, pl. LXIII, fig. 28. 1917. **Isotype:** On *Toddalia lanceolata* Lam. [= *Vepris lanceolata* (Lam.) G. Don, South Africa, Cape

Prov.], Kentani, 16 DEC 1914, leg. A. Pegler, ex PREM No. 8788, ILL 6402. **Holotype:** PREM [the single specimen cited with the Latin diagnosis on p. 746]. **Paratypes:** [1914], leg. A. Pegler No. 1960A, ILL 6401; Natal Prov., Henley, near Pietermaritzburg, 24 MAY 1915, leg. E.M. Doidge, ex PREM No. 8999, ILL 6403.

Meliola tomentosa H.G. Winter var. *calva* H. Rehm. Annales Mycologici 5:209. 1907. **Isotype:** An der Unterfläche der Blätter von *Styrax*, Brasilien [Brazil], Rio Grande do Sul, São Leopoldo, 1906, leg. J. Rick, ex H. Rehm, Ascomycetes Exsiccati No. 1707, ILL 4259 [only two microscopic preparations remain; the original specimen was obtained from B (comm. H. Sydow)]. ≡ *Irenina aberrans* F.L. Stevens. Annales Mycologici 25:462. 1927, nom. and stat. nov. [Stevens applied the term "n. sp." but evidently intended it as a new name; the epithet "*calva*" was preoccupied at the rank of species under both *Meliola* and *Irenina*]. ≡ *Asteridiella aberrans* (F.L. Stevens) C.G. Hansford. Beihefte zur Sydowia Annales Mycologici 2:516. 1961.

Meliola tonkinensis P.A. Karsten & C. Roumeguère. Revue Mycologique 12: 77–78. 1890. **Isotypes:** Hab. ad folia *Fici* [= *Ficus* sp., Tonkin (Vietnam)], Tu-Phap, JAN 1889, leg. B. Balansa No. 25, ex C. Roumeguère, Fungi Selecti Exsiccati No. 5944, ILL 3880 [leaf fragment and several microscopic preparations], and in bound exsiccati set. ≡ *Appendiculella tonkinensis* (P.A. Karsten & C. Roumeguère) R.A. Toro. Mycologia 19:71–72. 1927. ≡ *Irene tonkinensis* (P.A. Karsten & C. Roumeguère) F.L. Stevens. Annales Mycologici 25:427. 1927.

Meliola torta E.M. Doidge. Transactions of the Royal Society of South Africa 5:726, 744, pl. LX, fig. 16. 1917, char. emend., Transactions of the Royal Society of South Africa 7:193–194. 1919. **Isotype:** On *Trichocladus crinitus* (Thunb.) Persoon, [South Africa, Cape Prov.], Kingwilliamstown District, Izelini Forest, 8 JUN 1915 [as 14 JUN on the typed packet label], leg. Forester Emmett, ex PREM No. 9064, ILL 6400 [also an isotype of *Perisporina meliolicola* E.M. Doidge and an isolectotype of *Meliola scabra* E.M. Doidge, as well as an isotype of *Meliola neotorta* S.J. Hughes, nom. nov. (illegit.?)]. **Holotype:** PREM. ≡ *Leptomeliola torta* (E.M. Doidge) S.J. Hughes. Mycological Papers. International Mycological Institute No. 166:210. 1993 (illegit.?). [This name is based only on *M. torta* E.M. Doidge, 1917, not on the emended description of 1919.]. ≡ *Meliola neotorta* S.J. Hughes. Mycological Papers. International Mycological Institute No. 166:210. 1993, nom. nov. (illegit.?). [Hughes published this name as a substitute for *Meliola torta* E.M. Doidge, char. emend., 1919, non E.M. Doidge, 1917. If, however, the name *M. torta* is retained for the emended circumscription, as intended by Doidge and valid in our interpretation of the rules of nomenclature (the original type is not excluded), then the new name is superfluous (cf. Rec. 47A and Articles 47 & 52 of I.C.B.N.). Under this interpretation, *Leptomeliola torta* is to be treated as a new species, with priority starting in 1993, and it is in conflict with Articles 11.3 & 11.4 of I.C.B.N. The name *Perisporina meliolicola* E.M. Doidge, op. cit. 1919, placed by Hughes in synonymy under *Leptomeliola torta*, has priority as an available basionym for transfer of the taxon to *Leptomeliola*.].

Meliola tortuosa H.G. Winter in A. Gaillard. Le Genre *Meliola*, p. 67, pl. 21, fig. 2. 1892. **Residual isosyntype:** In foliis *Piperis* cujusdam [= *Piper* sp.], Brasilia [Brazil], Santa Catharina Prov., São Francisco, [JUL 1884], leg. E.H.G. Ule No. 202, ILL 4454, ex herb. H. Sydow. **Residual syntype:** S. **Assumed Lectotype:** E.H.G. Ule No. 1501, S [citation of this collection and institution for "type" by C.G. Hansford (1961:69) is interpreted as lectotypification]. ≡ *Irenopsis tortuosa* (H.G. Winter in A. Gaillard) F.L. Stevens. Annales Mycologici 25:439. 1927.

Meliola toruloidea F.L. Stevens. Illinois Biological Monographs 2:493, fig. 21. 1916. **Holotype:** On *Cassia quinquangulata* sensu auct. non L.C. Richard [misspelled "*quinquadrangulata*" = *Senna* cf. *nitida* (L.C. Richard) Irwin & Barneby], Porto Rico [Puerto Rico], Jajome Alto, 17 JUL 1915, leg. F.L. Stevens No. 8394, ILL 4491 [also a paratype of *Helminthosporium glabroides* F.L. Stevens]. **Paratypes:** Maricao, 10 JAN 1913, leg. F.L. Stevens No. 206, ILL 4496 and 4499; 20 JUL 1915, leg. F.L. Stevens No. 8980, ILL 4494; Aibonito, 5 NOV 1913, leg. F.L. Stevens No. 4015, ILL 4492; 16 JUL 1915, leg. F.L. Stevens No. 8468, ILL 4493; on *Inga laurina* (Sw.) Willd., Las Marias, 10 JUL 1915, leg. F.L. Stevens No. 8135, ILL 4495. ≡ *Irene toruloidea* (F.L. Stevens) F.L. Stevens & L.R. Tehon. Mycologia 18:18. 1926. ≡ *Irenopsis toruloidea* (F.L. Stevens) F.L. Stevens. Annales Mycologici 25:441. 1927.

Meliola tounateae F.L. Stevens. Annales Mycologici 26:204, pl. III, fig. 24. 1928. **Holotype:** On Leguminosae: *Tounatea* sp. [= *Swartzia* sp. (Fabaceae)], Panama, Baille Mona, 20 SEP 1924, leg. F.L. Stevens No. 675, ILL 6399.

Meliola trachelospermi H.S. Yates. Philippine Journal of Science, Section C (Botany) 13:370. (1918) 1919 [as *trachelospermae*, corrected in accordance with Articles 32.6 & 61 of I.C.B.N.]. **Isolectotype:** On leaves of *Trachelospermum* sp., [Philippines], Luzon, Rizal Prov., Mount Lumutan, 3 SEP [as JUL 1917 on the packet], leg. M. Ramos & G. Edaño, ex Phil. Bur. Sci. No. 29813, ILL 6398. **Assumed lectotype:** S [citation of this institution for location of "type" by C.G. Hansford (1961:562) is interpreted as lectotypification with the specimen at S].

Meliola trematis C.L. Spegazzini. Anales del Museo Nacional de Historia

Natural de Buenos Aires 23:45–46. 1912; Mycetes Argentinenses, Series 6, pp. 45–46, No. 1346. 1912 [as *tremae*, corrected by F.C. Deighton (1968) in accordance with current I.C.B.N.]. **Isotype**: Hab. ad folia viva *Tremae micranthae* [= *Trema micranthum* (L.) Blume] in silvis, [Argentina], Misiones, pr. Puerto León, JUL 1909, leg. A.L. Venturi No. 28, comm. C.L. Spegazzini [No. 1810 (cf. C.G. Hansford, 1961)], ILL 4265. **Holotype**: LPS. ≡ *Irenina trematis* (C.L. Spegazzini) F.L. Stevens. Annales Mycologici 25:457. 1927. ≡ *Asteridiella trematis* (C.L. Spegazzini) C.G. Hansford ex A.C. Batista & H. da Silva Maia. Anais da Sociedade de Biologia de Pernambuco 15(2): 453–454. 1957. [Epithets of the two synonyms were also published as *tremae* and are corrected herein.].

Meliola trichiliae M. Beeli. Bulletin du Jardin Botanique de L'État à Bruxelles 7:99–100. 1920. **Isotype**: In foliis *Trichilia retusa* Oliver, [C. Africa], Congo [Zaire], Bords de l'Aruwimi [River], 1913, leg. Bequaert No. 1495, ILL 6397. **Holotype**: BR.

Meliola trichostroma (G. Kunze) R.A. Toro var. *olecranoni* (F.L. Stevens & L.R. Tehon) C.G. Hansford, 1961. **See basionym**: *Meliola olecranoni* F.L. Stevens & L.R. Tehon, 1926.

Meliola trifurcata R. Ciferri var. *philippinensis* C.G. Hansford. Sydowia Annales Mycologici 11:60. (1957) 1958. **Holotype**: Hab. in foliis *Dysoxyli* sp. [= *Dysoxylum* sp.], Philippines, Luzon, Nueva Ecija [Prov., San José-Balete Road, 9 JAN 1931], leg. F.L. Stevens No. 1771 p.p., ILL 6535.

Meliola trinidadensis F.L. Stevens & L.R. Tehon. Mycologia 18:8, pl. 1, fig. 6. 1926. **Holotype**: On *Meibomia* sp. [= *Desmodium* sp., the packet labeled correctly], Trinidad, St. Augustine, 13 AUG 1922, leg. F.L. Stevens No. 825, ILL 6396.

Meliola triumfettae F.L. Stevens. Illinois Biological Monographs 2:498-499. 1916. **Holotype**: On *Triumfetta semitriloba* Jacquin, Porto Rico [Puerto Rico], Utuado, 8 NOV 1915 [as 1913 on the packet], leg. F.L. Stevens No. 4421, ILL 4314. **Paratypes**: Indiera Fria, 8 OCT 1913, leg. F.L. Stevens No. 3482, ILL 4315; on *Hibiscus tiliaceus* L., Dos Bocas, below Utuado, 8 JUL 1915, leg. F.L. Stevens No. 8073, ILL 4317; Arecibo-Lares Road, 21 JUN 1915, leg. F.L. Stevens No. 7249, ILL 4316; Maricao, 20 JUL 1915, leg. F.L. Stevens No. 8962, ILL 4318. ≡ *Irenopsis coronata* (C.L. Spegazzini) F.L. Stevens var. *triumfettae* (F.L. Stevens) F.L. Stevens. Annales Mycologici 25:435. 1927. ≡ *Irenopsis triumfettae* (F.L. Stevens) C.G. Hansford & F.C. Deighton. Mycological Papers. Commonwealth Mycological Institute 23:14. 1948.

Meliola tuberculata F.L. Stevens. Illinois Biological Monographs 2:490. 1916. **Holotype**: On unknown dicotyledonous plant, Porto Rico [Puerto Rico], Vega Baja, 2 JUL 1915, leg. F.L. Stevens No. 7742, ILL 3886.

Isotype: FH. ≡ *Appendiculella tuberculata* (F.L. Stevens) R.A. Toro. Mycologia 17:144. 1925. ≡ *Irene tuberculata* (F.L. Stevens) F.L. Stevens. Annales Mycologici 25:428. 1927.

Meliola tumor F.L. Stevens. Annales Mycologici 26:179, pl. II, fig. 6. 1928. **Holotype:** On Bignoniaceae sp. indet., British Guiana [Guyana], Rockstone, 16 JUL 1922, leg. F.L. Stevens No. 422, ILL 6395.

Meliola ulei C.G. Hansford. Proceedings of the Linnean Society of London 160: 132. (1948) 1949. **Isotypes:** Hab. in foliis Scrophulacearum spec. indet. [= Scrophulariaceae sp. indet.], Brazil, Amazonas, Rio Negro, [1902], leg. E.H.G. Ule, ex Mycotheca Brasiliensis No. 61, sub *Meliola microspora* N.T. Patouillard & A. Gaillard, ILL 5645 [as a microscopic preparation], and in bound exsiccati set. **Holotype:** PREM.

Meliola umirayensis H.S. Yates. Philippine Journal of Science, Section C (Botany) 13:370. (1918) 1919. **Isotype:** On leaves of *Ficus* sp., Philippines, Luzon, Tayabas Prov., Umiray, 2 JUN 1917, leg. M. Ramos & G. Edaño, ex Phil. Bur. Sci. No. 29081, ILL 6490. ≡ *Irenina umirayensis* (H.S. Yates) C.G. Hansford. Proceedings of the Linnean Society of London 157:170–171. 1946. ≡ *Asteridiella umirayensis* (H.S. Yates) C.G. Hansford ex C.G. Hansford. Beihefte zur Sydowia Annales Mycologici 2:323–324. 1961.

Meliola uncariae H. Rehm. Leaflets of Philippine Botany 6:2192. 1914. **Isotype:** Ad folia *Uncariae perrottetii* [misspelled "*perrottettii* " = *Uncaria perrottetii* (A. Richard) Merrill], Philippines, Luzon, Laguna Prov., Los Baños, [20] JUL 1913, leg. C.F. Baker No. 1280, ILL 6406. ≡ *Irenina uncariae* (H. Rehm) F.L. Stevens. Annales Mycologici 25:451. 1927. ≡ *Asteridiella uncariae* (H. Rehm) C.G. Hansford ex C.G. Hansford. Beihefte zur Sydowia Annales Mycologici 2:581. 1961.

Meliola uncinata H. Sydow. Leaflets of Philippine Botany 9:3120. 1925. **Isolectotype:** On leaves of *Horsfieldia gigantifolia* A.D.E. Elmer [nom. invalid. sine diagn. lat. = *Horsfieldia* sp. (Myristicaceae), Philippines], Luzon, Sorsogon Prov., Irosin, SEP 1916, leg. A.D.E. Elmer No. 17222, ILL 6410. **Assumed lectotype:** S [citation of this institution for location of "type" by C.G. Hansford (1961:61) is interpreted as lectotypification with the specimen at S].

Meliola uncitricha H. Sydow. Annales Mycologici 24:308–310. 1926. **Isolectotype:** Hab. in foliis vivis *Phoebes neurophyllae* [= *Phoebe neurophylla* Mez & Pittier], Costa Rica, Cerro de San Isidro pr. San Ramón, 9 FEB 1925, leg. H. Sydow, Fungi in Itinere Costaricensi Collecti No. 169e, ILL 6411. **Assumed lectotype :** S [citation of this institution for location of "type" by C.G. Hansford (1961:52) is interpreted as lectotypification]. **Residual isosyntype:** San Pedro de San Ramón, 10 FEB

1925, leg. H. Sydow, Fungi in Itinere Costaricensi Collecti No. 388, ILL 6412.

Meliola vaccinii F.L. Stevens. Bernice P. Bishop Museum Bulletin 19:30, text fig. 7b. 1925. **Lectotype** [designated herein]: On *Vaccinium reticulatum* J.E. Smith, Hawaii, [Island of] Hawaii, Hilo, [lava] flow of 1881, 8 JUL [1921], leg. F.L. Stevens No. 739, ILL 6416. **Isolectotypes**: ILL 6413 and 6418, BISH 499020 and 499957. **Paratypes**: [Island of] Hawaii, Kilauea, 16 JUL [1921], leg. F.L. Stevens No. 866, ILL 6417; 13 JUL [1921], leg. F.L. Stevens No. 821, ILL 4086 [also, p.p., the holotype of *Schiffnerula vaccinii* C.G. Hansford and a paratype of *Meliola alyxiae* F.L. Stevens]; Maui, Olinda Pipeline, 5 SEP [1921], leg. F.L. Stevens No. 1146, ILL 6415; 1916, leg. C.N. Forbes-F.L. Stevens No. 694, ILL 6414.

Meliola varia E.M. Doidge. Transactions of the Royal Society of South Africa 5:738, 747–748, pl. LXV, fig. 40. 1917. **Isotype**: On leaves of *Cissus rhomboidea* E. Meyer ex Harvey & Sonder, [South Africa], Natal [Prov.], Winter's Kloof, 26 JUN 1911 [as 12 JUL on the packet], leg. E.M. Doidge, ex PREM No. 1639, ILL 5605. **Holotype**: PREM.

Meliola variaseta F.L. Stevens. Annales Mycologici 26:204–205, pl. III, fig. 25. 1928. **Holotype**: On Sapindaceae sp. indet., Panama, Chagres [River], 2–3 miles off mouth, 23 AUG 1923, leg. F.L. Stevens No. 1299, ILL 6420.

Meliola varicuspis F.L. Stevens & L.R. Tehon. Mycologia 18:7, pl. 1, fig. 4. 1926. **Holotype**: On an undetermined Annonacea [= Annonaceae sp. indet.], Costa Rica, Aserri, 26 JUN 1923, leg. F.L. Stevens No. 132, ILL 6419.

Meliola varroniae F.C. Deighton. Mycological Papers. Commonwealth Mycological Institute 9:22–23, fig. 28. 1944. **Isotype**: On leaves of *Varronia* sp. [= *Cordia* sp.], Porto Rico [Puerto Rico], El Miradero, 3 AUG 1915, leg. F.L. Stevens No. 9133, ILL 4373. **Holotype**: K. ≡ *Irenopsis varroniae* (F.C. Deighton) C.G. Hansford. Beihefte zur Sydowia Annales Mycologici 2:628. 1961.

Meliola viburni H. & P. Sydow. Annales Mycologici 15:193. 1917. **Isotype**: Hab. in foliis *Viburni odoratissimi* [= *Viburnum odoratissimum* Ker Gawler], Philippines, Luzon, Subprov. Benguet, MAR–MAY 1916, leg. H.S. Yates, ex Phil. Bur. Sci. No. 25156, ILL 4281. ≡ *Irenina viburni* (H. & P. Sydow) F.L. Stevens. Annales Mycologici 25:457. 1927. ≡ *Asteridiella viburni* (H. & P. Sydow) C.G. Hansford ex C.G. Hansford. Beihefte zur Sydowia Annales Mycologici 2:612. 1961.

Meliola vicina H. Sydow. Annales Mycologici 21:95. 1923 [non H. Sydow (1926)]. **Assumed lectotype**: Hab. in foliis *Timonii ternifolii* [= *Timonius ternifolius* (Bartlet) Fernandez-Villar], Philippines, Palawan [Island],

Taytay, 15 MAY 1913, leg. E.D. Merrill No. 8886, ILL 6422 [annotation of this specimen and citation of F.L.S. herb. (ILL) for location of "type" by C.G. Hansford (1961:601) is interpreted as lectotypification].

Meliola vicina H. Sydow. Annales Mycologici 24:310. 1926, nom. illegit., non H. Sydow, 1923. **Lectotype** [designated herein]: Hab in foliis *Rauwolfiae nitidae* [= *Rauwolfia nitida* Jacquin], Costa Rica, Los Angeles de San Ramón, 30 JAN 1925, leg. H. Sydow, Fungi in Itinere Costaricensi Collecti No. 133, ILL 6421. ≡ *Meliola euopla* H. Sydow ex F.L. Stevens. Annales Mycologici 26:254. 1928, nom. nov.

Meliola vignae-gracilis C.G. Hansford & F.C. Deighton var. *panamensis* C.G. Hansford. Sydowia Annales Mycologici 9:78. 1955; Sydowia Annales Mycologici 10:95–96. (1956) 1957. **Isotype:** Hab. in foliis *Meibomiae* sp. [*Meibomia* = *Desmodium* sp.], Panama, Las Cruces trail, [2 SEP 1924], leg. F.L. Stevens No. 152, ILL 4736. **Holotype:** FH. **Paratypes:** [28 SEP 1924], leg. F.L. Stevens No. 869, ILL 4754; [Paitillia Point, 8 SEP 1924], leg. F.L. Stevens No. 370, ILL 4737; [Corozal, Trail No. 17, 30 AUG 1924], leg. F.L. Stevens No. 85, ILL 4739; in foliis *Canavaliae lasiocalycis* [= *Canavalia lasiocalyx* O. Kuntze], Ecuador [erroneously cited as Panama in the protologue], Barrn'nital, [17 SEP 1924], leg. F.L. Stevens No. 339, ILL 4758.

Meliola villaresiae P.C. Hennings. Hedwigia 36:218. 1897. **Isotype:** Auf Blättern von *Villaresia* sp. [= *Citronella* sp.], Brasilia [Brazil], Gayaz, leg. A. Glaziou No. 22713, ILL 6426 [as microscopic preparations].

Meliola visci F.L. Stevens. Bernice P. Bishop Museum Bulletin 19:38, text fig. 8e. 1925. **Holotype:** On *Viscum articulatum* sensu Hillebr. non N.L. Burman [= *Korthalsella complanata* (Tieghem) Engler], Hawaii, Oahu, Wahiawa, 31 MAY [1921], leg. F.L. Stevens No. 167, ILL 6429. **Isotypes:** BISH 499019 and 499956. **Paratype:** Maui, Olinda Pipeline, 5 SEP [1921], leg. F.L. Stevens No. 1149, ILL 6428.

Meliola walsurae F.L. Stevens ex C.G. Hansford. Sydowia Annales Mycologici 11:61. (1957) 1958. **Holotype:** Hab. in foliis *Walsurae* sp. [= *Walsura* sp.], Philippines, Mindanao, [Island, Zamboanga Distr.], Malangas, [OCT–NOV 1919], leg. M. Ramos & G. Edaño, ex Phil. Bur. Sci. No. 36500, ILL 6430.

Meliola wardii F.L. Stevens. Annales Mycologici 26:213–214, pl. IV, fig. 32. 1928. **Holotype:** On *Malouetia panamensis* Heurck & Muell. Arg., Panama, France Field, 2 SEP 1924, leg. F.L. Stevens No. 184, ILL 6432. **Paratypes:** Panama, Frijoles, 14 OCT 1924, leg. F.L. Stevens No. 1287 p.p., ILL 6434 [also the holotype of *Meliola wardii* F.L. Stevens var. *minor* C.G. Hansford]; on *Tabernaemontana* sp., British Guiana [Guyana], Kartabo, 21 JUL 1922, leg. F.L. Stevens No. 503, ILL 6435 [also the holotype of *Meliola wardii* F.L. Stevens var. *tabernaemontanae*

C.G. Hansford]; Rockstone, 17 JUL 1922, leg. F.L. Stevens No. 452, ILL 6431; leg. F.L. Stevens No. 474, ILL 6433 [ILL 6431 and 6433 are also paratypes of *Meliola wardii* F.L. Stevens var. *tabernaemontanae* C.G. Hansford].

Meliola wardii F.L. Stevens var. *minor* C.G. Hansford. Sydowia Annales Mycologici 11:61. (1957) 1958. **Holotype:** Hab. in foliis *Malouetiae panamensis* [= *Malouetia panamensis* Heurck & Muell. Arg.], Panama, [Frijoles, 14 OCT 1924], leg. F.L. Stevens No. 1287 p.p., ILL 6434 [originally cited as paratype of the species name].

Meliola wardii F.L. Stevens var. *tabernaemontanae* C.G. Hansford. Sydowia Annales Mycologici 11:61–62. (1957) 1958. **Holotype:** Hab. in foliis *Tabernaemontanae* sp. [= *Tabernaemontana* sp.], British Guiana [Guyana], Kartabo, [21 JUL 1922], leg. F.L. Stevens No. 503, ILL 6435. **Paratypes:** [Rockstone, 17 JUL 1922], leg. F.L. Stevens No. 452, ILL 6431; leg. F.L. Stevens No. 474, ILL 6433 [Stevens numbers 452, 474 and 503 were originally cited as paratypes of the species name].

Meliola wikstroemiicola C.G. Hansford. Sydowia Annales Mycologici 11:62. (1957) 1958. **Holotype:** Hab. in *Wikstroemiae* [= *Wikstroemia* sp.], Philippines, Luzon, Laguna [Prov.], Paete, [16 NOV 1930], leg. F.L. Stevens No. 987, ILL 6520. **Paratype:** *Wikstroemia* sp. [as *W. meyeniana* Warb. (misspelled "*megeniana*") on the packet, Rizal Prov., Antipolo, JUL 1917], leg. M. Ramos & G. Edaño, ex Phil. Bur. Sci. No. 29571, ILL 6437.

Meliola winteri C.L. Spegazzini. Anales de la Sociedad Cientifica Argentina 26:20–21. 1888; Fungi Guaranitici Pugillus II, pp. 18–19, No. 53. 1888. **Isotypes:** Hab. in folia viva *Solani verbascifolii* [probably *Solanum verbascifolium* sensu auct. non L. = *S.* cf. *erianthum* D. Don] in sylvis montanis in Sierra de Peribebuy [Paraguay], 15 SEP 1883, leg. B. Balansa No. 3986 [comm. C.L. Spegazzini No. 548 (cf. C.G. Hansford, 1961)], ILL 3887 and 3888. ≡ *Irene winteri* (C.L. Spegazzini) H. & P. Sydow. Annales Mycologici 15:194. 1917. ≡ *Asteridiella winteri* (C.L. Spegazzini) C.G. Hansford ex C.G. Hansford. Beihefte zur Sydowia Annales Mycologici 2:633–634. 1961.

Meliola wismarensis F.L. Stevens. Annales Mycologici 26:191–192, pl. III, fig. 14. 1928. **Holotype:** On *Solanum* sp., British Guiana [Guyana], Wismar, 14 JUL 1922, leg. F.L. Stevens No. 302, ILL 6436.

Meliola wrightiae H.S. Yates. Philippine Journal of Science, Section C (Botany) 13:371. (1918) 1919. **Isotype:** On leaves of *Wrightia laniti* (Blanco) Merrill, Philippines, Luzon, Rizal Prov., 28 NOV 1916, leg. M. Ramos, ex Phil. Bur. Sci. No. 26757, ILL 6438.

Meliola xenoderma H. Sydow. Annales Mycologici 24:311–313. 1926. **Isotypes:** Hab. in foliis vivis *Malpighiae glabrae* [= *Malpighia glabra* L.],

Costa Rica, San Pedro de San Ramón, 6 FEB 1925, leg. H. Sydow, Fungi in Itinere Costaricensi Collecti No. 184, ILL 6440; also in bound set of H. Sydow, Fungi Exotici Exsiccati, as No. 624.

Meliola xylopiae F.L. Stevens. Annales Mycologici 26:257, pl. V, fig. 66. 1928. **Lectotype:** On *Xylopia grandiflora* A. St. Hil., Panama, France Field, 3 OCT 1924, leg. F.L. Stevens No. 1102, ILL 6441 [citation of this collection as "type" by C.G. Hansford (1961:34) is accepted as lectotypfication]. **Residual syntype:** On *Xylopia frutescens* Aublet, leg. F.L. Stevens No. 988, ILL 6442.

Meliola xylosmae F.L. Stevens. Annales Mycologici 26:256, pl. V, fig. 65. 1928. **Holotype:** On Flacourtiaceae: *Myroxylon intermedium* (Seem.) O. Kuntze [= *Xylosma intermedia* (Seem.) Triana & Planchon], Panama, Pedro Miguel, 16 AUG 1923 [as 1924 on the packet], leg. F.L. Stevens No. 1103, ILL 6443.

Meliola yatesiana A. Trotter. Sylloge Fungorum 24:284. 1926, nom. nov. **Based on:** *Meliola diospyriae* H.S. Yates, (1917) 1918, nom. illegit., non *M. diospyri* H. & P. Sydow in H. & P. Sydow & E.J. Butler, 1911. **Isotypes:** On leaves of *Diospyros discolor* Willd., Philippines, Luzon, Tayabas Prov., Basiad, 8 DEC 1916, leg. H.S. Yates, ex Phil. Bur. Sci. No. 25711, ILL 5040, K. ≡ *Meliola diospyri* H. & P. Sydow in H. & P. Sydow & E.J. Butler var. *yatesiana* (A. Trotter) C.G. Hansford & F.C. Deighton. Mycological Papers. Commonwealth Mycological Institute 23:50. 1948.

Meliola zamboangensis C.G. Hansford. Beihefte zur Sydowia Annales Mycologici 1:119–120. 1957 [as *zamboagensis*, clearly a printer's error]. **Holotype:** Hab. in foliis *Dysoxyli* sp. [= *Dysoxylum* sp.], Philippines, Mindanao [Island], Zamboanga [District], Malangas, [OCT–NOV 1919], leg. M. Ramos & G. Edaño, ex Phil. Bur. Sci. No. 36345, ILL 5018.

Meliola zetekii F.L. Stevens. Annales Mycologici 26:192–193, pl. III, fig. 15. 1928. **Holotype:** On *Piper paulownifolium* C. DC. [= *P. carilloanum* C. DC.], Panama, Barro Colorado Island, 19 SEP 1924 [as 1923 on the packet], leg. F.L. Stevens No. 645, ILL 6444.

Meliola zollingeri A. Gaillard. Le Genre *Meliola*, pp. 105–106, pl. 1, fig. 4 [erroneously cited as "fig. 5" in the text]. 1892. **Isotype:** Ad paginam superiorem foliorum *Desmodii* cujusdam [= *Desmodium* sp.], in Ins. Java [Indonesia], leg. H. Zollinger No. 70, ILL 6453. **Holotype:** PC.

Meliola zollingeri A. Gaillard var. *minor* M. Beeli. Bulletin du Jardin Botanique de L'État à Bruxelles 7:100. 1920. **Isosyntype:** Épiphylle sur les feuilles d'une *Desmodium* [C. Africa, Zaire], Bokada, Kikwit, 1913–1914, leg. H.J.R. Vanderyst, ex PREM No. 2708, ILL 6458. **Syntype:** BR.

Meliolaster clavisporus (N.T. Patouillard) F. v. Höhnel, (1917) 1918. See **basionym:** *Meliola clavispora* N.T. Patouillard, 1890.

Meliolidium portoricense C.L. Spegazzini. Boletín de la Academia Nacional de Ciencias en Córdoba 26:336–337, illustr. Preprint 1923 [journal part issued in 1924]. **Probable isotype:** Hab. Abundante sobre la cara inferior del *Calophyllum calaba* sensu Jacquin non L. [= *C. antillanum* Britton in Britton & Wilson], Porto Rico [Puerto Rico], en los alrededores de Mayagüez, [15 JUN 1915], leg. F.L. Stevens No. 7059, ILL 3734 [also the holotype of *Meliola calophylli* F.L. Stevens]. **Holotype:** LPS 406.

Meliolina degeneri S. J. Hughes. Mycological Papers. International Mycological Institute 166:62–65, figs. 1D, 9C, 38, 39. 1993. **Paratype:** On *Metrosideros collina* (J.R. & G. Forster) A. Gray [ssp.] *polymorpha* (Gaudich.) Rock var. ? [= *M. polymorpha* Gaudich.], Hawaii, Molokai, Waialua ridge, SEP 1912, leg. C.N. Forbes No. 593, ILL 6499.

Meliolina haplochaeta H. & P. Sydow. Annales Mycologici 15:145. 1917. **Isotypes:** In foliis *Metrosideros polymorphae* [= *M. polymorpha* Gaudich.], Sandwicensium [Sandwich Islands, Hawaii], Oahu, Nuuanu Pali, 12 JAN 1909 [as 1907 on the packet], leg. H.L. Lyon No. 1, ILL 6497, BISH 499018. ≡ *Meliolinopsis haplochaeta* (H. & P. Sydow) M. Beeli. Bulletin du Jardin Botanique de L'État à Bruxelles 7:119. 1920.

Meliolina hawaiiensis S.J. Hughes. Mycological Papers. International Mycological Institute 166:78–81, figs. 2D, 49–51. 1993. **Holotype:** On *Metrosideros collina* (J.R. & G. Forster) A. Gray [ssp]. *polymorpha* (Gaudich.) Rock var. *incana* (H. Léveillé) St. John [= *M. polymorpha* Gaudich. var. *incana* (H. Léveillé) Scottsberg], Hawaii, [Island of] Hawaii, Kilauea, 11 JUL 1921, leg. F.L. Stevens No. 788, ILL 6515 [also a residual syntype of *Meliolina sydowiana* F.L. Stevens]. **Isotypes:** BISH 146043 and 499952 [also residual isosyntypes of *Meliolina sydowiana* F.L. Stevens].

Meliolina iquitosensis (P.C. Hennings) F.L. Stevens, 1927. **See basionym:** *Meliola iquitosensis* P.C. Hennings, 1904.

Meliolina malacensis (P.A. Saccardo) A. Trotter, 1926 [spelled malaccensis by later authors]. **See basionym:** *Meliola malacensis* P.A. Saccardo, 1918. [The above combination was also published by F.L. Stevens, 1927, as a superfluous later homonym.].

Meliolina megalospora (C.L. Spegazzini) F.L. Stevens, 1927. **See basionym:** *Meliola megalospora* C.L. Spegazzini, 1881.

Meliolina meliolae (F.L. Stevens) F.L. Stevens, 1927. **See basionym:** *Perisporium meliolae* F.L. Stevens, 1918.

Meliolina paulliniae (F.L. Stevens) F.L. Stevens, 1927. **See basionym:** *Perisporium paulliniae* F.L. Stevens, 1918.

Meliolina philippinensis F.L. Stevens. Annales Mycologici 25:417. 1927. **Holotype:** On Lauraceae, *Cryptocarya* sp., [Philippines], Samar [Island], Catubig River, FEB–MAR 1916, leg. M. Ramos, ex Phil Bur. Sci. No. 24720, ILL 6503. ≡ *Meliola philippinensis* (F.L. Stevens) C.G. Hansford, Beihefte zur Sydowia Annales Mycologici 2:59. 1961.

Meliolina pulcherrima (H. & P. Sydow) H. & P. Sydow, 1914. **See basionym:** *Meliola pulcherrima* H. & P. Sydow, 1913.

Meliolina radians H. & P. Sydow. Annales Mycologici 12:553. 1914. **Isotypes:** In foliis *Eugeniae xanthophyllae* [= *Eugenia xanthophylla* C.B. Robinson], Philippines, Luzon, Rizal Prov., Montalban, 23 FEB 1914, leg. M. Ramos, ex Phil. Bur. Sci. No. 17383, ILL 6510, BPI, FH, IMI. ≡ *Meliola radians* (H. & P. Sydow) M. Beeli. Bulletin du Jardin Botanique de L'État à Bruxelles 7:118, 156. 1920.

Meliolina saurauiae F.L. Stevens & E.F. Roldan. Philippine Journal of Science 56:48–49, fig. 1a. 1935, nom. invalid. sine diagn. lat. **Holotype:** On *Saurauia latibractea* Choisy, Philippines, Luzon, Benguet, Naguilian Road, 5 JAN 1931, leg. F.L. Stevens No. 1480, ILL 6512. ≡ *Toroa saurauiae* (F.L. Stevens & E.F. Roldan) C.G. Hansford. Mycological Papers. Commonwealth Mycological Institute 15:102. 1946, comb. nov. invalid. [in violation of Article 36.1 of I.C.B.N.].

Meliolina stevensii S.J. Hughes. Mycological Papers. International Mycological Institute 166:150–153, figs. 5C, 98, 99. 1993. **Holotype:** On *Metrosideros polymorpha* Gaudich. [as *M. collina* (J.R. & G. Forster) A. Gray (ssp.) *polymorpha* (Gaudich.) Rock var. (?) on the packet], Hawaii, [Island of] Hawaii, Keauhou, Kona, Bishop Estate Road, 25 JUL 1921, leg. F.L. Stevens No. 976, ILL 6514 [also an isolectotype of *Meliolina sydowiana* F.L. Stevens]. **Isotypes:** ILL 6519, BISH 146045 and 499954 [also the lectotype and isolectotypes, respectively, of *Meliolina sydowiana* F.L. Stevens]. **Paratypes:** [Between] Hilo and Kilauea, 10 JUL 1921, leg. F.L. Stevens No. 775, ILL 6494, BISH 598011.

Meliolina sydowiana F.L. Stevens. Bernice P. Bishop Museum Bulletin 19:46–47, text fig. 10a. 1925. **Lectotype** [designated by S.J. Hughes (1993)]: On *Metrosideros polymorpha* Gaudich. var. *incana* (H. Léveillé) Scottsberg [cited as *M. collina* (J.R. & G. Forster) A. Gray (ssp.) *polymorpha* (Gaudich.) Rock var. *incana* (H. Léveillé) St. John, Island of] Hawaii, Kealakehua, [as Keauhou, Kona, Bishop Estate Road on some of the packets], 25 JUL [1921], leg. F.L. Stevens No. 976, ILL 6519. **Isolectotypes:** ILL 6514, BISH 146045 and 499954 [Stevens No. 976, as ILL 6514, is also the holotype of *Meliolina stevensii* S.J. Hughes]. **Residual syntypes** [BISH numbers cited are residual isosyntypes]: On *Metrosideros polymorpha* Gaudich. var. *incana* (H. Léveillé) Scottsberg [cited as above], Kilauea, 11 JUL 1921, leg. F.L. Stevens No. 788, ILL 6515, BISH 146043 and 499952 [also the holotype and isotypes, respectively, of *Meliolina hawaiiensis* S.J. Hughes]; on *Metrosideros macropus* Hooker & Arnott, Hawaii, Oahu, Olympus, 24 JUN [1921], leg. F.L. Stevens No. 721, ILL 6517, BISH 146046 and 499951; Oahu, Tantalus, 22 JUN [1921], leg. F.L. Stevens No. 639, ILL 6513, BISH 499017 and 499953; Maui, Olinda pipe line, 5 SEP 1921, leg. F.L. Stevens No. 1144, ILL 6518, BISH 146042 and 499950; leg. F.L. Stevens No. 1145, ILL 6516, BISH 146044 and 499949.

Meliolina yatesii H. & P. Sydow. Annales Mycologici 15:195. 1917. **Isotype:** In foliis *Viburni* (?) spec. [= *Viburnum* sp. (?), as *Eugenia* sp. on the packet], Philippines, Luzon, Laguna Prov., Mt. Banahao, OCT 1915, leg. H.S. Yates, ex Phil. Bur. Sci. No. 25134, ILL 6491. ≡ *Meliolinopsis yatesii* (H. & P. Sydow) M. Beeli. Bulletin du Jardin Botanique de L'État à Bruxelles 7:119. 1920.

Meliolinopsis clavatispora (C.L. Spegazzini) M. Beeli, 1920. **See basionym:** *Meliola* ? *clavatispora* C.L. Spegazzini, 1889.

Meliolinopsis clavispora (N.T. Patouillard) M. Beeli, 1920. **See basionym:** *Meliola clavispora* N.T. Patouillard, 1890.

Meliolinopsis costaricensis (F.L. Stevens) F. Petrak, 1951. **See basionym:** *Hyalomeliolina costaricensis* F.L. Stevens, 1927.

Meliolinopsis guianensis (F.L. Stevens) F. Petrak, 1951. **See basionym:** *Hyalomeliolina guianensis* F.L. Stevens, 1923.

Meliolinopsis haplochaeta (H. & P. Sydow) M. Beeli, 1920. **See basionym:** *Meliolina haplochaeta* H. & P. Sydow, 1917.

Meliolinopsis hyalospora (J.H. Léveillé) M. Beeli, 1920. **See basionym:** *Meliola hyalospora* J.H. Léveillé, 1846.

Meliolinopsis iquitosensis (P.C. Hennings) M. Beeli, 1920. **See basionym:** *Meliola iquitosensis* P.C. Hennings, 1904.

Meliolinopsis manaosensis (P.C. Hennings) M. Beeli, 1920. **See basionym:** *Meliola manaosensis* P.C. Hennings, 1904.

Meliolinopsis meliolae (F.L. Stevens) F. Petrak, 1951. **See basionym:** *Perisporium meliolae* F.L. Stevens, 1918.

Meliolinopsis palmicola F.L. Stevens. Illinois Biological Monographs 8:193, pl. VIII, figs. 66–67. 1923. **Holotype:** On *Bactris* sp., Trinidad, Cumuto, 16 AUG 1922, leg. F.L. Stevens No. 1000, ILL 6539.

Meliolinopsis quercina (N.T. Patouillard) M. Beeli, 1920. **See basionym:** *Meliola quercina* N.T. Patouillard, 1890.

Meliolinopsis yatesii (H. & P. Sydow) M. Beeli, 1920. **See basionym:** *Meliolina yatesii* H. & P. Sydow, 1917.

Melioliphila graminicola (F.L. Stevens) C.L. Spegazzini, 1923. **See basionym:** *Calonectria graminicola* F.L. Stevens, 1918.

Mendoziopeltis ilicifolii A.C. Batista & G.E. Peres in A.C. Batista. Instituto de Micologia Universidade do Recife. Publicação 56:438, fig. 135.

1959 [as *ilexifolii*, corrected in accordance with Rec. 60H.1 and Articles 60 & 61 of I.C.B.N.]. **Isotypes:** Sôbre folhas de *Ilex nitida* (Vahl) Maxim., associado a *Trichomerium* sp. e *Atichia lopesii* [nom. ined., Puerto Rico], Rio Maricao, 20 SEP 1913, leg. F.L. Stevens No. 3613, ILL 33014, BPI. **Holotype:** URM 13370.

Merismella concinna H. Sydow. Annales Mycologici 25:115–116. 1927. **Isotype:** In foliis vivis *Caseariae silvestris* [= *Casearia sylvestris* Sw.], Costa Rica, Grecia, 13 JAN 1925, leg. H. Sydow, Fungi in Itinere Costaricensi Collecti No. 136, ILL 6737.

Metabotryon connatum H. Sydow. Annales Mycologici 24:412–413. 1926. **Isotype:** Hab. parasiticum in stromatibus *Cyclostomellae oncophorae* [= *Cyclostomella oncophora* H. Sydow] ad folia *Ocoteae veraguensis* [= *Ocotea veraguensis* (Meisner) Mez], Costa Rica, San Pedro de San Ramón, 22 JAN 1925, leg. H. Sydow, Fungi in Itinere Costaricensi Collecti No. 171b, ILL 8665 [also an isotype of *Parabotryon connatum* H. Sydow].

Metasphaeria abortiva F.L. Stevens. Transactions of the Illinois State Academy of Science 10:186, fig. 7. 1917. **Holotype:** On *Varronia alba* Jacquin [= *Cordia alba* (Jacquin) Roemer & J.A. Schultes], Porto Rico [Puerto Rico], Mayagüez, [30 JAN 1913], leg. F.L. Stevens No. 304, ILL 10061. **Isotype:** ILL 10056 . **Paratypes:** [14 JAN 1914], leg. F.L. Stevens No. 6782, ILL 10060; Maricao, [8 OCT 1913], leg. F.L. Stevens No. 3457, ILL 10057; leg. F.L. Stevens No. 3465, ILL 10058; Arecibo-Lares Road, [21 JUN 1915], leg. F.L. Stevens No. 7315, ILL 10059.

Metasphaeria abundans H. Rehm. Leaflets of Philippine Botany 6:2201. 1914. **Assumed isotype:** Ad emortuum *Alangium begoniifolium* (Roxb.) Baillon [= *A. chinense* (Lour.) Rehder], Philippines, Luzon, Laguna Prov., [Mount Maquiling], Los Baños, OCT 1913, leg. M.B. Raimundo, comm. C.F. Baker No. 1742 [ex Fungi Malayana No. 48], ILL 10062.

Metasphaeria hawaiiensis F.L. Stevens & P.A. Young in F.L. Stevens. Bernice P. Bishop Museum Bulletin 19:106. 1925. **Holotype:** On living leaves of *Metrosideros polymorpha* Gaudich., Hawaii, [Island of] Hawaii, Kilauea, 13 JUL [1921], leg. F.L. Stevens No. 826, ILL 10067. **Isotypes:** BISH 499947 and 499948.

Metasphaeria hibiscincola H. Rehm. Leaflets of Philippine Botany 6:2202. 1914. **Assumed isotype:** Ad ramulos emortuos *Hibisci rosa-sinensis* [= *Hibiscus rosa-sinensis* L.], Philippines, Luzon, Laguna Prov., [Mount Maquiling], near Los Baños, AUG 1913, leg. C.F. Baker No. 1424b [ex Fungi Malayana No. 161], ILL 10068.

Metasphaeria reyesii P.A. Saccardo. Annales Mycologici 12:305. 1914. **Assumed isotype:** Hab. in caulibus emortuis *Synedrellae nodiflorae* [=

Synedrella nodiflora (L.) Gaertner], Philippines, Luzon, [Laguna Prov.], Mount Maquiling, Los Baños, AUG 1913, leg. S.A. Reyes No. 1430 [ex C.F. Baker, Fungi Malayana No. 162], ILL 10071.

Microcallis amadelpha H. Sydow. Annales Mycologici 24:342–343. 1926. **Isosyntype**: Hab. in foliis *Roupalae veraguensis* [= *Roupala veraguensis* Klotzsch ex Meisner in Martius], Costa Rica, San Pedro de San Ramón, 28 JAN 1925, leg. H. Sydow, Fungi in Itinere Costaricensi Collecti No. 38h, ILL 6664.

Microcera curta P.A. Saccardo. Annales Mycologici 7:437. 1909. **Isotype**: Hab. in scutellia Coccorum ad ramos vivos *Tiliae platyphyllae* [= *Tilia platyphyllos* Scop., Brandenburg: Berganlagen], Tamsel [now Poland], 2 DEC 1908, leg. P. Vogel [ex H. Sydow, Mycotheca Germanica No. 849], ILL 16524.

Microclava coccolobae F.L. Stevens. Transactions of the Illinois State Academy of Science 10:206. 1917; Mycologia 11(1):7, pl. 3, fig. 10. DEC 1918 [as *coccolobiae*, corrected in accordance with Articles 32.6 & 61 of I.C.B.N.]. **Holotype**: On *Coccoloba diversifolia* Jacquin, Porto Rico [Puerto Rico], Maricao, [20 JUL 1915], leg. F.L. Stevens No. 8877, ILL 16236.

Microclava miconiae F.L. Stevens. Transactions of the Illinois State Academy of Science 10:206, fig. 11. 1917 [as *miconia*]. **Holotype**: On *Miconia laevigata* (L.) D. Don in Sweet, Porto Rico [Puerto Rico], Aguas Buenas, [9 JUL 1913], leg. F.L. Stevens No. 302, ILL 16237. **Isotype**: ILL 7254 [ILL 7254 is also the holotype and ILL 16237 an isotype of *Echidnodella miconiae* R.W. Ryan; both specimens are also syntypes of *Blastotrichum miconiae* F.L. Stevens].

Microdiplodia constrictula F. Bubák var. *crini* R. Ciferri. Sydowia Annales Mycologici 10:158. (1956) 1957. **Isotype**: Hab. in foliis siccis *Crini* sp. [= *Crinum* sp.], [Dominican Republic], Espaillat Prov., Moca, Valle del Cibao, [cult.] in the garden, Estación National Agronomica, SEP 1927, leg. R. Ciferri, Mycoflora Domingensis Exsiccata No. 382, ILL 33136.

Microdiplodia henningsii R. Staritz ex H. & P. Sydow. Annales Mycologici 2:192. 1904, nom. nud. **Isotype**: [Auf trockenen Stengeln von *Chenopodium album* L., Germany, Anhalt: Ziebigk bei Dessau, NOV 1903], leg. R. Staritz, ex H. & P. Sydow, Mycotheca Germanica No. 142, ILL 11335.

Micropeltella constricta F.L. Stevens & H.W. Manter. Botanical Gazette 79:281, figs. 40–43. 1925. **Syntypes**: On unknown members of the Annonaceae [as Anonaceae], British Guiana [Guyana], Kartabo, 22 JUL [1922], leg. F.L. Stevens No. 568, ILL 6740; Trinidad, Cumuto, 16 AUG 1922 [erroneously as JUL in the protologue], leg. F.L. Stevens No. 914 [renumbered on the packet as No. 1009], ILL 6741.

Micropeltella minima F.L. Stevens & H.W. Manter. Botanical Gazette 79:280–281, fig. 39. 1925. **Syntypes:** On unknown members of Annonaceae [as Anonaceae], British Guiana [Guyana], Tumatumari, 9 JUL [1922], leg. F.L. Stevens No. 1002, ILL 6743; Kartabo, 24 JUL [1922], leg. F.L. Stevens No. 1012, ILL 6742; on unknown host, Kartabo, 23 JUL [1922], leg. F.L. Stevens No. 1010, ILL 6744.

Micropeltella sparsa F.L. Stevens & H.W. Manter. Botanical Gazette 79:281–282, figs. 43–44. 1925. **Holotype:** On *Anacardium occidentale* L., British Guiana [Guyana], Rockstone, 13 JUL [1922], leg. F.L. Stevens No. 1013, ILL 6745.

Micropeltidium monense C.L. Spegazzini. Boletín de la Academia Nacional de Ciencias en Córdoba 26:351. Preprint 1923 [journal part issued in 1924]. **Possible isotypes:** Hab. Sobre las hojas vivas de *Amyris elemifera* L., Porto Rico [Puerto Rico], Isla Mona [Mona Island, 20–21 DEC 1913], leg. F.L. Stevens No. 6150, ILL 5657 and 5660 [also isotypes of *Stevensula monensis* C.L. Spegazzini, as well as paratypes of *Meliola monensis* F.L. Stevens and of *Helminthosporium glabroides* F.L. Stevens]. **Possible holotype:** LPS 743 p.p. ≡ *Parapeltella monensis* (C.L. Spegazzini) C.G. Orejuela in A.C. Batista. Instituto de Micologia. Universidade do Recife, Pernambuco. Publicação 56:293. 1959.

Micropeltidium portoricense C.L. Spegazzini. Boletín de la Academia Nacional de Ciencias en Córdoba 26:351-352. Preprint 1923 [journal part issued in 1924]. **Potential isotype** [not confirmed by LPS]: Hab. Sobre la cara superior de las hojas vivas de *Comocladia glabra* (J.A. Schultes) Sprengel, Porto Rico [Puerto Rico], en los alrededores del Rosario, [4 JUL 1915], leg. F.L. Stevens No. 9015 p.p., ILL 4334 [also the holotype of *Meliola comocladiae* F.L. Stevens]. ≡ *Parapeltella portoricensis* (C.L. Spegazzini) C.G. Orejuela. Mycologia 36:449. 1944.

Micropeltidium trigonostemonis F.L. Stevens & M. Schneider. Natural and Applied Science Bulletin. University of the Philippines 3:23–24, fig. 2. 1933. **Holotype:** On Euphorbiaceae: *Trigonostemon philippinensis* Stapf, Philippines, [Luzon], Laguna [Prov.], Mount Maquiling, 18 JAN 1931, leg. F.L. Stevens No. 2006, ILL 6738.

Micropeltis aroidicola F.L. Stevens & H.W. Manter. Botanical Gazette 79:279, figs. 34–35. 1925. **Holotype:** On *Philodendron* sp., British Guiana [Guyana], Kartabo, 23 JUL [1922], leg. F.L. Stevens No. 544, ILL 6748.

Micropeltis dispora F.L. Stevens & H.W. Manter. Botanical Gazette 79:278, figs. 31–33. 1925. **Holotype:** On unknown host, British Guiana [Guyana], Kartabo, 23 JUL [1922], leg. F.L. Stevens No. 593, ILL 6751.

Micropeltis dissociabilis F.L. Stevens & H.W. Manter. Botanical Gazette 79:278–279, fig. 38. 1925. **Holotype:** On unknown host, British Guiana [Guyana], Coverden, 8 AUG [1922], leg. F.L. Stevens No. 785, ILL 6752.

Micropeltis guianensis F.L. Stevens & H.W. Manter. Botanical Gazette 79:279, fig. 36. 1925. **Holotype:** On unknown host, British Guiana [Guyana], Tumatumari, 12 JUL [1922], leg. F.L. Stevens No. 241, ILL 6753.

Micropeltis tetraspora F.L. Stevens & H.W. Manter. Botanical Gazette 79:278, fig. 30. 1925. **Holotype:** On unknown host, British Guiana [Guyana], Kartabo, 24 JUL [1922], leg. F.L. Stevens No. 653, ILL 6759.

Micropeltis wildemanii M. Beeli. Bulletin du Jardin Botanique de L'État à Bruxelles 8:4, fig. 8. 1922 [as *wildemani*, corrected in accordance with Articles 32.6 & 61 of I.C.B.N.]. **Isotype:** Sur les feuilles glabres d'une Legumineuse [= Fabaceae], Congo [C. Africa, Zaire], Ipamu, leg. H.J.R. Vanderyst No. 9998, ILL 6761.

Microsphaeria elevata T.J. Burrill. Bulletin of the Illinois Museum of Natural History 1(1):58–59, pl. 2, fig. 4. 1876 [= Bulletin of the Illinois State Laboratory of Natural History, Volume 1]. **Holotype:** On upper sides of leaves of *Catalpa bignonioides* Walter, [Illinois, Champaign County, Urbana, 21 SEP–16 OCT 1876], leg. T.J. Burrill No. 21664, ILL 1088. **Probable isotype:** T.J. Burrill No. 21665, ILL 1089 [according to a footnote in the protologue, both specimens were collected in the vicinity of T.J. Burrill's residence, listed in the faculty directory of 1872–1878 as 1007 West Green Street, Urbana].

Microstroma ingicola E.M.R. Lamkey in F.L. Stevens. Mycologia 12:52. 1920 [as *ingaicola*, corrected according to Rec. 60H.1 and Articles 60 & 61 of I.C.B.N.]. **Holotype:** On *Inga laurina* (Sw.) Willd., Porto Rico [Puerto Rico], Mayagüez, JAN 1914, leg. F.L. Stevens No. 6711, ILL 28596. **Isotype:** ILL 28597.

Microstroma pithecellobii E.M.R. Lamkey in F.L. Stevens. Mycologia 12:52. 1920 [as *pithecolobii*, corrected in accordance with current I.C.B.N. (cf. F.C. Deighton, 1968)]. **Holotype:** On *Pithecellobium saman* (Jacquin) Bentham [as *Pithecolobium* = *Albizia saman* (Jacquin) F. Mueller], Porto Rico [Puerto Rico], Mayagüez, 5 JAN 1914 [as on the original label; erroneously as DEC 1913 in the protologue], leg. F.L. Stevens No. 6734, ILL 28665. **Isotypes:** ILL 28664, BPI 70896.

Microthyriella distincta F.L. Stevens & H.W. Manter. Botanical Gazette 79:290–291, fig. 63. 1925. **Holotype:** On unknown host, British Guiana [Guyana], Kartabo, 24 JUL [1922], leg. F.L. Stevens No. 675, ILL 6762.

Microthyriella domingensis R. Ciferri. Sydowia Annales Mycologici 10:144–145. (1956) 1957. **Isotype:** Hab. in foliis vivis vel semisiccis *Caesalpiniae coriariae* [= *Caesalpinia coriaria* (Jacquin) Willd., Dominican

Republic], Cordillera Septentrional, Monte Cristy Prov., Monte Cristy ad viam El Morro, MAR 1930, leg. R. Ciferri, Mycoflora Domingensis Exsiccata No. 356, ILL 33170.

Microthyriella guianensis F.L. Stevens & H.W. Manter. Botanical Gazette 79:290, fig. 62. 1925. **Holotype**: On unknown member of Ochnaceae, British Guiana [Guyana], Tumatumari, 10 JUL [1922], leg. F.L. Stevens No. 130, ILL 6764.

Microthyriella hibisci F.L. Stevens. Bernice P. Bishop Museum Bulletin 19:88–89, text fig. 20, pl. IX(E,F,G). 1925. **Syntypes** [BISH numbers cited are isosyntypes]: On *Hibiscus* sp. [cult], Hawaii, Oahu, Beretania St., Honolulu, 18 MAY [1921], leg. F.L. Stevens No. 5, ILL 6767, BISH 146048 and 499944; Honolulu, 2 JUN [1921], leg. F.L. Stevens No. 189, ILL 6766, BISH 146050 and 499945; leg. F.L. Stevens No. 193, ILL 6765, BISH 146049, 146051, 499943 and 499946.

Microthyriella roupalae H. Sydow. Annales Mycologici 25:95. 1927. **Isosyntype**: Hab. in foliis vivis *Roupalae veraguensis* [= *Roupala veraguensis* Klotzsch ex Meisner in Martius], Costa Rica, Mondongo pr. San Ramón, 3 FEB 1925, leg. H. Sydow, ex Fungi in Itinere Costaricensi Collecti No. 229c, ILL 6772.

Microthyrium calophylli R.W. Ryan. Mycologia 16:179. 1924. **Holotype**: On *Calophyllum* sp., Porto Rico [Puerto Rico], Maricao, [3 MAR 1913], leg. F.L. Stevens No. 881, ILL 7298.

Microthyrium elatum H. Rehm. Philippine Journal of Science, Section C (Botany) 8:254. 1913. **Assumed isotype**: Ad petiolos emortuos *Coryphae elatae* [= *Corypha elata* Roxb.], Philippines, Luzon, Laguna Prov., Los Baños, [Mount Maquiling], JAN 1913, C.F. Baker No. 28 [ex Fungi Malayana No. 53], ILL 7300.

Microxyphium americanum A.C. Batista in A.C. Batista and R. Ciferri. Quaderno Laboratorio Crittogamico Instituto Botanico Della Universita Pavia 31:183. 1963 — an error in the protologue, the name not adopted by the author in the text and caption of the illustration. **See**: *Scolecoxyphium americanum* A.C. Batista in A.C. Batista and R. Ciferri (loc. cit.).

Milesina lygodii H. Sydow. Mycologia 17:255. 1925. **Holotype**: On *Lygodium* sp., British Guiana [Guyana], Tumatumari, 11 JUL 1922, leg. F.L. Stevens No. 154, ILL 28260. **Isotype**: ILL 28259.

Milesina vogesiaca H. Sydow. Annales Mycologici 8:491–492. 1910. **Isotype**: Hab. in frondibus *Aspidii lobati* [*Aspidium lobatum* (C. Presl) Sw. = *Tectaria* sp.], Elsass [France, Alsace], regionis montis Hohneck Vogesorum, pr. Fischboedle, [12] JUL 1910, leg. H. Sydow, Mycotheca Germanica No. 878, ILL 28261.

Mollisia lithocarpi E.K. Cash. Mycologia 50:647–648. (1958) 1959. **Isotype**:

On dead leaves attached to fallen branches of *Lithocarpus densiflorus* (Hooker & Arnott) Rehder, California, Santa Cruz County, Big Basin State Park, 15 JUL 1954, leg. L. Bonar, [California Fungi No. 1178], ILL 33540.

Monilia crataegi H. Diedicke ex P.A. Saccardo. Sylloge Fungorum 18:502. 1906; originally published by H. Diedicke in H. & P. Sydow. Annales Mycologici 2:529. 1904, as nom. nud. **Isotype**: Auf lebenden Blättern von *Crataegus oxyacantha* L., [Germany, Thüringen], Erfurt, [Andreas-Glacis, MAY 1903/04], leg. H. Diedicke, ex H. & P. Sydow, Mycotheca Germanica No. 282, ILL 13972.

Monogrammia miconiae F.L. Stevens. Transactions of the Illinois State Academy of Science 10:202–203, fig. 9. 1917. **Assumed holotype**: On *Miconia* sp. [= *M. laevigata* (L.) D. Don in Sweet], associated with *Hyalosphaera miconiae* F.L. Stevens [misspelled *Hyalosphaeria* but correctly spelled on the packets], Porto Rico [Puerto Rico], Yabucoa, [31 DEC 1913], leg. F.L. Stevens No. 6705, ILL 7299 [also the assumed holotype of *Paranectria miconiae* F.L. Stevens, the associated teleomorph (cf. Rossman, 1987), and a paratype of *Hyalosphaera miconiae* F.L. Stevens]. **Probable isotypes**: F.L. Stevens No. 6705a, ILL 8288b, 8398 and NY. [The holotype was only recently located in the general collections under the generic name *Microthyrium*. The packet holds an ample specimen, including a microscopic preparation and some pencilled drawings by a worker other than F.L. Stevens (perhaps E. Young) that are not pertinent for the type of this name. It was not available to A.Y. Rossman (1987:28), who designated and annotated F.L. Stevens No. 6705a, ILL 8288b, as the neotype of both this species and *Paranectria miconiae* F.L. Stevens. She annotated ILL 8398 (misquoted as "8395" in her publication) as "possibly part of the type specimen of *Paranectria miconiae*." Both packets hold only small leaf fragments, and a microscopic preparation exists of ILL 8288b. ILL 8398, with an original label giving the correct date and locality (although misspelled), also holds notes and drawings for both the anamorph and the teleomorph, in part by Stevens but possibly placed there later (see also comments under *Paranectria miconiae*).]. ≡ *Titaea miconiae* (F.L. Stevens) S.C. Damon, Journal of the Washington Academy of Sciences 42:367. 1952.

Monosporium uredinicola F.L. Stevens. Transactions of the Illinois State Academy of Science 10:201. 1917 [as *uredicolum*, corrected in accordance with Articles 32.6 & 61 of I.C.B.N.]. **Holotype**: On *Coleosporium ipomoeae* (L.D. v. Schweinitz) T.J. Burrill on *Ipomoea batatas* (L.) Lam., Porto Rico [Puerto Rico, Preston's Ranch, 31 DEC 1913], leg. F.L. Stevens No. 6668, ILL 13993. **Isotypes**: ILL 14495 and 14496.

Morenoella cestri R.W. Ryan. Mycologia 16:192. 1924. **Holotype:** On unknown host [later identified as *Cestrum* sp.], Porto Rico [Puerto Rico], Quebradillas, [1915], leg. F.L. Stevens No. 4994, ILL 7349. **Probable isotype:** ILL 7307.

Morenoella decalvans (N.T. Patouillard) F. Theissen var. *laugeriae* R.W. Ryan. Mycologia 16:193. 1924 [as *langeriae*, corrected in accordance with Articles 60 & 61 of I.C.B.N.]. **Holotype:** On *Laugeria* sp. [misspelled "*Langeria*" = cf. *Neolaugeria* sp.], Porto Rico [Puerto Rico], Rio Tanamá, [6 JUL 1915], leg. F.L. Stevens No. 821, ILL 7310.

Morenoella decalvans (N.T. Patouillard) F. Theissen var. *rondeletiae* R.W. Ryan. Mycologia 16:192–193. 1924. **Holotype:** On *Rondeletia* sp., Porto Rico [Puerto Rico], Santa Ana, [31 DEC 1913], leg. F.L. Stevens No. 6689, ILL 7309 [for another specimen of the same collection, the host was identified as *Laugeria resinosa* Vahl = *Neolaugeria* (see holotype of *Morenoella laugeriae* R.W. Ryan)].

Morenoella decalvans (N.T. Patouillard) F. Theissen var. *stigmaphylli* R.W. Ryan. Mycologia 16:193. 1924. **Syntype:** On *Stigmaphyllon* sp. [cited as *Stigmatophyllum* (orth. var.)], Porto Rico [Puerto Rico], Santa Ana, [31 DEC 1913], leg. F.L. Stevens No. 6654, ILL 7312. **Probable syntype:** Leg. F.L. Stevens No. 6652 [as 6552 in the protologue— probably an error], ILL 7311.

Morenoella dothideoides (J.B. Ellis & B.M. Everhart) F. v. Höhnel var. *impetiolaris* R.W. Ryan. Mycologia 16:192. 1924. **Holotype:** On *Miconia impetiolaris* (Sw.) D. Don ex DC., Porto Rico [Puerto Rico], Mayagüez, [24 JUN 1915], leg. F.L. Stevens No. 7421, ILL 7314. **Paratype:** Consumo, [27 APR 1913], leg. F.L. Stevens No. 893, ILL 7318.

Morenoella gigantea R.W. Ryan. Mycologia 16:194. 1924 [as *giganteae*, corrected in accordance with Articles 32.6 & 61 of I.C.B.N.]. **Holotype:** On *Miconia laevigata* (L.) D. Don in Sweet, Porto Rico [Puerto Rico], Rio Maricao above Maricao, [20 SEP 1913], leg. F.L. Stevens No. 3645, ILL 7313.

Morenoella laugeriae R.W. Ryan. Mycologia 16:192. 1924 [as *langeriae*, corrected in accordance with Articles 60 & 61 of I.C.B.N.]. **Holotype:** On *Laugeria resinosa* Vahl [misspelled "*Langeria*" = *Neolaugeria resinosa* (Vahl) Nicolson], Porto Rico [Puerto Rico], Santa Ana, [31 DEC 1913], leg. F.L. Stevens No. 6689, ILL 7316 [for another specimen of the same collection, the host was identified as *Rondeletia* sp. (see holotype of *Morenoella decalvans* (N.T. Patouillard) F. Theissen var. *rondeletiae* R.W. Ryan)]. **Paratype:** Porto Rico [Puerto Rico], Rio Tanamá, [6 JUL 1915], leg. F.L. Stevens No. 7821, ILL 7315.

Morenoella melastomatacearum R.W. Ryan. Mycologia 16:194. 1924 [as *melastomacearum*, corrected in accordance with current I.C.B.N. (cf.

F.C. Deighton, 1968)]. **Holotype:** On Melastomataceae sp. indet. [cited as Melastomaceae], Porto Rico [Puerto Rico], Monte de Oro, [3 DEC 1913], leg. F.L. Stevens No. 544, ILL 7317.

Morenoella miconiae R.W. Ryan. Mycologia 16:191. 1924. **Holotype:** On *Miconia* [cf.] *splendens* sensu auct. non (Sw.) Griseb. [= *M. prasina* Sw.], Porto Rico [Puerto Rico], Las Marias, [10 JUL 1915], leg. F.L. Stevens No. 8154 [erroneously as 8145 in the protologue], ILL 7322. **Paratypes:** On *Miconia macrophylla* (D. Don) Triana [= *M. serrulata* (DC.) Naudin], Las Marias, [10 JUL 1915], leg. F.L. Stevens No. 8137, ILL 7321; on *Miconia prasina* (Sw.) DC. [as *prasiana*], Las Marias, [10 JUL 1915], leg. F.L. Stevens No. 8165, ILL 7320; leg. F.L. Stevens No. 8160, ILL 6930, 4411, 7326 and 7345 [also the lectotype and three isolectotypes, respectively, of *Echidnodella melastomatacearum* R.W. Ryan, as well as assumed paratypes of *Meliola miconiae* F.L. Stevens. Only the packet of ILL 4411 is marked with the latter name.].

Morenoella miconiicola R.W. Ryan. Mycologia 16:191. 1924 [as *miconicola*, corrected in accordance with Articles 60 & 61 of I.C.B.N.]. **Lectotype** [designated herein]: On *Miconia prasina* (Sw.) DC., Porto Rico [Puerto Rico], Mayagüez Mesa, [25 JUN 1915], leg. F.L. Stevens No. 7451, ILL 7347a. **Isolectotypes:** ILL 7347b, c, d. **Paratypes:** Leg. F.L. Stevens No. 7452, ILL 7323 and 7346.

Morenoella oxyanthi E.M. Doidge. Transactions of the Royal Society of South Africa 8:270, 281. 1920 [as *oxyanthae*, corrected in accordance with Articles 32.6 & 61 of I.C.B.N.]. **Paratype:** On *Oxyanthus gerrardii* Sonder in Harvey & Sonder, South Africa, [Natal Prov.], Maritzburg [Pietermaritzburg], Town Bush Valley, 21 MAR 1916, leg. E.M. Doidge, ex PREM No. 9719, ILL 7342.

Morenoella portoricensis C.L. Spegazzini. Boletín de la Academia Nacional de Ciencias en Córdoba 26:343–344. Preprint 1923 [journal part issued in 1924]. **Isotype:** Hab. Sobre las hojas vivas de *Ocotea leucoxylon* (Sw.) de Lanessan, Porto Rico [Puerto Rico], en los alrededores de Mayagüez [as Mayagüez on the packet, 24 JUN 1915], leg. F.L. Stevens No. 7393, ILL 7338. **Holotype:** LPS 1326. [Stevens No. 7393 is one of several collections listed for the name by R.W. Ryan (1924:193), but it is the one that best matches the citation and label data of the holotype. Ryan erroneously cited as "type" a collection from Trujillo Alto, 15 AUG 1915, leg. F.L. Stevens No. 9433, ILL 7336].

Morenoella pothodei (H. Rehm) F. Theissen var. *laevigatae* R.W. Ryan. Mycologia 16:193–194. 1924. **Holotype:** On *Miconia laevigata* (L.) D. Don in Sweet, Porto Rico [Puerto Rico], El Alto de la Bandera, [16 JUL 1915], leg. F.L. Stevens No. 8689, ILL 7344.

Morenoella psychotriae R.W. Ryan. Mycologia 16:194. 1924. **Holotype:** On unknown host, Porto Rico [Puerto Rico], Monte de Oro, [3 DEC 1913], leg. F.L. Stevens No. 5664, ILL 7348.

Morenoina africana E.M. Doidge. Transactions of the Royal Society of South Africa 8:242. 1920. **Isotype:** On pinnules of *Dryopteris inaequalis* (Schlecht.) O. Kuntze, South Africa, Natal [Prov.], Zwartkop, near Maritzburg [Pietermaritzburg], 19 JUL 1918, leg. E.M. Doidge, ex PREM No. 11605, ILL 7353. **Holotype:** PREM.

Mycoacia pinicola J. Eriksson. Svensk Botanisk Tidskrift 43:59–60, fig. 2. 1949. **Paratype:** On decaying trunk of *Pinus sylvestris* L., Sweden, Skåne, Halland: Ö. Karup Parish, Hemmeslöv, 5 JUN 1947, leg. Berit & J. Eriksson No. 1463, ILL 33202, ex UPS. **Holotype:** J. Eriksson No. 659, UPS.

Mycoleptodiscus terrestris (J.W. Gerdemann) S.A. Ostazeski, 1968. **See basionym:** *Leptodiscus terrestris* J.W. Gerdemann, 1953. [*Mycoleptodiscus* S.A. Ostazeski, 1968, is a substitute name for *Leptodiscus* J.W. Gerdemann, 1953 (nom. illegit.), non R. Hertwig, 1877].

Mycophaga guianensis F.L. Stevens. Illinois Biological Monographs 8:197–198. 1923. **Syntypes:** Growing as a parasite on undetermined hyphodiate mycelium on *Anacardium* sp. (Cashew), British Guiana [Guyana], Rockstone, 13 JUL 1922, leg. F.L. Stevens No. 253, ILL 6543; Tumatumari [as Tumtumari], 8 AUG 1922, leg. F.L. Stevens No. 65, ILL 6542.

Mycosphaerella andromedae S.M. Tracy & F.S. Earle ex L.E. Miles. Plant Disease Reporter 19:55. 1935, nom. invalid. sine diagn. lat. **Based on:** *Sphaerella andromedae* S.M. Tracy & F.S. Earle, 1895 (nom. illegit.), non B. Auerswald in G. Gonnermann & G.L. Rabenhorst, 1869. **Isotype:** On living leaves of *Andromeda nitida* Bartram ex Marshall [= *Lyonia lucida* (Lam.) K. Koch, the latter host name inscribed on the packet label], Mississippi, Ocean Springs, [10] MAR 1888, leg. F.S. Earle s.n., ILL 11975, ex BPI 71469. [The combination published in 1935 is legitimate by virtue of transfer of the taxon to another genus (cf. Article 58.3 of I.C.B.N.). It is treated, however, as the name of a new taxon and is not validly published (cf. Article 36.1)].

Mycosphaerella anthurii L.E. Miles. Transactions of the Illinois State Academy of Science 10:252. 1917. **Paratype:** On leaves of *Anthurium acaule* (Jacquin) Schott, Porto Rico [Puerto Rico], Monte Allegrillo [as Alleguillo, 10 MAY 1913], leg. [W.E. Hess] as F.L. Stevens No. 1420, ILL 9783. ≡ *Sphaerella anthurii* (L.E. Miles) A. Trotter. Sylloge Fungorum 24:850–851. 1928.

Mycosphaerella artocarpi F.L. Stevens & P.A. Young in F.L. Stevens. Bernice P. Bishop Museum Bulletin 19:101. 1925. **Syntypes:** On living leaves of *Artocarpus incisa* L.f. [= *A. altilis* (Parkinson) Fosb.], Hawaii,

Oahu, Hakipuu, Mr. Albert F. Judd's garden, 19 JUN [1921], leg. F.L. Stevens No. 566, ILL 13223; leg. F.L. Stevens No. 579c, ILL 9786.

Mycosphaerella brideliae H. & P. Sydow. Annales Mycologici 12:199–200. 1914. **Assumed isotype:** Hab. in foliis vivis *Brideliae stipularis* [= *Bridelia stipularis* (L.) Blume], Philippines, [Luzon], Laguna Prov., Los Baños, [Mount Maquiling], 10 JAN 1914, leg. C.F. Baker No. 2577, [ex Fungi Malayana No. 55], ILL 9787. ≡ *Sphaerella brideliae* (H. & P. Sydow) A. Trotter. Sylloge Fungorum 24:862. 1928.

Mycosphaerella caricae H. & P. Sydow. Annales Mycologici 11:403–404. 1913. **Assumed isotype:** Hab. in foliis vivis *Caricae papayae* [= *Carica papaya* L.], Philippines, [Luzon, Laguna Prov.], Los Baños, Mount Maquiling, 18 JUL 1913, leg. S.A. Reyes, comm. C.F. Baker No. 1512 [ex Fungi Malayana No. 56], ILL 9791.

Mycosphaerella cassiae F.L. Stevens. Illinois Biological Monographs 11:199. 1927, nom. illegit., non H. Sydow, 1925. **Syntypes:** On *Cassia* sp., Costa Rica, Siquirres, 1 AUG 1923, leg. F.L. Stevens No. 724, ILL 9793; Columbiana, 19 JUL 1923, leg. F.L. Stevens No. 909, ILL 8904a p.p. [associated with *Phyllachora lactea* F. Theissen & H. Sydow (as F.L. Stevens No. 568)]. ≡ *Mycosphaerella frauxii* M. Morelet. Bulletin de la Société des Sciences Naturelles et d'Archéologie de Toulon et du Var 175:5. 1968, nom. nov.

Mycosphaerella chrysobalani L.E. Miles. Transactions of the Illinois State Academy of Science 10:252. 1917. **Holotype:** On leaves of *Chrysobalanus icaco* L., Porto Rico [Puerto Rico], Rio Piedras, [3 NOV 1913], leg. F.L. Stevens No. 5699, ILL 9794. ≡ *Sphaerella chrysobalani* (L.E. Miles) A. Trotter. Sylloge Fungorum 24:882–883. 1928.

Mycosphaerella clusiae F.L. Stevens. Transactions of the Illinois State Academy of Science 10:181–182. 1917. **Holotype:** On *Clusia rosea* Jacquin, Porto Rico [Puerto Rico], Maricao [as Monte Alegrillo on the packet, 10 MAY 1913], leg. F.L. Stevens No. 1374, ILL 9799. **Paratypes:** Maricao, [19 JUL 1915], leg. F.L. Stevens No. 8829, ILL 9795; leg. F.L. Stevens No. 8849, ILL 9796; Lajas, [17 JUN 1915], leg. F.L. Stevens No. 7136, ILL 9797; Utuado, [8 NOV 1913], leg. F.L. Stevens No. 4587, ILL 9798. ≡ *Sphaerella clusiae* (F.L. Stevens) A. Trotter. Sylloge Fungorum 24:869. 1928.

Mycosphaerella cyaneae F.L. Stevens & P.A. Young in F.L. Stevens. Bernice P. Bishop Museum Bulletin 19:101. 1925. **Holotype:** On leaves of *Cyanea angustifolia* (Cham.) Hillebr., Hawaii, Oahu, Honolulu, 23 MAY [1921], leg. F.L. Stevens No. 723, ILL 9800 **Isotypes:** BISH 499077 and 499994.

Mycosphaerella dianellae F.L. Stevens & A.G. Weedon in F.L. Stevens. Bernice P. Bishop Museum Bulletin 19:102, text fig. 26c,d,e, pl. X(E).

1925. **Holotype:** On *Dianella odorata* sensu Hillebr. non Blume [= *D. sandwicensis* Hooker & Arnott], Hawaii, Kauai, Waimea Canyon, 15 JUN [1921], leg. F.L. Stevens No. 421, ILL 9805 [also the holotype of *Phaeosphaerella dianellae* F.L. Stevens]. **Isotypes:** BISH 499076 and 499993. **Paratypes:** Kauai, Kalalau trail, 16 JUN [1921], leg. F.L. Stevens No. 528, ILL 9804; Oahu, Wahiawa, 3 JUN [1921], leg. F.L. Stevens No. 253, ILL 9801; Waiahole ditch trail, 12 JUN [1921], leg. F.L. Stevens No. 405, ILL 9802; Maui, 1920, leg. C.N. Forbes No. 1999, ILL 9803 [F.L. Stevens numbers 528 and 253, as well as C.N. Forbes No. 1999 are also paratypes of *Phaeosphaerella dianellae* F.L. Stevens].

Mycosphaerella didymopanacis L.E. Miles. Transactions of the Illinois State Academy of Science 10:249–250. 1917. **Holotype:** On living leaves of *Didymopanax morototonii* (Aublet) Decaisne & Planchon [misspelled "*mortoni*"], Porto Rico [Puerto Rico], Añasco, [12 OCT 1913], leg. F.L. Stevens No. 3591, ILL 9824. **Isotypes:** ILL 9810, NY. **Paratypes:** [Mayagüez Distr.], Añasco, [FEB 1912], leg. F.L. Stevens No. 35, ILL 9819; [30 JAN 1913], leg. F.L. Stevens No. 297, ILL 9811 and 9818; [Rosario, 27 OCT 1913], leg. F.L. Stevens No. 3780, ILL 9823; [Monte de Oro, 3 DEC 1913], leg. F.L Stevens No. 5716, ILL 9821; leg. F.L. Stevens No. 5748, ILL 9808; [Las Marias, 10 JUL 1915], leg. F.L. Stevens No. 8140, ILL 9814; Utuado, [8 NOV 1913], leg. F.L. Stevens No. 4681, ILL 9822; [Arecibo-Lares Road, 21 JAN 1914], leg. F.L. Stevens No. 6829, ILL 9813. ≡ *Sphaerella didymopanacis* (L.E. Miles) A. Trotter. Sylloge Fungorum 24:852. 1928.

Mycosphaerella dubia L.E. Miles. Transactions of the Illinois State Academy of Science 10:250. 1917. **Holotype:** On living leaves of *Solanum*? sp., Porto Rico [Puerto Rico], Maricao, [4 APR 1913], leg. F.L. Stevens No. 750, ILL 9826. **Isotype:** NY. ≡ *Sphaerella dubia* (L.E. Miles) A. Trotter. Sylloge Fungorum 24:887–888. 1928.

Mycosphaerella erythrinae F.L. Stevens. Annales Mycologici 28:285. 1930, nom. illegit., non S.H. Koorders, 1907. **Holotype:** On *Erythrina* sp., Panama, Juan Diaz, 21 AUG 1923, leg. F.L. Stevens No. 1235, ILL 9827. ≡ *Mycosphaerella stevensii* B.A. Tomilin. Novosti Sistematiki Nizshikh Rastenij [Novitates Systematicae Plantarum Non Vascularium] 1968:167. 1968, nom. nov.

Mycosphaerella frauxii M. Morelet. Bulletin de la Société des Sciences Naturelles et d'Archéologie de Toulon et du Var 175:5. 1968, nom. nov. **Based on:** *Mycosphaerella cassiae* F.L. Stevens, 1927 (nom. illegit.), non H. Sydow, 1925. **Syntypes:** On *Cassia* sp., Costa Rica, Siquirres, 1 AUG 1923, leg. F.L. Stevens No. 724, ILL 9793; Columbiana, 19 JUL 1923, leg. F.L. Stevens No. 909, ILL 8904a p.p. [associated with *Phyllachora lactea* F. Theissen & H. Sydow (as F.L. Stevens No. 568)].

Mycosphaerella freycinetiae F.L. Stevens. Bernice P. Bishop Museum Bulletin 19:103, text fig. 27a. 1925. **Holotype:** On *Freycinetia arnotti* Gaudich. [= *F. arborea* Gaudich.], Hawaii, Oahu, Kalihi Valley, DEC 1908, leg. C.N. Forbes No. 3, ILL 9840. **Isotypes:** BISH 486899 and 499075.

Mycosphaerella guttiferae L.E. Miles. Transactions of the Illinois State Academy of Science 10:250–251. 1917. **Holotype:** On living leaves of *Clusia gundlachii* Stahl, Porto Rico [Puerto Rico], Maricao, [10 JAN 1912], leg. F.L. Stevens No. 286, ILL 9837. **Isotype:** NY. **Paratypes:** [19 JUL 1915], leg. F.L. Stevens No. 8906, ILL 9836; [3 APR 1913], leg. F.L. Stevens No. 809, ILL 9727 and 9838 [also the holotype and an isotype, respectively, of *Guignardia clusiae* F.L. Stevens]. ≡ *Sphaerella guttiferae* (L.E. Miles) A. Trotter. Sylloge Fungorum 24:869. 1928.

Mycosphaerella hawaiiensis F.L. Stevens & P.A. Young in F.L. Stevens. Bernice P. Bishop Museum Bulletin 19:103. 1925. **Holotype:** On living leaves of *Gunnera petaloidea* Gaudich., Hawaii, Maui, Olinda Pipeline, 5 SEP [1921], leg. F.L. Stevens No. 1143b, ILL 9841. **Possible isotypes:** BISH 499074 and 499992.

Mycosphaerella hedychii F.L. Stevens & P.A. Young in F.L. Stevens. Bernice P. Bishop Museum Bulletin 19:103. 1925. **Holotype:** On *Hedychium coronarium* J. Koenig, Hawaii, [Island of] Hawaii, Wailuku River, 8 JUL [1921], leg. F.L. Stevens No. 744, ILL 9842. **Isotypes:** BISH 499073 and 499991.

Mycosphaerella kaduae F.L. Stevens & P.A. Young in F.L. Stevens. Bernice P. Bishop Museum Bulletin 19:103–104. 1925. **Syntypes** [BISH numbers cited are isosyntypes]: On *Kadua* sp. [= *Hedyotis* sp.], Hawaii, Oahu, Konahuanui, [no date], leg. Bergman, as F.L. Stevens No. 112, ILL 9847, BISH 499072 and 499989; on *Kadua grandis* A. Gray [= *Hedyotis acuminata* (Cham. & Schlecht.) Steudel], Oahu, Tantalus, 25 MAY [1921], leg. F.L. Stevens No. 93, ILL 9846, BISH 146061 and 499990; on living leaves of *Gouldia* sp. [= *Hedyotis* sp.], Tantalus, 22 JUN [1921], leg. F.L. Stevens No. 602, ILL 9848, BISH 146059 and 146060 [both F.L. Stevens No. 93 and No. 602 are also cited as syntypes of *Septoria gouldiae* F.L. Stevens & P.A. Young in F.L. Stevens].

Mycosphaerella maxima L.E. Miles. Transactions of the Illinois State Academy of Science 10:251. 1917. **Holotype:** On living leaves of an undetermined host, probably a member of the Rubiaceae, Porto Rico [Puerto Rico], Maricao, [APR 1915], leg. F.L. Stevens No. 754, ILL 9853. ≡ *Sphaerella maxima* (L.E. Miles) A. Trotter. Sylloge Fungorum 24:884. 1928.

Mycosphaerella metrosideri F.L. Stevens & P.A. Young in F.L. Stevens. Bernice P. Bishop Museum Bulletin 19:104. 1925. **Syntypes** [BISH numbers cited are isosyntypes]: On living leaves of *Metrosideros polymorpha* Gaudich., Hawaii, Oahu, Wahiawa, 31 MAY [1921], leg. F.L. Stevens No. 159, ILL 9857, BISH 146064 and 499986; Kalihi Valley, 2 JUN [1921], leg. F.L. Stevens No. 183, ILL 9855, BISH 499071 and 499988; Olympus, 24 JUN [1921], leg. F.L. Stevens No. 716, ILL 9856, BISH 146062 and 499987; [Island of] Hawaii, Kohala, 2 JUL [1919], leg. H.L. Lyon No. 481, ILL 9858, BISH 146063.

Mycosphaerella mucunae F.L. Stevens. Transactions of the Illinois State Academy of Science 10:182. 1917. **Holotype:** On *Mucuna pruriens* (L.) DC., Porto Rico [Puerto Rico], Añasco, [12 OCT 1913], leg. F.L. Stevens No. 3535, ILL 14896. **Isotype:** ILL 14880. ≡ *Sphaerella mucunae* (F.L. Stevens) A. Trotter. Sylloge Fungorum 24:873. 1928.

Mycosphaerella palmae L.E. Miles. Transactions of the Illinois State Academy of Science 10:252. 1917. **Holotype:** On leaves of palms [Arecaceae], Porto Rico [Puerto Rico], Guanica, [29 JUL 1915], leg. F.L. Stevens No. 2107, ILL 9863. **Isotype:** NY. ≡ *Sphaerella palmae* (L.E. Miles) A. Trotter. Sylloge Fungorum 24:881–882. 1928.

Mycosphaerella perseae L.E. Miles. Transactions of the Illinois State Academy of Science 10:251. 1917. **Holotype:** On living leaves of *Persea americana* P. Miller, Porto Rico [Puerto Rico], Maricao, [18 NOV 1913], leg. F.L. Stevens No. 4486, ILL 9870. **Paratypes:** Maricao, leg. F.L. Stevens No. 4809, ILL 9867; San Germán, [8 NOV 1913], leg. F.L. Stevens No. 5797, ILL 9868; Dos Bocas, below Utuado, [30 DEC 1913], leg. F.L. Stevens No. 6601, ILL 9869. ≡ *Sphaerella perseae* (L.E. Miles) A. Trotter. Sylloge Fungorum 24:871–872. 1928.

Mycosphaerella reyesii H. & P. Sydow. Annales Mycologici 12:200. 1914 [as *reyesi*, corrected in accordance with Articles 32.6 & 61 of I.C.B.N.]. **Assumed isotype:** Hab. in foliis languidis vel subemortuis *Sapindi saponariae* [= *Sapindus saponaria* L.], Philippines, [Luzon], Laguna Prov., [Mount Maquiling, near] Los Baños, 1 OCT 1913, leg. S.A. Reyes, comm. C.F. Baker No. 2141 [ex Fungi Malayana No. 59], ILL 9882. ≡ *Sphaerella reyesii* (H. & P. Sydow) A. Trotter. Sylloge Fungorum 24:885. 1928 [as *reyesi*, corrected herein].

Mycosphaerella scaevolae F.L. Stevens & P.A. Young in F.L. Stevens. Bernice P. Bishop Museum Bulletin 19:104. 1925. **Holotype:** On living leaves of *Scaevola chamissoniana* Gaudich., Hawaii, Oahu, Tantalus, 22 JUN [1921], leg. F.L. Stevens No. 660, ILL 9894 [also the holotype of *Pleospora scaevolae* F.L. Stevens & P.A. Young in F.L. Stevens]. **Isotypes:** BISH 499070 and 499985. **Paratypes:** Tantalus, leg. F.L. Stevens No. 614, ILL 9899; Olympus, 24 JUN [1921], leg. F.L. Stevens No. 707, ILL 9901; leg. F.L. Stevens No. 722, ILL 9900; leg. F.L. Stevens No. 724, ILL 9895; leg. F.L. Stevens No. 700, ILL 9893; Kauai, Kalalau trail, 16

JUN [1921], leg. F.L. Stevens No. 522, ILL 9891; on *Scaevola mollis* Hooker & Arnott, Oahu, Wahiawa, 3 JUN [1921], leg. F.L. Stevens No. 215, ILL 9898; Konahuanui, 3 NOV [1921], leg. H.L. Lyon No. 166, ILL 9896; Tantalus, 22 JUN [1921], leg. F.L. Stevens No. 615, ILL 9890; leg. F.L. Stevens No. 646, ILL 9897; on *Scaevola glabra* Hooker & Arnott, Oahu, Wahiawa, 2 JUN [1921, as 22 JUN on the packet], leg. F.L. Stevens No. 204, ILL 9892.

Mycosphaerella sequoiae L. Bonar. Mycologia 34:184. 1942. **Isotype:** On leaves and twigs of *Sequoia sempervirens* (Lambert ex D. Don) Endl., California, Humboldt County, Trinidad, Spring 1932, leg. H.E. Parks, [California Fungi No. 651], ILL 9904. **Holotype:** UC 653844. **Paratype:** 1 FEB 1933, leg. H.E. Parks, [California Fungi No. 652], ILL 9905.

Mycosphaerella stevensii B.A. Tomilin. Novosti Sistematiki Nizshikh Rastenij [Novitates Systematicae Plantarum Non Vascularium] 1968:167. 1968, nom. nov. **Based on:** *M. erythrinae* F.L. Stevens, 1930 (nom. illegit.), non S.H. Koorders, 1907. **Holotype:** On *Erythrina* sp., Panama, Juan Diaz, 21 AUG 1923, leg. F.L. Stevens No. 1235, ILL 9827.

Mycosphaerella subastoma F.L. Stevens & N.E. Dalbey. Mycologia 11(1):8, pl. 3, figs. 13–15. DEC 1918. **Holotype:** On *Anemia adiantifolia* (L.) Sw. [as *Aneimia*], Porto Rico [Puerto Rico], Dos Bocas, [8 JUL 1915], leg. F.L. Stevens No. 8058, ILL 9906. ≡ *Sphaerella subastoma* (F.L. Stevens & N.E. Dalbey) A. Trotter. Sylloge Fungorum 24:865. 1928.

Mycosphaerella tabebuiae L.E. Miles. Transactions of the Illinois State Academy of Science 10:249. 1917. **Holotype:** On living leaves of *Tabebuia haemantha* (Bertolini ex Sprengel) DC., Porto Rico [Puerto Rico], Vega Baja, [18 MAY 1913], leg. F.L. Stevens No. 2021, ILL 9807. **Paratype:** Mona Island, [20–21 DEC 1913], leg. F.L. Stevens No. 6187, ILL 9806. ≡ *Sphaerella tabebuiae* (L.E. Miles) A. Trotter. Sylloge Fungorum 24:854. 1928.

Mycosphaerium anemones F.E. & E.S. Clements. Cryptogamae Formationum Coloradensium, Century 3, No. 214. Anno 1907, nom. nud. **Isotype:** ... in caulibus emortuis *Anemones globosae* [*Anemone globosa* (Torrey & Gray) Nutt. ex Pritzel = *A. multifida* Poiret var. *hudsoniana* DC.], Colorado, Palsgrove Canyon, alt. 2700 m, 19 JUL 1906, leg. F.E. & E.S. Clements, ILL 9397.

Mycosphaerium artemisiae F.E. & E.S. Clements. Cryptogamae Formationum Coloradensium, Century 3, No. 215. Anno 1907, nom. nud. **Isotype:** ... in caulibus emortuis stantibusque *Artemisiae canadensis* [*Artemisia canadensis* Michaux = *A. campestris* L., sensu lato], Colorado, Ruxton Dell, alt. 2800 m, 21 JUL 1906, leg. F.E. & E.S. Clements, ILL 9398.

Mycosphaerium calthae F.E. & E.S. Clements. Cryptogamae Formationum Coloradensium, Century 3, No. 216. Anno 1907, nom. nud. **Isotype:** ... in foliis petiolisque putrescentibus *Calthae leptosepalae* [= *Caltha leptosepala* DC.], Colorado, Cabin Canyon, alt. 2700 m, 10 JUL 1906, leg. F.E. & E.S. Clements, ILL 9400.

Mycosphaerium fendlerae F.E. & E.S. Clements. Cryptogamae Formationum Coloradensium, Century 5, No. 419. Anno 1908, nom. nud. **Isotype:** ... in foliis vetustis *Fendlerae rupicolae* [= *Fendlera rupicola* A. Gray], Colorado, Durango, alt. 2000 m, 29 JUL 1907, leg. F.E. & E.S. Clements, ILL 9405.

Mycosphaerium insidens F.E. & E.S. Clements. Cryptogamae Formationum Coloradensium, Century 3, No. 221. Anno 1907, nom. nud. **Isotype:** ... in caulibus emortuis stantibusque *Thalictri sparsiflori* [= *Thalictrum sparsiflorum* Turcz. ex Fischer & C.A. Meyer], Colorado, Minnehaha, alt. 2600 m, 3 JUL 1906, leg. F.E. & E.S. Clements, ILL 9407.

Mycosphaerium lini F.E. & E.S. Clements. Cryptogamae Formationum Coloradensium, Century 5, No. 420. Anno 1908, nom. nud. **Isotype:** ... in caulibus emortuis *Lini lewisii* [= *Linum lewisii* Pursh], Colorado, Sulphur Springs, alt. 2400 m, 22 JUL 1907, leg. F.E. & E.S. Clements, ILL 9410.

Mycosphaerium melaenodes F.E. & E.S. Clements. Cryptogamae Formationum Coloradensium, Century 5, No. 421. Anno 1908, nom. nud. **Isotype:** ... in caulibus emortuis *Astragali junciformis* [*Astragalus junciformis* A. Nelson = *A. convallarius* E. Greene], Colorado, Sulphur Springs, alt. 2400 m, 18 JUL 1907, leg. F.E. & E.S. Clements, ILL 9412.

Mycosphaerium octopetalae F.E. & E.S. Clements (nom. nud.) var. *majus* F.E. & E.S. Clements. Cryptogamae Formationum Coloradensium, Century 3, No. 223. Anno 1907, nom. nud. **Isotype:** ... in foliis emortuis *Dryadis octopetalae* [= *Dryas octopetala* L.], Colorado, Bottomless Pit, alt. 3600 m, 13 JUL 1906, leg. F.E. & E.S. Clements, ILL 9414.

Mycosphaerium orthosporum F.E. & E.S. Clements. Cryptogamae Formationum Coloradensium, Century 5, No. 423. Anno 1908, nom. nud. **Isotype:** ... in caulibus emortuis *Pentstemonis proceri* [= *Penstemon procerus* Douglas ex Graham], Colorado, Sulphur Springs, alt. 2400 m, 19 JUL 1907, leg. F.E. & E.S. Clements, ILL 9415.

Mycosphaerium primulae F.E. & E.S. Clements (nom. nud.) var. *majus* F.E. & E.S. Clements. Cryptogamae Formationum Coloradensium, Century 3, No. 225. Anno 1907, nom. nud. **Isotype:** ... in foliis putridis *Primulae parryi* [= *Primula parryi* A. Gray], Colorado, Bottomless Pit, alt. 3600 m, 1 AUG 1906, leg. F.E. & E.S. Clements, ILL 9419.

Mycosphaerium rudbeckiae F.E. & E.S. Clements. Cryptogamae Formationum Coloradensium, Century 3, No. 226. Anno 1907, nom. nud. **Iso-**

type: ... in caulibus emortuis *Rudbeckiae flavae* [*Rudbeckia flava* T.V. Moore = *R. hirta* L.], Colorado, Ruxton Dell, alt. 2800 m, 26 JUL 1906, leg. F.E. & E.S. Clements, ILL 9421.

Mycosphaerium tassianum F.E. & E.S. Clements (nom. nud.) var. *macrosporum* F.E. & E.S. Clements. Cryptogamae Formationum Coloradensium, Century 3, No. 228. Anno 1907, nom. nud. **Isotype:** ... in foliis vetustis *Calamagrostidis purpurascentis* [= *Calamagrostis purpurascens* R. Br.], Colorado, Minnehaha, alt. 2600 m, 19 JUL 1906, leg. F.E. & E.S. Clements, ILL 9424.

Mycosphaerium tassianum F.E. & E.S. Clements (nom. nud.) var. *vagnerae* F.E. & E.S. Clements. Cryptogamae Formationum Coloradensium, Century 3, No. 230. Anno 1907, nom. nud. **Isotype:** ... in foliis putridinis *Vagnerae stellatae* [*Vagnera stellata* (L.) Morong = *Smilacina stellata* (L.) Desf.], Colorado, Minnehaha, alt. 2600 m, 25 JUN 1906, leg. F.E. & E.S. Clements, ILL 9426.

Mycosphaerium veratri F.E. & E.S. Clements. Cryptogamae Formationum Coloradensium, Century 5, No. 426. Anno 1908, nom. nud. **Isotype:** ... in foliis vetustis *Veratri speciosi* [*Veratrum speciosum* Rydb. = *V. californicum* Durand], Colorado, Sierra Blanca, alt. 2800 m, 21 JUN 1907, leg. F.E. & E.S. Clements, ILL 9429.

Mycovellosiella cayoponiae (F.L. Stevens & W.G. Solheim in W.G. Solheim & F.L. Stevens) M. Muntañola, 1960. **See basionym:** *Cercospora cayaponiae* F.L. Stevens & W.G. Solheim in W.G. Solheim & F.L. Stevens, 1931.

Mycovellosiella mikaniae (F.L. Stevens) F.C. Deighton, 1974. **See basionym:** *Cladosporium mikaniae* F.L. Stevens, 1917.

Myiocopron conjunctum H. & P. Sydow. Annales Mycologici 12:200–201. 1914. **Assumed isotype:** Hab. in foliis *Daemonoropis* [= *Daemonorops* sp.], Philippines, [Luzon], Laguna Prov., Mount Maquiling, near Los Baños, 24 DEC 1913, leg. C.F. Baker No. 2228 [ex Fungi Malayana No. 165], ILL 7354.

Myiocopron freycinetiae (F.L. Stevens & R.W. Ryan in F.L. Stevens) G. Arnaud, 1931. **See basionym:** *Peltella freycinetiae* F.L. Stevens & R.W. Ryan in F.L. Stevens, 1925.

Myriangina miconiae F.L. Stevens & A.G. Weedon. Mycologia 15:201–202, pl. 18, fig. 5; pl. 20, figs. 16–18. 1923. **Syntypes:** On *Miconia* sp. indet., British Guiana [Guyana], Tumatumari, 11 JUL 1922, leg. F.L. Stevens No. 174, ILL 33017; Demerara-Essequibo R.R., 15 JUL 1922, leg. F.L. Stevens No. 332, ILL 33016.

Myrianginella costaricensis F.L. Stevens. Illinois Biological Monographs 11:165–166, pl. I, figs. 4–5. 1927. **Holotype:** On *Miconia* sp., Costa Rica, Siquirres, 31 JUL 1923, leg. F.L. Stevens No. 698, ILL 33018.

Myrianginella tapirirae F.L. Stevens & A.G. Weedon. Mycologia 15:197–199, figs. 1, 2, 6–9. 1923 [as *tapirae*, corrected in accordance with Articles 60 & 61 of I.C.B.N.]. **Holotype:** On *Tapirira* sp. [misspelled "*Tapira*," British Guiana [Guyana], Kartabo, 22 JUL 1922, leg. F.L. Stevens No. 525, ILL 32646 [also the holotype of *Endodothella tapirirae*]. **Isotype:** ILL 8600 [a microscopic preparation].

Myriangium sabaleos A.G. Weedon. Mycologia 18:218–219. 1926. **Holotype:** On *Sabal palmetto* (Walter) Lodd. ex J.A. & J.H. Schultes, Florida, St. Petersburg, 15 FEB 1923, leg. A.G. Weedon No. 2, ILL 33015.

Myrioconium comitatum J.J. Davis. Transactions of the Wisconsin Academy of Sciences, Arts and Letters 19(2):686. 1919. **Isosyntype:** On leaves of *Populus tremuloides* Michaux, Wisconsin, Mountain, [9 JUL 1915], leg. J.J. Davis [Fungi Wisconsinenses Exsiccati No. 61], ILL 13782.

Mytilinidion juniperi J.B. Ellis & B.M. Everhart. Journal of Mycology 4:57. 1888. **Isotype:** On outer bark of living *Juniperus virginiana* L. [red cedar], New Jersey, Newfield, APR 1888, leg. J.B. Ellis, [ex J.B. Ellis and B.M. Everhart, North American Fungi No. 2152], ILL 7642.

Myxomycidium flavum G.W. Martin. Mycologia 30:435–438, figs. 16–28. 1938. **Holotype:** Colombia, Dept. Magdalena, Sierra Nevada de Santa Marta, [vicinity of Dos Aguas, east of Hacienda Cincinati], alt. 1400–1500 m, 14 AUG 1935, leg. G.W. Martin No. 3371, ILL 32581.

Myxosporella ulmicola L. Bonar. Mycologia 59:599. 1967. **Isotypes:** On ... dead twigs of *Ulmus americana* L., West Virginia, Fort Hill Farm, near Burlington, 3 JUL 1963, leg. L. Bonar, ILL 33178, BPI. **Holotype:** UC 1318860.

Myxosporium diedickei H. & P. Sydow. Annales Mycologici 2:529. 1904. **Isotype:** Hab. in ramis *Mori albae* [= *Morus alba* L., Germany, Thuringia], pr. Erfurt, [26 JUL 1904], leg. H. Diedicke, ex H. & P. Sydow, Mycotheca Germanica No. 279, ILL 13788.

N

Naemosphaera hyptidicola F.L. Stevens. Botanical Gazette 65:233–234. 1918. **Holotype:** On *Meliola hyptidicola* F.L. Stevens on *Hyptis* sp., Porto Rico [Puerto Rico], Monte de Oro, [13 DEC 1913], leg. F.L. Stevens No. 5760, ILL 4099 [also a paratype of *Meliola hyptidicola* F.L. Stevens].

Nectria meliolicola F.L. Stevens. Botanical Gazette 65:231. 1918. **Holotype:**

Associated with an undetermined *Fusarium* sp. and *Meliola paulliniae* F.L. Stevens, on *Casearia sylvestris* Sw., Porto Rico [Puerto Rico], Mayagüez, [1 MAY 1915], leg. F.L. Stevens No. 1051, ILL 5884 [also the holotype of *Fusarium meliolicola* F.L. Stevens and a paratype of *Meliola paulliniae* F.L. Stevens].

Neohaplomyces neomedonalis R.K. Benjamin. El Aliso 3:192–193, figs. 13, 14. 1955. **Holotype:** On all parts of *Neomedon arizonense* Csy. (Coleoptera: Staphylinidae), Arizona, Pima County, Santa Catalina Mountains, northeast of Tuscon, 15 JUL 1938, leg. O. Bryant, as R.K. Benjamin No. 874, ILL 21947. **Paratypes:** 12 APR 1936, leg. R.K. Benjamin No. 572, ILL 21945; Graham County, Graham Mountains, 15 AUG 1933, leg. O. Bryant, as R.K. Benjamin No. 876, ILL 21948.

Neostomella tabernaemontanae H. Sydow. Annales Mycologici 25:39–41. 1927. **Isosyntypes:** In foliis *Tabernaemontanae longipedis* [= *Tabernaemontana longipes* J. Donnell Smith], Costa Rica, San Pedro de San Ramón, 5 FEB 1925, leg. H. Sydow, Fungi in Itinere Costaricensi Collecti No. 393, ILL 6792; in foliis *Tabernaemontanae sananho* [= *Tabernaemontana sananho* Ruiz & Pavon], Costa Rica, San Pedro de San Ramón, 10 FEB 1925, leg. H. Sydow, Fungi in Itinere Costaricensi Collecti No. 128 p.p., ILL 6791; Cerro de San Isidro pr. San Ramón, 9 FEB 1925, leg. H. Sydow, Fungi in Itinere Costaricensi Collecti No. 395 p.p., ILL 3658 [also an isosyntype of *Dimerium consimile* H. Sydow and of *Cicinnobella consimilis* H. Sydow].

Neottiopezis sclerothrix F.E. & E.S. Clements. Cryptogamae Formationum Coloradensium, Century 3, No. 295. Anno 1907, nom. nud. **Isotype:** ... in terra uda ... , Colorado, Minnehaha, alt. 2500 m, 1 AUG 1906, leg. F.E. & E.S. Clements, ILL 7676.

Nigredo graminicola (T.J. Burrill) J.C. Arthur, 1906. **See basionym:** *Uromyces graminicola* T.J. Burrill, 1884.

Niptera grewiae H. Rehm. Leaflets of Philippine Botany 8:2928–2929. 1915. **Assumed isotype:** Ad folia Grewiae [= *Grewia multiflora* Blanco], Philippines, [Luzon], Laguna Prov., [Mount Maquiling, near] Los Baños, FEB 1914, leg. C.F. Baker No. 2885 [ex Fungi Malayana No. 167], ILL 7967.

Niptera pella F.E. & E.S. Clements. Cryptogamae Formationum Coloradensium, Century 1, No. 88. Anno 1906, nom nud. **Isotype:** ... ad bases udas reliquasque *Streptopodi amplexicaulis* [*Streptopus amplexicaulis* Poiret = *S. amplexifolius* (L.) DC.], Colorado, Jack Brook, alt. 2600 m, 2 AUG 1905, leg. F.E. & E.S. Clements, ILL 7969.

Nowellia guianensis F.L. Stevens. Illinois Biological Monographs 8:177–179, pl. I, fig. 7; pl. II, figs. 2–14; pl. XIII, fig. 93. 1923. **Holotype:** On Celastraceae sp. indet., British Guiana [Guyana], Demerara-Essequibo

R.R., 15 JUL 1922, leg. F.L. Stevens No. 357, ILL 8636. **Isotypes:** ILL 8636a, b.

Nummularia fragillima H. Rehm. Leaflets of Philippine Botany 8:2959. 1916. **Assumed isotype:** Ad *Calamum* emortuum [= *Calamus* sp.], Philippines, [Luzon, Laguna Prov., Mount Maquiling], APR 1914, leg. C.F. Baker No. 3187, [ex C.F. Baker, Fungi Malayana No. 169], ILL 10689.

O

Odontia crustula L.W. Miller. Mycologia 26:29, pl. 3, fig. 7. 1934. **Isotype:** On linden [= *Tilia* sp.], Iowa, Milford, [Little Sioux River], 16 JUN 1931, leg. L.W. Miller [No. 16], ILL 32589. **Holotype:** IA [currently at ISC].

Odontia sacchari E.A. Burt. Annals of the Missouri Botanical Garden 4:233–235, fig. 1. 1917. **Paratype:** On dead sheath bases and cane trash of sugar cane [= *Saccharum* sp.], Porto Rico [Puerto Rico], Rio Piedras, [17 JUL 1916], leg. J.A. Stevenson No. 5628, ILL 29241.

Oidium cococarpum F.L. Stevens & A.S. Peirce. Indian Journal of Agricultural Science 3:915. 1933. **Assumed holotype:** On a seed of *Cocos nucifera* L., [India, Poona, Bombay, 1932], leg. B.N. Uppal No. 65, ILL 14000. **Isotype:** AMH.

Oidium erysiphoides E.M. Fries:E.M. Fries forma *tagetes* R. Ciferri. Mycoflora Domingensis Exsiccata No. 332. Anno 1956, nom. nud. **Isotype:** On *Tagetes patula* L. (cult.), [Dominican Republic], Santiago Prov., Valle del Cibao, Santiago, in gardens at Hato del Yaque, 24 DEC 1929, leg. R. Ciferri, ILL 33100.

Oligostroma suttoniae F.L. Stevens. Bernice P. Bishop Museum Bulletin 19:22–23, text fig. 6b,c. 1925. **Holotype:** On *Suttonia lessertiana* (A. DC.) Mez [= *Myrsine lessertiana* A. DC.], Hawaii, [Island of] Hawaii, Kilauea, 16 JUL [1921], leg. F.L. Stevens No. 868a, ILL 8638. **Isotypes:** BISH 499039 and 499981.

Omphalia albo-flava A.P. Morgan var. *longipes* F.E. & E.S. Clements. Cryptogamae Formationum Coloradensium, Century 2, No. 179. Anno 1906, nom. nud. [as *albiflava longipes*]. **Isotype:** ... ad terram muscosam ..., Colorado, Mount Palsgrove, alt. 2700 m, 5 SEP 1904, leg. F.E. & E.S. Clements, ILL 30340.

Ophiodothella panamensis F.L. Stevens. Illinois Biological Monographs 11:196–197, pl. IX, figs. 69–71. 1927. **Syntypes:** On *Cordia heterophylla* Roemer & J.A. Schultes, Panama, Juan Mina, 18 AUG 1923, leg. F.L.

Stevens No. 1163, ILL 8652; Panama [City], 21 AUG 1923, leg. F.L. Stevens No. 1217, ILL 8644.

Oplothecium palmae F.L. Stevens. Illinois Biological Monographs 8:194. 1923. **Holotype:** On palm [Arecaceae sp. indet.], British Guiana [Guyana], Tumatumari, 11 JUL 1922, leg. F.L. Stevens & R.I. Dowell No. 134, ILL 6665.

Otthia panici F.L. Stevens. Transactions of the Illinois State Academy of Science 10:185. 1917. **Holotype:** On *Panicum maximum* Jacquin, Porto Rico [Puerto Rico], Preston's Ranch, [31 DEC 1913], leg. F.L. Stevens No. 6659, ILL 9630. **Paratype:** Jayuya, [17 DEC 1913], leg. F.L. Stevens No. 5994, ILL 9631.

Otthiella furcraeae F.L. Stevens & de Coursey. Illinois Biological Monographs 11:198–199, pl. IX, fig. 74; pl. XVII, fig. 119. 1927 [as *fourcroyae*, corrected in accordance with Rec. 60H.1 and Articles 60 & 61 of I.C.B.N.]. **Holotype:** On *Furcraea* sp. [as *Fourcroya*], Costa Rica, El Alto, 6 JUL 1923, leg. F.L. Stevens No. 251, ILL 9370.

Ovularia edwiniae F.E. & E.S. Clements. Cryptogamae Formationum Coloradensium, Century 3, No. 260. Anno 1907, nom. nud. **Isotype:** ... in foliis *Edwiniae americanae* [*Edwinia americana* (Torrey & Gray) A.A. Heller = *Jamesia americana* Torrey & Gray], Colorado, Larkspur Dell, alt. 2500 m, 15 AUG 1906, leg. F.E. & E.S. Clements, ILL 14090.

Ovularia minutissima H. Sydow. Annales Mycologici 6:481. 1908. **Isotype:** Hab. in foliis *Hyperici quadranguli* [*Hypericum quadrangulum* L., nomen ambiguum = *Hypericum* sp., Europe, probably now Poland], Riesengebirge: Weisswassergrund [bei Spindelmühl], 1 SEP 1908, leg. H. Sydow, Mycotheca Germanica No. 732, ILL 14094.

Ovularia vogeliana P.A. Saccardo and H. & P. Sydow. Annales Mycologici 2:194. 1904. **Isotype:** Hab. in foliis vivis *Coluteae arborescentis* [= *Colutea arborescens* L.], Germaniae [now Poland], pr. Tamsel [16 OCT 1903], leg. P. Vogel, ex H. & P. Sydow, Mycotheca Germanica No. 190, ILL 14136.

P

?*Paepalopsis deformans* H. Sydow. Annales Mycologici 5:398–399. 1907. **Assumed isotype:** Hab. in antheris *Ruborum* [*Rubus* sp.] pluribus locis in Thuringia [Germany, Horba bei Paulinzella und im Steiger bei Erfurt, JUN 1907], leg. H. Diedicke, ex H. Sydow, Mycotheca Germanica No. 633, ILL 14138.

Palawaniella eucleae E.M. Doidge. Bothalia 1(1):16–17. 1921. **Isotype**: On living leaves of *Euclea macrophylla* E. Meyer, [South Africa, Cape Prov.], Howiesons Poort, near Grahamstown, 12 JUL 1919, leg. E.M. Doidge, ex PREM No. 12375, ILL 8664. **Holotype**: PREM.

Parabotryon connatum H. Sydow. Annales Mycologici 24:374–377. 1926. **Isotype**: Parasiticum in stromatibus *Cyclostomella oncophorae* H. Sydow ad folia *Ocoteae veraguensis* [= *Ocotea veraguensis* (Meisner) Mez], Costa Rica, San Pedro de San Ramón, 22 JAN 1925, leg. H. Sydow, Fungi in Itinere Costaricensi Collecti No. 171b, ILL 8665 [also an isotype of *Metabotryon connatum* H. Sydow].

Paranectria luxurians H. Rehm. Leaflets of Philippine Botany 8:2924. 1915. **Isosyntype**: Ad Meliolam Maesae [= *Meliola maesae* H. Rehm], Philippines, Luzon, [Laguna Prov., Mount Maquiling, near] Los Baños, APR 1913, leg. Eladio Sablan, comm. C.F. Baker No. 2882b, [ex Fungi Malayana, No.171], ILL 8396.

Paranectria meliolicola F.L. Stevens. Botanical Gazette 65:232–233, text fig. 3. 1918. **Holotype**: On *Meliola tortuosa* H.G. Winter on *Piper umbellatum* L. [= *Lepianthes umbellata* (L.) Raf.], Porto Rico [Puerto Rico], Rio Maricao above Maricao, [20 SEP 1913], leg. F.L. Stevens No. 3634, ILL 4451. **Paratype**: On *Meliola glabroides* F.L. Stevens on *Piper aduncum* L., Lares, [22 NOV 1913], leg. F.L. Stevens No. 4930, ILL 4031.

Paranectria miconiae F.L. Stevens. Botanical Gazette 65:233, text fig. 4. 1918. **Assumed holotype**: On *Miconia* sp. [= *M. laevigata* (L.) D. Don in Sweet] on microthyriaceous fungus, Porto Rico [Puerto Rico], Yabucoa, [31 DEC 1913], leg. F.L. Stevens No. 6705, ILL 7299 [also the assumed holotype of *Monogrammia miconiae* F.L. Stevens (the associated anamorph), and a paratype of *Hyalosphaera miconiae* F.L. Stevens]. **Probable isotypes**: F.L. Stevens No. 6705a, ILL 8288b and 8398 . [The holotype was only recently located in the general collections under the generic name *Microthyrium*. The packet holds an ample specimen, including a microscopic preparation and some pencilled drawings by a worker other than F.L. Stevens (perhaps E. Young) that are not pertinent for the type of this name. It was not available to A.Y. Rossman (1987:27–28), who designated and annotated F.L. Stevens No. 6705a, ILL 8288b, as the neotype of both *Paranectria miconiae* and *Monogrammia miconiae* F.L. Stevens. A microscopic preparation exists of ILL 8288b, and Rossman annotated ILL 8398 (misquoted as 8395 in her publication) as "possibly part of the type specimen of *Paranectria miconiae*." Both packets hold only small leaf fragments, and that of ILL 8398, with an original label giving the correct date and locality (although misspelled), also holds notes and drawings, in part by Stevens but possibly placed there later, for both the teleomorph and the anamorph (see also comments under *Monogrammia miconiae*).]. ≡ *Paranectriella miconiae* (F.L. Stevens) A.Y. Rossman. Mycological Papers. International Mycological Institute 157:27–28, figs. 15, 16. 1987.

Paranectriella miconiae (F.L. Stevens) A.Y. Rossman, 1987. **See Basionym**: *Paranectria miconiae* F.L. Stevens, 1918.

Parapeltella monensis (C.L. Spegazzini) C.G. Orejuela in A.C. Batista, 1959. **See basionym**: *Micropeltidium monense* C.L. Spegazzini, 1923.

Parapeltella portoricensis (C.L. Spegazzini) C.G. Orejuela, 1944. **See basionym**: *Micropeltidium portoricense* C.L. Spegazzini, 1923 (1924).

Parascorias byrsonimae J.M. Mendoza in F.L. Stevens. Annales Mycologici 28:366. 1930. **Holotype**: On *Byrsonima* sp., British Guiana [Guyana], Tumatumari, 11 JUL [1922], leg. F.L. Stevens No. 109, ILL 6666 [also the holotype of *Petrakiopeltis byrsonimae* A.C. Batista, A.F. Vital & R. Ciferri].

Parastenella magnoliae (A.G. Weedon) J.C. David, 1991. **See basionym**: *Heterosporium magnoliae* A.G. Weedon, 1926.

Parasterina brachystoma (H. Rehm) F. Theissen var. *laxa* E.M. Doidge. Transactions of the Royal Society of South Africa 8:245. 1920 [listed on p. 273 of this article as *Asterina brachystoma* (H. Rehm) F. Theissen var. *laxa* E.M. Doidge (nom. invalid. — cf. Article 34.1 of I.C.B.N.)]. **Isosyntype**: On *Oxyanthus gerrardii* Sonder in Harvey & Sonder, South Africa, Durban, Berea, 28 JAN 1917, leg. P.A. van der Bijl, ex PREM No. 11017, ILL 7356.

Parenglerula macowaniana (F. v. Thümen) F. v. Höhnel, 1910. **See basionym**: *Meliola macowaniana* F. v. Thümen, 1876.

Parksia libocedri E.K. Cash. Mycologia 37:312–314, figs 1, 4. 1945. **Paratype**: On *Libocedrus decurrens* Torrey [= *Calocedrus decurrens* (Torrey) Florin], California, Del Norte County, [Darlingtonia], along the Middle Fork of the Smith River, MAR 1936, leg. H.E. & S.T. Parks No. 5604, ILL 30668.

Parodiopsis clusiicola C.G. Hansford. Sydowia Annales Mycologici 11:44. (1957) 1958. **Holotype**: Hab. in foliis *Clusiae krugianae* [*Clusia krugiana* Urban = *C. clusioides* (Griseb.) D'Arcy], Porto Rico [Puerto Rico, Maricao, 3 APR 1913], leg. F.L. Stevens No. 816, ILL 6546. **Assumed isotypes**: ILL 3690, BPI. **Paratypes**: [Rio Maricao above Maricao, 20 SEP 1913], leg. F.L. Stevens & W.E. Hess No. 3615, ILL 6547; [Maricao, 10 JAN 1912], leg. F.L. Stevens No. 285a, ILL 6548; [19 JUL 1915], leg. F.L. Stevens No. 8826, ILL 3692 and 6549; [20 JUL 1915], leg. F.L. Stevens No. 8862, ILL 3693 and 6550 [F.L. Stevens numbers 8826 and 8862, p.p., are also paratypes of *Parodiopsis portoricensis* C.G. Hansford].

Parodiopsis portoricensis C.G. Hansford. Beihefte zur Sydowia Annales Mycologici 1:86.1957. **Holotype:** Hab. in foliis *Clusiae roseae* [*Clusia rosea* Jacquin, an erroneous identification = *C. clusioides* (Griseb.) D'Arcy], Porto Rico [Puerto Rico], Maricao, [3 APR 1913], leg. F.L. Stevens No. 816, ILL 3690. **Assumed isotype:** ILL 6546. **Paratypes:** leg. F.L. Stevens No. 746, ILL 3691; [10 JAN 1912], leg. F.L. Stevens No. 285a, ILL 3687; Rio Maricao above Maricao, [20 SEP 1913], leg. F.L. Stevens No. 3615, ILL 3688 and 3689; in foliis *Clusiae krugianae* [*Clusia krugiana* Urban = *C. clusioides* (Griseb.) D'Arcy, 19 JUL 1915], leg. F.L. Stevens No. 8826, ILL 3692 and 6549; [20 JUL 1915], leg. F.L. Stevens No. 8862, ILL 3693 and 6550 [Stevens numbers 8826 and 8862, p.p., are also paratypes of *Parodiopsis clusiicola* C.G. Hansford].

Passalora cecropiae F.L. Stevens. Transactions of the Illinois State Academy of Science 10:207. 1917 [as *cercropiae*, corrected in accordance with Articles 60 & 61 of I.C.B.N.]. **Holotype:** On *Cecropia peltata* L., Porto Rico [Puerto Rico, Rio] Arecibo, [8 JUL 1915], leg. F.L. Stevens No. 7790, ILL 16245.

Patellaria cyanea J.B. Ellis & G.W. Martin. Journal of Mycology 1:97. 1885. **Isotype:** On living leaves of *Quercus* (*laurifolia* Michaux?) [as *Q. virens* Aiton? var. on the packet = *Q.* cf. *virginiana* P. Miller], FEB 1885, Florida, [Green Cove Springs], leg. G.W. Martin s.n., ILL 8042.

Patouillardina clavispora (N.T. Patouillard) G. Arnaud, 1917. **See basionym:** *Meliola clavispora* N.T. Patouillard, 1890.

Pauahia sideroxyli F.L. Stevens. Bernice P. Bishop Museum Bulletin 19:17–18, text fig. 2a, b. 1925. **Holotype:** On *Sideroxylon rhynchospermum* Rock [misspelled "*rhyncospermum*" = *Pouteria sandwicensis* (A. Gray) Baehni & Degener], Hawaii, Maui, Nahiku, JAN 1909, leg. H.L. Lyon No. 61, ILL 8666. **Isotype:** BISH 499038.

Paullicorticium jacksonii A.E. Liberta. Brittonia 14:223, fig. 5. 1962. **Isotype:** [On *Tsuga* sp.], Canada, Ontario, woods west of Maple, 11 SEP 1943, leg. H.S. Jackson s.n., ILL 33544. **Holotype:** TRTC 18722.

Pellicularia ansosa H.S. Jackson & D.P. Rogers in D.P. Rogers. Farlowia 1:103–104, fig. 6. 1943. **Isotypes:** On conifer wood and bark of *Picea sitchensis* (Bong.) Carrière, Washington, Olympic Mountains, Deer Lake Trail, 13 JUN 1939, leg. A.H. Smith No. 14344, ILL 32518, TRTC. **Holotype:** MICH.

Pellicularia asperula D.P. Rogers. Farlowia 1:100–101, fig. 2. 1943. **Isotype:** On fallen decayed hardwood limb, Cuba, Santa Clara Prov., Cienfuegos, Soledad, Blanco's Woods, 1 JUL 1941, leg. W.L. White No. 603, ILL 32519. **Holotype:** NY.

Pellicularia biapiculata D.P. Rogers in G.W. Martin. Lloydia 7:71–72, fig. 6. 1944. **Isotype:** Brazil, Estado do São Paulo, São Leopoldo, DEC

1939, leg. J. Rick [as SUI 1555], ILL 32631. **Holotype:** IA [currently at ISC].

Pellicularia chordulata D.P. Rogers. Farlowia 1:98–99, fig. 1. 1943. **Holotype:** On dead bark [of *Salix* sp.], Ohio, Ten Mile Creek, west of Toledo, 12 AUG 1935, leg. D.P. Rogers No. 946, ILL 32632.

Pellicularia cystidiata D.P. Rogers. Farlowia 1:101–102, fig. 4. 1943. **Isotype:** On [?]*Picea* sp. and log of undetermined conifer, Canada, Ontario, Constance Bay, JUN 1933, leg. J.W. Grove [OTB F6335, currently DAOM], ILL 32633. **Holotype:** DAOM.

Pellicularia digitata D.P. Rogers in G.W. Martin. Lloydia 7:72, fig. 10. 1944. **Isotype:** On completely charred wood, Panama, Chiriqui Prov., valley of upper Rio Chiriqui Viejo, alt. 1600–1800 m, 3 AUG 1935, leg. G.W. Martin No. 2387, ILL 32634.

Pellicularia lembospora D.P. Rogers. Farlowia 1:109–110, fig. 8. 1943. **Isotype:** On wood and bark of unidentified dicotyledonous species, and on *Bambusa vulgaris* Schrader ex J.C. Wendl., British Guiana [Guyana], Bartica, 18 JAN 1924, leg. D.H. Linder No. 731, ILL 32623. **Holotype:** FH.

Pellioniella macrospora E.R. Spencer. Botanical Gazette 72:276. 1921. **Holotype:** Parasitic on the endosperm of seed of *Bertholletia nobilis* Miers and *B. excelsa* Humb. & Bonpl., [from wholesale firms in Chicago and a retail grocery in Champaign], Illinois, 1920, leg. E.R. Spencer s.n., ILL 11337 [a single packet holding three unmarked vials of material].

Peltella freycinetiae F.L. Stevens & R.W. Ryan in F.L. Stevens. Bernice P. Bishop Museum Bulletin 19:69. 1925. **Holotype:** On *Freycinetia arnottii* Gaudich. [= *F. arborea* Gaudich.], Hawaii, Oahu, Wahiawa, 3 JUN 1921, leg. F.L. Stevens No. 977, ILL 7358. **Isotypes:** BISH 499979 and 499880. ≡ *Myiocopron freycinetiae* (F.L. Stevens & R.W. Ryan in F.L. Stevens) G. Arnaud. Annales de Cryptogamie Exotique 4:88. 1931.

Peniophora admirabilis E.A. Burt. Annals of the Missouri Botanical Garden 12:304–305. 1925. **Isotypes:** On decaying wood of stump of *Ulmus* sp., New York, Oneonta, [3] MAY 1914, leg. E.A. Burt, ILL 32841, TRTC. **Holotype:** FH.

Peniophora compta H.S. Jackson. Canadian Journal of Research, Section C (Botanical Sciences) 26:138, fig. 8. 1948. **Paratype:** [On *Pinus* sp.], Canada, Ontario, Timagami, Paradis' Bay, 26 AUG 1936, leg. R. Biggs [No. 732], ILL 32628, ex TRTC 16675. **Holotype:** TRTC.

Peniophora cymosa D.P. Rogers & H.S. Jackson in H.S. Jackson. Canadian Journal of Research, Section C (Botanical Sciences) 26:133–134, fig. 4. 1948. **Isotypes:** [On coniferous wood], North Carolina, Highlands, 17 AUG 1933, leg. G.W. Martin No. 1321, ILL 32629, TRTC. **Holotype:** IA [currently at ISC].

Peniophora delectans L.O. Overholts. Mycologia 26:513. 1934. **Isotypes:** On dead conifer wood, Pennsylvania, Clarion County, Cook Forest, 23 JUN 1932, leg. L.O. Overholts No. 16260, ILL 32627, TRTC. **Holotype:** PAC.

Peniophora hamata H.S. Jackson. Canadian Journal of Research, Section C (Botanical Sciences) 26:133, fig. 3. 1948. **Paratype:** On *Abies balsamea* (L.) P. Miller, Canada, Ontario, Lake Timagami, East Mainland, Long Point, 12 AUG 1936, leg. R. Biggs, ex TRTC No. 9877, ILL 32630. **Holotype:** TRTC 9871.

Peniophora heterobasidioides D.P. Rogers. University of Iowa Studies in Natural History 17:30–31, fig. 15. 1935. **Holotype:** On a sodden log of aspen [= *Populus* sp.], [Iowa], Iowa City, 8 JUL 1934, leg. D.P. Rogers No. 329, ILL 32608.

Peniophora heterocystidia E.A. Burt. Annals of the Missouri Botanical Garden 12:293–295. 1925. **Isotype:** On fallen limbs of ... [*Fagus* sp.], Vermont, Middlebury, [3 OCT 1901], leg. E.A. Burt s.n., ILL 32609. **Holotype:** FH.

Peniophora hiulca E.A. Burt. Annals of the Missouri Botanical Garden 12:272–273. 1925. **Isotype:** On bark and decaying wood of frondose species, Jamaica, Castleton Gardens, [14–15 DEC 1908], leg. W.A. & E.L. Murrill No. 71, ILL 32604. **Holotype:** NY.

Peniophora inornata H.S. Jackson & D.P. Rogers in H.S. Jackson. Canadian Journal of Research, Section C (Botanical Sciences) 26:139, fig. 9. 1948. **Paratype:** On wood and bark of *Pseudotsuga taxifolia* (Lambert) Britton [= *P. menziesii* (Mirbel) Franco], Oregon, [Curry County], Chetco River, about 15 miles above Brookings, 12 JUL 1939, leg. A.M. & D.P. Rogers No. 764, ILL 32610, ex TRTC. **Holotype:** TRTC 15004.

Peniophora lauta H.S. Jackson. Canadian Journal of Research, Section C (Botanical Sciences) 26:129–132, fig. 1. 1948. **Isotype:** On bark of *Thuja occidentalis* L., Canada, Ontario, York County, Toronto, Don Valley, near Sunnybrook Park, 17 SEP 1942, leg. H.S. Jackson, ex TRTC No. 17581, ILL 32611. **Holotype:** TRTC.

Peniophora leiocystis F.E. & E.S. Clements. Cryptogamae Formationum Coloradensium, Century 4, No. 334. Anno 1907, nom. nud. **Isotype:** ... ad lignum *Pseudotsugae* [= *Pseudotsuga* sp.], Colorado, Minnehaha, alt. 2500 m, 19 JUN 1906, leg. F.E. & E.S. Clements, ILL 28875.

Peniophora phosphorescens E.A. Burt. Annals of the Missouri Botanical Garden 12:273–274. 1925. **Isotype:** On rotten wood, Jamaica, leg. A.E. Wright s.n., comm. W.G. Farlow, ILL 32614. **Holotype:** FH.

Peniophora piceina L.O. Overholts. Mycologia 22:238–239, pl. 28, fig. 3. 1930. **Isotype:** On bark of limbs of *Picea rubens* Sargent, New Hampshire, Cherry Mountain, 7 JUL 1926 [as 1927 on the packet], leg. P.

Spaulding No. 43890, comm. L.O. Overholts No. 11263, ILL 32535, ex TRTC. **Holotype:** PAC.

Peniophora probata H.S. Jackson. Canadian Journal of Research, Section C (Botanical Sciences) 26:134–136, fig. 5. 1948. **Paratype:** On decaying wood of *Pinus* sp., Canada, Ontario, Lake Timagami, Bear Island, 14 AUG 1936, leg. R. Biggs, ex TRTC No. 12903, ILL 32615.

Peniophora ralla H.S. Jackson. Canadian Journal of Research, Section C (Botanical Sciences) 26:136–137, fig. 6. 1948. **Paratype:** On bark and wood of *Abies balsamea* (L.) P. Miller [as *Alnus?* on the packet], Canada, Ontario, Lake Timagami, Portage to Gull Lake, 10 AUG 1936, leg. H.S. Jackson, ex TRTC No. 9770, ILL 32617. **Holotype:** TRTC 9771.

Peniophora subcalcea V. Litschauer. Oesterreichische Botanische Zeitschrift 88:119–120, fig. 3. 1939. **Isotypes:** Hab. in cortice putrida arborum frondosarum. In Austria inferiore: apud pagum "Lunz" pr. lacum "Lunzer See," 14 SEP 1930, leg. V. Litschauer s.n., ILL 32621, TRTC.

Peniophora tenuissima C.H. Peck. Bulletin of the New York State Museum 157:30,114. (1911) 1912. **Isotype:** On bark of yellow birch, *Betula lutea* Michaux f. [= *B. alleghaniensis* Britton], New York, Adirondack Mountains, Mt. McIntyre, North Elba, [21] JUN [1911], leg. C.H. Peck s.n., ILL 32616. **Holotype:** NYS.

Peridermium gracile J.C. Carter & F.D. Kern. Bulletin of the Torrey Botanical Club 33:417–418. 1906. **Isotypes:** On leaves of *Pinus filifolia* Lindley, Mexico, mountains above Oaxaca, 28 MAY 1894, leg. C.G. Pringle s.n., ILL 28238, FH.

Perisporina dentritica F.L. Stevens. Illinois Biological Monographs 11:171–172, pl. III, figs. 21–23. 1927. **Holotype:** On *Inga* sp., Costa Rica, Experiencia Farm, 18 JUL 1927, leg. F.L. Stevens No. 517, ILL 6560.

Perisporina lantanae F.L. Stevens. Transactions of the Illinois State Academy of Science 10:170. 1917 [as *Perisporium lantanae* on the packet]. **Holotype:** On *Lantana camara* L., Porto Rico [Puerto Rico], Lares, [22 NOV 1913], leg. F.L. Stevens No. 4924, ILL 6589.

Perisporina meliolae (F.L. Stevens) C.L. Spegazzini, 1923 (1924). **See basionym:** *Periosporium meliolae* F.L. Stevens, 1918.

Perisporina meliolicola E.M. Doidge. Transactions of the Royal Society of South Africa 7:195–196, fig. 3. 1919. **Isotype:** Hab. in mycelio *Meliolae glabrae* et *Meliolae tortae* in foliis *Trichocladi criniti* [= *Trichocladus crinitus* (Thunb.) Persoon, South Africa, Cape Prov.], Izelini Forest, Kingwilliamstown [District], 14 JUN 1915 [as 8 JUN in the type citations of *Meliola scabra* E.M. Doidge and *M. torta* E.M. Doidge], leg. Emmett, ex PREM No. 9064, ILL 6400 [also an isotype of *Meliola torta* E.M. Doidge, 1917, char. emend., 1919, and of *M. neotorta* S.J.

Hughes, nom. nov. (illegit.?), as well as an isolectotype of *M. scabra* E.M. Doidge]. **Holotype**: PREM.

Perisporina paulliniae (F.L. Stevens) C.L. Spegazzini, 1923. **See basionym**: *Perisporium paulliniae* F.L. Stevens, 1918.

Perisporina portoricensis (F.L. Stevens & R. Higley ex F.L. Stevens) C.G. Hansford, 1958. **See basionym**: *Perisporium portoricense* F.L. Stevens & R. Higley ex F.L. Stevens, 1917.

Perisporium bromeliae F.L. Stevens. Transactions of the Illinois State Academy of Science 10:168. 1917. **Holotype**: On *Bromelia pinguin* L., Porto Rico [Puerto Rico], Manati, 5 NOV 1913, leg. F.L. Stevens No. 4329, ILL 6565. **Paratypes**: [20 APR 1913], leg. F.L. Stevens No. 1832, ILL 6568; [Dos Bocas, below] Utuado, [30 DEC 1913], leg. F.L. Stevens No. 6577, ILL 6567; [8 JUL 1915], leg. F.L. Stevens No. 8081, ILL 6583; Mayagüez, [31 OCT 1913], leg. F.L. Stevens No. 3912, ILL 6566; [29 JUN 1915], leg. F.L. Stevens No. 7573, ILL 6578; [14 MAY 1915], leg. F.L. Stevens No. 7094, ILL Nos. 6572 and 6575; [25 JUN 1915], leg. F.L. Stevens No. 7426, ILL 6588; Rio Tanamá, [7 JUL 1915], leg. F.L. Stevens No. 7999, ILL 6573; leg. F.L. Stevens No. 8106, ILL 6585; Santa Ana, [1 JUL 1915], leg. F.L. Stevens No. 7613, ILL 6584; Cataño, [2 JUL 1915], leg. F.L. Stevens No. 7708, ILL 6574; Vega Baja, [2 JUL 1915], leg. F.L. Stevens No. 7719, ILL 6579; Florida Adentro, [1 JUL 1915], leg. F.L. Stevens No. 7679, ILL 6580; Lajas, [17 JUN 1915], leg. F.L. Stevens No. 7150, ILL 6577; Hormigueros, [23 JUN 1915], leg. F.L. Stevens No. 7370, ILL 6581; Coamo, [16 JUL 1915], leg. F.L. Stevens No. 8355, ILL 6587; [6 JUL 1915], leg. F.L. Stevens No. 8356, ILL 6586; Maricao, [19 JUL 1915], leg. F.L. Stevens No. 8925, ILL 6582; Añasco, [20 JUL 1915], leg. F.L. Stevens No. 8751, ILL 6571.

Perisporium meliolae F.L. Stevens. Botanical Gazette 65:228–229, text fig. 2. 1918. **Holotype**: On *Meliola compositarum* F.S. Earle var. *portoricensis* F.L. Stevens, on *Eupatorium portoricense* Urban, Porto Rico [Puerto Rico], Dos Bocas, near Utuada, [16 DEC 1913], leg. F.L. Stevens No. 6032, ILL 3777 [also a paratype of *Coniothyrium glabroides* F.L. Stevens and of *Helminthosporium glabroides* F.L. Stevens. Furthermore, this collection and all paratypes cited below are also paratypes of *Meliola compositarum* F.S. Earle var. *portoricensis* F.L. Stevens]. **Isotype**: ILL 11033. **Paratypes** [on the same host as the holotype]: San Sebastian, [22 NOV 1913], leg. F.L. Stevens No. 5192, ILL 3773, 4955 and 6590; Dos Bocas, near Utuado, [16 DEC 1913], leg. F.L. Stevens No. 6031, ILL 3796 [also a paratype of *Helminthosporium glabroides* F.L. Stevens]; [30 DEC 1913], leg. F.L. Stevens No. 6861, ILL 3795 ; leg. F.L. Stevens No. 6866, ILL 3794 and 8673 [also an isotype and the holotype, respectively, of *Phaeodothiopsis eupatorii*

F.L. Stevens]; leg. F.L. Stevens 6557, ILL 3790; [River junction below Utuado, 16 DEC 1913], leg. F.L. Stevens No. 6056, ILL 3775 and 3858 [also paratypes of *Helminthosporium glabroides* F.L. Stevens]; leg. F.L. Stevens No. 6003, ILL 3787 and 3792. ≡ *Perisporina meliolae* (F.L. Stevens) C.L. Spegazzini. Boletín de la Academia Nacional de Ciencias en Córdoba 26:339. Preprint 1923 [journal part issued in 1924]. ≡ *Meliolina meliolae* (F.L. Stevens) F.L. Stevens. Annales Mycologici 25:416. 1927. ≡ *Phaeophragmeriella meliolae* (F.L. Stevens) C.G. Hansford. Mycological Papers. Commonweath Mycological Institute 15:96. 1946. ≡ *Meliolinopsis meliolae* (F.L. Stevens) F. Petrak. Sydowia Annales Mycologici 5:335. 1951. ≡ *Stevensula meliolae* (F.L. Stevens) R.A. Toro. The Journal of Agriculture of the University of Puerto Rico 36:80-81. 1952.

Perisporium paulliniae F.L. Stevens. Botanical Gazette 65:228, text fig. 1. 1918. **Holotype**: On *Meliola hessii* F.L. Stevens on *Paullinia pinnata* L., Porto Rico [Puerto Rico], Mayagüez, [4 MAY 1913], leg. F.L. Stevens No. 1207, ILL 6222 [also an isotype of *Dexteria pulchella* F.L. Stevens]. **Isotypes**: ILL 8249 [also the holotype of *Dexteria pulchella* F.L. Stevens], and perhaps F.L. Stevens No. 1207b, ILL 6222b [a microscopic preparation of ILL 6222, cited as a paratype of *Meliola hessii* F.L. Stevens and of *Helminthosporium glabroides* F.L. Stevens]. ≡ *Perisporina paulliniae* (F.L. Stevens) C.L. Spegazzini. Boletín de la Academia Nacional de Ciencias en Córdoba 26:339. Preprint 1923 [journal part issued in 1924]. ≡ *Meliolina paulliniae* (F.L. Stevens) F.L. Stevens. Annales Mycologici 25:416. 1927.

Perisporium portoricense F.L. Stevens & R. Higley ex F.L. Stevens. Transactions of the Illinois State Academy of Science 10:169, fig. 1. 1917. **Holotype**: On *Calophyllum calaba* sensu Jacquin non L. [= *C. antillanum* Britton in Britton & Wilson], Porto Rico [Puerto Rico], Mayagüez Mesa, [25 JUN 1915], leg. F.L. Stevens No. 7489 p.p., ILL 6591 [also the holotype of *Trichasterina calophylli* C.G. Hansford and a possible isosyntype of *Meliolidium portoricense* C.L. Spegazzini]. **Paratype**: Vega Baja, [5 NOV 1913], leg. F.L. Stevens No. 4310, ILL 6592 [also a paratype of *Meliola calophylli* F.L. Stevens]. ≡ *Perisporina portoricensis* (F.L. Stevens & R. Higley ex F.L. Stevens) C.G. Hansford, Sydowia Annales Mycologici 11:44–45. (1957) 1958.

Perisporium truncatum F.L. Stevens. Transactions of the Illinois State Academy of Science 10:167–168. 1917. **Holotype**: On *Inga laurina* (Sw.) Willd., Porto Rico [Puerto Rico], Mayagüez, [14 JUN 1915], leg. F.L. Stevens No. 7049, ILL 6603. **Isotype**: ILL 6598. **Paratypes**: Maricao, [20 SEP 1913], leg. F.L. Stevens No. 3657, ILL 6601; [14 JUN 1915], F.L. Stevens No. 7023, ILL 6606; Mayagüez, [25 JUN 1925], leg.

F.L. Stevens No. 7477, ILL 6602; [14 JUN 1915], leg. F.L. Stevens No. 7038, ILL 6605; [25 JUN 1915], leg. F.L. Stevens No. 7474, ILL 6607; [30 APR 1913], leg. F.L. Stevens No. 974, ILL 6595; [21 OCT 1913], leg. F.L. Stevens No. 3905, ILL 6593; [1 MAY 1913], leg. F.L. Stevens No. 1076, ILL 6597; Mayagüez, [4 AUG 1915], leg. F.L. Stevens No. 9137, ILL 6599; El Alto de la Bandera, [14 JUL 1915], leg. F.L. Stevens No. 8273, ILL 6604; [JUN 1915], leg. F.L. Stevens No. 7559, ILL 6600; Coamo, [6 APR 1913], leg. F.L. Stevens No. 605, ILL 6594; on *Inga vera* Willd., Maricao, [3 APR 1913], leg. F.L. Stevens No. 762, ILL 6596.

Perizomella inquinans H. Sydow. Annales Mycologici 25:106–108. 1927. **Isosyntypes**: Hab. parasitica in stromatibus *Phyllachorae* spec. [parasitans] ad folia *Phoebes costaricanae* [= *Phoebe costaricana* Mez & Pittier], Costa Rica, San Pedro de San Ramón, 23 JAN 1925, leg. H. Sydow, Fungi in Itinere Costaricensi Collecti No. 170d, ILL 11338; 28 JAN 1925, leg. H. Sydow, Fungi in Itinere Costaricensi Collecti No. 406, ILL 11339.

Peronoplasmopara portoricensis E.M.R. Lamkey in F.L. Stevens. Mycologia 12:52–53. 1920. **Syntype**: On *Melia azedarach* L., Porto Rico [Puerto Rico], Florida Adentro, [1 JUL] 1915, leg. F.L. Stevens No. 7687, ILL 2238.

Peronospora celtidis M.B. Waite. Journal of Mycology 7:105–106, pl. 17, figs. 1–16. 1894. **Isosyntypes**: On *Celtis occidentalis* L., Washington, D.C., 7 OCT 1891, leg. M.B. Waite No. 556, ILL 2343; 9 OCT 1891, leg. M.B. Waite No. 557, ILL 2345, ex BPI 71218.

Peronospora cynoglossi T.J. Burrill ex W.T. Swingle. Transactions of the Kansas Academy of Science 11:77. 1889. **Isotype**: [On *Cynoglossum officinale* L.], Illinois, [Tunnel Hill], MAY 1882, leg. A.B. Seymour], ex J.B. Ellis & B.M. Everhart, North American Fungi No. 2206. Anno 1889 [the host cited as *C. virginianum* L. on the Exsiccatum label], ILL 2379.

Peronospora hedeomae W.A. Kellerman & W.T. Swingle. Transactions of the Kansas Academy of Science 11:81. 1889. **Isosyntype**: On *Hedeoma hispida* Pursh, Kansas, Riley County, Manhattan, 20 MAY 1889, leg. W.A. Kellerman & W.T. Swingle No. 1671, ILL 2554.

Peronospora illinoensis W.G. Farlow. Botanical Gazette 8:332–333. 1883, nom. invalid. [provisional name, cf. Article 34.1 of I.C.B.N.]. **Isosyntypes**: On *Parietaria pensylvanica* Muhl. ex Willd, Illinois, [Camp Point, 27 JUN 1882], leg. A.B. Seymour [No. 5302], ILL 2561 and 2982, BPI; [Quincy, 29 JUN 1882], leg. A.B. Seymour [No. 5354], ILL 2562 and 2985, BPI; leg. A.B. Seymour [No. 5355], ILL 2563 and 2981, BPI. **Syntypes**: FH.

Peronospora swinglei J.B. Ellis & W.A. Kellerman. Journal of Mycology 3:104. 1887. **Isotype**: On *Salvia lanceolata* Willd. non Lam. [= *S. reflexa* Hornem.], Kansas, Manhattan, JUN [& JUL] 1887, leg. W.T. Swingle [ex J.B. Ellis & B.M. Everhart, North American Fungi No. 2203], ILL 2780.

Pestalotia foliorum J.B. Ellis & B.M. Everhart. Fungi Columbiani, Century 16, No. 1551. Anno 1901, nom nud. [published as *Pestalozzia* (orth. var.), the generic name corrected in accordance with current I.C.B.N.]. **Isotype**: On leaves of *Diospyros virginiana* L., Alabama, Tuskegee, 15 SEP 1900, leg. G.W. Carver, ILL 13801.

Pestalotia foliorum J.B. Ellis & B.M. Everhart (nom. nud.) var. *rosae* J.B. Ellis & B.M. Everhart. Fungi Columbiani, Century 16, No. 1552. Anno 1901, nom. nud. [published as *Pestalozzia* (orth. var.), the generic name corrected in accordance with current I.C.B.N.]. **Isotype**: On leaves of *Rosa* (sp. cult.), Alabama, Tuskegee, 14 SEP 1900, leg. G.W. Carver, ILL 13802.

Pestalotia lucumae L.R. Tehon. Botanical Gazette 67:508–509. 1919 [cited as *Pestalozzia* (orth. var.), the generic name corrected in accordance with current I.C.B.N.]. **Holotype**: On leaves of *Lucuma multiflora* A. DC. [= *Pouteria multiflora* (A. DC.) Eyma], Porto Rico [Puerto Rico], Monte Alegrillo, 20 JUN 1913, leg. F.L. Stevens No. 2301, ILL 13817.

Pestalotia peregrina J.B. Ellis & G.W. Martin. Journal of Mycology 1:100–101. 1885 [cited as *Pestalozzia* (orth. var.), the generic name corrected in accordance with current I.C.B.N.]. **Isotype**: On dead leaves of *Pinus austriaca* Hoess [= *P. nigra* Arnold], New Jersey, Newfield, MAY 1885, leg. J.B. Ellis [ex J.B. Ellis & B.M. Everhart, North American Fungi No. 1627], ILL 13826.

Petrakiopeltis byrsonimae A.C. Batista, A.F. Vital & R. Ciferri in A.C. Batista, C.A.A. Costa & R. Ciferri. Instituto de Micologia, Universidade do Recife, Brasil. Publicação 90:10–12, figs. 2, 3. 1957; Atti dell' Istituto Botanico dell' Università e Laboratorio Crittogamica di Pavia, Series 5, 15:42–44, figs. 2, 3. 1958. **Holotype**: In foliis *Byrsonimae* sp. [= *Byrsonima* sp.], British Guiana [Guyana], Tumatumari, 11 JUL 1922, leg. F.L. Stevens [No. 109], ILL 6666 [also the holotype of *Parascorias byrsonimae* M. Mendoza in F.L. Stevens].

Pezizella albo-tincta H. Rehm in H. & P. Sydow. Annales Mycologici 2:191. 1904. **Isotype**: Hab. in caulibus putridis *Artemisiae campestris* [= *Artemisia campestris* L., Germany, Brandenburg], Zehlendorf pr. Berlin, [15 OCT 1903], leg. H. Sydow, ex H. & P. Sydow, Mycotheca Germanica No. 127, ILL 7872.

Pezizella aristospora L. Bonar. Mycologia 34:183, fig. 1c, d. 1942. **Isotype**: On dead leaves and small fallen twigs of *Sequoia gigantea* (Lindley)

Decaisne [= *Sequoiadendron giganteum* (Lindley) Buchholz], California, Yosemite National Park, Tuolumne Grove, 23 JUL 1933, leg. L. Bonar, California Fungi No. 660, ILL 7685. **Holotype:** UC 653847.

Pezizella salmonea F.E & E.S. Clements. Cryptogamae Formationum Coloradensium, Century 1, No. 76. Anno 1905, nom. nud. **Isotype:** ... in rimis corticis *Pseudotsugae mucronatae* [*Pseudotsuga mucronata* (Raf.) Sudw. in Holz. = *P. menziesii* (Mirbel) Franco], Colorado, Minnehaha, alt. 2600 m, 8 SEP 1904, leg. F.E. & E.S. Clements, ILL 7881.

Pezoloma griseum F.E. & E.S. Clements ex F.E. Clements. Minnesota Botanical Studies 4:186, pl. 25, fig. 2. 1911; originally distributed in Cryptogamae Formationum Coloradensium, Century 3, No. 292. Anno 1907, as nom. nud. **Isotype:** Ad et inter radiculas udas *Betulae occidentalis* [= *Betula occidentalis* Hooker], Colorado, Minnehaha, [Ruxton Brook], alt. 2700 m, [4] JUL 1906, leg. F.E. & E.S. Clements, ILL 7882.

Phacellula gouaniae H. Sydow. Annales Mycologici 25:139–140. 1927. **Isosyntype:** Hab. in foliis vivis *Gouaniae tomentosae* [= *Gouania tomentosa* Jacquin], ad fluv. Rio Poas inter Sabanilla de Alajuela et San Pedro, 10 JAN 1925, leg. H. Sydow, Fungi in Itinere Costaricensi Collecti No. 392, ILL 8399 [also an isotype of *Dendrodochium gouaniae* H. Sydow].

Phaeocapnias mucronata (J.P.F.C. Montagne) R. Ciferri & A.C. Batista in A.C. Batista & R. Ciferri, 1963. **See basionym:** *Capnodium mucronatum* J.P.F.C. Montagne, 1860.

Phaeociboria garryae E.K. Cash. Mycologia 50:652, fig. 7. (1958) 1959. **Paratype:** On dry leaves from cut brush of *Garrya elliptica* Douglas ex Lindley, California, [Humboldt County], Trinidad, Spruce Cove, JAN 1947, leg. H.E. Parks No. 6943, [ex California Fungi No. 1189], ILL 33053. **Holotype:** BPI. **Isotype:** UC.

Phaeodimeriella asperula H. Sydow. Annales Mycologici 23:333–334. 1925. **Isotype:** Hab. parasitica in mycelio et thyriotheciis *Asterinae acalyphae* [= *Asterina acalyphae* H. Sydow] ad folia *Acalyphae macrostachyae* var. *hirsutissimae* [= *Acalypha macrostachya* Jacquin var. *hirsutissima* (Willd.) Muell. Arg.], Costa Rica, San Ramón, 22 JAN 1925, leg. H. Sydow, Fungi in Itinere Costaricensi Collecti No. 349, ILL 6610 [also an isotype of *Cicinnobella asperula* H. Sydow].

Phaeodimeriella exigua H. Sydow. Annales Mycologici 24:327–329. 1926. **Isotype:** Hab. parasitica in mycelio *Asterinae* spec. indet. [= *Asterina* sp.] ad folia *Roupalae veraguensis* [= *Roupala veraguensis* Klotzsch ex Meisner], Costa Rica, Mondongo pr. San Ramón, 3 FEB 1925, leg. H. Sydow, Fungi in Itinere Costaricensi Collecti No. 229b, ILL 6612 [also an isotype of *Cicinnobella exigua* H. Sydow]. **Paratype:** Parasitica in

Asterinae guaraniticae [= *Asterina guaranitica* C.L. Spegazzini], ad folia *Trichiliae oerstedianae* [= *Trichilia oerstediana* C. DC. in DC.], Costa Rica, San Pedro de San Ramón, 5 FEB 1925, leg. H. Sydow, Fungi in Itinere Costaricensi Collecti No. 124a, ILL 6613.

Phaeodothiopsis eupatorii F.L. Stevens. Botanical Gazette 69:252–253, figs. 16, 17. 1920 [as *Phaeodothopsis*, corrected in accordance with Articles 60 & 61 of I.C.B.N.]. **Holotype:** On *Eupatorium portoricense* Urban, Porto Rico [Puerto Rico], Dos Bocas, below Utuado, [30 DEC 1913], leg. F.L. Stevens No. 6866, ILL 8673 [also a paratype of *Meliola compositarum* F.S. Earle var. *portoricensis* F.L. Stevens and of *Perisporium meliolae* F.L. Stevens]. **Isotype:** ILL 3794. **Paratypes:** Leg. F.L. Stevens No. 6537, ILL 8672; [16 DEC 1913], leg. F.L. Stevens No. 6034, ILL 8671 and 33023; [29 DEC 1913], leg. F.L. Stevens No. 6830, ILL 3515 and 8670 [also paratypes of *Meliola compositarum* F.S. Earle var. *portoricensis* F.L. Stevens].

Phaeodothis costaricensis F.L. Stevens. Illinois Biological Monographs 11:195–196, pl. VIII, figs. 67–68. 1927 [as *Phaedothis*, corrected in accordance with Articles 60 & 61 of I.C.B.N.]. **Holotype:** On unknown member of Rubiaceae, Costa Rica, Siquirres, 31 JUL 1923, leg. F.L. Stevens No. 673, ILL 8674.

Phaeodothis gigantochloae H. Rehm. Leaflets of Philippine Botany 6:2223. 1914. **Possible isosyntype:** Ad *Gigantochloam* emortuam [as *Gigantochloa scribneriana* Merrill on the packet = *G. levis* (Blanco) Merrill], Philippines, Luzon, Laguna Prov., [Mount Maquiling, near] Los Baños, JUL 1913, leg. S.E. Reyes, comm. C.F. Baker [ex Fungi Malayana No. 66, ILL 8675; cited in the protologue as Baker Nos. 1256 and 1257].

Phaeopeltosphaeria panamensis F.L. Stevens & C.M. King in F.L. Stevens. Illinois Biological Monographs 11:202–203, pl. X, figs. 81–83. 1927. **Holotype:** On *Chaetochloa vulpiseta* (Lam.) Hitchc. & Chase [= *Setaria vulpiseta* (Lam.) Roemer & J.A. Schultes], Panama, Gamboa, 16 AUG 1923, leg. F.L. Stevens No. 1350, ILL 10348.

Phaeophragmeriella meliolae (F.L. Stevens) C.G. Hansford, 1946. **See basionym:** *Perisporium meliolae* F.L. Stevens, 1918.

Phaeophragmocauma buddlejae F.L. Stevens. Annales Mycologici 29:103–104, figs. 1–3. 1931 [as *buddleyae*, corrected in accordance with Rec. 60H.1 and Articles 60 & 61 of I.C.B.N.]. **Holotype:** On *Buddleja incana* Ruiz & Pavon [cited as *Buddleya*], Peru, Tarma, 3 DEC 1924, leg. F.L. Stevens No. 35, ILL 8676.

Phaeoramularia manihotis (F.L. Stevens & W.G. Solheim) M.B. Ellis, 1976. **See basionym:** *Ragnhildiana manihotis* F.L. Stevens & W.G. Solheim in W.G. Solheim & F.L. Stevens, 1931.

Phaeosaccardinula dematia V.M. Miller & L. Bonar. University of California Publications in Botany 19:411–412, pl. 67, figs. 8, 9; pl. 68, fig. 2. 1941. **Isotype**: On *Baccharis pilularis* DC. ssp. *consanguinea* (DC.) C.B. Wolf (erect form of *B. pilularis*), California, San Mateo County, Moss Beach, [17] APR 1924, leg. H.E. Parks No. 2139 [ex California Fungi No. 661], ILL 31509. **Holotype**: UC 617322. **Paratypes**: On *Umbellularia californica* (Hooker & Arnott) Nutt., California, Monterey County, Big Sur, [14] AUG 1937, leg. L. Bonar, [California Fungi No. 662], ILL 31510; on *Baccharis pilularis* DC. *typica* C.B. Wolf [= *B. pilularis* ssp. *pilularis*] (prostrate form of *B. pilularis*), Marin County, east of Point Reyes Lighthouse, [14] JUL 1939, leg. L. Bonar, ILL 31401; on *Sequoia sempervirens* (Lambert ex D. Don) Endl., Monterey County, Big Sur, [14] AUG 1937, leg. L. Bonar, [California Fungi No. 608], ILL 33494 [also a paratype of *Chaetasbolisia falcata* V.M. Miller & L. Bonar].

Phaeosaccardinula morindae J.M. Mendoza in F.L. Stevens. Bernice P. Bishop Museum Bulletin 19:59, pl. V(45–48). 1925. **Holotype**: On *Morinda citrifolia* L., Hawaii, Oahu, Hakipuu, on Albert F. Judd's property, 6 JUN [1921], leg. F.L. Stevens No. 572, ILL 7359. **Isotypes**: BISH 499977 and 499978.

Phaeosphaerella dianellae F.L. Stevens. Bernice P. Bishop Museum Bulletin 19:105. 1925. **Holotype**: On *Dianella odorata* sensu Hillebr. non Blume [= *D. sandwicensis* Hooker & Arnott], Hawaii, Kauai, Waimea Canyon, 15 JUN [1921], leg. F.L. Stevens No. 421, ILL 9805 [also the holotype of *Mycosphaerella dianellae* F.L. Stevens & A.G. Weedon in F.L. Stevens]. **Paratypes**: Oahu, Wahiawa, 3 JUN [1921], leg. F.L. Stevens No. 253, ILL 9801; Waiahole ditch trail, 12 JUN [1921], leg. F.L. Stevens No. 405, ILL 9802; Kauai, Kalalau Trail, 16 JUN [1921], leg. F.L. Stevens No. 528, ILL 9804; Maui, 1920, leg. C.N. Forbes No. 1999, ILL 9803 [F.L. Stevens Nos. 253, 405 and 528, as well as C.N. Forbes No. 1999, are also paratypes of *Mycosphaerella dianellae* F.L. Stevens & A.G. Weedon in F.L. Stevens].

Phaeosphaerella hawaiiensis F.L. Stevens & McMunn in F.L. Stevens. Bernice P. Bishop Museum Bulletin 19:105. 1925. **Holotype**: On unknown dicotyledonous host, Hawaii, [Island of] Hawaii, Waimea, 30 JUL [1921], leg. F.L. Stevens No. 1040 [as 1040a on the packet], ILL 9910.

Phaeosphaerella mangiferae F.L. Stevens & A.G. Weedon in F.L. Stevens. Bernice P. Bishop Museum Bulletin 19:105. 1925. **Holotype**: On *Mangifera indica* L. (mango), Hawaii, Oahu, Hakipuu, 19 JUN [1921], leg. F.L. Stevens No. 583, ILL 9912. **Isotypes**: BISH 499037 and 499976.

Phaeospora cacticola F.L. Stevens. Transactions of the Illinois State Academy of Science 10:177, 181. 1917. **Holotype:** On *Rhipsalis cassutha* Gaertner [cited as *cassytha* (orth. var.) = *R. baccifera* (Sol. ex J.S. Miller) Stearn], Porto Rico [Puerto Rico], Monte de Oro, near Cayey, [3 DEC 1913], leg. F.L. Stevens No. 5662, ILL 9436.

Phaeoxyphiella rondeletiae A.C. Batista, M.L. Nascimento & R. Ciferri in A.C. Batista & R. Ciferri. Quaderno Laboratorio Crittogamico Istituto Botanico Della Universita Pavia 31:150–151, pl. 14, fig. 78. 1963. **Isotypes:** On leaves of *Rondeletia* sp., with *Antennariella perseae* A.C. Batista, M.L. Nascimento & R. Ciferri, *Microxyphium columnatum* A.C. Batista, R. Ciferri & M.L. Nascimento, and *Achaetobotrys* sp., Porto Rico [Puerto Rico], Luquillo Forest, 12 FEB 1913 [as 2 DEC 1913 on the packet], leg. F.L. Stevens No. 5569, ILL 33503, BPI. **Holotype:** URM [I.M.U.R.] 13101.

Phialea incertella H. Rehm in H. & P. Sydow. Annales Mycologici 4:485. (1906) JAN 1907. **Isotype:** Hab. ad folia emortua *Koeleriae cristatae* [*Koeleria cristata* (L.) Persoon p.p. (as *K. ciliata* Kerner in Baenitz on the packet) = *K. macrantha* (Ledeb.) J.A. Schultes (sensu lato), Germany, Thüringen: Steiger], pr. Erfurt, [JUN–JUL 1906], leg. H. Diedicke, ex H. & P. Sydow, Mycotheca Germanica No. 505, ILL 7890.

Phialea turbinata H. Sydow in H. & P. Sydow. Annales Mycologici 5:397. 1907. **Isotype:** Hab. in caulibus putrescentibus *Ranunculi* sp. [= *Ranunculus* sp., Germany], Hessen-Nassau, Eube pr. Gersfeld, 6 JUL 1907, leg. H. Sydow, Mycotheca Germanica No. 599, ILL 7894.

Phialetea aerospora A.C. Batista & M.L. Nascimento in A.C. Batista, H. da Silva Maia, J.A. de Lima & E.A.F. da Matta. Atas do Instituto de Micologia 1:264–266, figs. 14, 15. 1960. **Assumed isotypes:** In foliis *Amomis caryophyllatae* [*Amomis caryophyllata* (Jacquin) Krug & Urban = *Pimenta racemosa* (P. Miller) J.W. Moore] with *Meliola amomicola* F.L. Stevens and *Stomiopeltella machadoi* A.C. Batista & J.A. de Lima in A.C. Batista, Porto Rico [Puerto Rico], Monte Alegrillo, 14 NOV 1913, leg. F.L. Stevens [No. 4757], ILL 33502, BPI [also assumed isotypes of *Meliola amomicola* F.L. Stevens var. *longispora* A.C. Batista]. **Holotype:** URM 14099.

Philonectria insignis (F. Petrak & R. Ciferri) J.A. v. Arx var. *macrospora* A.C. Batista & A.F. Vital in A.C. Batista, C.A.A. Costa & A.F. Vital. Anais da Sociedade de Biologia de Pernambuco 15:502–504, fig. 9. 1957. **Holotype:** In mycelium *Sporochismi* sp. [= *Sporochisma* sp.] in folii *Rollandii racemosae* [*Rollandia racemosa* (H. Mann) Hillebr. = *R. humboldtiana* Gaudich.], sub *Limaciniopsis rollandiae* J.M. Mendoza, Hawaii, Oahu, Waiahole ditch trail, [12 JUN 1921], leg. F.L. Stevens [No. 407], ILL 6663 [erroneously cited as 663; also the holotype of *Limaciniopsis rollandiae* J.M. Mendoza in F.L. Stevens].

Phleospora callistea H. Sydow. Annales Mycologici 7:439–440. 1909. **Isotype:** Hab. ad folia viva *Osmundae regalis* [= *Osmunda regalis* L.], Germaniae [Mecklenburg: Moorwald], pr. Müritz, 14 AUG 1909, leg. H. Sydow, Mycotheca Germanica No. 785, ILL 9467 [anamorph (as Pycnidiumform) and also an isotype of *Sphaerella callistea* H. Sydow].

Phleospora muhlenbergiae R. Sprague & W.G. Solheim in W.G. Solheim. Mycologia 41:628. 1949. **Isotype:** On living leaf blades and sheaths of *Muhlenbergia arizonica* Scribner, Arizona, [Pima County], Santa Catalina Mountains, east of Tucson, Oak Flats Picnic Grounds, 12 NOV 1948, leg. W.G. Solheim & R. Solheim, ex Mycoflora Saximontanensis Exsiccata No. 478, ILL 20892. **Holotype:** RMS 2448.

Phlyctaena bromi F.E. & E.S. Clements. Cryptogamae Formationum Coloradensium, Century 3, No. 258. Anno 1907, nom. nud. **Isotype:** ... in vaginulis et culmis emortuis *Bromi ciliati* [= *Bromus ciliatus* L.], Colorado, Halfway, alt. 2700 m, 25 JUN 1906, leg. F.E. & E.S. Clements, ILL 11378.

Phlyctaena campanulae F.E. & E.S. Clements. Cryptogamae Formationum Coloradensium, Century 3, No. 257. Anno 1907, nom. nud. **Isotype:** ... in caulibus emortuis *Campanulae petiolatae* [*Campanula petiolata* A. DC. = *C. rotundifolia* L.], Colorado, Cross Ruxton, alt. 2600 m, 13 AUG 1906, leg. F.E. & E.S. Clements, ILL 11379.

Phoenicostroma chamaedoreae H. Sydow. Annales Mycologici 23:345–348, fig. 5. 1925. **Isotype:** Hab. in foliis vivis vel languidis *Chamaedoreae bifurcatae* [= *Chamaedorea bifurcata* Oersted], Costa Rica, Monte Poas, pr. Grecia, 15 JAN 1925, leg. H. Sydow, Fungi in Itinere Costaricensi Collecti No. 196, ILL 8677 [also an isotype of *Lembosia poasensis* H. Sydow].

Pholidota subpraecox F.E. & E.S. Clements. Cryptogamae Formationum Coloradensium, Century 4, No. 372. Anno 1907, nom. nud. **Isotype:** ... ad terram ..., Colorado, Mushroom Park, alt. 2800 m, 21 JUL 1906, leg. F.E. & E.S. Clements, ILL 30303.

Pholidota trachyspora F.E. & E.S. Clements. Cryptogamae Formationum Coloradensium, Century 4, No. 373. Anno 1907, nom. nud. **Isotype:** ... ad terram ..., Colorado, Sugar Loaf Park, alt. 2500 m, 13 AUG 1906, leg. F.E. & E.S. Clements, ILL 30304.

Phoma antennariae F.E. & E.S. Clements. Cryptogamae Formationum Coloradensium, Century 3, No. 243. Anno 1907, nom. nud. **Isotype:** ... gregarius in foliis vivis languidisque *Antennariae anaphaloidis* [= *Antennaria anaphaloides* Rydb.], Colorado, Sugar Loaf Park, alt. 2600 m, 13 AUG 1906, leg. F.E. & E.S. Clements, ILL 11395.

Phoma conigena P.A. Karsten var. *abieticola* P.A. Saccardo in H. & P. Sydow. Annales Mycologici 3:233. 1905. **Isotype:** Hab. in squamis emortuis conorum *Abietis excelsae* [*Abies excelsa* (Lam.) Poiret = *Picea abies* (L.) H. Karsten, Germany], Thüringen: [Forstbezirk Bürgerholz], pr. Sondershausen, 10 MAR 1905, leg. G. Oertel, ex H. & P. Sydow, Mycotheca Germanica No. 333, ILL 11404.

Phoma frigida P.A. Saccardo. Annales Mycologici 6:561. 1908. **Isotype:** Hab. in ramis junioribus *Populi tremulae* [= *Populus tremula* L., Germany, Brandenburg]: Rüdnitz pr. Bernau, [26] MAY 1907, leg. H. Sydow, [Mycotheca Germanica No. 710], ILL 11412.

Phoma herbarum G.D. Westendorp var. *grindeliae* F.E. & E.S. Clements. Cryptogamae Formationum Coloradensium, Century 5, No. 471. Anno 1908, nom. nud. **Isotype:** ... in caulibus vetustis *Grindeliae squarrosae* [= *Grindelia squarrosa* (Pursh) Dunal], Colorado, Boulder, alt. 1600 m, 31 JUL 1907, leg. F.E. & E.S. Clements, ILL 11420.

Phoma oleracea P.A. Saccardo var. *bryoniae* P.A. Saccardo. Annales Mycologici 7:435. 1909. **Isotype:** Hab. in caulibus emortuis *Bryoniae albae* [= *Bryonia alba* L., Brandenburg: Baumschulen] Tamsel [now Poland, 15 MAY 1909], leg. P. Vogel No. 73, [ex H. Sydow, Mycotheca Germanica No. 809], ILL 11434.

Phoma torilis H. Sydow. Annales Mycologici 8:492. 1910. **Isotype:** Hab. in caulibus siccis *Torilis anthrisci* [*Torilis anthriscus* (L.) C.C. Gmelin non Gaertner = *T. japonica* (Houtt.) DC., Germany, Brandenburg]: Tiefensee pr. Werneuchen, 30 MAY 1909, leg. H. Sydow, Mycotheca Germanica No. 914, ILL 11457.

Phoma tremulae P.A. Saccardo in H. & P. Sydow. Annales Mycologici 2:529. 1904. **Isotype:** Hab. in ramis *Populi tremulae* [= *Populus tremula* L., Germany, Brandenburg]: Zehlendorf pr. Berlin, [8 NOV 1903], leg. H. Sydow, ex H. & P. Sydow, Mycotheca Germanica No. 264, ILL 11458.

Phoma ulicis H. & P. Sydow. Annales Mycologici 3:420. 1905. **Isotype:** Hab. in spinis *Ulicis europaeae* [= *Ulex europaeus* L., Germany, Pommern: zwischen] Lobbe [und Göhren], Insel Rügen, 24 AUG 1905, leg. H. & P. Sydow, Mycotheca Germanica No. 411, ILL 11460.

Phomatospora migrans H. Rehm. Leaflets of Philippine Botany 8:2936. 1916. **Assumed isotype:** Ad *Arengam sacchariferam* [*Arenga saccharifera* Labill. = *A. pinnata* (Wurmb.) Merrill], Philippines, [Luzon, Laguna Prov.], Los Baños, AUG 1913, leg. S.A. Reyes, comm. C.F. Baker No. 1455 [ex Fungi Malayana No. 177], ILL 10296.

Phomopsis bertholletianum E.R. Spencer. Botanical Gazette 72:288. 1921. **Holotype:** On Brazil nuts [= *Bertholletia* sp.], Illinois, [from wholesale firms in Chicago and retail grocery stores in Champaign and Urbana — a single packet with two unmarked vials], 1920, leg. E.R. Spencer s.n., ILL 11475.

Phomopsis loti J. Upadhyay. Phytopathology 56:764–765, figs. 1–5. 1966.
Holotype: In seminibus *Loti corniculati* culta [= *Lotus corniculatus* L.], Illinois, Champaign County, Urbana, Agricultural Experiment Station, Agronomy South Farm, [SEP 1958], leg. J.W. Gerdemann & J. Upadhyay s.n., ILL 31866. **Isotypes:** ILL 31866a and 31866b.

Phomopsis oblita P.A. Saccardo. Annales Mycologici 8:343. 1910. **Isotype:** Hab. in caulibus *Artemisiae absinthii* [= *Artemisia absinthium* L.] morientibus, [Germany, Mecklenburg]: pr. Graal, 19 AUG 1909 [as 9 AUG on the packet], leg. H. Sydow, [Mycotheca Germanica No. 915], ILL 11480.

Phragmocapnias smilacina J.M. Mendoza in F.L. Stevens. Bernice P. Bishop Museum Bulletin 19:58–59, pl. IV(41–44). 1925. **Syntypes** [BISH numbers cited are isosyntypes]: On *Smilax* sp., Hawaii, Oahu, Olympus, 24 JUN [1921], leg. F.L. Stevens No. 981, ILL 6667, BISH 499975; leg. F.L. Stevens No. 670, ILL 6668, BISH 499972, 499973 and 499974 [the host erroneously cited in the protologue and also labeled on the packet as *Pelea* sp. ILL 6667 is also the holotype of *Ainsworthia smilacina* A.C. Batista & A.F. Vital and of *Trichopeltum hawaiiense* A.C. Batista & C.A.A. Costa in A.C. Batista, C.A.A. Costa & R. Ciferri (the latter erroneously cited as being in NY). ILL 6668 is also the holotype of *Plochmopeltidella smilacina* J.M. Mendoza in F.L. Stevens & H.W. Manter and of *Pycnidiopeltis smilacina* A.C. Batista & C.A.A. Costa in A.C. Batista & R. Ciferri.].

Phragmopeltis callista H. Sydow. Annales Mycologici 25:109–110. 1927. **Isotype:** Hab. in foliis vivis *Ocoteae insularis* [= *Ocotea insularis* (Meisner) Mez], Costa Rica, Piedades de San Ramón, 26 JAN 1925, leg. H. Sydow, Fungi in Itinere Costaricensi Collecti No. 168a, ILL 13184.

Phragmopeltis phoebes H. Sydow. Annales Mycologici 25:111–113. 1927. **Isotype:** In foliis vivis *Phoebes costaricanae* [= *Phoebe costaricana* Mez & Pittier], Costa Rica, San Pedro de San Ramón, 23 JAN 1925, leg. H. Sydow, Fungi in Itinere Costaricensi Collecti No. 170a, ILL 7636 [also an isotype of *Hysterostomella phoebes* H. Sydow].

Phragmothyriella bakeri H. Rehm. Leaflets of Philippine Botany 6:2230–2231. 1914. **Assumed isosyntype:** Ad *Schizostachyum* sp. emortuum, Philippines, Luzon, Laguna Prov., Los Baños, NOV 1913, leg. S.A. Reyes, comm. C.F. Baker No. 1968b [ex Fungi Malayana No. 68], ILL 33195.

Phragmothyriella luzonensis F.L. Stevens & M. Schneider. University of the Philippines, Natural and Applied Science Bulletin 3:24–25, fig. 3. 1933. **Holotype:** On Celastraceae: *Euonymus javanicus* Blume, Philippines, [Luzon], Laguna [Prov.], Mount Maquiling, 18 JAN 1931, leg. F.L. Stevens No. 1929, ILL 6773.

Phyllachora aegiphilae F.L. Stevens. Illinois Biological Monographs 8:184, pl. V, fig. 35. 1923. **Holotype:** On *Aegiphila* sp., British Guiana [Guyana], Rockstone, 17 JUL 1922, leg. F.L. Stevens No. 458, ILL 8682.

Phyllachora amphibola H. Sydow. Annales Mycologici 24:383–385. 1926. **Isotype:** Hab. in foliis *Ingae* spec. [= *Inga* sp.], Costa Rica, Aserri, 26 DEC 1924, leg. H. Sydow, Fungi in Itinere Costaricensi Collecti No. 15, ILL 8685.

Phyllachora aserriensis H. Sydow. Annales Mycologici 25:4–5. 1927. **Isotype:** In foliis *Paspali paniculati* [= *Paspalum paniculatum* L.], Costa Rica, Aserri, 26 DEC 1924, leg. H. Sydow, Fungi in Itinere Costaricensi Collecti No. 175, ILL 8698.

Phyllachora banisteriae F.L. Stevens & N.E. Dalbey. Botanical Gazette 68:54, pl. VI, figs. 1, 2. 1919. **Holotype:** On *Banisteria tomentosa* Schlecht. [= *Heteropteris* sp.], Porto Rico [Puerto Rico], Vega Baja, [1915], leg. F.L. Stevens No. 8341, ILL 8702 [the specimen is missing from the packet].

Phyllachora baphispora H. Sydow. Annales Mycologici 24:386–387. 1926. **Isotype:** Hab. in foliis *Verbesinae myriocephalae* [= *Verbesina myriocephala* Schultz-Bip.], Costa Rica, Aserri, 1 JAN 1925, leg. H. Sydow, Fungi in Itinere Costaricensi Collecti No. 227, ILL 8849.

Phyllachora bourreriae F.L. Stevens & N.E. Dalbey. Botanical Gazette 68:54, pl. VI, figs. 3–4. 1919. **Holotype and isotypes:** On *Bourreria succulenta* Jacquin, Porto Rico [Puerto Rico], Vega Alta, [NOV 1913], leg. F.L. Stevens No. 4149, ILL 8703 [represented by three packets]. **Paratype:** [Alta] Jayuda [as Joyuda, 31 MAR 1913], leg. F.L. Stevens No. 4770 [as 4770Y in the protologue], ILL 8704.

Phyllachora byttneriae F.L. Stevens. Annales Mycologici 29:104. 1931 [as *buettneriae*, corrected in accordance with Rec. 60H.1 and Articles 60 & 61 of I.C.B.N. . **Holotype:** On *Byttneria brevipes* Bentham [cited as *Buettneria*, nom. rej.], Ecuador, Barrn'nital, 17 NOV 1924, leg. F.L. Stevens No. 341, ILL 8708.

Phyllachora canafistulae F.L. Stevens & N.E. Dalbey. Botanical Gazette 68:55, pl. VI, figs. 5, 6. 1919. **Holotype:** On *Cassia fistula* L., Porto Rico [Puerto Rico], Mayagüez, [14 JUN 1915], leg. F.L. Stevens No. 7022, ILL 8710.

Phyllachora casimiroae F.L. Stevens & G.E. King in F.L. Stevens. Illinois Biological Monographs 11:185–186, pl. VII, fig. 57; pl. VIII, fig. 58; pl. XVI, fig. 111. 1927. **Syntypes:** On *Casimiroa tetrameria* Millsp., Costa Rica, El Alto, 6 JUL 1923, leg. F.L. Stevens No. 233, ILL 4002 [also the holotype of *Asteridiella casimiroae* C.G. Hansford]; leg. F.L. Stevens No. 234, ILL 8712.

Phyllachora chaetochloae F.L. Stevens. Illinois Biological Monographs

8:184, pl. V, figs. 36–38; pl. XIV, fig. 97. 1923. **Holotype:** On *Chaetochloa tenax* (L.C. Richard) A.S. Hitchc. [= *Setaria tenax* (L.C. Richard) Desv.], Trinidad, Cumuto, 15 AUG 1922, leg. F.L. Stevens No. 882, ILL 8724.

Phyllachora circinata H. Sydow var. *sanguinea* H. Rehm. Leaflets of Philippine Botany 6:2274–2275. 1914. **Assumed isotype:** Ad *Ficum heterophyllam* [= *Ficus heterophylla* L.f.], Philippines, [Luzon, Laguna Prov.], Los Baños, JAN 1914, leg. C.F. Baker No. 2606, [ex Fungi Malayana No. 70], ILL 8736.

Phyllachora congruens H. Rehm. Leaflets of Philippine Botany 6:2220–2221. 1914. **Assumed isosyntype:** Ad folia *Panici carinati* [= *Panicum carinatum* J. & C. Presl], Philippines, Luzon, Laguna Prov., Los Baños, SEP 1913, leg. M.B. Raimundo, comm. C.F. Baker Nos. 1725 and 1815 [ex Fungi Malayana No. 71], ILL 8737.

Phyllachora dasylirii (C.H. Peck) P.A. Saccardo, 1883. **See basionym:** *Dothidea dasylirii* C.H. Peck, 1882.

Phyllachora davillae F.L. Stevens. Annales Mycologici 28:282. 1930. **Holotype:** On *Davilla rugosa* Poiret, Panama, Sweetwater, Fort Sherman, 6 OCT 1924, leg. F.L. Stevens No. 1075, ILL 8860.

Phyllachora dimorphandrae F.L. Stevens. Illinois Biological Monographs 8:185, pl. V, figs. 39–41; pl. XIV, fig. 96. 1923. **Holotype:** On *Dimorphandra* sp., British Guiana [Guyana], Demerara-Essequibo R.R., 15 JUL 1922, leg. F.L. Stevens No. 333, ILL 8864. **Paratypes:** Wismar, 14 JUL 1922, leg. F.L. Stevens No. 291, ILL 8865; Kartabo, 22 JUL 1922, leg. F.L. Stevens No. 629, ILL 8866.

Phyllachora distichlidis F.L. Stevens. Annales Mycologici 29:104–105. 1931. **Holotype:** On *Distichlis thalassica* E. Desv., Peru, Callao, 18 DEC 1924, leg. F.L. Stevens No. 243, ILL 8868.

Phyllachora dolichogena (M.J. Berkeley & C.E. Broome) P.A. Saccardo ssp. *phaseolina* (H. Sydow in H. &. P. Sydow) P.F. Cannon, 1991. **See basionym:** *Phyllachora phaseolina* H. Sydow in H. & P. Sydow, 1913.

Phyllachora donacina H. Rehm. Leaflets of Philippine Botany 6:2222. 1914. **Assumed isosyntype:** Ad culmum *Donacis cannaeformis* [= *Donax cannaeformis* (G. Forster) K. Schum., Philippines, Luzon, Laguna Prov., Los Baños, JUL 1913, leg. S.A. Reyes, comm. C.F. Baker No. 1271, ILL 8869.

Phyllachora drypeticola F.L. Stevens & N.E. Dalbey. Botanical Gazette 68:55, pl. VI, figs. 7, 8. 1919. **Holotype:** On leaves of *Drypetes* sp., Porto Rico [Puerto Rico], Rio Tanamá, Arecibo, [6 JUL 1915], leg. F.L. Stevens No. 7828, ILL 8879. **Paratypes:** On leaves of *Drypetes glauca* Vahl, Maricao [as Maracao in the protologue, 13 NOV 1913], leg. F.L. Stevens No. 4508, ILL 8870 and 8872; [4 NOV 1913], leg. F.L. Stevens

No. 4472, ILL 8877; [10 MAY 1913], leg. F.L. Stevens No. 1353, ILL 8876; [4 MAR 1913], leg. F.L. Stevens No. 730, ILL 8875; [13 NOV 1913], leg. F.L. Stevens No. 8558D, ILL 8871; [as Monte Alegrillo on the packet, 14 NOV 1913], leg. F.L. Stevens No. 4746, ILL 8880; El Gigante, [10 JUL 1915], leg. F.L. Stevens No. 8558, ILL 8878; Utuado, [8 NOV 1913], leg. F.L. Stevens No. 4387, ILL 8873; Mayagüez, [20 MAY 1913], leg. F.L. Stevens No. 1834, ILL 8874.

Phyllachora elmeri H. Sydow in H. Rehm. Leaflets of Philippine Botany 6:2219. 1914, nom. nud. **Possible Isotype:** Ad *Ficum ulmifoliam* [= *Ficus ulmifolia* Lam.], Philippines, Luzon, Laguna Prov., [Mount Maquiling], near Los Baños, OCT 1913, leg. M.B. Raimundo, comm. C.F. Baker No. 1973 [ex Fungi Malayana No. 74], ILL 8881.

Phyllachora freycinetiae F.L. Stevens. Bernice P. Bishop Museum Bulletin 19:22, pl. II(C). 1925. **Holotype:** On *Freycinetia arnottii* Gaudich. [= *F. arborea* Gaudich.], Hawaii, Oahu, Kalihi Valley, 2 JUN [1921], leg. F.L. Stevens No. 184, ILL 8898. **Isotypes:** BISH 500003 and 500342.

Phyllachora genipae F.L. Stevens & N.E. Dalbey. Botanical Gazette 68:55–56, pl. VIII, figs. 27, 28. 1919 [as *gnipae*, corrected in accordance with Articles 60 & 61 of I.C.B.N.]. **Holotype:** On *Genipa americana* L. [misspelled "*Gnipa*"], Porto Rico [Puerto Rico], El Gigante, [10 JUL 1915], leg. F.L. Stevens No. 8520, ILL 8754.

Phyllachora greciana H. Sydow. Annales Mycologici 24:390–391. 1926. **Isotype:** Hab. in foliis *Cyperi feracis* [= *Cyperus ferax* L.C. Richard], Costa Rica, Grecia, 20 JAN 1924, leg. H. Sydow, Fungi in Itinere Costaricensi Collecti No. 263, ILL 8764.

Phyllachora guianensis F.L. Stevens. Illinois Biological Monographs 8:185–186. 1923. **Syntypes:** On *Paspalum virgatum* L., British Guiana [Guyana], Tumatumari, 9 JUL 1922, leg. F.L. Stevens No. 32, ILL 8842; 10 JUL 1922, leg. F.L. Stevens No. 142, ILL 8845; Georgetown, Lamada canal, 2 AUG 1922 [as 22 AUG on the packet], leg. F.L. Stevens No. 712, ILL 8847; Coverden, 4 AUG 1922, leg. F.L. Stevens No. 730, ILL 8844.

Phyllachora heterotrichi F.L. Stevens & N.E. Dalbey. Botanical Gazette 68:56, pl. VI, figs. 9, 10. 1919 [as *heterotrichae*, corrected in accordance with Articles 32.6 & 61 of I.C.B.N.]. **Holotype:** On *Heterotrichum cymosum* (Wendl. ex Sprengel) Urban, Porto Rico [Puerto Rico], Villa Alba, [3 JAN 1912], leg. F.L. Stevens No. 116, ILL 8853. **Isotype:** ILL 9731. [Stevens No. 116 is also cited as a paratype of *Guignardia heterotrichi* F.L. Stevens.].

Phyllachora icacoreae F.L. Stevens. Illinois Biological Monographs 11:187, pl. VIII, fig. 59; pl. XVI, fig. 112. 1927. **Holotype:** On *Icacorea* sp. [= *Ardisia* sp.], Costa Rica, Cartago, 23 JUL 1923, leg. F.L. Stevens No.

46, ILL 8855. **Paratypes:** Leg. F.L. Stevens No. 72, ILL 8857; leg. F.L. Stevens No. 89, ILL 8856.

Phyllachora ingicola H. Sydow. Annales Mycologici 24:391–392. 1926. **Isotype:** Hab. in foliis *Ingae verae* [=*Inga vera* Willd.; as *I.* pr. *goldmanii* Pittier on the packet], Costa Rica, San Pedro de San Ramón, 23 JAN 1925, leg. H. Sydow, Fungi in Itinere Costaricensi Collecti No. 108d, ILL 8858.

Phyllachora ischaemi L.R. Tehon. Botanical Gazette 67:507, pl. XVIII, figs. 2, 3. 1919 [as *ischmaemi*, corrected by A. Trotter (1926) in accordance with current I.C.B.N.], nom. illegit., non H. Sydow, 1915. **Holotype:** On *Ischaemum latifolium* Kunth [misspelled "*Ischmaemum*"], Martinique, St. Pierre, [4 AUG 1913], leg. F.L. Stevens No. 2972, ILL 8859. ≡ *Phyllachora tehonis* A. Trotter. Sylloge Fungorum 24:580. 1926, nom. nov.

Phyllachora juraensis P.C. Hennings var. *minima* F.L. Stevens. Illinois Biological Monographs 11:187. 1927. **Holotype:** On *Brownea* sp., Panama, Tapia, 15 AUG 1923, leg. F.L. Stevens No. 1025, ILL 8902. **Isotype:** ILL 8903.

Phyllachora lasiacis H. Sydow. Annales Mycologici 23:374–375. 1925. **Isotype:** In foliis *Lasiacis divaricata* (L.) A.S. Hitchc., Costa Rica, La Caja pr. San José, 24 DEC 1924, leg. H. Sydow, Fungi in Itinere Costaricensi Collecti No. 203, ILL 8907.

Phyllachora mauriae H. Sydow. Annales Mycologici 23:376–378. 1925. **Isosyntype:** Hab. in foliis *Mauriae glaucae* [= *Mauria glauca* J. Donnell Smith], Costa Rica, La Caja pr. San José, 14 FEB 1925, leg. H. Sydow, Fungi in Itinere Costaricensi Collecti No. 327, ILL 8926.

Phyllachora mayepeae F.L. Stevens & N.E. Dalbey. Botanical Gazette 68:56–57, pl. VII, fig. 14. 1919. **Holotype:** On *Mayepea domingensis* (Lam.) Krug & Urban [= *Chionanthus domingensis* Lam.], Porto Rico [Puerto Rico], Maricao, [3 APR 1913], leg. F.L. Stevens No. 785, ILL 8938. **Paratypes:** Maricao, [3 APR 1913], leg. F.L. Stevens No. 775, ILL 8939; [4 MAR 1913], leg. F.L. Stevens No. 731, ILL 8935; [3 APR 1913], leg. F.L. Stevens No. 765, ILL 8937; [14 NOV 1913], leg. F.L. Stevens No. 4751, ILL 8936; [10 JAN 1913], leg. F.L. Stevens No. 196, ILL 8934; [19 JUL 1915], leg. F.L. Stevens No. 8787, ILL 8932; [Monte Alegrillo, 1915], leg. F.L. Stevens No. 4720, ILL 8941; Mayagüez Mesa, [25 JUN 1915], leg. F.L. Stevens No. 7471, ILL 8940; 29 JUN 1915, leg. F.L. Stevens No. 7585, ILL 8930; Coamo, 1 JAN 1913, leg. F.L. Stevens No. 148, ILL 8931.

Phyllachora meibomiae F.L. Stevens. Illinois Biological Monographs 11:188–189, pl. VIII, fig. 60; pl. XVI, fig. 113. 1927. **Holotype:** On *Meibomia* sp. [= *Desmodium* sp.], Costa Rica, Santa Cecelia [as San

Cecilia], 7 AUG 1923, leg. F.L. Stevens No. 756, ILL 8943. **Paratype:** Leg. F.L. Stevens No. 745, ILL 8942.

Phyllachora metastelmatis F.L. Stevens & N.E. Dalbey. Botanical Gazette 68:57, pl. VII, figs. 15, 16. 1919 [as *metastelmae*, corrected in accordance with Rec. 60H.1 and Articles 60 & 61 of I.C.B.N.]. **Holotype:** On stems of *Metastelma* sp., Porto Rico [Puerto Rico], El Alto de la Bandera, [16 JUL 1915], leg. F.L. Stevens No. 8715, ILL 8944.

Phyllachora myrsinicola E.M. Doidge. Bothalia 1(2):81. 1922. **Isotype:** Hab. in foliis *Myrsine melanophleos* (L.) R. Br. [= *Rapanea melanophloeos* (L.) Mez], South Africa, Natal Prov., Duncairn, near Maritzburg [Pietermaritzburg], 13 JUL 1921, leg. E.M. Doidge, ex PREM No. 15015, ILL 8947. **Holotype:** PREM.

Phyllachora nectandrae F.L. Stevens & N.E. Dalbey. Botanical Gazette 68:57, pl. VIII, figs. 23, 24. 1919. **Holotype:** On *Nectandra patens* (Sw.) Griseb. [= *Ocotea patens* (Sw.) Nees], Porto Rico [Puerto Rico], Maricao, [20 SEP 1913], leg. F.L. Stevens No. 3608, ILL 8949. **Paratypes:** [8 OCT 1913], leg. F.L. Stevens No. 3435, ILL 8950; [20 SEP 1913], leg. F.L. Stevens No. 3730, ILL 8948; [19 JUL 1915], leg. F.L. Stevens No. 8949, ILL 8951.

Phyllachora nitens P. Garman. Mycologia 7:339. 1915, nom. illegit., non (J.H. Léveillé) M.C. Cooke, 1885. **Holotype:** On leaves of *Schlegelia brachyantha* Griseb., Porto Rico [Puerto Rico], Maricao, [3 APR 1913], leg. F.L. Stevens No. 873, ILL 8957. **Paratypes:** Leg. F.L. Stevens No. 857, ILL 8958; Ponce, [8 NOV 1913], leg. F.L. Stevens No. 4352, ILL 8953; Monte Alegrillo, [NOV 1913], leg. F.L. Stevens No. 4501, ILL 8954; Rio Grande, [13 NOV 1913,] leg. F.L. Stevens No. 4502, ILL 8955 and 8959; Preston's Ranch, [JAN 1914], leg. F.L. Stevens No. 6776, ILL 8961.

Phyllachora nitens (J.H. Léveillé) M.C. Cooke ssp. *isthmea* P.F. Cannon. Mycological Papers. International Mycological Institute 163:147. 1991, nom. nov. **Based on:** *Catacauma galactiae* F.L. Stevens. Annales Mycologici 29:102-103. 1931, non *Phyllachora nitens* (J.H. Léveillé) M.C. Cooke ssp. *galactiae* (F.S. Earle) P.F. Cannon, 1991. **Holotype:** On *Galactia speciosa* (DC.) Britton, Peru, Palca, 6 DEC 1924, leg. F.L. Stevens No. 36, ILL 8499.

Phyllachora ocoteicola F.L. Stevens & N.E. Dalbey. Botanical Gazette 68:57, pl. VIII, figs. 25, 26. 1919. **Holotype:** On *Ocotea leucoxylon* (Sw.) de Lanessan, Porto Rico [Puerto Rico], Monte Alegrillo, [4 NOV 1913], leg. F.L. Stevens No. 4768, ILL 8965. **Paratypes:** [14 NOV 1913], leg. F.L. Stevens No. 4767, ILL 8964; leg. F.L. Stevens No. 4725, ILL 8506 and 8963 [also the holotype and an isotype, respectively, of *Catacauma ocoteae* F.L. Stevens]; Monte de Oro, [3 DEC 1913], leg. F.L. Stevens No. 5669, ILL 8966.

Phyllachora ocoteicola F.L. Stevens & N.E. Dalbey [as Dalby] var. *costaricensis* F.L. Stevens. Illinois Biological Monographs 11:189. 1927. **Holotype:** On *Ocotea* sp., Costa Rica, Peralta, 12 JUL 1923, leg. F.L. Stevens No. 390, ILL 8967.

Phyllachora orbicula H. Rehm. Leaflets of Philippine Botany 6:2221–2222. 1914. **Assumed isotype:** Ad folia *Bambusae blumeanae* [= *Bambusa blumeana* Blume ex J.H. Schultes], Philippines, Luzon, Laguna Prov., Los Baños, AUG 1913, leg. S.A. Reyes, comm. C.F. Baker No. 1603, [ex Fungi Malayana No. 79], ILL 8970.

Phyllachora orbicularis C.L. Spegazzini. Boletín de la Academia Nacional de Ciencias en Córdoba 26:357–358, illustr. Preprint 1923 [journal part issued in 1924]. **Probable isotype:** Hab. En las hojas vivas de *Cordia nitida* Vahl in West [= *C. laevigata* Lam.], Porto Rico [Puerto Rico], Martín Peña, [11 AUG 1915], leg. F.L. Stevens No. 9329, ILL 4127 [also the assumed holotype of *Asteridiella longipoda* (A. Gaillard) C.G. Hansford ex C.G. Hansford var. *minor* C.G. Hansford ex C.G. Hansford]. **Holotype:** LPS 237.

Phyllachora panamensis F.L. Stevens. Illinois Biological Monographs 11:189–190. 1927. **Syntypes:** On *Rourea glabra* Kunth, Panama, Gamboa, 16 AUG 1923, leg. F.L. Stevens No. 1094, ILL 8974; Juan Mina, 18 AUG 1923, leg. F.L. Stevens No. 1351, ILL 8973.

Phyllachora parilis H. Sydow. Annales Mycologici 25:3–4. 1927. **Isotype:** Hab. in foliis *Paspali candidi* [= *Paspalum candidum* (Humb. & Bonpl.) Kunth], Costa Rica, Aserri, 26 DEC 1924, leg. H. Sydow, Fungi in Itinere Costaricensi Collecti No. 202 p.p., ILL 8975.

Phyllachora phaseolina H. Sydow in H. & P. Sydow. Philippine Journal of Science. Section C (Botany) 8:494. 1913. **Isoneotypes:** On *Phaseolus calcaratus* Roxb. [= *Vigna umbellata* Thunb.) Ohwi & Ohashi], Philippines, Luzon, Laguna Prov. [Mount Maquiling], Los Baños., SEP 1913, leg. C.F. Baker, ex Fungi Malayana No. 81, ILL 9041, BPI, K, NY. **Neotype:** S [designated by P.F. Cannon (1991)]. ≡ *Phyllachora dolichogena* (M.J. Berkeley & C.E. Broome) P.A. Saccardo ssp. *phaseolina* (H. Sydow in H. &. P. Sydow) P.F. Cannon. Mycological Papers. International Mycological Institute 163:108–109. 1991.

Phyllachora phoebes H. Sydow. Annales Mycologici 24:397–398. 1926. **Isotype:** Hab. in foliis *Phoebes tonduzii* [= *Phoebe tonduzii* Mez], Costa Rica, Grecia, 19 JAN 1925, leg. H. Sydow, Fungi in Itinere Costaricensi Collecti No. 160b, ILL 9043.

Phyllachora picramniae F.L. Stevens. Illinois Biological Monographs 11:190–191, pl. VIII, figs. 61–62; pl. XVII, fig. 121. 1927. **Holotype:** On *Picramnia bonplandiana* L.R. Tulasne [= *P.* cf. *antidesma* Sw.], Costa Rica, Aserri, 26 JUN 1923, leg. F.L. Stevens No. 119, ILL 9046.

Paratype: Desamparados, 27 JUN 1923, leg. F.L. Stevens No. 138, ILL 9045; on *Picramnia antidesma* Sw., El Alto, 6 JUL 1923, leg. F.L. Stevens No. 242, ILL 9044.

Phyllachora psuedes H. Rehm. Leaflets of Philippine Botany 6:2219. 1914, nom. nud. **Possible isotype:** Ad folia *Fici notae* [= *Ficus nota* (Blanco) Merrill], Philippines, Luzon, Laguna Prov., [Mount Maquiling], Los Baños, OCT 1913, leg. S.A. Reyes, comm. C.F. Baker No. 1983 [ex Fungi Malayana No. 82], ILL 9052.

Phyllachora quadraspora L.R. Tehon. Botanical Gazette 67:507, fig. 4. 1919. **Holotype:** On *Paspalum glabrum* Poiret [= *P. laxum* Lam.], Porto Rico [Puerto Rico], Maricao, [19 JUL 1915], leg. F.L. Stevens No. 8803, ILL 9070.

Phyllachora ramonensis H. Sydow. Annales Mycologici 23:379–380. 1925. **Isotype:** Hab. in foliis *Nectandrae reticulatae* [*Nectandra reticulata* (Ruiz & Pavon) Mez = *Ocotea* sp.], Costa Rica, San Pedro de San Ramón, 25 JAN 1925, leg. H. Sydow, Fungi in Itinere Costaricensi Collecti No. 371, ILL 9073.

Phyllachora sancta F. Petrak & R. Ciferri. Annales Mycologici 30:247–248. 1932. **Isotype:** In foliis vivis *Guajaci sancti* [= *Guajacum sanctum* L., Dominican Republic], Cordillera Septentrional, Santiago Prov., Las Lagunas, Santiago, hills at Arroyo Haranguillo, alt. ca. 400 m, 20 OCT 1930, leg. E.L. Ekman No. 4047, ex R. Ciferri, Mycoflora Domingensis Exsiccata No. 323, ILL 33203, ex herb. R. Ciferri.

Phyllachora sapindacearum F.L. Stevens. Illinois Biological Monographs 11:191. 1927. **Holotype:** On *Serjania mexicana* Willd., Panama, France Field, 24 AUG 1923, leg. F.L. Stevens No. 1327, ILL 9091.

Phyllachora sarcomphali R. Ciferri. Sydowia Annales Mycologici 10:135–136. (1956) 1957. **Isotype:** Hab. in foliis vivis *Sarcomphali* sp. [*Sarcomphalus* = *Ziziphus* sp., Dominican Republic], Espaillat Prov., Valle del Cibao, Moca, pr. via Salcedo, FEB 1928, leg. R. Ciferri, Mycoflora Domingensis Exsiccata No. 379, ILL 33064.

Phyllachora stena H. Sydow. Annales Mycologici 23:380–381. 1925. **Isosyntype:** Hab. in foliis *Mauriae biringo* [= *Mauria biringo* L.R. Tulasne], Costa Rica, Grecia, 17 JAN 1925, leg. H. Sydow, Fungi in Itinere Costaricensi Collecti No. 248 p.p., ILL 9128.

Phyllachora stenocarpa H. Sydow. Annales Mycologici 24:402–403. 1926. **Isotype:** Hab. in foliis vivis *Topobeae durandianae* [= *Topobea durandiana* Cogn.], Costa Rica, Cerro de San Isidro pr. San Ramón, 9 FEB 1925, leg. H. Sydow, Fungi in Itinere Costaricensi Collecti No. 151, ILL 9129.

Phyllachora stevensii H. Sydow in F.L. Stevens. Illinois Biological Monographs 11:191, pl. XVI, fig. 115. 1927. **Holotype:** On *Meibomia* sp. [=

Desmodium sp.], Panama, Gamboa, 17 AUG 1923, leg. F.L. Stevens No. 1007, ILL 9121. **Paratype**: Gatuncillo, 18 AUG 1923, leg. F.L. Stevens No. 1150, ILL 9122.

Phyllachora tabernaemontanae F.L. Stevens. Illinois Biological Monographs 8:186–187, pl. V, figs. 42, 43; pl. XIV, fig. 98. 1923. **Holotype**: On *Tabernaemontana* sp., British Guiana [Guyana], Kartabo, 22 JUL 1922, leg. F.L. Stevens No. 564, ILL 9130.

Phyllachora tehonis A. Trotter. Sylloge Fungorum 24:580. 1926, nom. nov. **Based on**: *Phyllachora ischaemi* L.R. Tehon, 1919 (nom. illegit.), non H. Sydow, 1915. **Holotype**: On *Ischaemum latifolium* Kunth, Martinique, St. Pierre, [4 AUG 1913], leg. F.L. Stevens No. 2972, ILL 8859.

Phyllachora tiliae F.L. Stevens. Illinois Biological Monographs 8:187, pl. VI, figs. 44–46. 1923. **Holotype**: On unknown species of Tiliaceae, British Guiana [Guyana], Tumatumari, 12 JUL 1922, leg. F.L. Stevens No. 227, ILL 9131. **Isotype**: ILL 9132.

Phyllachora trophis F.L. Stevens. Illinois Biological Monographs 11:192–193, pl. VIII, figs. 63–65. 1927. **Holotype**: On *Trophis racemosa* (L.) Urban, Costa Rica, Cartago, 23 JUN 1923, leg. F.L. Stevens No. 85, ILL 9157.

Phyllachora tumatumariana P.F. Cannon. Mycological Papers. International Mycological Institute 163:181–182, fig. 42. 1991. **Holotype**: On unknown legume [= Fabaceae sp. indet.], Guyana [as British Guiana on the packet], Tumatumari, 10 JUL 1922, leg. F.L. Stevens No. 138, ILL 9212.

Phyllachora veraguensis H. Sydow. Annales Mycologici 24:396–397. 1926. **Isotype**: Hab. in foliis vivis *Ocoteae veraguensis* [= *Ocotea veraguensis* (Meisner) Mez], Costa Rica, Grecia, 13 JAN 1925, leg. H. Sydow, Fungi in Itinere Costaricensi Collecti No. 155, ILL 9184.

Phyllachora vismiae F.L. Stevens. Illinois Biological Monographs 11:193. 1927. **Holotype**: On *Vismia guianensis* (Aublet) Choisy, Costa Rica, Escasu, 29 JUL 1923, leg. F.L. Stevens No. 155, ILL 9186.

Phyllachora wismarensis F.L. Stevens. Illinois Biological Monographs 8:187–188, pl. VI, figs. 47–50; pl. XV, fig. 99. 1923. **Holotype**: On *Ficus* sp., British Guiana [Guyana], Demerara-Essequibo R.R., 15 JUL 1922, leg. F.L. Stevens No. 397, ILL 9200. **Paratype**: Leg. F.L. Stevens No. 334, ILL 9201.

Phyllachorella schistocarphae F.L. Stevens. Illinois Biological Monographs 11:194, pl. VIII, fig. 66; pl. XVII, fig. 116. 1927. **Syntypes**: On *Schistocarpha hoffmannii* O. Kuntze [= *S. oppositifolia* (O. Kuntze) Rydb.], Costa Rica, Peralta, 13 JUL 1923, leg. F.L. Stevens No. 434, ILL 9219; leg. F.L. Stevens No. 454, ILL 9220.

Phyllosticta adianticola E. Young. Mycologia 7:144. 1915. **Holotype**: On leaves of *Adiantum tenerum* Sw., Porto Rico [Puerto Rico], Manati,

[5 NOV 1913], leg. F.L. Stevens No. 4299, ILL 11565. **Paratypes**: Utuado, [8 NOV 1913], leg. F.L. Stevens No. 4588, ILL 11567; Quebradillos, [22 NOV 1913], leg. F.L. Stevens No. 5000, ILL 11566.

Phyllosticta adoxae F.E. & E.S. Clements. Cryptogamae Formationum Coloradensium, Century 3, No. 242. Anno 1907, nom. nud. **Isotype**: ... in foliis languidis emortuisque *Adoxae moschatellinae* [= *Adoxa moschatellina* L.], Colorado, Mount Garfield, alt. 3700 m, 21 AUG 1906, leg. F.E. & E.S. Clements, ILL 11495.

Phyllosticta aesculi J.B. Ellis & G.W. Martin. Journal of Mycology 2:130. 1886. **Isotype**: On living leaves of *Aesculus glabra* Willd., Missouri, [Columbia, SEP 1886], leg. B.T. Galloway No. 76, ILL 11497. **Holotype**: NY.

Phyllosticta allantospora J.B. Ellis & B.M. Everhart. Proceedings of the Academy of Natural Sciences of Philadelphia 1894:355. 1894. **Isotype**: On leaves of *Cakile americana* Nutt. [= *C. edentula* (Bigelow) Hooker], New Jersey, Sandy Hook, [29] JUN 1892, leg. F.L. Stevens [No. 7], ILL 11508.

Phyllosticta alpinicola W.G. Solheim. Mycologia 41:625–626. 1949. **Isotype**: On living leaves of *Trifolium parryi* A. Gray, Wyoming, Albany County, Medicine Bow Mountains, roadside below Brooklyn Lake, [alt. 10,300 ft.], 16 AUG 1930, leg. W.G. Solheim No. 50, ex Mycoflora Saximontanenses Exsiccati No. 454, ILL 20916.

Phyllosticta araliana E. Young. Mycologia 7:148–149. 1915. **Holotype**: On leaves of *Dendropanax arboreus* (L.) Decaisne & Planchon, Porto Rico [Puerto Rico], Maricao, [4 APR 1913], leg. F.L. Stevens No. 755, ILL 11525.

Phyllosticta begoniicola F.L. Stevens & Baechler. Illinois Biological Monographs 11:204. 1927. **Holotype**: On *Begonia* sp., Costa Rica, Peralta, 13 JUL 1923, leg. F.L. Stevens No. 429, ILL 11551.

Phyllosticta bixina E. Young. Mycologia 7:148. 1915. **Holotype**: On *Bixa orellana* L., Porto Rico [Puerto Rico], Maricao, [2 OCT 1913], leg. F.L. Stevens No. 174, ILL 11562. **Paratypes**: San Germán, [8 NOV 1913], leg. F.L. Stevens No. 5794, ILL 11558; Rosario, [14 NOV 1913], leg. F.L. Stevens No. 4844, ILL 11559; Mayagüez, [30 JAN 1913], leg. F.L. Stevens No. 298, ILL 11557; Coamo, [4 JAN 1913], leg. F.L. Stevens No. 53, ILL 11556; Punta Santiago, [23 JUN 1913], leg. F.L. Stevens No. 2458, ILL 11556a; Añasco, 21 [SEP 1913], leg. F.L. Stevens No. 3208, ILL 11560; Adjuntas, [22 NOV 1913], leg. F.L. Stevens No. 4975a, ILL 11561.

Phyllosticta bonduc F.L. Stevens. Botanical Gazette 69:256. 1920. **Holotype**: On *Caesalpinia bonduc* Roxb., Porto Rico [Puerto Rico], Guanica, [2 MAR 1913], leg. F.L. Stevens No. 360, ILL 11563.

Phyllosticta borinquensis E. Young. Mycologia 7:147–148. 1915. **Holotype:** On leaves of *Helicteres jamaicensis* Jacquin, Porto Rico [Puerto Rico], San Germán, [12 AUG 1913], leg. F.L. Stevens No. 5672, ILL 11564. **Isotype:** ILL 11564a.

Phyllosticta casimiroae F.L. Stevens & A.G. Weedon in F.L. Stevens. Bernice P. Bishop Museum Bulletin 19:129. 1925. **Syntypes:** On *Casimiroa edulis* La Llave, tree No. 176, Hawaii, [Oahu, Honolulu], Agricultural Experiment Station, [9 MAY 1913], leg. H.L. Lyon No. 320, ILL 11571; leg. H.L. Lyon No. 329, ILL 11570.

Phyllosticta chelonanthi F.L. Stevens. Illinois Biological Monographs 11:204. 1927. **Holotype:** On *Chelonanthus acutangulus* (Ruiz & Pavon) Gilg [= *Irlbachia alata* (Aublet) Maas ssp. *alata*], Costa Rica, La Palma, 8 JUL 1923, leg. F.L. Stevens No. 284, ILL 11582.

Phyllosticta clusiae F.L. Stevens. Transactions of the Illinois Academy of Science 10:195. 1917. **Holotype:** On dead leaves of *Clusia rosea* Jacquin, Porto Rico [Puerto Rico], Maricao, [4 MAR 1913], leg. F.L. Stevens No. 739a, ILL 11594.

Phyllosticta codiaei F.L. Stevens & P.A. Young in F.L. Stevens. Bernice P. Bishop Museum Bulletin 19:129. 1925. **Holotype:** On living leaves of *Codiaeum moluccanum* sensu auct., non (L.) Decaisne [= *Codiaeum variegatum* (L.) A.H.L. Jussieu], Hawaii, Oahu, Honolulu, 20 MAY [1921], leg. F.L. Stevens No. 31, ILL 11597. **Isotype:** BISH 500002.

Phyllosticta colocasiophila A.G. Weedon in F.L. Stevens. Bernice P. Bishop Museum Bulletin 19:129–131, text fig. 28a, pl. X(F). 1925. **Syntypes** [BISH numbers cited are isosyntypes]: On *Colocasia* sp. (taro), Hawaii, [Island of] Hawaii, Kona, Keauhou, Bishop Estate Road, 23 JUL [1921], leg. F.L. Stevens No. 943, ILL 11602, BISH 500341; Kilauea, 16 JUL [1921], leg. F.L. Stevens No. 873, ILL 11601, BISH 500340.

Phyllosticta commelinicola E. Young. Mycologia 7:144–145. 1915. **Holotype:** On leaves of *Commelina nudiflora* L., Porto Rico [Puerto Rico], Hormigueros, [14 JAN 1913], leg. F.L. Stevens No. 214, ILL 11604.

Phyllosticta cordillerana R. Ciferri. Sydowia Annales Mycologici 10:159–160 (1956) 1957. **Isotype:** Hab. in foliis vivis *Bambusearum* ?cujusdam [= *Bambusa* sp., Dominican Republic], Cordillera Central, La Vega Prov., ad viam Jarabacoa-Constanza, in foresta montana, MAR 1929, leg. R. Ciferri, Mycoflora Domingensis Exsiccata No. 410, ILL 33078.

Phyllosticta cordylinophila P.A. Young in F.L. Stevens. Bernice P. Bishop Museum Bulletin 19:133. 1925. **Holotype:** On living leaves of *Cordyline terminalis* (L.) Kunth [= *C. fruticosa* (L.) A. Chev.] (the Hawaiian ti plant), Hawaii, Oahu, Hawaii Sugar Planter's Station, SEP 1921, leg. F.L. Stevens No. 1132, ILL 11613. **Isotypes:** BISH 146157,

500001 and 500339. **Paratypes:** Honolulu, Manoa Valley, 23 MAY [1921], leg. F.L. Stevens No. 63, ILL 11612; Honolulu, [Nuanu St.], 18 AUG [1921], leg. F.L. Stevens No. 1133, ILL 11614.

Phyllosticta delphinii F.E. & E.S. Clements. Cryptogamae Formationum Coloradensium, Century 5, No. 468. Anno 1908, nom. nud. **Isotype:** ... in foliis vivis *Delphinii multiflori* [*Delphinium multiflorum* Rydb. = *D. occidentale* (S. Watson) S. Watson], Colorado, Sulphur Springs, alt. 2400 m, 22 JUL 1907, leg. F.E. & E.S. Clements, ILL 11632.

Phyllosticta dircae J.B. Ellis & J. Dearness. Canadian Record of Science 5:267. 1893. **Isotype:** On leaves of *Dirca palustris* L., Canada, [Ontario], London, JUL 1892, leg. J. Dearness No. 1934, ex J.B. Ellis & B.M. Everhart, North American Fungi No. 2838, ILL 11638.

Phyllosticta erechtitis F.L. Stevens & P.A. Young in F.L. Stevens. Bernice P. Bishop Museum Bulletin 19:131. 1925. **Holotype:** On living leaves of *Erechtites* sp. [= *E. hieraciifolia* (L.) Raf. ex DC.], Hawaii, Kauai, Waimea, 15 JUN [1921], leg. F.L. Stevens No. 543, ILL 11643. **Isotypes:** BISH 146159, 146160, 486966 and 499090.

Phyllosticta erythrinicola E. Young. Mycologia 7:146. 1915. **Holotype:** On leaves of *Erythrina micropteryx* Poeppig in Urban [= *E. poeppigiana* (Walpers) O.F. Cook], Porto Rico [Puerto Rico], Villa Alba, [3 JAN 1913], leg. F.L. Stevens No. 110, ILL 11648. **Paratypes:** Jajome Alta, [3 DEC 1913], leg. F.L. Stevens No. 5633, ILL 11650; Yauco, [3 OCT 1913], leg. F.L. Stevens No. 3157, ILL 11652; Mayagüez, [8 JUL 1912], leg. F.L. Stevens No. 68, ILL 11651 and 20405.

Phyllosticta eugeniae E. Young. Mycologia 7:148. 1915. **Holotype:** On *Eugenia buxifolia* (Sw.) Willd. [= *E. foetida* Persoon], Porto Rico [Puerto Rico], Mona Island, [20–21 NOV 1913], leg. F.L. Stevens No. 6230, ILL 11656. **Paratype:** Leg. F.L. Stevens No. 6091, ILL 11655; leg. F.L. Stevens No. 6127, ILL 11654.

Phyllosticta heliconiae F.L. Stevens & P.A. Young in F.L. Stevens. Bernice P. Bishop Museum Bulletin 19:131. 1925. **Holotype:** On living leaves of *Heliconia* sp., Hawaii, Oahu, Hakipuu, 19 JUN [1921], leg. F.L. Stevens No. 574, ILL 11674. **Isotypes:** BISH 499089 and 500338.

Phyllosticta lantanae F.L. Stevens. Transactions of the Illinois State Academy of Science 10:195. 1918. **Holotype:** On *Lantana odorata* L. [= *L. involucrata* L. var. *odorata* (L.) Moldenke], Porto Rico [Puerto Rico], Desecheo Island, [31 MAY 1913], leg. F.L. Stevens No. 1681 [erroneously as No. 168 in the protologue], ILL 11704. **Paratypes:** Mona Island, [20–21 DEC 1913], leg. F.L. Stevens No. 6416, ILL 11703; leg. F.L. Stevens No. 6440, ILL 11706; Utuado, [30 DEC 1913], leg. F.L. Stevens No. 6592, ILL 11705; Guanica, [3 FEB 1915], leg. F.L. Stevens No. 332, ILL 11702.

Phyllosticta lycii J.B. Ellis & W.A. Kellerman. American Naturalist 17:1166. 1883. **Assumed isotype**: On leaves of *Lycium vulgare* Dunal in DC. [= *L. barbarum* L., Ohio, Fairfield County, Lancaster, 30 MAY 1883], leg. W.A. Kellerman s.n., ILL 11725.

Phyllosticta momisiana E. Young. Mycologia 7:145. 1915. **Holotype**: On leaves of *Momisia iguanaea* (Jacquin) Rose & Standley [= *Celtis iguanaea* (Jacquin) Sargent], Porto Rico [Puerto Rico], Coamo, 6 APR 1913, leg. F.L. Stevens No. 834, ILL 11742.

Phyllosticta musae F.L. Stevens & P.A. Young in F.L. Stevens. Bernice P. Bishop Museum Bulletin 19:132. 1925. **Holotype**: On living leaves of *Musa* sp. (banana), Hawaii, Oahu, Honolulu, Manoa Valley, 24 MAY [1921], leg. F.L. Stevens No. 76, ILL 11745. **Isotypes**: BISH 499088 and 500337.

Phyllosticta musicola F.L. Stevens & P.A. Young in F.L. Stevens. Bernice P. Bishop Museum Bulletin 19:132. 1925. **Holotype**: On living leaves of *Musa* sp. (banana), Hawaii, Oahu, Honolulu, MAY 1919, leg. H.L. Lyon s.n., ILL 11746. **Isotype**: BISH 499087.

Phyllosticta oleae J.B. Ellis & G.W. Martin. Journal of Mycology 2:17. 1886. **Isotype**: On leaves of *Olea americana* L. [= *Osmanthus americanus* (L.) Bentham & J.D. Hooker ex A. Gray], Florida, [Green Cove Springs, FEB 1885], leg. G.W. Martin s.n., ILL 11752.

Phyllosticta pandanicola E. Young. Mycologia 7:150. 1915. **Holotype**: On leaves of *Pandanus* sp. indet., Porto Rico [Puerto Rico], Santurce, [17 JAN 1913], leg. F.L. Stevens No. 240, ILL 11762.

Phyllosticta panici E. Young. Mycologia 7:144. 1915. **Holotype**: On leaves of *Panicum maximum* Jacquin, Porto Rico [Puerto Rico], Coamo, [6 APR 1913], leg. F.L. Stevens No. 830, ILL 11764. **Paratype**: Martin Peña, [21 JAN 1899], leg. A.A. Heller No. 377, ILL 11765.

Phyllosticta phaea P.A. Saccardo. Annales Mycologici 11:559. 1913. **Assumed isotype**: Hab. in foliis nondum emortuis *Crataegi oxyacanthae* [= *Crataegus oxyacantha* L., Czech Republic], M[ähr.]-Weisskirchen [= Hranice], SEP 1913, leg. F. Petrak, ILL 11774.

Phyllosticta pithecellobii E. Young. Mycologia 7:145. 1915 [as *pithecolobii*, corrected in accordance with Rec. 60H.1 and Articles 60 & 61 of I.C.B.N.]. **Holotype**: On leaves of *Pithecellobium unguis-cati* (L.) Bentham [as *Pithecolobium*], Porto Rico [Puerto Rico], Desecheo [Island, 31 MAY 1913], leg. F.L. Stevens No. 1576, ILL 11782. **Paratype**: Yauco, [3 OCT 1913], leg. F.L. Stevens No. 3267, ILL 11781.

Phyllosticta pithecellobii E. Young var. *monensis* E. Young. Mycologia 7:145–146. 1915. **Holotype**: On leaves of *Pithecellobium unguis-cati* (L.) Bentham [as *Pithecolobium*], Porto Rico [Puerto Rico], Mona Island, [20–21 DEC 1913], leg. F.L. Stevens No. 6137, ILL 11779.

Phyllosticta portoricensis E. Young. Mycologia 7:147. 1915. **Holotype**: On leaves of *Croton lucidus* L., Porto Rico [Puerto Rico], Guanica, [3 FEB 1913], leg. F.L. Stevens No. 325, ILL 11799.

Phyllosticta pothicola A.G. Weedon in F.L. Stevens. Bernice P. Bishop Museum Bulletin 19:132–133. 1925. **Syntypes**: On *Pothos* sp., Hawaii, Oahu, Waikiki, 18 MAY [1921], leg. F.L. Stevens No. 3, ILL 11800; Honolulu, 20 MAY [1921], leg. F.L. Stevens No. 25, ILL 11801. **Isosyntypes**: BISH 499086 and 500336.

Phyllosticta punctata J.B. Ellis & J. Dearness. Canadian Record of Science 5:268. 1893. **Isotype**: On leaves of *Viburnum opulus* L., Canada, [Ontario], London, 15 AUG 1892, leg. J. Dearness No. 1982, ex J.B. Ellis & B.M. Everhart, North American Fungi No. 2832, ILL 11807.

Phyllosticta ragnhildae W.G. Solheim. Mycologia 41:626–627. 1949. **Isotype**: On living leaves of *Antennaria pulcherima* (Hooker) E. Greene, Wyoming, Albany County, Laramie Mountains, Happy Jack Picnic Area, [alt. 8400 ft.], 6 AUG 1942, leg. W.G. & R. Solheim No. 2093, ex Mycoflora Saximontanensis Exsiccata No. 462, ILL 20908.

Phyllosticta sancti-iosephi R. Ciferri. Sydowia Annales Mycologici 10:161. (1956) 1957. **Isotype**: Hab. in foliis vivis *Cassiae* sp. [= *Cassia* sp., Dominican Republic], Cordillera Central, Santiago Prov., S. José de Las Matas, in sylvis, FEB 1926, leg. R. Ciferri, Mycoflora Domingensis Exsiccata No. 388, ILL 33140.

Phyllosticta sechii E. Young. Mycologia 7:149. 1915. **Holotype**: On leaves of *Sechium edule* (Jacquin) Sw., Porto Rico [Puerto Rico], Mayagüez, [9 OCT 1913], leg. F.L. Stevens No. 3357, ILL 11827.

Phyllosticta smilacinae W.G. Solheim. Mycologia 41:627–628. 1949. **Isotype**: On yellowing leaves of *Smilacina amplexicaulis* Nutt. ex Baker [= *S. racemosa* (L.) Desf. var. *amplexicaulis* (Nutt. ex Baker) S. Watson], Colorado, Ouray County, Ouray Picnic Grounds, 12 OCT 1948, leg. W.G. & R. Solheim No. 2250, ex Mycoflora Saximontanensis Exsiccata No. 464, ILL 20906.

Phyllosticta stevensii E. Young. Mycologia 7:147. 1915. **Holotype**: On leaves of *Triumfetta semitriloba* Jacquin, Porto Rico [Puerto Rico], Coamo, [1 JAN 1913], leg. F.L. Stevens No. 119, ILL 11858. **Paratype**: Villa Alba, [3 JAN 1913], leg. F.L. Stevens No. 94, ILL 11857.

Phyllosticta superficiale F.L. Stevens. Transactions of the Illinois State Academy of Science 10:195–196. 1917. **Holotype**: On *Passiflora sexflora* Jussieu, Porto Rico [Puerto Rico], Ponce, [8 NOV 1913], leg. F.L. Stevens No. 4337, ILL 11889. **Paratypes**: Leg. F.L. Stevens No. 4377, ILL 11885; Monte de Oro, [3 DEC 1913], leg. F.L. Stevens No. 5736, ILL 11887.

Phyllosticta yaguarum R. Ciferri. Sydowia Annales Mycologici 10:162.

(1956) 1957. **Isotype**: Hab. in foliis vivis *Caseariae* sp. [= *Casearia* sp.], "palo de yagua" dicta, [Dominican Republic], Cordillera Central, La Vega Prov., Bonao, in sylva, MAR 1926, leg. R. Ciferri, Mycoflora Domingensis Exsiccata No. 402, ILL 33085.

Phyllosticta yunaensis R. Ciferri. Sydowia Annales Mycologici 10:162. (1956) 1957. **Isotype**: Hab. in foliis vivis *Trichiliae pallidae* [= *Trichilia pallida* Sw.] (Meliaceae), [Dominican Republic], Cordillera Central, La Vega Prov., Bonao, secus Rio Yuna, in forest, JUN 1926, leg. E. & R. Ciferri, Mycoflora Domingensis Exsiccata No. 365, ILL 33154.

Phyllosticta zingiberis F.L. Stevens & R.W. Ryan in F.L. Stevens. Bernice P. Bishop Museum Bulletin 19:133. 1925. **Syntypes** [BISH numbers cited are isosyntypes]: On leaves of *Zingiber zerumbet* (L.) J.E. Smith, Hawaii, Oahu, [Honolulu], Olympus, 24 JUN [as 22 JUN 1921 on the packet], leg. F.L. Stevens No. 655, ILL 11917, BISH 499085 and 500335; leg. F.L. Stevens No. 961, ILL 11918, BISH 146168 and 500335.

Phyllostictina murrayae H. & P. Sydow in H. & P. Sydow & E.J. Butler. Annales Mycologici 14:186. 1916. **Isotypes**: Hab. in foliis vivis *Murrayae koenigii* [= *Murraya koenigii* (L.) Sprengel], India, Dehra, 30 AUG 1905, leg. E.J. Butler No. 1689, ILL 11744 and 11927.

Physalacria solida F.E. & E.S. Clements. Cryptogamae Formationum Coloradensium, Century 4, No. 333. Anno 1907, nom. nud. **Isotype**: ... ad terram et ad acus *Pseudotsugae* [= *Pseudotsuga* sp.], Colorado, Minnehaha, alt. 2600 m, 7 SEP 1906, leg. F.E. & E.S. Clements, ILL 20015.

Physalospora andirae F.L. Stevens. Transactions of the Illinois State Academy of Science 10:184–185. 1917. **Possible holotype**: On *Andira jamaicensis* (W. Wright) Urban [= *A. inermis* (W. Wright) Kunth ex DC.], Porto Rico [Puerto Rico], Camuy, leg. F.L. Stevens No. 7277, ILL 10090 [no data on the packet label except species name, marked as "sp. nov.," and name of the host]. **Paratypes**: Mayagüez, [1 MAY 1913], leg. F.L. Stevens No. 1037, ILL 10104 and 10105; [14 MAR 1913], leg. F.L. Stevens No. 1479, ILL 10106, 10107 and 10109; [31 OCT 1913], leg. F.L. Stevens No. 3939, ILL 10099; leg. F.L. Stevens No. 3950, ILL 10115; Vega Baja, [20 FEB 1913], leg. F.L. Stevens No. 492, ILL 10101; 22 FEB 1913, leg. F.L. Stevens No. 465, ILL 10091 and 10095; San Sebastian, [22 NOV 1913], leg. F.L. Stevens No. 5198, ILL 10098; Maricao, [20 SEP 1913], leg. F.L. Stevens No. 3628, ILL 10102 and 10108 [also probable isotypes of *Dothidella andiricola* C.L. Spegazzini]; Cabo Rojo, [27 DEC 1913], leg. F.L. Stevens No. 6485, ILL 10097; Coamo, [6 APR 1913], leg. F.L. Stevens No. 842 [this number cited twice in the protologue but for different collections], ILL 10092 and 10103; 16 JUL 1915, leg. F.L. Stevens No. 8357, ILL 10089; [16 JUL 1915], leg. F.L. Stevens No. 8478, ILL 10116; Quebradillas, [12 NOV 1913], leg. F.L. Stevens No. 4999, ILL 10096; Hormigueros, [14 JAN 1913], leg. F.L. Stevens No. 218, ILL 10117; San Germán, [8 DEC 1913], leg. F.L. Stevens No. 5808, ILL 10118; [12 DEC 1913], leg. F.L. Stevens No. 842, ILL 10093; Lajas, [18 JUN 1915], leg. F.L. Stevens No. 7178, ILL 10088; Rio Tanamá, [6 JUL 1915], leg. F.L. Stevens No. 7835, ILL 10094; Arecibo-Lares Road, [21 JUN 1915], leg. F.L. Stevens No. 7294, ILL 10114; Martin Peña, [11 AUG 1915], leg. F.L. Stevens No. 9315, ILL 10112; Peñuelas, [no date], leg. F.L. Stevens No. 9163, ILL 10110 and 10113; Santa Catalina, [28 AUG 1913], leg. F.L. Stevens No. 2721, ILL 10100.

Physalospora caryophyllinicola F.L. Stevens. Transactions of the Illinois State Academy of Science 10:184. 1917. **Holotype:** On *Drymaria cordata* (L.) Willd. ex J.A. Schultes, Porto Rico [Puerto Rico], Jayuya, [17 DEC 1913], leg. F.L. Stevens No. 5937, ILL 10119.

Physalospora cestri F.L. Stevens. Illinois Biological Monographs 11:200, pl. IX, fig. 75. 1927. **Holotype:** On *Cestrum* sp., Costa Rica, Port Limon, 10 AUG 1923, leg. F.L. Stevens No. 865, ILL 10126.

Physalospora dinochloae H. Rehm. Leaflets of Philippine Botany 8:2937. 1916. **Assumed isotype:** Ad *Dinochloam* sp. [= *Dinochloa* sp.], Philippines, [Luzon, Laguna Prov., Mount Maquiling, near] Los Baños, [DEC 1913], leg. C.F. Baker No. 2189a [ex Fungi Malayana No. 181a (in exsiccati set as 181a, b, on the same packet label)], ILL 10128 [also, under No. 181b, an assumed isotype of *Guignardia dinochloae* H. Rehm].

Physalospora lepachydis J.B. Ellis & B.M. Everhart var. *major* F.E. & E.S. Clements. Cryptogamae Formationum Coloradensium, Century 5, No. 412. Anno 1908, nom. nud. **Isotype:** ... in foliis vivis *Ratibidae tagetis* [= *Ratibida tagetes* (James) Barnhart], Colorado, Yuma, alt. 1250 m, 25 JUL 1907, leg. F.E. & E.S. Clements, ILL 10134.

Physalospora pandani F.L. Stevens & A.S. Peirce. Indian Journal of Agricultural Science 3:913. 1933. **Assumed holotype:** On *Pandanus* sp., [India, Poona, Bombay, 1932], leg. B.N. Uppal [No. 33], ILL 10142. **Isotype:** AMH.

Physalospora peribambusina H. Rehm. Leaflets of Philippine Botany 8:2937. 1916. **Possible isosyntype:** Ad emortuam *Bambusam vulgarem* [= *Bambusa vulgaris* Schrader ex J.C. Wendl.], Philippines, Luzon, [Laguna Prov., Mount Maquiling], Los Baños, OCT 1913, leg. S.A. Reyes, comm. C.F. Baker Nos. 1896, 1901 [ex Fungi Malayana No. 183 — neither of the cited Baker numbers is indicated on the specimen label in the exsiccati set], ILL 10143.

Physalospora polygoni J.B. Ellis & B.M. Everhart. Fungi Columbiani, Century 16, No. 1562. Anno 1901, nom. nud. **Isotype:** On dead stems of *Polygonum pensylvanicum* L., Alabama, Tuskegee, 29 NOV 1900, leg. G.W. Carver, ILL 10145.

Physalospora psidii F.L. Stevens & A.S. Peirce. Indian Journal of Agricultural Science 3:913. 1933. **Assumed holotype:** On *Psidium guajava* L., [India, Poona, Bombay, 1932], leg. B.M. Uppal [No. 6], ILL 10140. **Isotype:** AMH.

Physalospora quadraspora F.L. Stevens & W.G. Solheim in F.L. Stevens. Annales Mycologici 28:367–368. 1930. **Holotype:** On *Commelina* sp., Costa Rica, Siquirres, 31 JUL 1923, leg. F.L. Stevens No. 722, ILL 10147.

Physarum tenerum G.A. Rex. Proceedings of the Academy of Natural Sciences of Philadelphia 1890:192. 1890. **Assumed isotype:** On rotten wood, Adirondack Mountains, New York, leg. A.P. Morgan, ex J.B. Ellis & B.M. Everhart, North American Fungi No. 2489, ILL 33556 [although "Adirondack Mts., N.Y." is indicated on the label of No. 2489, two locations are cited and discussed in the protologue: Philadelphia and Ohio].

Phytophthora phaseoli R. Thaxter. Botanical Gazette 14:274. 1889. **Isosyntype:** On parts of stems and leaves of the lima bean [= *Phaseolus lunatus* L.], Connecticut, New Haven, SEP 1889, leg. R. Thaxter [No. 21886], ILL 2891.

Pileolaria domingensis R. Ciferri. Sydowia Annales Mycologici 10:151. (1956) 1957. **Isotype:** Hab. in foliis vivis *Comocladiae* sp. [= *Comocladia* sp., Dominican Republic], Cordillera Septentrional, Puerto Plata Prov., in sylva costiera, JUN 1929, leg. R. Ciferri, Mycoflora Domingensis Exsiccata No. 423, ILL 33196.

Pirostoma dianellae F.L. Stevens & P.A. Young in F.L. Stevens. Bernice P. Bishop Museum Bulletin 19:143–144. 1925. **Holotype:** On leaves of *Dianella odorata* sensu Hillebr. non Blume [= *D. sandwicensis* Hooker & Arnott], Hawaiian Islands, [summer] 1921, leg. F.L. Stevens s.n., ILL 13199. **Isotypes:** BISH 499084 and 500333.

Pistillaria fusiformis C.H. Kauffman. Papers of the Michigan Academy of Science, Arts, and Letters 5:119, pl. 2, fig. 2. 1926. **Isotype:** On rotten wood of conifer, Oregon, Mt. Hood, 7 OCT [1922], leg. L.E. Wehmeyer s.n., ILL 32582. **Holotype:** MICH.

Placosphaerella silvatica P.A. Saccardo. Annales Mycologici 8:344. 1910. **Isotype:** Hab. ad folia *Festucae heterophyllae* subviva [= *Festuca heterophylla* Lam., Germany, Thüringen], Steiger pr. Erfurt, AUG 1906, leg. H. Diedicke, [ex H. Sydow, Mycotheca Germanica No. 924], ILL 11929.

Plasmopara cephalophora J.J. Davis. Transactions of the Wisconsin Academy of Sciences, Arts and Letters 19:709–710, figs. 1–3. 1919. **Isosyntype**: On leaves of *Physostegia parviflora* Nutt. ex A. Gray, Wisconsin, Shiocton, 21 AUG 1917, leg. J.J. Davis [Fungi Wisconsinenses Exsiccati No. 67], ILL 2904.

Plasmopara galinsogae L. Campbell. Mycologia 24:333, fig. 1b. 1932. **Holotype**: On *Galinsoga parviflora* Cav., Guatemala, 1930, leg. Bjorn Palm s.n., ILL 2919.

Plasmopara palmii L. Campbell. Mycologia 24:332–333, fig. 1a. 1932. **Holotype**: On *Eupatorium ariolare* DC., [Guatemala], Antigua, 1930, leg. Bjorn Palm s.n., ILL 2995. **Isotype**: WSP 31288.

Platygloea sphaerospora G.W. Martin. Mycologia 26:261–262, pl. 31, figs. 1, 2. 1934. **Isotype**: On rotten wood of *Quercus rubra* L. (*Q. falcata* Michaux), New Jersey, Cape May County, Dias Creek, 10 SEP 1932, leg. G.W. Martin No. 1222, ILL 32805.

Pleiostomella halleriae E.M. Doidge. Bothalia 1(1):17–18. 1921. **Paratype**: On *Halleria lucida* L., [South Africa, Cape Prov.], Grahamstown, Howiesons Poort, 17 NOV 1917, leg. E.M. Doidge, ex PREM No. 10959, ILL 9222.

Pleospora balsamorhizae S.M. Tracy & F.S. Earle [as *balsamorrhizae*] var. *perseptis* F.E. & E.S. Clements. Cryptogamae Formationum Coloradensium, Century 5, No. 439. Anno 1908, nom. nud. [The variant spelling of the specific epithet, introduced by Tracy & Earle, 1888, is changed to the currently accepted spelling in accordance with Rec. 60H.1 and Articles 60 & 61 of I.C.B.N.]. **Isotype**: ... in caulibus vetustis *Balsamorrhizae sagittatae* [= *Balsamorhiza sagittata* (Pursh) Nutt.], Colorado, Sulphur Springs, alt. 2400 m, 22 JUL 1907, leg. F.E. & E.S. Clements, ILL 10153.

Pleospora colla F.E. & E.S. Clements. Cryptogamae Formationum Coloradensium, Century 5, No. 440. Anno 1908, nom. nud. **Isotype**: ... in caulibus vetustis *Asteris fremontii* [*Aster fremontii* Torrey & Gray = *A. occidentalis* (Nutt.) Torrey & Gray var. *fremontii* (Torrey & Gray) A.G. Jones], Colorado, Ouray, alt. 2400 m, 13 JUL 1907, leg. F.E. & E.S. Clements, ILL 10153.

Pleospora cybospora F.E. & E.S. Clements. Cryptogamae Formationum Coloradensium, Century 1, No. 34. Anno 1906, nom. nud. **Isotype**: ... in caulibus emortuis *Drymocallis fissae* [*D. fissa* (Nutt. in Torrey & Gray) Rydb. = *Potentilla fissa* Nutt. in Torrey & Gray], Colorado, Golf Links, alt. 2700 m, 20 AUG 1905, leg. F.E. & E.S. Clements, ILL 10155.

Pleospora cytisi L. Fuckel var. *lineata* F.E. & E.S. Clements. Cryptogamae Formationum Coloradensium, Century 1, No. 35. Anno 1906, nom. nud. **Isotype**: ... in ramis emortuis *Opulastri monogynae* [*Opulaster* *monogynus* (Torrey) O. Kuntze = *Physocarpus monogynus* (Torrey) Coulter], Colorado, Minnehaha, alt. 2600 m, 7 AUG 1905, leg. F.E. & E.S. Clements, ILL 10154.

Pleospora niessleana J. Kunze. Fungi Selecti Exsiccati, Century 1, No. 71. Anno 1876, nom. nud. **Isosyntype:** Ad *Meliloti albi* [= *Melilotus albus* Medik.], ... caules aridos, humi stratos, [Germany], "Oberrissdorfer Thal pr. Islebiam" [Eisleben], (Sax. Bor.), MAY 1875, leg. J. Kunze, ILL 10167 [for explanation on type status of collection see L.E. Wehmeyer (1961)].

Pleospora scaevolae F.L. Stevens & P.A. Young in F.L. Stevens. Bernice P. Bishop Museum Bulletin 19:107. 1925. **Holotype:** On living leaves of *Scaevola chamissoniana* Gaudich., Hawaii, Oahu, Tantalus, 22 JUN [1921], leg. F.L. Stevens No. 660, ILL 9894 [also the holotype of *Mycosphaerella scaevolae* F.L. Stevens & P.A. Young in F.L. Stevens]. **Isotypes:** BISH 499070 and 499985.

Plicaria fuckelii H. Rehm var. *caerulescens* F.E. & E.S. Clements. Cryptogamae Formationum Coloradensium, Century 2, No.124. Anno 1906, nom. nud. **Isotype:** ... ad terram ..., Colorado, Beaver Dam, alt. 2700 m, 18 AUG 1904, leg. F.E. & E.S. Clements, ILL 7737.

Plochmopeltidella smilacina J.M. Mendoza in F.L. Stevens & H.W. Manter. Botanical Gazette 79:292, figs. 64–66. 1925. **Holotype:** On *Smilax* sp. [erroneously identified on the packet as *Pelea* sp.], Hawaii, Oahu, Olympus, 24 JUN 1921 [erroneously as 6 JUN in the protologue]; leg. F.L. Stevens No. 670, ILL 6668 [also the holotype of *Pycnidiopeltis smilacina* A.C. Batista & C.A.A. Costa in A.C. Batista & R. Ciferri, and a syntype of *Phragmocapnias smilacina* J.M. Mendoza in F.L. Stevens]. **Isotypes:** ILL 6668c, BISH 499972, 499973 and 499974.

Pluriporus gouldiae F.L. Stevens & R.W. Ryan in F.L. Stevens. Bernice P. Bishop Museum Bulletin 19:65–66, text fig. 13, pl. VI(E). 1925. **Assumed holotype:** On *Gouldia coriacea* (Hooker & Arnott) Hillebr. [= *Hedyotis terminalis* (Hooker & Arnott) W.L. Wagner & Herbst], Hawaii, Kauai, Waimea Canyon pipe trail, 15 JUN 1921 [as Oahu, etc., 31 MAY 1921 on the packet— almost certainly an error in transcription of the label data], leg. F.L. Stevens No. 455, ILL 7363 [including four microscopic preparations used for the published illustrations]. **Isotypes:** BISH 500331 and 500332. **Paratypes:** Kauai, Waimea Canyon pipe trail, 15 JUN 1921, leg. F.L. Stevens No. 454, ILL 6262 [including two microscopic preparations], and perhaps ILL 7362 [with the correct number and fungus name but erroneous label data transcribed to the packet].

Polystigma adenostomatis W.G. Farlow in J.B. Ellis & B.M. Everhart. Fungi Columbiani, Century 21, No. 2049. Anno 1905. **Isotype:** On living leaves of *Adenostoma fasciculatum* Hooker & Arnott, California, San Mateo County, 12 APR 1903, leg. C.H. Thompson, ILL 8405.

Polystomella costaricensis F.L. Stevens. Illinois Biological Monographs 11:175–176, pl. III, figs. 25–26; pl. XIV, figs. 101–102. 1927. **Holotype:** On *Struthanthus* sp., Costa Rica, El Alto, 6 JUN 1923, leg. F.L. Stevens No. 255, ILL 6805. **Paratype:** Cartago, 23 JUN 1923, leg. F.L. Stevens No. 97, ILL 6806.

Polystomella kaduae F.L. Stevens & R.W. Ryan in F.L. Stevens. Bernice P. Bishop Museum Bulletin 19:65. 1925. **Holotype:** On *Kadua glomerata* Hooker & Arnott [= *Hedyotis centranthoides* (Hooker & Arnott) Steudel], Hawaii, [Island of] Hawaii, Kealakekua, 25 JUL [1921], leg. F.L. Stevens No. 1005, ILL 6807. **Isotypes:** BISH 500328, 500329 and 500330.

Polystomella pulcherrima C.L. Spegazzini. Anales de la Sociedad Cientifica Argentina 26:53. 1888; Fungi Guaranitici, Pugillus II, p. 51, No. 137. 1888. **Isotype:** Ad folia viva *Solani boerhaviaefolii* [= *Solanum boerhaviifolium* Sendtner in Martius, Paraguay], pr. Guarapi, NOV 1883, leg. J.I. Puiggari No. 2704, comm. C.L. Spegazzini No. 4056, ILL 6808.

Polystomella trichiliae H. Sydow. Annales Mycologici 23:389–392. 1925. **Isosyntypes:** In foliis vivis *Trichiliae havanensis* [= *Trichilia havanensis* Jacquin], Costa Rica, La Caja pr. San José, 3 JAN 1925, leg. H. Sydow, Fungi in Itinere Costaricensi Collecti, No. 374, ILL 6809; in foliis *Trichiliae oerstedianae* [= *Trichilia oerstediana* C. DC. in DC.], Costa Rica, San Pedro de San Ramón, 5 FEB 1925, leg. H. Sydow No. 124a, ILL 6810 and 7004.

Polystomellopsis mirabilis F.L. Stevens. Illinois Biological Monographs 8:198–200, pl. X, figs. 80–84; pl. XI, figs. 85–87; pl. XVIII, figs. 109–110; pl. XIX, figs. 111–113. 1923. **Holotype:** On *Hirtella* sp., British Guiana (Guyana), Kartabo, 24 JUL 1922, leg. F.L. Stevens No. 647, ILL 6816. **Paratypes:** Kartabo, 21 JUL 1922 [as Rockstone, 17 JUL on the packet], leg. F.L. Stevens No. 484, ILL 6812; 23 JUL 1922, leg. F.L. Stevens No. 595, ILL 6815; Trinidad, Cumuto, 16 AUG 1922, leg. F.L. Stevens No. 943, ILL 6811 and 6818; on *Coccoloba* sp., British Guiana [= Guyana], Tumatumari, 8 JUL 1922, leg. F.L. Stevens No. 40, ILL 6814; on unknown host, Tumatumari, 8 JUL 1922, leg. F.L. Stevens No. 45, ILL 6813.

Poria oleagina L.O. Overholts. Bulletin of the Pennsylvania Agricultural Experiment Station 418:34. 1942. **Isotype:** On prostrate log [of *Picea rubens* Sargent], New Hampshire, North Conway, 10 AUG 1920 [as 8 AUG on the packet], leg. W.H. Snell No. 577, L.O. Overholts herb. No. 6077, ILL 32570, ex. herb. Snell. **Holotype:** PACMA.

Prospodium couraliae H. Sydow. Annales Mycologici 23:320. 1925. **Isosyntype: Hab. in foliis *Couraliae roseae* [*Couralia rosea* (Bertolini) J. Donnell Smith = *Tabebuia rosea* (Bertolini) DC.], Costa Rica, La Caja pr. San José, 4 JAN 1925, leg. H. Sydow, Fungi in Itinere Costaricensi Collecti No. 338, ILL 19122.

Protocoronospora phoradendri L. Darling. Madroño 5:241–245, figs. 1–3. 1940. **Isotype:** On *Phoradendron flavescens* (Pursh) Nutt. var. *macrophyllum* Engelm. [= *P. macrophyllum* (Engelm.) Cockerell], California, Lake County, north shore of Clear Lake, [as Lakeport on the packet], 30 MAR 1934, leg. L. Bonar [California Fungi No. 688], ILL 13834. **Holotype:** UC 615587.

Protoscypha pulla H. Sydow. Annales Mycologici 23:403–405, fig. 12. 1925. **Isotype:** Hab. in foliis *Miconiae thomasianae* [= *Miconia thomasiana* DC.], Costa Rica, Los Angeles de San Ramón, 30 JAN 1925, leg. H. Sydow, Fungi in Itinere Costaricensi Collecti No.150, ILL 7972.

Protostegia pini F.E. & E.S. Clements. Cryptogamae Formationum Coloradensium, Century 5, No. 498. Anno 1908, nom. nud. **Isotype:** ... ad acus emortuas *Pini murrayana* [*Pinus murrayana* Grev. & Balf. = *P. contorta* Douglas ex Loud. var. *murrayana* (Grev. & Balf.) Engelm.], Colorado, Sulphur Springs, alt. 2400 m, 20 JUL 1907, leg. F.E. & E.S. Clements, ILL 13210.

Pseudographis opulastri F.E. & E.S. Clements. Cryptogamae Formationum Coloradensium, Century 3, No. 282. Anno 1907, nom. nud. **Isotype:** ... in ramulis dejectis *Opulastri monogynae* [*Opulaster monogynus* (Torrey) O. Kuntze = *Physocarpus monogynus* (Torrey) Coulter], Colorado, Minnehaha, alt. 2600 m, 25 JUN 1906, leg. F.E. & E.S. Clements, ILL 7473.

Pseudonectria pipericola F.L. Stevens. Botanical Gazette 65:230. 1918. **Holotype:** On *Meliola tortuosa* H.G. Winter on *Piper umbellatum* L. [= *Lepianthes umbellata* (L.) Raf.], Porto Rico [Puerto Rico], Jajome Alto, [3 DEC 1913], leg. F.L. Stevens No. 5656, ILL 4470. **Paratypes:** Rio Tanamá, [near Arecibo, 6 JUL 1915], leg. F.L. Stevens No. 7916, ILL 8412, ex BPI 70994 and ILL 4452, ex BPI 70996; leg. F.L. Stevens No. 7848, ILL 4482; Añasco [12 OCT 1913], leg. F.L. Stevens No. 3578, ILL 4445; on *Piper marginatum* Jacquin, Rio Arecibo, [8 JUL 1915], leg. F.L. Stevens No. 7777, ILL 4449; Rio Tanamá, [6 JUL 1915], leg. F.L. Stevens No. 7842, ILL 4448.

Pseudoparodiella vernoniae F.L. Stevens. Illinois Biological Monographs 11:166–167, pl. I, figs. 6–8; pl. II, figs. 9–15. 1927. **Holotype:** On *Vernonia canescens* Kunth in H.B.K., Costa Rica, Peralta, 12 JUL 1923, leg. F.L. Stevens No. 352, ILL 6614.

Pseudovalsa tropicalis L.E. Weymeyer in G.W. Martin. Lloydia 7:67–68, figs. 1–3. 1944. **Isotype:** On bark of large prostrate trunk, Panama, valley of Upper Rio Chiriquí Viejo, alt. 1600–1800 m, 3 JUL 1935, leg. G.W. Martin No. 2416, ILL 32558. **Holotype:** IA [currently at ISC].

Ptychopeltis roupalae H. Sydow. Annales Mycologici 25:78–82, fig. 1. 1927. **Isotype:** Hab. in foliis *Roupalae veraguensis* [= *Roupala veraguensis* Klotzsch ex Meisner in Martius], Costa Rica, Mondongo pr. San Ramón, 3 FEB 1925, leg. H. Sydow, Fungi in Itinere Costaricensi Collecti No. 229a, ILL 7361.

Puccinia amphispilusa P. Dietel & E.W.D. Holway in P. Dietel. Erythea 3:79. 1895. **Isotype:** On *Polygonum* sp., California, Lassen County, 28 JUL 1894, leg. F.P. Nutting s.n., ILL 19210.

Puccinia andropogonis L.D. v. Schweinitz var. *onobrychidis* (T.J. Burrill) J.C. Arthur, 1934. **See basionym:** *Aecidium onobrychidis* T.J. Burrill, 1884.

Puccinia bicolor J.B. Ellis & B.M. Everhart. Bulletin of the Torrey Botanical Club 27:572. 1900. **Isotype:** On leaves of *Hieracium scouleri* Hooker, Washington, Waitsburg, 7 MAY 1900, leg. R.M. Horner [No. 1433, ex J.B. Ellis & B.M. Everhart, Fungi Columbiani No. 1570], ILL 19588. **Holotype:** NY.

Puccinia caricis-shepherdiae J.J. Davis. Transactions of the Wisconsin Academy of Sciences, Arts and Letters 21:301–302. 1924. **Isotype:** On leaves and culms of *Carex eburnea* Boott ex Hooker, Wisconsin, Fish Creek, [28 SEP 1919], leg. J.J. Davis [Fungi Wisconsinenses Exsiccati No. 129], ILL 19770.

Puccinia caulicola S.M. Tracy & B.T. Galloway. Journal of Mycology 4:20. 1888. **Isotype:** On *Salvia lanceolata* Willd. non Lam. [= *S. reflexa* Hornem.], Colorado, Canon City, 21 AUG 1887, leg. S.M. Tracy & A.W. Evans s.n., ILL 19788.

Puccinia claytoniicola G.B. Cummins. Bulletin of the Torrey Botanical Club 79:218, fig. 7. 1952. **Isotypes:** On leaves, stems, and inflorescences of *Claytonia lanceolata* Pursh, Wyoming, [Carbon County], Medicine Bow Mountains, Park Headquarters, 8 JUL 1947, leg. W.G. Solheim No. 2155 [ex Mycoflora Saximontanensis Exsiccata No. 541], ILL 31296, BPI, PUR, RMS.

Puccinia crepidis-grandiflorae A. Hasler. Zentralblatt für Bakteriologie, Parasitenkunde und Infektionskrankheiten, Abt. 2, 21:510. 1908. **Isotype:** Auf Blättern von *Crepis grandiflora* (Allioni) Tausch non Willd. [= *C.* cf. *conyzifolia* (Gouan) A. Kerner, Europe, now Poland], Riesengebirge: am kleinen Teiche bei Krummkübel, 24 AUG 1908, leg. H. Sydow, Mycotheca Germanica No. 759, ILL 20135.

Puccinia cymbopogonis G.E. Massee. Kew Bulletin 1911:224. 1911. **Isotypes:** On *Cymbopogon citratus* (DC. ex Nees) Stapf, Africa, Uganda, Entebbe, leg. R. Fyffe s.n., ILL 20160, PUR. **Holotype:** K.

Puccinia cymopteri P. Dietel & E.W.D. Holway in P. Dietel. Botanical Gazette 18:255. 1893. **Isotype:** On *Cymopterus terebinthinus* (Hooker) Torrey & Gray [= *Pteryxia terebinthina* (Hooker) Coulter & Rose], California, Kings River Canyon, JUL 1892, leg. E.W.D. Holway, ILL 20161.

Puccinia distichlidis J.B. Ellis & B.M. Everhart. Proceedings of the Academy of Natural Sciences of Philadelphia 1893:152. 1893 [as *distichlydis*, corrected in accordance with Articles 60 & 61 of I.C.B.N.]. **Isotypes:** On *Distichlis maritima* Raf. [as *Distichlys* — an error for *Spartina gracilis* Trin. (fide G.B. Cummins, 1971)], Montana, Helena, 21 SEP 1891, leg. F.D. Kelsey No. 23 [ex North American Fungi, No. 2890], ILL 20236, PUR. **Holotype:** NY.

Puccinia eulobi P. Dietel & E.W.D. Holway in P. Dietel. Erythea 1:249–250. 1893. **Isotype:** On all green parts of *Eulobus californicus* Nutt. ex Torrey & Gray [= *Camissonia californica* (Nutt. ex Torrey & Gray) Raven], California, Pasadena, JUL 1893, leg. A.J. McClatchie No. 295, ILL 20368.

Puccinia fimbristylidis J.C. Arthur. Bulletin of the Torrey Botanical Club 33:28–29. 1906. **Paratypes:** On *Fimbristylis holwayana* Fernald, Mexico, State of Jalisco, Chapala, 9 SEP 1899 [as 18 SEP on the packet], leg. E.W.D. Holway No. 3443, ILL 20443, PUR 26941 and 57435. **Holotype:** PUR 26943.

Puccinia impressa H. Sydow. Annales Mycologici 24:289. 1926. **Isosyntype:** Hab. in foliis *Solani salviifolii* [= *Solanum salviifolium* Lam.], Costa Rica, La Caja pr. San José, 24 DEC 1924, leg. H. Sydow, Fungi in Itinere Costaricensi Collecti No. 58, ILL 22697.

Puccinia raunkiaeri C.C.F. Ferdinandsen & Ø. Winge. Botanisk Tidsskrift 29:8–9, pl. 1, fig. 1. 1909. **Isosyntype:** In caulibus, petiolis foliisque *Rivinae humilis* [= *Rivina humilis* L.], U.S. Virgin Islands, St. Thomas, Løvenlund, [leg. C. Raunkiaer, 10 MAY 1906, Museum Botanicum Hauniense No. 1819], ILL 23761.

Puccinia ruelliae-bourgaei P. Dietel & E.W.D. Holway in E.W.D. Holway. Botanical Gazette 31:329. 1901. **Isotype:** On *Ruellia bourgaei* Hemsley, [Mexico, State of Jalisco], Chapala, 20 SEP 1899, leg. E.W.D. Holway No. 3471, ILL 23918.

Puccinia scleriicola J.C. Arthur. Mycologia 7:232–234. 1915. **Paratypes:** On Cyperaceae: *Scleria* sp., Porto Rico [Puerto Rico], Preston's Ranch, near Naguabo, 31 DEC 1912 [as 1915 on the packet], leg. F.L. Stevens No. 6684, ILL 24305, 24304 and 24306. **Holotype:** PUR.

Puccinia seymeriae T.J. Burrill. Botanical Gazette 9:189. DEC 1884; Bulletin of the Illinois State Laboratory of Natural History 2(3):188–189. AUG 1885. **Holotype:** On *Seymeria macrophylla* Nutt. [= *Dasistoma macrophylla* (Nutt.) Raf.], Illinois, [McLean County], Bloomington, 2 SEP 1879, leg. A.B. Seymour [No. 2762], ILL 24330. **Isotype:** BPI.

Puccinia seymouriana J.C. Arthur. Botanical Gazette 34:11–13, fig. 3. 1902. **Paratypes**: On *Spartina cynosuroides* (L.) Roth, Illinois, [McHenry County], English Prairie, [27] AUG 1881, leg. A.B. Seymour No. 1346, ILL 24326a, PUR 22720; [Fulton County], Canton, [1] OCT 1881, leg. A.B. Seymour No. 1771, ILL 24326b, PUR 22719; [McLean County], Normal, [11] OCT 1881, leg. A.B. Seymour No. 1829, ILL 24327a; [26] MAY 1882, leg. A.B. Seymour No. 4794, ILL 24325, PUR 22718; SEP 1882, leg. A.B. Seymour No. 6262, ILL 24322, PUR 22717; 19 JUN 1882, leg. C.A. Hart No. 5272, ILL 24324, PUR 22716. **Holotype**: PUR 22724.

Puccinia tenuis T.J. Burrill. Botanical Gazette 9:188–189. DEC 1884; Bulletin of the Illinois State Laboratory of Natural History 2(3):181. AUG 1885. **Holotype**: On leaves of *Eupatorium ageratoides* L.f. [= *E. rugosum* Houtt.], Illinois, [McLean County], Bloomington, 3 SEP 1879, leg. A.B. Seymour s.n., ILL 24724 [specimen marked "type" on the packet; confirmed by G.B. Cummins (1978)]. **Isotype**: PUR 61219.

Puccinia tetramerii A.B. Seymour in C.G. Pringle. Mexican Fungi, Decade 1, No. 9. Anno 1896 [cited as *tetrameri*, corrected in accordance with Articles 32.6 and 61 of I.C.B.N.]. **Isotypes**: On leaves of *Tetramerium aureum* Rose [= *T. glandulosum* Oersted], Mexico, Oaxaca, Tomellin Cañon, 30 NOV 1895, leg. C.G. Pringle, ILL 24741, 24742 and 33197, DAOM, FH.

Pucciniastrum wikstroemiae J.C. Arthur in F.L. Stevens. Bernice P. Bishop Museum Bulletin 19:115. 1925. **Isotypes**: On *Wikstroemia uva-ursi* A. Gray, Hawaii, [Island of] Hawaii, Kapapala Ranch, 18 JUL [1921], leg. F.L. Stevens No. 892, ILL 25127, BISH 146257 and 500327. **Holotype**: PUR F-2100.

Pycnidiopeltis smilacina A.C. Batista & C.A.A. Costa in A.C. Batista & R. Ciferri. Mycopathologia 11:83–85, figs. 66, 67. 1959. **Holotype**: In foliis *Smilacis* sp. [= *Smilax* sp.; erroneously as *Pelea* sp. on the packet], Hawaii, Oahu, Olympus, 24 JUN 1921, leg. F.L. Stevens [No. 670], ILL 6668 [also a syntype of *Phragmocapsias smilacina* J.M. Mendoza in F.L. Stevens, and the holotype of *Plochmopeltidella smilacina* J.M. Mendoza in F.L. Stevens & H.W. Manter]. **Isotypes**: BISH 499972, 499973 and 499974.

Pycnidiostroma eugeniae F.L. Stevens. Illinois Biological Monographs 11:198, pl. IX, fig. 73; pl. XVII, figs. 117–118. 1927. **Holotype**: On *Eugenia oerstediana* Berg [= *Calyptranthes costaricensis* Berg], Costa Rica, Cartago, 2 JUL 1923, leg. F.L. Stevens No. 187, ILL 13108. **Paratypes**: El Alto, 6 JUL 1923, leg. F.L. Stevens No. 244 [erroneously cited as 224 in the protologue], ILL 13110; on *Calyptranthes costaricensis* Berg [as *Eugenia oerstediana* Berg on the packet], El Alto, 6 JUL 1923, leg. F.L. Stevens No. 237, ILL 13111; leg. F.L. Stevens No. 250, ILL 13109.

Pycnodothis tetracerae F.L. Stevens. Illinois Biological Monographs 8:198, pl. IX, figs. 78–79; pl. XIX, fig. 108. 1923. **Holotype:** On *Tetracera* sp., British Guiana [Guyana], Demerara-Essequibo R.R., 15 JUL 1922, leg. F.L. Stevens No. 408, ILL 6615. **Isotype:** ILL 6616.

Pyrenopeziza doryphora F.E. & E.S. Clements. Cryptogamae Formationum Coloradensium, Century 3, No. 288. Anno 1907, nom. nud. **Isotype:** ... ad folia emortua *Caricis tolmiei* [*Carex tolmiei* Bailey non Boott ex Hooker = *Carex* sp.], Colorado, Bottomless Pit, alt. 3600 m, 13 JUL 1906, leg. F.E. & E.S. Clements, ILL 8011.

Pyrenopeziza doryphora F.E. & E.S. Clements [nom. nud.] var. *heleocharidis* F.E. & E.S. Clements. Cryptogamae Formationum Coloradensium, Century 6, No. 525. Anno 1908, nom. nud. **Isotype:** ... in culmis emortuis *Heleocharidis palustris* [= *Eleocharis palustris* (L.) Roemer & J.A. Schultes] ..., Colorado, Sulphur Springs, alt. 2400 m, 19 JUL 1907, leg. F.E. & E.S. Clements, ILL 8010.

Pyrenophora ciliata (J.B. Ellis) P.A. Saccardo var *ecoronis* F.E. & E.S. Clements. Cryptogamae Formationum Coloradensium, Century 5, No. 446. Anno 1908, nom. nud. **Isotype:** ... in caulibus foliisque vetustis *Giliae pungentis* [*Gilia pungens* (Torrey) Bentham in A. DC. = *Leptodactylon pungens* (Torrey) Torrey ex Nutt.], Colorado, La Veta, alt. 2100 m, 19 JUN 1907, leg. F.E. & E.S. Clements, ILL 10179.

Pyrenophora corynis F.E. & E.S. Clements. Cryptogamae Formationum Coloradensium, Century 5, No. 447. Anno 1908, nom. nud. **Isotype:** ... in caulibus vetustis *Anaphalidis margaritaceae* [= *Anaphalis margaritacea* (L.) Bentham & J.D. Hooker], Colorado, Long's Peak Inn, alt. 2700 m, 16 AUG 1907, leg. F.E. & E.S. Clements, ILL 10182.

Pyrenophora gigantis F.E. & E.S. Clements. Cryptogamae Formationum Coloradensium, Century 5, No. 448. Anno 1908, nom. nud. **Isotype:** ... in caulibus vetustis *Lupini ammophili* [= *Lupinus ammophilus* E. Greene], Colorado, Sulphur Springs, alt. 2400 m, 23 JUL 1907, leg. F.E. & E.S. Clements, ILL 10184.

Pyrenophora hispida (G. Niessl) P.A. Saccardo var. *pericomes* F.E. & E.S. Clements. Cryptogamae Formationum Coloradensium, Century 3, No. 240. Anno 1907, nom. nud. **Isotype:** ... in caulibus emortuis *Pericomes caudatae* [= *Pericome caudata* A. Gray], Colorado, Engelmann Canyon, alt. 2300 m, 17 JUL 1906, leg. F.E. & E.S. Clements, ILL 10188.

Pyrenophora ipomoeae F.E. & E.S. Clements. Cryptogamae Formationum Coloradensium, Century 5, No. 450. Anno 1907, nom. nud. **Isotype:** In caulibus vetustis *Ipomoeae leptophyllae* [= *Ipomoea leptophylla* Torrey], Colorado, Wray, alt. 1100 m, 25 AUG 1907, leg. F.E. & E.S. Clements, ILL 10189. ≡ *Comoclathris ipomoeae* F.E. Clements. Minnesota Botanical Studies 4:185–186. 1911 [Clements did not validate the original name but cited it, pro syn., under the new, validly published name].

Pyrenophora oedospora F.E. & E.S. Clements. Cryptogamae Formationum Coloradensium, Century 5, No. 451. Anno 1908, nom. nud. **Isotype:** ... in caulibus emortuis *Machaerantherae variantis* [*Machaeranthera varians* E. Greene = *M. bigelovii* (A. Gray) E. Greene], Colorado, Sulphur Springs, alt. 2400 m, 22 JUL 1907, leg. F.E. & E.S. Clements, ILL 10190.

Pyrenostigme siparunae H. Sydow. Annales Mycologici 24:370–372. 1926. **Isosyntypes:** Hab. in foliis *Siparunae patelliformis* [= *Siparuna patelliformis* Perkins], Costa Rica, Mondongo pr. San Ramón, 3 FEB 1925, leg. H. Sydow, Fungi in Itinere Costaricensi Collecti No. 211, ILL 9232; San Pedro de San Ramón, 29 JAN 1925, leg. H. Sydow, Fungi in Itinere Costaricensi Collecti No. 401, ILL 9233.

R

Ragnhildiana agerati (F.L. Stevens) F.L. Stevens & W.G. Solheim in W.G. Solheim & F.L. Stevens, 1931. **See basionym:** *Cercospora agerati* F.L. Stevens, 1925.

Ragnhildiana cyathulae F.L. Stevens & W.G. Solheim in W.G. Solheim & F.L. Stevens. Mycologia 23:403, fig. 12. 1931. **Holotype:** On leaves of *Cyathula achyranthoides* (Kunth) Moq. in DC., British Guiana [Guyana], Coverden, [4 AUG 1922], leg. F.L. Stevens No. 743, ILL 16298.

Ragnhildiana manihotis F.L. Stevens & W.G. Solheim in W.G. Solheim & F.L. Stevens. Mycologia 23:404–405. 1931. **Holotype:** On *Manihot utilissima* Pohl [= *M. esculenta* Crantz], British Guiana [Guyana], Penal Settlement, [25 JUL 1922], leg. F.L. Stevens No. 683, ILL 16302. **Paratypes:** Panama, Frijoles, [20 AUG 1923], leg. F.L. Stevens No. 1181, ILL 16303; Porto Rico [Puerto Rico], Bayamon, [24 FEB 1916], leg. J.A. Stevenson No. 3932, ILL 16300; Dos Bocas, [below Utuado, 30 DEC 1913], leg. F.L. Stevens No. 6557, ILL 16301. ≡ *Phaeoramularia manihotis* (F.L. Stevens & W.G. Solheim) M.B. Ellis. More Dematiaceous Hyphomycetes, p. 321. 1976.

Ragnhildiana trematis F.L. Stevens & W.G. Solheim in W.G. Solheim & F.L. Stevens. Mycologia 23:405. 1931 [as *tremae*, corrected in accordance with current I.C.B.N. (cf. F.C. Deighton, 1968)]. **Holotype:** On *Trema micranthum* (L.) Blume, Trinidad, St. Clair, [19 AUG 1922], leg. F.L.

Stevens No. 889, ILL 16306. ≡ *Cercospora trematis* (F.L. Stevens & W.G. Solheim in W.G. Solheim & F.L. Stevens) C. Chupp in C.E. Chardon & R.A. Toro. Monographs of the University of Porto Rico, Series B, 2: 253. 1934, non C.G. Hansford, 1944, nec K. Sawada, 1944 [the epithet also published as *tremae* and corrected herein].

Ramularia angelicae F.E. & E.S. Clements. Cryptogamae Formationum Coloradensium, Century 3, No. 262. Anno 1907, nom. nud. **Isotype:** ... in foliis vivis *Angelicae grayi* [= *Angelica grayi* (Coulter & Rose) Coulter & Rose], Colorado, Garfield Glen, alt. 3600 m, 21 AUG 1906, leg. F.E. & E.S. Clements, ILL 14175.

Ramularia arnicae J.B. Ellis & B.M. Everhart. Proceedings of the Academy of Natural Sciences of Philadelphia 1891:85. 1891 [as *arnicalis*, corrected in accordance with Articles 32.6 & 61 of I.C.B.N.]. **Isotype:** On *Arnica cordifolia* Hooker, Montana, Rimini, [24] JUN 1889, leg. F.D. Kelsey No. 88, ILL 14200. **Holotype:** NY.

Ramularia artemisiae J.J. Davis. Transactions of the Wisconsin Academy of Sciences, Arts and Letters 22:173. 1926. **Isotype:** On leaves of *Artemisia caudata* Michaux [= *A. campestris* L. ssp. *caudata* (Michaux) H.M. Hall & F.E Clements], Wisconsin, Lewis, 2 AUG 1924, leg. J.J. Davis, [Fungi Wisconsinenses Exsiccati No. 135], ILL 14203.

Ramularia cardamines H. & P. Sydow. Annales Mycologici 1:538. 1903. **Isotype:** Hab. in foliis vivis vel languidis *Cardamines amarae* [= *Cardamine amara* L.], ad fines Saxoniae et Bohemiae, [Germany], Dürrkamnitzschlucht [bei Herrnskretschen, 16 AUG 1903], leg. H. & P. Sydow, Mycotheca Germanica No. 92, ILL 14225.

Ramularia cercosporoides J.B. Ellis & B.M. Everhart. Proceedings of the Academy of Natural Sciences of Philadelphia 1895:437. 1895. **Isotype:** On leaves of *Epilobium spicatum* Lam. [= *E. angustifolium* L.], Washington, Seattle, AUG 1894, leg. C.V. Piper No. 290, ILL 14234. **Holotype:** NY.

Ramularia conspicua H. & P. Sydow. Annales Mycologici 1:538. 1903. **Isotype:** Hab. in pagina super. foliorum vivorum *Hieracii murorum* [= *Hieracium murorum* L.], ad fines Saxoniae et Bohemiae, [Germany], Dürrkamnitzschlucht [bei Herrnskretschen, 19 AUG 1903], leg. H. & P. Sydow, ex Mycotheca Germanica No. 94, ILL 14243.

Ramularia dispar J.J. Davis. Transactions of the Wisconsin Academy of Sciences, Arts and Letters 19:702. 1919. **Isotype:** On leaves of *Eupatorium purpureum* L., Wisconsin, Danbury, 30 AUG 1916, leg. J.J. Davis, [Fungi Wisconsinenses Exsiccati No. 87], ILL 14267.

Ramularia gracilipes J.J. Davis. Transactions of the Wisconsin Academy of Sciences, Arts and Letters 22:173. 1926. **Isotype:** On leaves of *Cornus alternifolia* L.f., Wisconsin, Bruce, 4 SEP 1924, leg. J.J. Davis, [Fungi Wisconsinenses Exsiccati No. 132], ILL 14284.

Ramularia grindeliae J.B. Ellis & W.A. Kellerman. Bulletin of the Torrey Botanical Club 11:122. 1884. **Isotype:** On leaves of *Grindelia squarrosa* (Pursh) Dunal, 15 AUG 1884, Kansas, [Riley County, Manhattan], leg. W.A. Kellerman No. 616, ILL 14286. **Holotype:** NY.

Ramularia ipomoeae F.L. Stevens. Bernice P. Bishop Museum Bulletin 19:150. 1925. **Holotype:** On *Ipomoea bona-nox* L. (moonflower, cult.), Hawaii, [Island of] Hawaii, Kealakekua, 21 JUL [1921], leg. F.L. Stevens No. 908, ILL 14311 [associated with and also the holotype of *Sphaerulina ipomoeae* F.L. Stevens]. **Isotypes:** BISH 499083 and 500326.

Ramularia microlepiae F.L. Stevens. Bernice P. Bishop Museum Bulletin 19:151. 1925. **Syntypes** [BISH numbers cited are isosyntypes]: On *Microlepia* sp., Hawaii, Kauai, Kalalau trail, 16 JUN [1921], leg. F.L. Stevens No. 500, ILL 14341, BISH 146265 and 500323; Maui, Pogue's ditch trail, 6 SEP [1921], leg. F.L. Stevens No. 1155, ILL 14336, BISH 146260 and 500322; Oahu, Wahiawa, 31 MAY [1921], leg. F.L. Stevens No. 169, ILL 14339, BISH 146266 and 500318; 3 JUN [1921], leg. F.L. Stevens No. 255, ILL 14333, BISH 146261 and 500317; Kalihi Valley, 2 JUN [1921], leg. F.L. Stevens No. 175, ILL 14335, BISH 146264 and 500316; Ahren's ditch trail, 8 JUN [1921], leg. F.L. Stevens No. 282, ILL 14334, BISH 146263 and 500325; Palolo Valley, 10 JUN [1921], leg. F.L. Stevens No. 336, ILL 14338, BISH 146258 and 500320; Waiahole ditch trail, 12 JUN [1921], leg. F.L. Stevens No. 391, ILL 14340, BISH 146259 and 500319; Tantalus, 22 JUN [1921], leg. F.L. Stevens No. 606, ILL 14342, BISH 146262 and 500324; Kolekole Pass, 27 JUN [1921], leg. F.L. Stevens No. 729, ILL 14337, BISH 500321 and 499082.

Ramularia mimosae F.L. Stevens & N.E. Dalbey. Mycologia 11(1):6, pl. 3, fig. 8. DEC 1918. **Holotype:** On *Mimosa pudica* L., Porto Rico [Puerto Rico], Coamo Springs, [16 JUL 1915], leg. F.L. Stevens No. 8367, ILL 14343. **Paratypes:** Porto Rico [Puerto Rico], Peñuelas, [JUL 1915], leg. F.L. Stevens No. 7215, ILL 14344; Arecibo-Lares Road, [21 JUN 1915], leg. F.L. Stevens No. 7298, ILL 14346; Mayagüez, [10 JUN 1915], leg. F.L. Stevens No. 7110, ILL 14345; Lajas, [17 JUN 1915], leg. F.L. Stevens No. 7158, ILL 14347.

Ramularia nephrolepis F.L. Stevens. Bernice P. Bishop Museum Bulletin 19:150, text fig. 31b. 1925. **Syntypes:** On *Nephrolepis exaltata* (L.) Schott, Hawaii, Oahu, Ahren's ditch trail, 8 JUN [1921], leg. F.L. Stevens No. 287, ILL 14356; Hawaii, Oahu, Palolo Valley, 10 JUN [1921], leg. F.L. Stevens No. 311, ILL 14357.

Ramularia phaceliae L. Bonar. Mycologia 38:344. 1946. **Isotype:** On living leaves of *Phacelia procera* A. Gray, California, Plumas County, Gold Lake, 31 JUL 1942, leg. L. Bonar, California Fungi No. 824, ILL 31463, ex UC. **Holotype:** UC 697788.

Ramularia serotina J.B. Ellis & B.M. Everhart. Journal of Mycology 5:69. 1889. **Assumed isotype:** On leaves of *Solidago serotina* Aiton non Retz. [= *S. gigantea* Aiton], Illinois, Lake County, JUL 1888, leg. J.J. Davis No. 39, ILL 14412 [no collection number indicated on the packet].

Ramularia tenuis J.J. Davis. Transactions of the Wisconsin Academy of Sciences, Arts and Letters 21:261. 1924. **Isotype:** On leaves of *Solidago latifolia* L. [= *S. flexicaulis* L.], Wisconsin, Holcombe, 9 AUG 1920, leg. J.J. Davis, [Fungi Wisconsinenses Exsiccati No. 165], ILL 14426.

Ramularia thelypodii F.E. & E.S. Clements. Cryptogamae Formationum Coloradensium, Century 6, No. 504. Anno 1908, nom. nud. **Isotype:** ... ad folia viva *Thelypodii integrifolii* [= *Thelypodium integrifolium* (Nutt.) Endl. ex Walp.], Colorado, Sulphur Springs, alt. 2400 m, 19 JUL 1907, leg. F.E. & E.S. Clements, ILL 14425.

Ramularia variata J.J. Davis. Transactions of the Wisconsin Academy of Sciences, Arts and Letters 19:688. 1919. **Isotype:** On leaves of *Monarda fistulosa* L., Wisconsin, Lynxville, 3 SEP 1915, leg. J.J. Davis, [Fungi Wisconsinenses Exsiccati No. 62], ILL 14468.

Ravenelia farlowiana P. Dietel. Hedwigia 33:369. 1894. **Isosyntype:** On *Acacia anisophylla* S. Watson, Mexico, [Jimulco], MAY 1885, leg. C.G. Pringle, ILL 25003, ex FH. **Syntype:** S [fide G.B. Cummins, 1975]. ≡ *Dendroecia farlowiana* (P. Dietel) J.C. Arthur. Résultats Scientifiques du Congrès Botanique Vienne 1905. p. 340. 1906.

Ravenelia igualica J.C. Arthur. North American Flora 7:136–137. 1907. **Isotype:** On Mimosaceae [= Fabaceae subfam. Mimosoideae]: *Acacia filicina* Willd., Mexico, State of Guerrero, Iguala, 3 NOV 1903, leg. E.W.D. Holway No. 5312, ILL 25007. **Holotype:** PUR.

Ravenelia opaca P. Dietel. Hedwigia 34:291. 1895; originally distributed as *Ravenelia indica* M.J. Berkeley forma *opaca* A.B. Seymour & F.S. Earle. Economic Fungi Fasc. 5, No. 203, nom. nud. **Isotype:** On living leaves of honey locust, *Gleditsia triacanthos* L. [cited as *Gleditschia*], Illinois, Union County, Clear Creek, [13 SEP 1890], leg. F.S. Earle, ex Economic Fungi No. 203, ILL 33555. [The specimen is filed in the exsiccati set; the name is included herein because apparently this fungus has not been found since the type collection was made.]. ≡ *Dendroecia opaca* (P. Dietel) J.C. Arthur. North American Flora 7(2): 145-146. 1907.

Ravenelia stevensii J.C. Arthur. Mycologia 7:178. 1915. **Assumed isotype:** On *Acacia riparia* sensu auct. non Kunth [= *A. retusa* (Jacquin) Howard], Porto Rico [Puerto Rico], Guayanilla, 13 NOV 1913, leg. F.L. Stevens No. 5881, ILL 25030. **Holotype:** PUR 6216. **Paratype:** Vega Baja, 22 FEB [1913], leg. F.L. Stevens No. 366, ILL 25029.

Ravenelia versatilis sensu G.B. Cummins. Mycologia 67:1042. 1975, non

P. Dietel. Hedwigia 33:64 (15 APR 1894), emend., op. cit. pp. 368–369. 20 DEC, 1894. **Isolectotype:** On *Acacia anisophylla* S. Watson, Mexico, Jimulco, MAY 1885, leg. C.G. Pringle, ILL 25003, ex FH [also an isosyntype of *R. farlowiana* P. Dietel]. **Lectotype** [designated by G.B. Cummins (1975:1042)]: S. [The name was originally published as *Ravenelia versatilis* (C.H. Peck) P. Dietel, and the type of *Uromyces versatilis* C.H. Peck (on *Acacia greggii* A. Gray, Arizona, leg. C.G. Pringle) was included as one of several collections (syntypes) cited for *R. versatilis*. In the discussion preceding his emendation (pp. 368–369), Dietel explicitly replaces the original diagnosis (p. 64) with the emended diagnosis. He also explicitly excludes (among others) the specimen of *Acacia anisophylla* from the collections he considers representative of the species. Of the original material cited (p. 64), therefore, only the Pringle collection of *Acacia greggia* from Arizona remains attached to Dietel's name. Implied typification of *Ravenelia versatilis* P. Dietel with a specimen of the latter was published by J.C. Arthur, North American Flora 7(2):136. 1907, who cited Arizona as the type locality, a decision upheld in J.C. Arthur (1934). The lectotypification by Cummins is contrary to the intentions of the original author and is, in our interpretation of I.C.B.N., unacceptable. Dietel explicitly assigned his specimen of *Acacia anisophylla* from Mexico, leg. C.G. Pringle, to represent one of several collections (syntypes) of *R. farlowiana* P. Dietel, sp. nov. (p. 369). Our specimen (ILL 25003) is a duplicate of the above collection, i.e., an isosyntype of the latter name, not an isolectotype of *R. versatilis* P. Dietel.].

Rhabdospora brunellae J.B. Ellis & B.M. Everhart. North American Fungi, Century 26, No. 2580. Anno 1891, nom. nud. **Isotype:** On dead stems of *Prunella vulgaris* L. [as *Brunella*], Canada, [Ontario], London, JUL 1899, leg. J. Dearness s.n., ILL 11938.

Rhabdospora pittospori F.L. Stevens & P.A. Young in F.L. Stevens. Bernice P. Bishop Museum Bulletin 19:141. 1925. **Holotype:** On dead capsules of *Pittosporum* sp., Hawaii, [Island of] Hawaii, Kona, 23 JUL 1911, leg. C.N. Forbes No. 21, ILL 11948.

Rheumatopeltis quercus F.L. Stevens. Illinois Biological Monographs 11:176–177, pl. IV, figs. 27–31; pl. XIV, figs. 103–104. 1927 [as *querci*, corrected in accordance with Articles 32.6 and 61 of I.C.B.N.]. **Holotype:** On *Quercus eugeniifolia* Liebm., Costa Rica, Cartago, 23 JUN 1923, leg. F.L. Stevens No. 68, ILL 9234.

Rhizopogon separabilis S.M. Zeller. Mycologia 31:3, figs. 17, 18. 1939. **Paratypes:** [On sandy soil] under pine and ericads, Oregon, Lane County, [south of] Sutton Lake, 26 NOV 1937, leg. A.M. & D.P. Rogers No. 415, ILL 32563; leg. A.M. & D.P. Rogers No. 451, ILL 32564.

Roestelia harknessiana J.B. Ellis & B.M. Everhart in F.D. Kern. Bulletin of the Torrey Botanical Club 34:462–463. 1907. **Isotype**: On *Amelanchier alnifolia* (Nutt.) Nutt., California, Klamath River, JUL 1887, collector unknown, comm. H.W. Harkness, (ex J.B. Ellis & B.M. Everhart, North American Fungi No. 2714), ILL 28166. **Holotype**: NY.

Rosellinia aquila (E.M. Fries:E.M. Fries) G. de Notaris var. *minor* F.E. & E.S. Clements. Cryptogamae Formationum Coloradensium, Century 1, No. 11. Anno 1906, nom. nud. **Isotype**: ... in ramis vetustis *Sambuci microbotryae* ... [*Sambucus microbotrys* Rydb. = *S. racemosa* L. var. *microbotrys* (Rydb.) Kearney & Peebles], Colorado, Larkspur Dell, alt. 2500 m, 20 JUL 1905, leg. F.E. & E.S. Clements, ILL 9443.

Rosellinia citriformis F.L. Stevens & A.G. Weedon in F.L. Stevens. Bernice P. Bishop Museum Bulletin 19:95–96, text fig. 23b. 1925. **Holotype**: On dead twig, Hawaii, Molokai, 20 OCT 1913, leg. L.D. Larsen, comm. H.L. Lyon No. 75, ILL 9446.

Rosellinia (Tassiella) crustacea H. Rehm. Leaflets of Philippine Botany 8:2941–2942. 1916. **Assumed isotype**: Ad calamos vivos *Schizostachyi* [= *Schizostachyum* sp.], Philippines, [Luzon, Laguna Prov.], Los Baños, MAY 1914, leg. S.A. Reyes, comm. C.F. Baker No. 3372 [ex Fungi Malayana No. 187], ILL 9447.

Rosellinia (Tassiella) horrida H. Rehm. Leaflets of Philippine Botany 8:2941. 1916. **Assumed isotype**: Ad corticem emortuum, Philippines, [Luzon, Laguna Prov., Los Baños], Mount Maquiling, MAR 1914, leg. C.F. Baker No. 2909 [ex Fungi Malayana No. 188], ILL 9448.

Rosellinia (Coniomela) maquilingiana H. Rehm. Leaflets of Philippine Botany 8:2942. 1916. **Assumed isotype**: Ad ramum corticatum deciduum, Philippines, [Luzon, Laguna Prov., Los Baños], Mount Maquiling, MAY 1914, leg. S.A. Reyes, comm. C.F. Baker No. 3347 [ex Fungi Malayana No. 189 (as *R. makilingiana*)], ILL 9449.

Rosellinia metachroa C.C.F. Ferdinandsen & Ø. Winge. Botanisk Tidsskrift 29:16, pl. 2, fig. 3. 1909. **Isosyntype**: Ad lignum corticatum vel nudum in insulis St. Croix et St. Jan [St. John], Indiae occidentalis, [U.S. Virgin Islands], 17 MAR 1906, leg. C. Raunkiaer No. 1740, ILL 9450.

S

Schiffnerula lisianthii C.G. Hansford. Beihefte zur Sydowia Annales Mycologici 1:88. 1957 [as *lisianthi*, corrected in accordance with Articles 32.6 and 61 of I.C.B.N.]. **Isotypes**: In foliis *Lisianthi grandiflori* [*Li-*

sianthius grandiflorus Aublet = *Irlbachia alata* (Aublet) Maas ssp. *alata*], British Guiana [Guyana], Wismar, [14 JUL 1922], leg. F.L. Stevens No. 316, ILL 5388 and 5398 [also the holotype and an isotype, respectively, of *Meliola lisianthii* F.L. Stevens & L.R. Tehon]. **Holotype:** FH.

Schiffnerula vaccinii C.G. Hansford. Beihefte zur Sydowia Annales Mycologici 1:88. 1957. **Holotype:** Hab. in foliis *Vaccinii reticulati* [= *Vaccinium reticulatum* J.E. Smith], Hawaii, [Island of] Hawaii, Kilauea, leg. F.L. Stevens No. 821 p.p., ILL 4086 p.p. [also a paratype of *Meliola alyxiae* F.L. Stevens and of *Meliola vaccinii* F.L. Stevens].

Schizochora pandani F.L. Stevens. Bernice P. Bishop Museum Bulletin 19:20, text fig 5a–c, pl. II(D). 1925. **Syntypes:** On *Pandanus odoratissimus* [prob. sensu auct., non L.f. = *Pandanus tectorius* Parkinson ex Zucc.], Hawaii, Oahu, Waiahole ditch trail, 6 JUN [1921], leg. F.L. Stevens No. 408, ILL 9242; Oahu, Kalihi Valley, 2 JUN [1922?], leg. F.L. Stevens No. 187, ILL 9243. **Isosyntypes:** BISH 146276, 499081, 500314, and 500315.

Schizothyrella sydowiana P.A. Saccardo in H. & P. Sydow. Annales Mycologici 3:233. 1905. **Isotype:** Hab. in culmis et foliis emortuis *Phragmitis communis* [*Phragmites communis* Trin. = *P. australis* (Cav.) Trin. ex Steudel, Germany, Brandenburg], Wannsee bei Berlin, 2 OCT 1904, leg. H. & P. Sydow, Mycotheca Germanica No. 341, ILL 13212.

Schweinitziella palmigena F.L. Stevens. Illinois Biological Monographs 11:177–178, pl. IV, figs. 32–33. 1927. **Holotype:** On *Chamaedorea* sp., Costa Rica, Peralta, 13 JUL 1923, leg. F.L. Stevens No. 417, ILL 9244.

Scirrhia gigantochloae H. Rehm. Leaflets of Philippine Botany 6:2223–2224. 1914. **Assumed isotype:** Ad folia *Gigantochloae scribnerianae* [= *Gigantochloa scribneriana* Merrill], Philippines, Luzon, Laguna Prov., Los Baños, JUL 1913, leg. S.A. Reyes, comm. C.F. Baker No. 1519 [ex Fungi Malayana No. 86], ILL 9245.

Sclerotium crustuliforme M.R. Roberge ex J.B. Desmazières. Annales des Sciences Naturelles, Paris, Botanique, Series 3, 10:346. 1848. **Assumed isosyntype:** Ad petiolos siccos *Aceris negundinis* [= *Acer negundo* L.], leg. M.R. Roberge s.n., ILL 16621.

Sclerotium portoricense F.L. Stevens. Transactions of the Illinois State Academy of Science 10:215–216, fig. 13 (p. 217). 1917. **Holotype:** On *Cynodon dactylon* (L.) Persoon, Porto Rico [Puerto Rico], Santurce, [23 FEB 1913], leg. F.L. Stevens No. 378, ILL 16607. **Paratype:** DEC 1916, leg. J.A. Stevenson [No. 18000], ILL 16608.

Scolecoccoidea costaricensis F.L. Stevens. Illinois Biological Monographs 11:178–179, pl. IV, fig. 39. 1927. **Holotype:** On *Miconia* sp., Costa Rica, Parismina Junction, 20 JUL 1923, leg. F.L. Stevens No. 603, ILL 9250.

Scolecodothopsis ingae F.L. Stevens. Illinois Biological Monographs 8:183, pl. IV, figs. 32–34; pl. XIII, fig. 95. 1923. **Holotype**: On *Inga* sp., British Guiana [Guyana], Demerara-Essequibo R.R., 15 JUL 1922, leg. F.L. Stevens No. 406, ILL 9253. **Paratypes**: Tumatumari, 8 JUL 1922, leg. F.L. Stevens No. 58, ILL 9255; Kartabo, 21 JUL 1922, leg. F.L. Stevens No. 510, ILL 9251 and 9254 [in a handwriting other than that of F.L. Stevens, ILL 9254 and 9255 are erroneously inscribed on the packet as "Demerara-Essequibo R.R., 15 JUL 1922"].

Scolecopeltella microcarpa C.L. Spegazzini. Boletín de la Academia Nacional de Ciencias en Córdoba 26:354. Preprint 1923 [journal part issued in 1924]. **Potential isotype** [not confirmed by LPS]: Hab. sobre las hojas vivas de *Philodendron krebsii* Schott [= *P. consanguineum* Schott], Porto Rico [Puerto Rico], en las cercanías de Río Arecibo [as Arecibo-Lares Road on the packets, 21 JUN 1915], leg. F.L. Stevens No. 7225 p.p., ILL 5991 [also the holotype of *Meliola philodendri* F.L. Stevens].

Scolecopeltella portoricensis C.L. Spegazzini. Boletín de la Academia Nacional de Ciencias en Córdoba 26:354–355, illustr. Preprint 1923 [journal part issued in 1924]. **Potential isotype** [not confirmed by LPS]: Hab. Sobre las hojas vivas de *Dipholis salicifolia* (L.) A. DC. [= *Sideroxylon salicifolium* (L.) Lam.], Porto Rico [Puerto Rico], en los alrededores de Guayanillas [Guayanilla, 14 JUL 1915], leg. F.L. Stevens No. 8549 p.p., ILL 5037 [also the holotype of *Meliola dipholidis* F.L. Stevens and a paratype of *Helminthosporium helleri* F.L. Stevens].

Scolecopeltidella palmarum J.M. Mendoza in F.L. Stevens & H.W. Manter. Botanical Gazette 79:293, figs. 73–75. 1925. **Holotype**: On palm [Arecaceae], British Guiana [Guyana], Rockstone, 17 JUL [1922], leg. F.L. Stevens No. 442, ILL 6669.

Scolecopeltidium costi F.L. Stevens & H.W. Manter. Botanical Gazette 79:284, figs. 58–60. 1925. **Syntypes**: On *Costus* sp., British Guiana [Guyana], Rockstone, 16 JUL [1922], leg. F.L. Stevens No. 425, ILL 6778; on *Serjania paucidentata* DC., Kartabo, 23 JUL [1922], leg. F.L. Stevens No. 585, ILL 6777; on unknown host plant, Tumatumari, 12 JUL [1922], leg. F.L. Stevens No. 234, ILL 6776.

Scolecopeltidium hormosporum F.L. Stevens & H.W. Manter. Botanical Gazette 79:283, figs. 45–46. 1925. **Holotype**: On unknown host, British Guiana [Guyana], Kartabo, 23 JUL [1922], leg. F.L. Stevens No. 581, ILL 6779.

Scolecopeltidium liciniae F.L. Stevens & H.W. Manter. Botanical Gazette 79:283–284, fig. 54. 1925. **Syntypes**: On *Licinia* sp. [= *Anthericum* sp.], British Guiana [Guyana], Rockstone, 17 JUL [1922], leg. F.L. Stevens No. 479, ILL 6780; on Marantaceae sp. indet., British Guiana [Guyana], Rockstone, 16 JUL [1922], leg. F.L. Stevens No. 423, ILL 6781.

Scolecopeltidium mirabile F.L. Stevens & H.W. Manter. Botanical Gazette 79:283, figs. 47–50. 1925. **Holotype**: On unknown member of Simaroubaceae [as Simarubaceae], British Guiana [Guyana], Kartabo, 23 JUL [1922], leg. F.L. Stevens No. 678, ILL 6782.

Scolecopeltidium multiseptatum F.L. Stevens & H.W. Manter. Botanical Gazette 79:283, figs. 51–53. 1925. **Syntypes**: On *Philodendron* sp., British Guiana [Guyana], Wismar [as Kartabo on the packet], 24 JUL [1922], leg. F.L. Stevens No. 1004, ILL 6786; 14 JUL 1922, leg. F.L. Stevens No. 265 [erroneously as 264 in the protologue], ILL 6784; Kartabo, 22 JUL 1922, leg. F.L. Stevens No. 544, ILL 6785.

Scolecopeltis cestri R.A. Toro. Mycologia 17:137. 1925. **Holotype**: On *Cestrum* sp., Porto Rico [Puerto Rico, Mayagüez, 4 JAN 1918], leg. F.L. Stevens No. 7576, ILL 7213 [also the holotype of *Aulographum cestri* R.W. Ryan].

Scolecopeltis pachyasca C.L. Spegazzini. Boletín de la Academia Nacional de Ciencias en Córdoba 26:353. Preprint 1923 [journal part issued in 1924]. **Potential isotypes** [not confirmed by LPS]: Hab. Sobre las hojas vivas de *Coccoloba laurifolia* sensu auct. non Jacquin [as *Coccolobis* = *Coccoloba diversifolia* Jacquin], Porto Rico [Puerto Rico], cerca Río Arecibo [as Arecibo-Lares Road on the packets, 21 JUN 1915], leg. F.L. Stevens No. 7292 p.p., ILL 4398 and 7371 [also the holotype and an isotype, respectively, of *Meliola rectangularis* F.L. Stevens. In addition, Stevens No. 7292 is also cited as a syntype of *Seynesia coccolobae* R.W. Ryan].

Scolecotrichum punctulatum S.M. Tracy & F.S. Earle. Bulletin of the Torrey Botanical Club 22:178 1895. **Isotypes**: On *Iris pabularia* Hort. ex Hasselbr. [= *Iris* cf. *ensata* Thunb. var.], Mississippi, Starkville, [1] JAN 1894, [leg. S.M. Tracy s.n.], ILL 16330, 16331 and 33497, BPI.

Scolecoxyphium americanum A.C. Batista in A.C. Batista & R. Ciferri. Quaderno Laboratorio Crittogamico Instituto Botanico Della Universita Pavia 31:183–184, pl. 12, fig. 67. 1963 [erroneously as *Microxyphium* in the protologue, a name not adopted by the author]. **Paratypes**: On *Melastoma* sp. [associated with *Achaetobotrys* sp.], Porto Rico [Puerto Rico], Maricao, 8 OCT 1913, leg. F.L. Stevens [No. 3404], ILL 33498, BPI, URM 13323 . **Holotype**: URM 5566.

Scoriadopsis miconiae J.M. Mendoza in F.L. Stevens. Annales Mycologici 28:365–366. 1930. **Holotype**: On *Miconia* sp., British Guiana [Guyana], Tumatumari, 10 JUL [1922], leg. F.L. Stevens No. 94, ILL 6670.

Scutellinia livida H.C.F. Schumacher var. *sphaerophysa* F.E. & E.S. Clements. Cryptogamae Formationum Coloradensium, Century 3, No. 294. Anno 1907, nom. nud. **Isotype**: Colorado, Fern Glen, alt. 2600 m, leg. F.E. & E.S. Clements, 8 AUG 1906, ILL 7744.

Sebacina atra J.M. McGuire. Lloydia 4:27–28, pl. 4, figs. 67–72. 1941. **Isotype:** On a sodden deciduous log, Iowa, Iowa City, 17 MAR 1933, [leg. D.P. Rogers No. 582], ILL 32829. **Holotype:** IA [currently at ISC].

Sebacina farinacea D.P. Rogers. Pacific Science 1:97–99, fig. 1. 1947. **Holotype:** On dead leaf bases and sheaths of *Cocos nucifera* L., Hawaii, [Oahu, Honolulu], Manoa, University of Hawaii campus, 29 OCT 1946, leg. D.P. Rogers No. 1884, ILL 32818. **Isotypes:** BISH 499080, IA [currently at ISC].

Sebacina lactescens E.A. Burt. Annals of the Missouri Botanical Garden 13:336–337. 1926. **Isotypes:** Grenada, Grand Etang, B.W.2, 1912, 1913, leg. R. Thaxter, comm. W.G. Farlow No. 153, ILL 32848, FH, IA [currently at ISC]. **Holotype:** MO 55236 [currently at BPI].

Sebacina (Bourdotia) megaspora G.W. Martin. Mycologia 28:214–216, pl. 1, figs. 1–12. 1936. **Isotype:** South Australia National Park, 28 JUL 1923, leg. J.B. Cleland No. 43, ILL 32822. **Holotype:** IA [currently at ISC].

Sebacina molybdea J.M. McGuire. Lloydia 4:17–20, pl. 2, figs. 22–25. 1941. **Isotype:** On the lower sides of logs ... of *Populus* sp. [identified as *Populus grandidentata* Michaux on the packet], Iowa, North Liberty, 26 APR 1939, leg. G.W. Martin No. 4664, ILL 32823. **Holotype:** IA [currently at ISC].

Sebacina murina E.A. Burt. Annals of the Missouri Botanical Garden 13:337. 1926. **Isotype:** On decorticated, weathered, badly decayed wood on mountain side, alt. 800–1500 ft., Mexico, Motzorongo, near Córdoba, [15] JAN [1910], leg. W.A. & E.L. Murrill No. 986, ILL 32832, ex NY. **Holotype:** MO 54609 [currently at BPI].

Sebacina obscura G.W. Martin. Lloydia 7:70, fig. 5. 1944. **Isotypes:** On dead leaf-stalk of date palm [= *Phoenix dactylifera* L.], Panama, Canal Zone, Summit, 19 JUL 1935, leg. G.W. Martin No. 2873, ILL 32827, BPI. **Holotype:** IA [currently at ISC].

Sebacina petiolata D.P. Rogers. Pacific Science 1:99–100, fig. 2. 1947. **Holotype:** On trunk of *Cocos nucifera* L., Marshall Islands, Likiep Island, Likiep, 28 AUG 1946, leg. D.P. Rogers No. 1475, ILL 32811. **Isotypes:** BISH 499079, FH , IA [currently at ISC].

Sebacina prolifera D.P. Rogers. Mycologia 28:347–362, figs. 1–34. 1936. **Holotype:** On lower side of a sodden, well-rotted, decorticated log of *Ulmus* sp., Iowa, Iowa City, Linder's Wood, 20 MAR 1935, leg. D.P. Rogers No. 80, ILL 32821. **Isotypes:** FH, IA [currently at ISC], K, US.

Sebacina (Bourdotia) rimosa H.S. Jackson & G.W. Martin in G.W. Martin. Mycologia 32:684, fig. 1. 1940. **Isotypes:** On *Thuja occidentalis* L., [Canada], Ontario, Maple, 13 NOV 1938 [as DEC on the packet], leg. H.S. Jackson, ex TRTC 13086, ILL 32812, IA [currently at ISC]. **Holotype:** TRTC.

Sebacina umbrina D.P. Rogers. University of Iowa Studies in Natural History 17:39–40, fig. 19. 1935. **Holotype**: On bark of dead branch of *Fraxinus* sp., Iowa, West Okoboji Lake, [Elmcrest], Miller's Bay, 9 AUG 1933, leg. A.M. & D.P. Rogers [No. 278], ILL 32826.

Septobasidium bakeri N.T. Patouillard. Leaflets of Philippine Botany 6:2239–2240. 1914. **Assumed isotypes**: Sur les colonies de *Odonaspis* nov. sp. (*Coccides*), de chaumes de *Schizostachyum*, [Philippines, Luzon, Laguna Prov.], Los Baños, [AUG 1913], leg. C.F. Baker No. 73 [ex Fungi Malayana No. 87], ILL 28852, B, FH, NCU, NY.

Septobasidium sinuosum J.N. Couch. Journal of the Elisha Mitchell Scientific Society 51:65–67, plates 8, 29. 1935, nom. invalid. sine diagn. lat. **Paratype**: On *Liquidambar styraciflua* L., Louisiana, [Saint Tammany Parish, Honey Island, near] New Orleans, 31 DEC 1932 [as 1931 on the packet], leg. [J.N. Couch &] D.P. Rogers No. 9935, ILL 32807.

Septogloeum hercynicum H. & P. Sydow. Annales Mycologici 3:233–234. 1905. **Isotype**: Hab. auf lebenden Blättern von *Acer* spec. (*A.* ? *dasycarpum* Ehrh.) [= *A. saccharinum* L., Germany], am Rehberger Graben zwischen Oderteich und St. Andreasberg im Harz, [24 AUG 1904], leg. P. Sydow, ex H. & P. Sydow, Mycotheca Germanica No. 343, ILL 13843.

Septogloeum sulphureum H. Sydow. Annales Mycologici 8:493. 1910. **Isotype**: Hab. in ramis *Abietis pectinatae* [*Abies pectinata* (Lam.) DC. = *A. alba* P. Miller] in silva, France: [Elsass], region of Mount Hohneck, [Central] Vogesen, Schiessrotried, [near Metzeral, 14] JUL 1910, leg. H. Sydow, Mycotheca Germanica No. 934, ILL 13854.

Septoria aurea J.B. Ellis & B.M. Everhart. Proceedings of the Academy of Natural Sciences of Philadelphia 1893:163–164. 1893. **Isosyntype**: On leaves of *Ribes aureum* Pursh, Kansas, Rockport, [22 JUL 1892], leg. E. Bartholomew No. 49 [ex J.B. Ellis & B.M. Everhart, North American Fungi No. 2844], ILL 12005.

Septoria canavaliae H.L. Lyon ex H. Sydow. Fungi Exotici Exsiccati, Fascicle 4:191. Anno 1913; Annales Mycologici 11:388. 1913. **Isotypes**: In foliis *Canavaliae ensiformis* [= *Canavalia ensiformis* (L.) DC.], Hawaii, Oahu, Honolulu, 4 FEB 1913, leg. H.L. Lyon No. 264, ILL 12059, BISH 500313.

Septoria cavendishiae F.L. Stevens. Illinois Biological Monographs 11:206, pl. XI, fig. 87. 1927. **Holotype**: On *Cavendishia* sp., Costa Rica, Cartago, 23 JUN 1923, leg. F.L. Stevens No. 51, ILL 12072.

Septoria clermontiae F.L. Stevens & P.A. Young in F.L. Stevens. Bernice P. Bishop Museum Bulletin 19:138. 1925. **Lectotype**: On living leaves of *Clermontia* sp., Hawaii, Oahu, Tantalus, 22 JUN [1921], leg. F.L. Stevens No. 659, ILL 12109 [marked "type" by F.L. Stevens on the packet; accepted and confirmed as lectotypification herein]. **Isolectotypes**: BISH 500012 and 500372. **Residual syntype**: On *Clermontia kakeana* Meyen (?), 25 MAY [1921], leg. F.L. Stevens No. 98, ILL 12110. **Residual isosyntypes**: BISH 146286 and 500371.

Septoria drummondii J.B. Ellis & B.M. Everhart. Journal of Mycology 7:133. 1894. **Isotype**: On leaves of *Phlox drummondii* Hooker, Canada, [Ontario], London, SEP 1891, leg. J. Dearness No. 820 [ex J.B. Ellis & B.M. Everhart, North American Fungi No. 2847], ILL 12198.

Septoria galeobdoli H. Diedicke. Hedwigia 42:(166)–(167). 1903. **Isotype**: Auf überwinterten Blättern von *Galeobdolon luteum* Hudson [= *Lamiastrum galeobdolon* (L.) Ehrend. & Polatschek, Germany], Sachsen, Steiger bei Erfurt, 15 APR 1903, leg. H. Diedicke [ex H. & P. Sydow, Mycotheca Germanica No. 41], ILL 12242.

Septoria gouldiae F.L. Stevens & P.A. Young in F.L. Stevens. Bernice P. Bishop Museum Bulletin 19:138–139. 1925. **Syntypes** [BISH numbers cited are isosyntypes]: On living leaves of *Gouldia lanceolata* (Wawra) A.A. Heller [= *Hedyotis terminalis* (Hooker & Arnott) W.L. Wagner & Herbst], Hawaii, Oahu, Tantalus, 22 JUN [1921, as 25 MAY on the packet], leg. F.L. Stevens No. 602, ILL 9848, BISH 500011 and 500369; on *Kadua grandis* A. Gray [= *Hedyotis acuminata* (Cham. & Schlecht.) Steudel], 29 MAY [1921], leg. F.L. Stevens No. 93, ILL 9846, BISH 146061 and 499990 [both ILL 9846 and 9848 are also syntypes of *Mycosphaerella kaduae* F.L. Stevens & P.A. Young in F.L. Stevens].

Septoria guettardae P. Garman. Mycologia 7:334, pl. 171, fig. 3. 1915. **Lectotype** [designated herein]: On leaves of *Guettarda ovalifolia* Urban, [Puerto Rico], Monte Alegrillo, NOV 1913, leg. F.L. Stevens No. 4741 [as 9759 in the protologue; probably in error since the packet of No. 4741 is clearly labeled type], ILL 12273. **Isolectotypes**: ILL 9933 and 9941 [F.L. Stevens No. 4741 is also cited as paratype of *Stigmatea guettardae* L.R. Tehon].

Septoria hawaiiensis F.L. Stevens & O.A. Plunkett in F.L. Stevens. Bernice P. Bishop Museum Bulletin 19:139. 1925. **Holotype**: On living leaves of *Gouldia* sp. [= *Hedyotis* sp.], Hawaii, [Island of] Hawaii, Kohala Mountains, Waimea, [18] SEP 1911, leg. C.N. Forbes No. 500, ILL 12275 [also the holotype of *Sphaeropsis gouldiae* F.L. Stevens & O.A. Plunkett in F.L. Stevens].

Septoria jackmanii J.B. Ellis & B.M. Everhart. Journal of Mycology 7:132. 1894 [as *jackmani*, corrected in accordance with Articles 32.6 & 61 of I.C.B.N.]. **Isotype**: On leaves of *Clematis* × *jackmanii* Th. Moore (pro sp., cult.) [= *C. lanuginosa* Lindley × *C. viticella* L.], New York, Geneva, AUG 1891, leg. D.G. Fairchild [ex J.B. Ellis & B.M. Everhart, North American Fungi No. 2853], ILL 12320.

Septoria lantanae P. Garman. Mycologia 7:334, pl. 171, fig. 4. 1915. **Holotype**: On leaves of *Lantana camara* L., Porto Rico [Puerto Rico], leg. F.L. Stevens No. 221X, ILL 12370.

Septoria ligustici E.F. Guba. Rhodora 41:519. 1939. **Isotype**: On blighted leaves of *Ligusticum scothicum* L., Massachusetts, Nantucket County, about Capaum Pond, 1 OCT 1936, leg. E.F. Guba, List of Second Hundred Fungi of Nantucket No. 192, ILL 21169.

Septoria macrostoma F.E. & E.S. Clements. Cryptogamae Formationum Coloradensium, Century 1, No. 55. Anno 1906, nom. nud. **Isotype**: ... in foliis dejectis *Araliae nudicaulis* [= *Aralia nudicaulis* L.], Colorado, Jack Brook, alt. 2600 m, 2 AUG 1905, leg. F.E. & E.S. Clements, ILL 12429.

Septoria miconiae P. Garman. Mycologia 7:333, pl. 171, fig. 2. 1915. **Holotype**: On leaves of *Miconia laevigata* (L.) D. Don in Sweet, Porto Rico [Puerto Rico], Las Marias, [22 JAN 1913], leg. F.L. Stevens No. 357, ILL 12449. **Paratypes**: Leg. F.L. Stevens No. 369, ILL 12448; [3 JAN 1913], leg. F.L. Stevens No. 117, ILL 12447.

Septoria negundinis J.B. Ellis & B.M. Everhart. Proceedings of the Academy of Natural Sciences of Philadelphia 1893:165. 1893. **Isotype**: On leaves of *Negundo aceroides* Moench [= *Acer negundo* L.], Canada, [Ontario], London, AUG 1892, leg. J. Dearness No. 19 [ex J.B. Ellis & B.M. Everhart, North American Fungi No. 2859], ILL 12473.

Septoria oedospora F.E. & E.S. Clements. Cryptogamae Formationum Coloradensium, Century 5, No. 493. Anno 1908, nom. nud. **Isotype**: ... in foliis vivis *Symphoricarpi oreophili* [= *Symphoricarpos oreophilus* A. Gray], Colorado, Sulphur Springs, alt. 2400 m, 20 JUL 1907, leg. F.E. & E.S. Clements, ILL 12484.

Septoria petitiae P. Garman. Mycologia 7:333, pl. 171, fig. 1. 1915. **Holotype**: On leaves of *Petitia domingensis* Jacquin, Porto Rico [Puerto Rico], Cabo Rojo, [27 DEC 1913], leg. F.L. Stevens No. 6470, ILL 12552. **Paratype**: Leg. F.L. Stevens No. 9756, ILL 12553.

Septoria phylloptosica R. Ciferri. Sydowia Annales Mycologici 10:163–164. (1956) 1957. **Isotype**: Hab. in foliis vivis *Metopii* sp. [= *Metopium* sp.] (Anacardiaceae), [Dominican Republic], Cordillera Septentrional, Puerto Plata Prov., Foresta Costiera pr. Sosua, 25 MAR 1930, leg. R. Ciferri, Mycoflora Domingensis Exsiccata No. 419, ILL 33119.

Septoria pityrogrammae P. Garman. Mycologia 7:334. 1915. **Holotype**: On leaves of *Pityrogramma calomelanos* (L.) Link, Porto Rico [Puerto Rico], Maricao, Indiera Fria, [8 OCT 1913], leg. F.L. Stevens No. 3484, ILL 12588. **Isotype**: ILL 12585.

Septoria plucheae E.F. Guba. Rhodora 41:518–519. 1939. **Isotype**: On living leaves of *Pluchea camphorata* (L.) DC., Massachusetts, Nantucket County, Coskata Pond, 16 AUG and 1 OCT 1936, leg. E.F. Guba, List of Second Hundred Fungi of Nantucket No. 191, ILL 21171.

Septoria prosopidis F.L. Stevens & A.S. Peirce. Indian Journal of Agricultural Science 3:914. 1933. **Assumed holotype**: On *Prosopis juliflora* (Sw.) DC. (sensu lato), [India, Poona, Bombay, Anno 1932], leg. B.N. Uppal [No. 47], ILL 12674. **Isotype**: AMH.

Septoria rollandiae F.L. Stevens & P.A. Young in F.L. Stevens. Bernice P. Bishop Museum Bulletin 19:140, pl. X(D). 1925. **Holotype**: On leaves of *Rollandia crispa* Gaudich., Hawaii, Oahu, Olympus, 24 JUN [1921], leg. F.L. Stevens No. 706, ILL 12698. **Isotypes**: BISH 500010 and 500368.

Septoriopsis chamaesyces F.L. Stevens & N.E. Dalbey. Mycologia 11(1):4, pl. 2, figs. 1–2. DEC 1918 [as *chamaesyceae*, corrected in accordance with Articles 32.6 & 61 of I.C.B.N.]. **Holotype**: On *Chamaesyce hypericifolia* Millsp., Porto Rico [Puerto Rico], Rio Piedras, leg. F.L. Stevens No. 9445, ILL 16535. **Paratype**: Trujillo, leg. F.L. Stevens No. 9438, ILL 16534. ≡ *Cercoseptoria chamaesyces* (F.L. Stevens & N.E. Dalbey) F. Petrak. Annales Mycologici 23:69. 1925 [the epithet also cited as *chamaesyceae* and corrected herein].

Septoriopsis piperis F.L. Stevens & N.E. Dalbey. Mycologia 11(1):5, pl. 2, figs. 3–4. DEC 1918. **Holotype**: On *Piper medium* Jacquin [= *P. amalago* L.], Porto Rico [Puerto Rico], San Germán, leg. F.L. Stevens No. 5792, ILL 16536. ≡ *Cercoseptoria piperis* (F.L. Stevens & N.E. Dalbey) F. Petrak. Annales Mycologici 23:69. 1925.

Seynesia alstoniae H. Rehm. Leaflets of Philippine Botany 6:2227. 1914. **Assumed isosyntype**: Ad folia *Alstoniae macrophyllae* [= *Alstonia macrophylla* Wallich ex G. Don], Philippines, Luzon, Laguna Prov., Los Baños, OCT 1913, leg. S.A. Reyes, comm. C.F. Baker No. 1748 [ex Fungi Malayana No. 88], ILL 7364.

Seynesia atkinsonii F.L. Stevens & R.W. Ryan in F.L. Stevens. Bernice P. Bishop Museum Bulletin 19:69. 1925. **Holotype**: On *Freycinetia arnottii* [= *F. arborea* Gaudich.], Hawaii, Oahu, Mt. Olympus, Palolo Valley, 16 JUN 1921 [as 10 JUN on the packet], leg. F.L. Stevens No. 300, ILL 7367. **Isotype**: BISH 500367. **Paratypes**: Tantalus, [5 SEP] 1909, leg. H.L. Lyon No. 87, ILL 7366 and 7375; [Palolo, 12 SEP] 1909, leg. H.L. Lyon No. 92, ILL 7365.

Seynesia coccolobae R.W. Ryan. Mycologia 16:178. 1924. **Syntypes**: On *Coccoloba laurifolia* sensu auct. non Jacquin [= *C. diversifolia* Jacquin], Porto Rico [Puerto Rico], Arecibo and Lares Road, [21 JUN 1915], leg. F.L. Stevens No. 7292, ILL 7371; Maricao, [13 APR 1913], leg. F.L. Stevens No. 813, ILL 7373; [probably Santa Ana, 1 JUL 1915], leg. F.L. Stevens No. 7611 [as 7611a on the packet], ILL 7372. **Assumed iso-**

syntype: F.L. Stevens No. 7292, ILL 4398 [also the holotype of *Meliola rectangularis* F.L. Stevens. The specimens of Stevens No. 7292 at ILL are also potential isotypes of *Scolecopeltis pachyasca* C.L. Spegazzini and paratypes of *Helminthosporium panici* F.L. Stevens.].

Seynesia cordiae R.W. Ryan. Mycologia 16:178. 1924. **Holotype:** On *Cordia sulcata* DC., Porto Rico [Puerto Rico], Mayagüez, College grounds, [30 MAR 1913], leg. F.L. Stevens No. 975, ILL 7374.

Seynesiopeltis tetraplasandrae F.L. Stevens & R.W. Ryan in F.L. Stevens. Bernice P. Bishop Museum Bulletin 19:70, text fig. 14a, pl. VI(F). 1925. **Holotype:** On *Tetraplasandra hawaiensis* A. Gray, Hawaii, [Island of] Hawaii, Hamakua, upper ditch trail, 31 MAY 1921 [as 31 JUL on the packet], leg. F.L. Stevens No. 1089, ILL 7377. **Isotypes:** ILL 7378, BISH 500366. **Paratype:** On *Tetraplasandra meiandra* (Hillebr.) H.A.T. Harms [= *T. oahuensis* (A. Gray) H.A.T. Harms], Maui, Kenohuau, 1908, leg. C.N. Forbes s.n., ILL 7379.

Shropshiria chusqueae F.L. Stevens. Mycologia 19:231. 1927. **Holotype:** On *Chusquea simpliciflora* Munro, Panama, Brazos Brook Reservoir, 22 SEP 1924, leg. F.L. Stevens No. 697, ILL 17663. **Paratypes:** On *Chusquea* sp., Culebra, 2 SEP 1924 [as Las Cruces trail, 28 SEP on the packet], leg. F.L. Stevens No. 939, ILL 17664; Las Cruces trail, 28 SEP 1924, leg. F.L. Stevens No. 876, ILL 17665.

Sirocyphis nivea F.E. & E.S. Clements. Cryptogamae Formationum Coloradensium, Century 5, No. 497. Anno 1908, nom. nud. **Isotype:** ... ad caules vetustos *Pedicularis racemosae* [= *P. racemosa* Douglas ex Bentham in Hooker], Colorado, Long's Peak, alt. 3000 m, 10 AUG 1907, leg. F.E. & E.S. Clements, ILL 12974.

Sirodothis populi F.E. & E.S. Clements. Minnesota Botanical Studies 4:187–188, pl. 25, fig. 4. 1911. **Isotype:** In ramulis vetustis *Populi tremuloidis* [= *Populus tremuloides* Michaux], Colorado, Long's Peak Inn, alt. 2700 m, [6] AUG 1907, leg. F.E. & E.S. Clements, ex Cryptogamae Formationum Coloradensium No. 478, ILL 12975.

Sistotrema subtrigonospermum D.P. Rogers. University of Iowa Studies in Natural History 17:22, fig. 10. 1935. **Holotype:** On a log of *Ulmus* sp., Iowa, Iowa City, [Linder's Woods], 9 JUN 1934, leg. D.P. Rogers No. 275, ILL 32599.

Sphaceloma krugii A.A. Bitancourt & A.E. Jenkins. Arquivos do Instituto Biologico. São Paulo 19:103–104. 1949. **Paratypes:** On *Euphorbia prunifolia* Jacquin var. *repanda* Muell. Arg. [= *E.* cf. *heterophylla* L.], Brazil, Est. São Paulo, Jacareí, 23 FEB 1937, leg. A.A. Bitancourt No. 476, [ex Myriangiales Selecti Exsiccati No. 422], ILL 31364, USM 74153 [currently BPI], and IBI 2420. **Holotype:** IBI 2185.

Sphaceloma magnoliae A.E. Jenkins & J.H. Miller. Journal of the Washington Academy of Sciences 42:323–325, fig. 1. 1952. **Paratypes**: On leaves of *Magnolia grandiflora* L., Georgia, Augusta, Goshen Plantation, 7 APR 1943, leg. Mrs. J. McK. Speer, ex A.E. Jenkins & A.A. Bitancourt, Myriangiales Selecti Exsiccati No. 428, ILL 33499 and 33500, USM 74508 [currently BPI], and IBI 5306.

Sphaerella andromedae S.M. Tracy & F.S. Earle. Bulletin of the Torrey Botanical Club 22:176. APR 1895; Mississippi Agricultural Experiment Station Bulletin 34:102. MAY 1895, nom. illegit., non B. Auerswald in G. Gonnermann & G.L. Rabenhorst, 1869. **Isotype**: On living leaves of *Pieris nitida* [unpubl. name, as *Andromeda nitida* Bartram ex Marshall in the second reference = *Lyonia lucida* (Lam.) K. Koch, the latter host name inscribed on the packet label], Mississippi, Ocean Springs, [10] MAR 1888, leg. F.S. Earle s.n., ILL 11975, ex BPI 71469. ≡ *Mycosphaerella andromedae* S.M. Tracy & F.S. Earle ex L.E. Miles. Plant Disease Reporter 19:55. 1935, nom. invalid sine diagn. lat. [The latter name is legitimate by virtue of transfer of the taxon from another genus (cf. Article 58.3 of I.C.B.N.). It is to be treated, however, as the name of a new taxon dating from 1935, and it is not validly published (cf. Article 36.1).].

Sphaerella anthurii (L.E. Miles) A. Trotter, 1928. **See basionym**: *Mycosphaerella anthurii* L.E. Miles, 1917.

Sphaerella aquilina (E.M. Fries) B. Auerswald forma *aspidiorum* P.A. Saccardo. Annales Mycologici 7:435. 1909. **Assumed isotype**: Hab. in frondibus languidis *Aspidii filicis-maris* [*Aspidium filix-mas* (L.) Sw. = *Dryopteris filix-mas* (L.) Schott, Germany, Brandenburg], Tiefensee pr. Werneuchen, [30 MAY 1909], leg. H. Sydow No. 78, ex Mycotheca Germanica No. 784, ILL 9465.

Sphaerella brideliae (H. & P. Sydow) A. Trotter, 1928. **See basionym**: *Mycosphaerella brideliae* H. & P. Sydow, 1914.

Sphaerella callistea H. Sydow. Annales Mycologici 7:439. 1909. **Isotype**: Hab. ad folia viva *Osmundae regalis* [= *Osmunda regalis* L.], Germaniae, [Mecklenburg: Moorwald] pr. Müritz, 14 AUG 1909, leg. H. Sydow, Mycotheca Germanica No. 785, ILL 9467 [associated with and also an isotype of the anamorph: *Phleospora callistea* H. Sydow].

Sphaerella chrysobalani (L.E. Miles) A. Trotter, 1928. **See basionym**: *Mycosphaerella chrysobalani* L.E. Miles, 1917.

Sphaerella didymopanacis (L.E. Miles) A. Trotter, 1928. **See basionym**: *Mycosphaerella didymopanacis* L.E. Miles, 1917.

Sphaerella dubia (L.E. Miles) A. Trotter, 1928. **See basionym**: *Mycosphaerella dubia* L.E. Miles, 1917.

Sphaerella guttiferae (L.E. Miles) A. Trotter, 1928. **See basionym**: *Mycosphaerella guttiferae* L.E. Miles, 1917.

Sphaerella iridis B. Auerswald var. *anceps* P.A. Saccardo. Annales Mycologici 7:435. 1909. **Isotype:** Hab. in foliis morientibus *Iridis pseudacori* [= *Iris pseudacorus* L., Germany, Brandenburg]: Eichwalde pr. Berolinum [Berlin, 20 MAR 1909], leg. H. Sydow No. 79 [ex Mycotheca Germanica No. 786], ILL 9494.

Sphaerella maxima (L.E. Miles) A. Trotter, 1928. **See basionym:** *Mycosphaerella maxima* L.E. Miles, 1917.

Sphaerella mucunae (F.L. Stevens) A. Trotter, 1928. **See basionym:** *Mycosphaerella mucunae* F.L. Stevens, 1917.

Sphaerella oerteliana P.A. Saccardo in H. & P. Sydow. Annales Mycologici 2:528. 1904. **Isotype:** Hab. in caulibus emortuis *Coronillae montanae* [*Coronilla montana* Jacquin non Scop. = *C. coronata* L., Germany], Thüringen: [am Göldner] pr. Sondershausen, [10 JUN 1904], leg. G. Oertel, ex H. & P. Sydow, Mycotheca Germanica No. 237, ILL 9502.

Sphaerella palmae (L.E. Miles) A. Trotter, 1928. **See basionym:** *Mycosphaerella palmae* L.E. Miles, 1917.

Sphaerella perseae (L.E. Miles) A. Trotter, 1928. **See basionym:** *Mycosphaerella perseae* L.E. Miles, 1917.

Sphaerella reyesii (H. & P. Sydow) A. Trotter, 1928. **See basionym:** *Mycosphaerella reyesii* H. & P. Sydow 1914.

Sphaerella rhoina P.A. Saccardo. Annales Mycologici 6:561. 1908. **Assumed isotype:** Hab. in foliis putrescentibus *Rhois toxicodendri* [*Rhus toxicodendron* L. = *Toxicodendron radicans* (L.) O. Kuntze], Germaniae, [Brandenburg: Baumschulen] pr. Tamsel [now Poland, 11 MAY 1908], leg. P. Vogel [ex H. Sydow, Mycotheca Germanica No. 680], ILL 9516.

Sphaerella subastoma (F.L. Stevens & N.E. Dalbey) A. Trotter, 1928. **See basionym:** *Mycosphaerella subastoma* F.L. Stevens & N.E. Dalbey, 1918.

Sphaerella tabebuiae (L.E. Miles) A. Trotter, 1928. **See basionym:** *Mycosphaerella tabebuiae* L.E. Miles, 1917.

Sphaerella vogelii H. Sydow. Annales Mycologici 6:480. 1908. **Isotype:** Hab. in foliis putrescentibus *Rhamni catharticae* [= *Rhamnus cathartica* L., Brandenburg: Schlucht am Kirschberge bei] Tamsel [now Poland], 22 MAY 1908, leg. P. Vogel, ex H. Sydow, Mycotheca Germanica No. 681, ILL 9524.

Sphaeronaema helianthi F.E. & E.S. Clements. Cryptogamae Formationum Coloradensium, Century 5, No. 475. Anno 1908, nom. nud. [as *Sphaeronema*. The generic name sanctioned as *Sphaeronaema* and corrected in accordance with current I.C.B.N.]. **Isotype:** in caulibus emortuis *Helianthi petiolaris* [= *Helianthus petiolaris* Nutt.], Colorado, Fort Garland, alt. 2400 m, 23 JUN 1907, leg. F.E. & E.S. Clements, ILL 12984.

Sphaeronaema leonuri F.E. & E.S. Clements. Cryptogamae Formationum Coloradensium, Century 5, No. 476. Anno 1908, nom. nud. [as *Sphaeronema*. The generic name sanctioned as *Sphaeronaema* and corrected in accordance with current I.C.B.N.]. **Isotype:** ... in caulibus vetustis *Leonuri cardiacae* [= *Leonurus cardiaca* L.], Colorado, Boulder, alt. 1600 m, 29 JUL 1907, leg. F.E. & E.S. Clements, ILL 12986.

Sphaeronaema negundinis J.B. Ellis & B.M. Everhart. Proceedings of the Academy of Natural Sciences of Philadelphia 1893:158. 1893 [as *Sphaeronema*. The generic name sanctioned as *Sphaeronaema* and corrected in accordance with current I.C.B.N.]. **Isotype:** On dead limbs of *Negundo aceroides* Moench [= *Acer negundo* L.], Pennsylvania, Philadelphia, Fairmount Park, JUN 1890, leg. H. Bilgram s.n., comm. W.C. Stevenson, Jr., ex J.B. Ellis & B.M. Everhart, North American Fungi No. 2775, ILL 12987.

Sphaeronaema pilosum F.E. & E.S. Clements. Cryptogamae Formationum Coloradensium, Century 1, No. 52. Anno 1906, nom. nud. [as *Sphaeronema*. The generic name sanctioned as *Sphaeronaema* and corrected in accordance with current I.C.B.N.]. **Isotype:** ... in caulibus emortuis *Senecionis eremophili* [= *Senecio eremophilus* J. Richardson], Colorado, Minnehaha, alt. 2600 m, 3 AUG 1905, leg. F.E. & E.S. Clements, ILL 12989.

Sphaeronaema stellatum J.B. Ellis. Bulletin of the Torrey Botanical Club 6:107. 1876 [as *Sphaeronema*. The generic name sanctioned as *Sphaeronaema* and corrected in accordance with current I.C.B.N.]. **Isotype:** On dead stems of *Ilex glabra* (L.) A. Gray, [New Jersey, Newfield], MAY, leg. J.B. Ellis [ex J.B. Ellis & B.M. Everhart, North American Fungi No. 2170], ILL 12993.

Sphaeronaemella fragariae F.L. Stevens & A. Peterson. Phytopathology 6:260. 1916. **Holotype:** Hab. on fruit of cultivated *Fragaria* sp., Louisiana, ILL 12992 [see also J.L. Maas, Compendium of Strawberry Diseases, 1984]. **Assumed paratype:** ILL 12996.

Sphaeropsis gouldiae F.L. Stevens & O.A. Plunkett in F.L. Stevens. Bernice P. Bishop Museum Bulletin 19:136. 1925. **Holotype:** On living leaves of *Gouldia* sp. [= *Hedyotis* sp.], Hawaii, [Island of] Hawaii, Kohala Mountains, Waimea, [18] SEP 1911, leg. C.N. Forbes No. 500, ILL 12275 [also the holotype of *Septoria hawaiiensis* F.L. Stevens & O.A. Plunkett in F.L. Stevens].

Sphaeropsis lyndonvillae P.A. Saccardo. Annales Mycologici 4:275–276. 1906. **Isotype:** Hab. in ramulis *Hibisci syriaci* [= *Hibiscus syriacus* L.] culti in hortis, [New York], Lyndonville, 1 JUN 1906 [erroneously as JAN in the protologue], leg. C.E. Fairman s.n. [ex E. Bartholomew, Fungi Columbiani No. 2281], ILL 13006.

Sphaeropsis magnoliae J.B. Ellis & J. Dearness ex E. Bartholomew. Fungi Columbiani, Century 21, No. 2087. Anno 1905, nom. nud. **Isotype:** On *Magnolia* [*acuminata* (L.) L.], Canada, [Ontario], London, NOV 1903, leg. J. Dearness, ILL 13008.

Sphaeropsis persicae J.B. Ellis & E. Bartholomew. Journal of Mycology 8:175. 1902. **Isotype:** On dead branches of *Amygdalus persica* L. [= *Prunus persica* (L.) Batsch], Kansas, Rooks County, 2 OCT 1901, leg. E. Bartholomew, Fungi Columbiani No. 1590, ILL 13038.

Sphaerulina cibotii F.L. Stevens & E.F. Guba in F.L. Stevens. Bernice P. Bishop Museum Bulletin 19:105, text fig. 27b. 1925. **Lectotype** [designated herein]: On the pinnae of *Cibotium menziesii* Hooker [= *C. chamissoi* Kaulf.], Hawaii, Kauai, Waimea, 17 JUN [1921], leg. F.L. Stevens No. 545, ILL 9917 [text fig. 27b on p. 103 represents this specimen of Stevens No. 545]. **Isolectotypes:** BISH 146321, 146324, 146325 and 500364. **Residual syntype:** [As *Cibotium* sp. on the packet], Island of] Hawaii, Kealakekua [as Kona Keauhou, Bishop Estate Road on the packet], 25 JUL [1921], leg. F.L. Stevens No. 1003, ILL 9918. **Residual isosyntypes:** BISH 500009 and 500365.

Sphaerulina ipomoeae F.L. Stevens. Bernice P. Bishop Museum Bulletin 19:105–106. 1925. **Holotype:** On *Ipomoea bona-nox* L. (moonflower), Hawaii, [Island of] Hawaii, Kealakekua, 21 JUL [1921], leg. F.L. Stevens No. 908, ILL 14311 [associated with and also the holotype of *Ramularia ipomoeae* F.L. Stevens]. **Isotypes:** BISH 499083 and 500326.

Sphenospora smilacina H. Sydow. Annales Mycologici 23:318–319. 1925. **Paratype:** Hab. in foliis *Smilacis* spec. [= *Smilax* sp.], Costa Rica, San Pedro de San Ramón, 23 JAN 1925, leg. H. Sydow, Fungi in Itinere Costaricensi Collecti No. 141, ILL 24987.

Spirogramma boergesenii C.C.F. Ferdinandsen & Ø. Winge, Videnskabelige Meddelelser fra Dansk Naturhistorisk Forening i Kjøbenhavn [Copenhagen] 1908:143–144, pl. 41, fig. 3. 1909. **Isosyntypes:** Ad ramos siccos arborum in insula St. Jan [St. John, U.S. Virgin Islands]: Foy Gut, 18 MAR 1906, leg. F. Börgesen, ILL 10697, CP.

Spiropes guareicola (F.L. Stevens) R. Ciferri, 1955. **See basionym:** *Helminthosporium guareicola* F.L. Stevens, 1918.

Sporidesmium tabacinum J.B. Ellis & B.M. Everhart. Proceedings of the Academy of Natural Sciences of Philadelphia 1891:92. 1891. **Isotype:** On decaying wood of *Populus tremuloides* Michaux, Montana, [Cascade County], Sand Coulee, 31 May 1889, leg. F.W. Anderson No. 503, ILL 16342. **Holotype:** NY.

Sporidesmium vogelianum H. Sydow. Annales Mycologici 8:493. 1910 [as *Sporodesmium*, corrected in accordance with Articles 60 & 61 of

I.C.B.N.]. **Isotype:** Hab. in ramis emortuis junioribus et pedunculis *Celtidis occidentalis* [= *Celtis occidentalis* L., Poland], Tamsel, 2 JUN 1910, leg. P. Vogel, ex H. Sydow, Mycotheca Germanica No. 947, ILL 16343.

Stagonospora erythrinae F.L. Stevens & P.A. Young in F.L. Stevens. Bernice P. Bishop Museum Bulletin 19:137, text fig. 28c, d. 1925. **Holotype:** On dead leaves of *Erythrina monosperma* Gaudich., Hawaii, [Island of] Hawaii, between Kona and Waimea, 27 JUL [1921], leg. F.L. Stevens No. 1019, ILL 13048. **Isotypes:** BISH 500008 and 500363.

Stagonospora petasitidis J.B. Ellis & B.M. Everhart. Proceedings of the Academy of Natural Sciences of Philadelphia 1891:81. 1891. **Isotype:** On living leaves of *Petasites palmatus* (Aiton) A. Gray [cited as *palmata* = *P. frigidus* (L.) E.M. Fries var. *palmatus* (Aiton) Cronq.], Canada, [Ontario], London, 1 JUL 1890, leg. J. Dearness No. 1767 [ex J.B. Ellis & B.M. Everhart, North American Fungi No. 2467], ILL 13055. **Holotype:** NY.

Stegastroma guianense F.L. Stevens. Illinois Biological Monographs 8:200–201, pl. XI, figs. 88, 89. 1923. **Holotype:** On a mimosa-like legume [Fabaceae subfam. Mimosoideae], British Guiana [Guyana], Tumatumari, 11 JUL 1922, leg. F.L. Stevens No. 164, ILL 10349.

Stenellopsis magnoliae (A.G. Weedon) G. Morgan-Jones, 1980. **See basionym:** *Heterosporium magnoliae* A.G. Weedon, 1926.

Stephanoma meliolae F.L. Stevens & N.E. Dalbey. Mycologia 11(1):9. DEC 1918 [as *melioliae*, corrected in accordance with Articles 32.6 & 61 of I.C.B.N.]. **Holotype:** On *Meliola tortuosa* H.G. Winter on *Piper umbellatum* L. [= *Lepianthes umbellata* (L.) Raf.], Porto Rico [Puerto Rico], Lares, [22 NOV 1913], leg. F.L. Stevens No. 4843, ILL 4483.

Stereum beigehimenium A.R. Teixeira. Bragantia 5:403. 1945. **Isotypes:** On dead wood, Brazil, Est. do Rio Grande do Sul, São Leopoldo, NOV 1909, leg. J. Rick, Instituto Agronomico de Campinas (IACM) No. 5069, ILL 32931 [ex IACM], NY and S.

Stereum carpaticum A. Pilát. Hedwigia 70:78–79. 1930; Ceskoslovenska Akademie Zemedelska, Prague. Sbornik 5:392–394. 1930. **Isosyntype:** Ad truncos prostratos *Piceae excelsae* [*Picea excelsa* (Lam.) Link = *P. abies* (L.) H. Karsten] in carpaticus orientalibus in Rossia Subcarpatica [near Slovak Republ.—Carpatho-Ukraine border]: in valle [rivis] Balzatul (alt. 700 m), Distr. Rachovo, [AUG 1928], leg. A. Pilát s.n., ILL 32593, ex herb. V. Litschauer.

Stereum durum E.A. Burt. Annals of the Missouri Botanical Garden 7:226, text fig. 46; pl. 6, fig. 75. 1920. **Isotype:** On dead wood, Mexico, Jalapa, leg. C.L. Smith, Central American Fungi No. 147, ILL 32594.

Stereum underwoodii E.A. Burt. Annals of the Missouri Botanical Garden

13:327. 1926. **Isotype:** On bark of *Xolisma* [misspelled *Xolisima* = *Lyonia* sp. (Ericaceae)], Jamaica, base of John Crow Peak [as Jim Crow Peak on the packet, 18 APR 1903], leg. L.M. Underwood No. 2432, ILL 32592. **Holotype:** NY.

Stevensula meliolae (F.L. Stevens) R.A. Toro, 1952. **See basionym:** *Perisporium meliolae* F.L. Stevens, 1918.

Stevensula monensis C.L. Spegazzini. Boletín de la Academia Nacional de Ciencias en Córdoba 26:339–341, illustr. Preprint 1923 [journal part issued in 1924]. **Isotypes:** Hab. En el subículo de la *Meliola monensis* F.L. Stevens sobre las hojas de *Amyris elemifera* L., Porto Rico [Puerto Rico], Isla de Mona, [Mona Island, 20–21 DEC 1913], leg. F.L. Stevens No. 6150, ILL 5657 and 5660 [also possible isotypes of *Micropeltidium monense* C.L. Spegazzini, as well as paratypes of *Meliola monensis* F.L. Stevens and of *Helminthosporium glabroides* F.L. Stevens]. **Holotype:** LPS 743 [fide S.J. Hughes, 1993].

Stictis myricae E.K. Cash. Mycologia 50:655, fig. 9. 1958. **Isotypes:** On dead leaves of *Myrica californica* Cham., California, Trinidad, Spruce Cove, 24 JAN 1947, leg. H.E. Parks No. 7005 [ex California Fungi No. 1209], ILL 33027, UC. **Holotype:** BPI.

Stigmatea cinereo-maculans H. Rehm. Philippine Journal of Science, Section C (Botany) 8:257. 1913. **Assumed isotype:** Ad vaginam foliorum *Pandani* [= *Pandanus* sp.], Philippines, Luzon, Laguna Prov., Los Baños, [Mount Maquiling, SEP 1913], leg. C.F. Baker No. 622, [ex Fungi Malayana No. 90], ILL 9929.

Stigmatea guettardae L.R. Tehon. Botanical Gazette 67:508. 1919. **Holotype:** On *Guettarda ovalifolia* Urban, Porto Rico [Puerto Rico], Maricao, 10 JAN 1913, leg. F.L. Stevens No. 191, ILL 9932. **Isotype:** ILL 9941. **Paratypes:** Barros, 2 JAN 1913, leg. F.L. Stevens No. 164, ILL 9938 and 9940; Maricao, 5 APR 1913, leg. F.L. Stevens No. 771, ILL 9934; 19 JUL 1915, leg. F.L. Stevens No. 8804, ILL 9939 and 9943; Monte Alegrillo, [no date], leg. F.L. Stevens No. 4741, ILL 9933 and 9944; Indiera Fria, 8 OCT 1913, leg. F.L. Stevens No. 3338, ILL 9936, 9937 and 9942; on *Guettarda scabra* (L.) Vent., Tanamá River, 6 JUL 1915, leg. F.L. Stevens No. 7851, ILL 9945.

Stilbella proliferans F.L. Stevens. Illinois Biological Monographs 11:211–212, fig. 88. 1927. **Holotype:** On *Theobroma cacao* L., Costa Rica, Indiana branch, 18 JUL 1923, leg. F.L. Stevens No. 541, ILL 16441.

Stomatogene yuccae C.G. Hansford. Sydowia Annales Mycologici 11:68. (1957) 1958. **Holotype:** Hab. in foliis *Yuccae mohavensis* [*Yucca mohavensis* Sargent = *Y. schidigera* Roezl ex Ortgies], California, [San Diego County, Camp Kearny, 17 FEB 1929], leg. H.E. Parks No. 3379 (California Fungi No. 576), ILL 6617. **Isotype:** UC. **Paratype:** In foliis

Yuccae whipplei [= *Yucca whipplei* Torrey], La Jolla, FEB 1929, leg. H.E. Parks No. 3400b [ex California Fungi No. 577], ILL 6618.

Stomiopeltella suttoniae J.M. Mendoza in F.L. Stevens & H.W. Manter. Botanical Gazette 79:292–293, figs. 69–72. 1925. **Holotype**: On *Suttonia lessertiana* (A. DC.) Mez [= *Myrsine lessertiana* A. DC.], Hawaii, Oahu, Honolulu, Hamakua, upper ditch trail, 28 JUL [1921], leg. F.L. Stevens No. 1032, ILL 6671.

Stomiopeltis cassiae J.M. Mendoza in F.L. Stevens & H.W. Manter. Botanical Gazette 79:292, figs. 67–68. 1925. **Holotype**: On *Cassia* sp., British Guiana [Guyana], Tumatumari, 10 JUL [1922], leg. F.L. Stevens No. 115, ILL 6787.

Stomiopeltis heteromeris H. Sydow. Annales Mycologici 25:84–85. 1927. **Isotype**: Hab. in foliis *Phoebes neurophyllae* [= *Phoebe neurophylla* Mez & Pittier], Costa Rica, Cerro de San Isidro pr. San Ramón, 9 FEB 1925, leg. H. Sydow, Fungi in Itinere Costaricensi Collecti No. 169d, ILL 6788.

Stomiopeltis polyloculatis E.S. Luttrell. Mycologia 38:574–575. 1946. **Isotypes**: Hab. in culmis vivis *Arundinariae tectae* [*Arundinaria tecta* (Walter) Muhl. = *A. gigantea* (Walter) Muhl. ssp. *tecta* (Walter) McClure], Georgia, Spalding County, Experiment, 24 MAY 1943, leg. E.S. Luttrell, ILL 6789 [labeled co-type on the packet], BPI, NY. **Holotype**: FH.

Sucinaria minuta H. Sydow. Annales Mycologici 23:363–364, fig. 9. 1925. **Isotype**: Hab. foliis *Miconiae thomasianae* [= *Miconia thomasiana* DC.], Costa Rica, Los Angeles de San Ramón, 30 JAN 1925, leg. H. Sydow, Fungi in Itinere Costaricensi Collecti No. 272, ILL 8344.

Synchytrium boerhaviae F.L. Stevens. Illinois Biological Monographs 11:163, pl. I, figs. 1–2. 1927 [as *boerhaaviae*, corrected in accordance with Rec. 60H.1 and Articles 60 & 61 of I.C.B.N.]. **Holotype**: On *Boerhavia erecta* L. [as *Boerhaavia*], Costa Rica, El Roble, 25 SEP 1923 [as 25 JUN on the packet], leg. F.L. Stevens No. 621, ILL 3115.

Synchytrium cellulare J.J. Davis. Transactions of the Wisconsin Academy of Sciences, Arts and Letters 19(2):681. 1919. **Assumed isotype**: On leaves and petioles of *Boehmeria cylindrica* (L.) Sw., Wisconsin, Devil's Lake, [7 AUG 1913], leg. J.J. Davis [Fungi Wisconsinenses Exsiccati No. 66], ILL 3116.

Synchytrium chiltonii M.T. Cook. Mycologia 37:288, fig. 1c. 1945. **Isosyntype**: Hab. in *Stellaria media* (L.) Cyrillo, Louisiana, Baton Rouge, [27 MAR 1947], leg. M.T. Cook s.n., ILL 21265.

Synchytrium pulvereum J.J. Davis. Transactions of the Wisconsin Academy of Sciences, Arts and Letters 20:407–408, fig. 1. 1921. **Isosyntype**: On leaves of *Laportea canadensis* (L.) Weddell, Wisconsin, Hawkins,

[21] AUG [1918], leg. J.J. Davis [Fungi Wisconsinenses Exsiccati No. 105], ILL 3176.
Synchytrium sambuci M.T. Cook. Mycologia 41:24–27, figs. 1–9. 1949. **Lectotype** [designated herein]: On *Sambucus canadensis* L., Louisiana, [Baton Rouge, 12 MAR 1947], leg. M.T. Cook & W.J. Dickson s.n., ILL 21267.
Synchytrium scirpi J.J. Davis. Journal of Mycology 11:154–156, figs. 1, 2. 1905. **Isosyntype**: On leaves of *Scirpus atrovirens* Muhl. ex Willd., Wisconsin, Kenosha County, AUG and SEP 1905, leg. J.J. Davis [ex J.B. Ellis & B.M. Everhart, Fungi Columbiani No. 2178], ILL 3178.
Synchytrium smilacis M.T. Cook. Mycologia 43:105–107, figs. 7–12. 1951. **Isotype**: On *Smilax* sp., Louisiana, Baton Rouge, [27 APR 1948], leg. M.T. Cook s.n., ILL 21266.
Systremma pterocarpi E.M. Doidge. Bothalia 1(2):70. 1922. **Isotype**: On leaves and twigs of *Pterocarpus sericeus* Bentham [misspelled "*ceriseus*"], Rhodesia [Zimbabwe], Khami Ruins, 14 JUL 1920, leg. A.M. Bottomley, ex PREM No. 14101, ILL 9268. **Holotype**: PREM.
Syzygospora alba G.W. Martin. Journal of the Washington Academy of Sciences 27:112–114, fig. 1. 1937. **Isotype**: On fallen trunks and stumps, in mountain forest, Panama, Chiriqui Prov., valley of upper Rio Chiriqui Viejo, alt. 1600–1800 m, 4 JUL 1935, leg. G.W. Martin No. 2449, ILL 32808. **Paratype**: G.W. Martin No. 2167, ILL 32809. **Holotype**: IA [currently at ISC].

T

Tetraploa divergens S.M. Tracy & F.S. Earle. Bulletin of the Torrey Botanical Club 22:179. 1895. **Isotype**: On living or languishing leaves of *Panicum agrostidiforme* Lam. [= *P. laxum* Sw.], Mississippi, Starkville, OCT 1894, leg. S.M. Tracy s.n., ILL 13864, ex BPI. **Holotype**: BPI.
Theciopeltis guianensis F.L. Stevens & H.W. Manter. Botanical Gazette 79:285, figs. 6, 7, 61. 1925. **Holotype**: On unknown host, British Guiana [Guyana], Rockstone, 17 JUL [1922], leg. F.L. Stevens No. 470, ILL 6790.
Theissenula clavispora H. & P. Sydow. Annales Mycologici 12:198–199. 1914. **Assumed isotype**: In foliis subvivis *Schizostachyi acutiflori* [*Schizostachyum acutiflorum* Munro = *S. diffusum* (Blanco) Merrill], Philippines, [Luzon], Laguna Prov., Los Baños, [Mount Maquiling], 25 OCT 1913, leg. S.A. Reyes, comm. C.F. Baker No. 1937 [ex Fungi Malayana No. 91], ILL 6624.

Tichospora praestipa F.E. & E.S. Clements. Cryptogamae Formationum Coloradensium, Century 5, No. 456. Anno 1908, nom. nud. **Isotype:** ... ad caules emortuos *Amelanchieris oreophilae* [*Amelanchier oreophila* A. Nelson = *A. utahensis* Koehne], Colorado, Sulphur Springs, alt. 2400 m, 23 JUL 1907, leg. F.E. & E.S. Clements, ILL 9572.

Titaea miconiae (F.L. Stevens) S.C. Damon, 1952. **See basionym:** *Monogrammia miconiae* F.L. Stevens, 1917.

Tomentella bambusina A.P. Viegas. Jornal de Agronomia São Paulo 2:324. 1939. **Isotypes:** On *Bambusa vulgaris* Schrader ex J.C. Wendl., Brazil, Est. São Paulo, [Rio] Piracicaba, Rosario, 5 SEP 1938, leg. A.P. Viegas No. 2481 [also as C.T. White No. 3440], ILL 32583, CUP 27984. **Holotype:** Herb. Inst. Agronomico, Campinas, São Paulo.

Tonduzia psychotriae F.L. Stevens. Illinois Biological Monographs 11:168, pl. II, figs. 16, 17a, b. 1927. **Holotype:** On *Psychotria brachiata* Sw., Costa Rica, Columbiana, 19 JUL 1923, leg. F.L. Stevens No. 570, ILL 6619.

Toroa saurauiae (F.L. Stevens & E.F. Roldan) C.G. Hansford, 1946, comb. nov. invalid. [cf. Article 36.1 of I.C.B.N.]. **See basionym:** *Meliolina saurauiae* F.L. Stevens & E.F. Roldan, 1935 (nom. invalid. sine diagn. lat.).

Torula epistromata R. Ciferri. Sydowia Annales Mycologici 10:176–177. (1956) 1957. **Isotype:** Hab. in stromate probab. *Phyllachorae guazumae* [= *Phyllachora guazumae* P.C. Hennings], praecipue in pagina inferiore foliorum sicca *Guazumae ulmifoliae* [= *Guazuma ulmifolia* Lam., Dominican Republic], Santiago Prov., Valle del Cibao, Hato del Yaque, FEB 1930, leg. R. Ciferri, Mycoflora Domingensis No. 370, ILL 33545.

Trabutia minima F.L. Stevens & A.G. Weedon in F.L. Stevens. Bernice P. Bishop Museum Bulletin 19:18, text fig. 3a, b, pl. II(A). 1925. **Holotype:** On unknown dicotyledonous host, Hawaii, Kauai, 15 JUN [1921], leg. F.L. Stevens No. 445, ILL 9273. **Isotypes:** BISH 500007 and 500362.

Trabutia portoricensis F.L. Stevens. Botanical Gazette 70:401, fig. 3. 1920. **Holotype:** On *Coccoloba nivea* Jacquin [cited as *Cocolobis* = *Coccoloba venosa* L.], Porto Rico [Puerto Rico], Mayagüez, [31 OCT 1913], leg. F.L. Stevens No. 3907a, ILL 8543. **Paratype:** [College grounds, 30 APR 1913], leg. F.L. Stevens No. 976, ILL 8542.

Trabutia xylosmae F.L. Stevens. Illinois Biological Monographs 11:182–183, pl. VI, fig. 46; pl. XV, fig. 106. 1927. **Holotype:** On *Xylosma salzmannii* Eichler in Martius [= *X. seemannii* Triana & Planchon], Costa Rica, El Alto, 6 JUL 1923, leg. F.L. Stevens No. 243, ILL 9275. **Paratypes:** Cartago, 2 JUL 1923, leg. F.L. Stevens No. 186, ILL 9278;

on *Myroxylon ellipticum* (D. Clos) O. Kuntze [= *Xylosma* sp.], La Palma, 8 JUL 1923, leg. F.L. Stevens No. 276, ILL 9276.

Trabutiella cordiae F.L. Stevens. Botanical Gazette 70:401, fig. 4. 1920. **Holotype**: On *Cordia collococca* L., Porto Rico [Puerto Rico], Añasco, [28 JAN 1913], leg. F.L. Stevens No. 276, ILL 9281.

Tracya lemnae (W.A. Setchell) H. & P. Sydow, 1901. **See basionym**: *Cornuella lemnae* W.A. Setchell, 1891. [*Tracya* H. & P. Sydow, 1901, is a substitute name for *Cornuella* W.A. Setchell, MAY 1891 (nom. illegit.), non J.B.L. Pierre, JAN 1891].

Trematosphaeria corynis F.E. & E.S. Clements. Cryptogamae Formationum Coloradensium, Century 5, No. 437. Anno 1908, nom. nud. **Isotype**: ... in caulibus emortuis *Chrysothamni graveolentis* [*Chrysothamnus graveolens* (Nutt.) E. Greene = *C. nauseosus* (Pallas ex Pursh) Britton ssp. *graveolens* (Nutt.) Piper], Colorado, Fort Garland, alt. 2400 m, 23 JUN 1907, leg. F.E. & E.S. Clements, ILL 9578.

Trematosphaeria maquilingiana H. Rehm var. *schizostachyi* H. Rehm. Leaflets of Philippine Botany 8:2952. 1916. **Assumed isotype**: Ad *Schizostachyum emortuum*, Philippines, [Luzon, Laguna Prov.], in e acumine Mount Maquiling, JUN 1914, leg. C.F. Baker No. 3426 [ex Fungi Malayana No. 195], ILL 9577.

Trichasterina calophylli C.G. Hansford. Sydowia Annales Mycologici 11:63–64. (1957) 1958. **Holotype**: Hab. in foliis *Calophylli calabae* [*Calophyllum calaba* sensu Jacquin non L. = *C. antillanum* Britton in Britton & Wilson], Porto Rico [Puerto Rico], Mayagüez, [25 JUN 1915], leg. F.L. Stevens No. 7489 p.p., ILL 6591 [also the holotype of *Perisporium portoricense* F.L. Stevens & R. Higley ex F.L. Stevens and a possible isosyntype of *Meliolidium portoricense* C.L. Spegazzini].

Trichobelonium distinguendum H. Sydow. Annales Mycologici 6:480. 1908. **Isotype**: Hab. in culmis emortuis *Phragmitis communis* [*Phragmites communis* Trin. = *P. australis* (Cav.) Trin. ex Steudel, Germany, Marchia [Mark Brandenburg], Schmöckwitz pr. Berlin, AUG 1908 [as 9 AUG on the packet], leg. H. Sydow, Mycotheca Germanica No. 704, ILL 8029.

Trichobelonium melioloides H. Rehm. Leaflets of Philippine Botany 8:2929. 1915. **Assumed isotype**: Ad folia *Gigantochloae scribnerianae* [*Gigantochloa scribneriana* Merrill = *G. levis* (Blanco) Merrill], Philippines, [Luzon, Laguna Prov.], hills back of Paete, APR 1914, leg. C.F. Baker No. 3115 [ex Fungi Malayana No. 196], ILL 8031.

Trichomerium clitoriae A.C. Batista & R. Ciferri. Saccardoa 2:198–200, figs. 77–79. 1963. **Paratype**: On Compositae [= Asteraceae sp. indet.], associated with *Podoxyphium yuccae* A.C. Batista & M.L. Nascimento, Porto Rico [Puerto Rico], Rosario [El Rosario, 14] NOV 1914, leg. F.L. Stevens No. 4847, ILL 33025.

Trichomerium coccolobae A.C. Batista & R. Ciferri. Saccardoa 2:201, fig. 80. 1963. **Isotypes:** On *Coccoloba uvifera* L., Porto Rico [Puerto Rico], Maricao, 4 MAR 1913 [as 3 MAR on the packet], leg. F.L. Stevens No. 819, ILL 33026, BPI. **Holotype:** URM [I.M.U.R.] 13080.

Trichomerium coffeicola (A. Puttemans) C.L. Spegazzini var. *macrosporum* R. Ciferri & A.C. Batista. Saccardoa 2:202–203, fig. 82. 1963. **Paratype:** On *Citrus* sp., Porto Rico [Puerto Rico], Utuado, 8 NOV 1913, leg. F.L. Stevens No. 4592, ILL 33028, BPI. **Holotype:** BPI.

Trichomerium portoricense C.L Spegazzini. Boletín de la Academia Nacional de Ciencias en Córdoba 26:341. Preprint 1923 [journal part issued in 1924]. **Isotype:** Hab. En la cara inferior de las hojas vivas de *Psidium guajava* L., Porto Rico [Puerto Rico], en los alrededores de Mayagüez, [as Yauco on the packets (Mayagüez Distr.), 3 OCT 1913], leg. F.L. Stevens No. 3120, ILL 6064 [also the holotype of *Isthmospora spinosa* F.L. Stevens. The collection was originally cited for *Meliola psidii* E.M. Fries:E.M. Fries (cf. F.L. Stevens, 1916).]. **Holotype:** LPS 661.

Trichonectria bambusicola H. Rehm. Leaflets of Philippine Botany 6:2226. 1914. **Assumed isotype:** Ad folia *Bambusae blumeanae* [= *Bambusa blumeana* Blume ex J.H. Schultes], Philippines, Luzon, Laguna Prov., Los Baños, SEP 1913, leg. S.A. Reyes No. 1655 [ex C.F. Baker, Fungi Malayana No. 92], ILL 8419.

Trichopeltis pulchella C.L. Spegazzini. Boletín de la Academia Nacional de Ciencias en Córdoba 11:571–572. 1889; Fungi Puiggariani, Pugillus I, pp. 193–194, No. 364. 1889. **Isosyntype:** Hab. ad folia viva Myrtacearum [= Myrtaceae sp. indet., Brazil], in dumetis pr. Apiahy [Apiai], Hiem, 1888, leg. J.I. Puiggari No. 2365, comm. C.L. Spegazzini, ILL 6878.

Trichopeltis rhyacoides F.L. Stevens. Bernice P. Bishop Museum Bulletin 19:84, pl. VIII(G). 1925. **Holotype:** On *Alyxia oliviformis* Gaudich. [cited as *olivaeformis*], Hawaii, [Island of] Hawaii, Kealakekua, 25 JUL [1921], leg. F.L. Stevens No. 985, ILL 6877 [also a paratype of *Meliola alyxiae* F.L. Stevens and of *Trichothallus hawaiiensis* F.L Stevens]. **Isotypes:** BISH 500360 and 500361.

Trichopeltum hawaiiense A.C. Batista & C.A.A. Costa in A.C. Batista, C.A.A. Costa & R. Ciferri. Instituto de Micologia, Universidade do Recife, Pernambuco. Publicação 90:21–22, fig. 10. 1957; Atti Instituto Botanico della Università Laboratorio Crittogamico Pavia, Series 5, 15:53–54, fig. 10. 1958 [as *hawaiiensis*, corrected in accordance with Articles 32.6 & 61 of I.C.B.N.]. **Holotype:** In foliis *Smilacis* sp. [= *Smilax* sp.], Hawaii, Oahu, Olympus, 24 JUN 1921, leg. F.L. Stevens [No. 981], ILL 6667 [the herbarium acronym erroneously cited as NY; also the holotype of

Ainsworthia smilacina A.C. Batista & A.F. Vital, and a syntype of *Phragmocapnias smilacina* J.M. Mendoza in F.L. Stevens].
Trichothallus hawaiiensis F.L. Stevens. Bernice P. Bishop Museum Bulletin 19:85–86, text figs. 18b, c, 19a–d. 1925. **Holotype:** On *Scaevola* sp., Hawaii, Kauai, Kalalau trail, 16 JUN [1921], leg. F.L. Stevens No. 492, ILL 6879 [also a paratype of *Irene scaevolicola* F.L. Stevens]. **Isotype:** ILL 4415. **Paratypes:** On *Pelea* sp. [= *Melicope* sp.], Maui, Olinda Pipeline, 5 SEP [1921], leg. F.L. Stevens No. 1137, ILL 6888; [Island of] Hawaii, Kealakekua, 25 JUL [1921], leg. F.L. Stevens No. 986a, ILL 6880; 1919, leg. Fullaway & Giffard [No. 111? on the packet], ILL 5326; on *Metrosideros polymorpha* Gaudich., [Island of] Hawaii, between Hilo and Kilauea, 10 JUL [1921], leg. F.L. Stevens No. 777, ILL 6889; on *Broussaisia* sp., [Island of] Hawaii, Kilauea, 16 JUL [1921], leg. F.L. Stevens No. 862, ILL 6882; on *Clermontia* sp., Maui, Iao Valley, 7 SEP [1921], leg. F.L. Stevens No. 1154a, ILL 6881 [also the lectotype of *Asterina clermontiae* F.L. Stevens & R.W. Ryan in F.L. Stevens and an isotype of *Meliola lobeliae* F.L. Stevens]; on *Alyxia oliviformis* Gaudich. [cited as *olivaeformis*, Island of] Hawaii, Kealakekua, 25 JUL [1921], leg. F.L. Stevens No. 985, ILL 6877, BISH 500360, 500361 [also the holotype and two isotypes, respectively, of *Trichopeltis rhyacoides* F.L. Stevens, as well as paratypes of *Meliola alyxiae* F.L. Stevens]; on *Smilax* sp., leg. F.L. Stevens No. 1163, ILL 6886; on *Vincentia angustifolia* Gaudich. [= *Machaerina angustifolia* (Gaudich.) T. Koyama, Island of] Hawaii, Kealakekua, 25 JUL [1921], leg. F.L. Stevens No. 1007, ILL 6883; on sedge [= Cyperaceae sp. indet.], Kealakekua, 25 JUL [1921], leg. F.L. Stevens No. 998, ILL 6884; on *Elaphoglossum* sp., Oahu, Tantalus, 22 JUN [1921], leg. F.L. Stevens No. 662, ILL 6885; on *Freycinetia arnottii* Gaudich. [= *F. arborea* Gaudich.], 24 JUL [1921], Oahu, leg. F.L.Stevens No. 674, ILL 6887.
Trichothecium fusarioides F.L. Stevens. Transactions of the Illinois State Academy of Science 10:201–202. 1917. **Holotype:** On *Phyllachora peribebuyensis* C.L. Spegazzini on *Miconia* sp., Porto Rico [Puerto Rico], Maricao, [20 SEP 1913], leg. F.L. Stevens No. 3610, ILL 14497.
Trichothyrium fimbriatum C.L. Spegazzini. Revista Argentina de Historia Natural 1:418–419. 1891; Fungi Guaranitici Nonnulli Novi Vel Critici No. 124. **Isotype:** Ad folia plantae cujusdam indet. in sylvae, [Paraguay], Caáguazú, JAN 1882, C.L. Spegazzini No. 3577, ILL 6795. **Holotype:** LPS.
Trichothyrium sarciniferum C.L. Spegazzini. Boletín de la Academia Nacional de Ciencias en Córdoba 11:556–557. 1889; Fungi Puiggariani, Pugillus I, pp. 178–179, No. 342. 1889. **Isotype:** Ad folia coriacea Myrtaceae cujusdam, [Brazil], in sylvis pr. Apiahy [Apiai], Aut. 1881, leg. J.I. Puiggari No. 1626, comm. C.L. Spegazzini, ILL 6796.

Trichothyrium? serratum C.L. Spegazzini. Boletín de la Academia Nacional de Ciencias en Córdoba 11:557–558. 1889; Fungi Puiggariani, Pugillus I, pp. 179–180, No. 343. 1889. **Isotype:** Ad folia viva Convolvulaceae? cujusdam, [Brazil], in dumetis pr. Apiahy [Apiai], Aut. 1888, leg. J.I. Puiggari No. 2772, comm. C.L. Spegazzini, ILL 6797.

Tubercularia coccicola J.A. Stevenson. Annual Report of the Insular Experiment Station of Porto Rico 1917: 91–92. 1917. **Isotype:** On *Lepidosaphes beckii* and *Hemichionaspis minor* [Insecta] on leaves and twigs of *Citrus decumana* (L.) L. [= *C. maxima* (N.L. Burman) Merrill], Porto Rico [Puerto Rico], Espinosa, [27] MAR 1917, leg. J.A. Stevenson No. 6366, ILL 33562. **Holotype:** BPI.

Tubercularia nigra F.L. Stevens. Annales Mycologici 28:371. 1930. **Holotype:** On *Eupatorium* sp., British Guiana [Guyana], Rockstone, 13 JUL 1922, leg. F.L. Stevens No. 260, ILL 16546.

Tuberculina argillacea J.J. Davis. Transactions of the Wisconsin Academy of Sciences, Arts and Letters 21:293. 1923. **Isosyntype:** On *Caeoma*-infected leaves of *Rubus alleghaniensis* Porter ex Bailey, Wisconsin, Madison, [2 JUN 1921], leg. J.J. Davis [Fungi Wisconsinenses Exsiccati No. 117], ILL 16565.

Tuberculina costaricana H. Sydow. Annales Mycologici 25:154–155. 1927. **Isosyntypes:** Hab in uredosoris *Pucciniae hodgsonianae* [= *Puccinia hodgsoniana* F.D. Kern], ad folia *Eupatorii schultzii* [= *Eupatorium schultzii* Schnittspahn], Costa Rica, La Caja pr. San José, 21 DEC 1924, leg. H. Sydow, Fungi in Itinere Costaricensi Collecti No. 86, ILL 16567; in uredosoris *Pucciniae impeditae* [= *Puccinia impedita* E.B. Mains & E.W.D. Holway], ad folia *Salviae tiliaefoliae* [= *Salvia tiliifolia* Vahl], 18 DEC 1924, leg. H. Sydow, Fungi in Itinere Costaricensi Collecti No. 278, ILL 16566.

Tuberculina microstigma P.A. Saccardo. Annales Mycologici 6:563. 1908. **Isotype:** Hab. in foliis vivis *Achilleae millefolii* [= *Achillea millefolium* L.], Germaniae, [Brandenburg]: Dahlewitz pr. Zossen, [10] SEP 1906, leg. H. Sydow, [Mycotheca Germanica No. 750], ILL 16569.

Tulasnella allantospora E.M. Wakefield & A.A. Pearson. Transactions of the British Mycological Society 8:220, fig. 7. 1923. **Isotype:** On decorticated coniferous wood, [England], East Horsley, APR 1922, leg. A.A. Pearson s.n., ILL 32840. **Holotype:** K.

Tulasnella anceps G. Bresadola & H. Sydow in H. Sydow. Annales Mycologici 8:490–491. 1910. **Isotypes:** Hab. in frondibus *Pteridis aquilinae* [*Pteris aquilina* L. = *Pteridium aquilinum* (L.) Kuhn, Germany], Mecklenburg, pr. Graal, 15 AUG 1908 [as 1909 on the packet], leg. H. Sydow, Mycotheca Germanica No. 858, ILL 28701 and 32834.

Tulasnella cinchonae M. Raciborski. Bulletin International de l'Académie des Sciences de Cracovie, Classe des Sciences Mathématiques et Naturelles 1909(1):369. 1909. **Assumed isosyntype:** Auf jungen Ästen ... kult. *Cinchona* sp. [as *C. ledgeriana* Bern. Moens ex Trimen on the packet, Indonesia, Java, Soekanegara], leg. M. Raciborski s.n., ILL 32842. **Syntype:** KRA.

Tulasnella eichleriana G. Bresadola var. *lilaceo-cinerea* H. Bourdot & M.A. Donk. Nederlandsch Kruidkundig Archief. Verslagen en Mededeelingen der Nederlandsche Botanische Vereeniging 1930:69, 83. 1930. **Isotype:** ... in cortice *Quercus* ramuli decidui [= fallen branches of *Q. robur* L.], Netherlands, Bilthoven, [JUL 1928], leg. M.A. Donk No. 1272, ILL 32843, ex herb. M.A. Donk.

Tulasnella guttulata L.S. Olive. Mycologia 49:666–667, figs. 15–33. 1957. **Isotype:** On dead [decorticated] wood, [French Polynesia], Tahiti, Paea District, 5 JUL 1956, leg. L.S. Olive No. T404, ILL 32837. **Holotype:** NY.

Tulasnella microspora E.M. Wakefield & A.A. Pearson. Transactions of the British Mycological Society 8:220, fig. 8. 1923. **Isosyntype:** On rotten coniferous wood, [England], East Horsley, APR 1922, leg. A.A. Pearson s.n., ILL 32844. **Syntype:** K.

Tulasnella tremelloides E.M. Wakefield & A.A. Pearson. Transactions of the British Mycological Society 6:70. 1917. **Isotype:** On pine needles covering a nest of *Formica rufa*, [England], Surrey, Weybridge, NOV 1917, leg. A.A. Pearson s.n., ILL 32839. **Holotype:** K.

Tylostoma excentricum W.H. Long. Mycologia 36:332–333, fig. 5. 1944. **Isotype:** On low sand dunes, New Mexico, Bernalillo County, 3.5 miles from Albuquerque, on Highway 85, south end of Sandia Plaza Addition, alt. 4950 ft. [ca. 1650 m], 30 MAY 1941, leg. W.H. Long No. 9335 [erroneously as No. 9395 in the protologue], ILL 33554.

Tylostoma involucratum W.H. Long. Mycologia 36:330–332, fig. 4. 1944. **Isotype:** New Mexico, Luna County, 10 miles west of Deming, on Highway 80, alt. 4300 ft. [ca. 1430 m], 13 SEP 1941, leg. W.H. Long & D.J. Stouffer No. 9650, ILL 33541.

Tylostoma lysocephalum W.H. Long. Mycologia 36:325–327, fig. 2. 1944. **Isotype:** New Mexico, Luna County, 10 miles west of Deming, on Route 70, alt. 4300 ft. [ca. 1430 m], 12 SEP 1941, leg. W.H. Long & D.J. Stouffer No. 9639, ILL 33542.

Tylostoma polymorphum W.H. Long. Lloydia 10:129–131, fig. 12. 1947. **Isosyntype:** New Mexico, Bernalillo County, Albuquerque, Lippett Sandia Plaza Addition ... , about 4 miles from city on North 4th St., 16 MAY 1941, leg. W.H. Long No. 9308, ILL 32859.

U

Uleodothis paspali F.L. Stevens. Illinois Biological Monographs 8:181, pl. III, figs. 20–23. 1923. **Holotype:** On *Paspalum conjugatum* P. Bergius, British Guiana [Guyana], Coverden, 8 AUG 1922, leg. F.L. Stevens No. 759, ILL 9282.

Uleodothis pteridis F.L. Stevens. Botanical Gazette 69:248–249, figs. 6, 7. 1920. **Holotype:** On *Pteridium caudatum* (L.) Maxon, Porto Rico [Puerto Rico], Maricao, [18 NOV 1913], leg. F.L. Stevens No. 4814, ILL 9285. **Isotypes:** ILL 9283 and 9284, BPI.

Uleomyces comedens H. Sydow. Annales Mycologici 24:356–358. 1926. **Isotype:** Hab. parasiticus in stromatibus *Hysterostomellae phoebes* [= *Hysterostomella phoebes* H. Sydow], ad folia *Phoebes costaricanae* [= *Phoebe costaricana* Mez & Pittier], Costa Rica, San Pedro de San Ramón, 23 JAN 1925, leg. H. Sydow, Fungi in Itinere Costaricensi Collecti No. 170b, ILL 8421.

Uredo ammophilae H. & P. Sydow in J.T.C. Vestergren. Botaniska Notiser 1900:42–43. 1900. **Assumed isotype:** In foliis *Ammophilae arenariae* [= *Ammophila arenaria* (L.) Link, Germany, Pommern]: am Strande bei Thiessow, Insel Rügen, 28 JUL 1899, leg. H. & P. Sydow, [Mycotheca Germanica No. 14], ILL 28104.

Uredo bixae J.C. Arthur. Mycologia 7:327. 1915. **Isotype:** On *Bixa orellana* L., Porto Rico [Puerto Rico], Adjuntas, 2 MAR [1913], leg. F.L. Stevens No. 462, ILL 25179. **Holotype:** PUR 42881.

Uredo campeliae H. Sydow. Annales Mycologici 24:294. 1926. **Isotype:** Hab. in foliis *Campeliae zanoniae* [*Campelia zanonia* (L.) Kunth = *Tradescantia zanonia* (L.) Sw.], Costa Rica, San Pedro de San Ramón, 5 FEB 1925, leg. H. Sydow, Fungi in Itinere Costaricensi Collecti No. 20, ILL 28124.

Uredo concors J.C. Arthur. Mycologia 7:330–331. 1915. **Isotype:** On *Dolichos lablab* L. [= *Lablab purpureus* (L.) Sweet], Porto Rico [Puerto Rico], Jayuya, 17 DEC [1913], leg. F.L. Stevens No. 6042, ILL 28134. **Holotype:** PUR 3228. **Paratypes:** On *Teramnus uncinatus* (L.) Sw., Jayuya, 17 DEC [1913], leg. F.L. Stevens No. 5998, ILL 28135 and 28136.

Uredo fallaciosa J.C. Arthur. Mycologia 7:323. 1915. **Isotype:** On *Psychotria patens* sensu auct. non Sw. [= *P. deflexa* DC.], Porto Rico [Puerto Rico], Maricao, 3 APR [1913], leg. F.L. Stevens No. 774, ILL 28078. **Holotype:** PUR 42536. **Paratype:** Ponce, 8 NOV [1913], leg. F.L. Stevens No. 4341, ILL 28077, PUR 42537, and perhaps ILL 28076 [this specimen has a retyped label combining No. 4341 with the locality and date of F.L. Stevens No. 774; it may be, in fact, another isotype].

Uredo fenestrala J.C. Arthur. Mycologia 7:332. 1915. **Paratypes:** On *Phyllanthus grandifolius* sensu auct. non L. [= *P. juglandifolius* Willd.], Porto Rico [Puerto Rico], Bayamon, 19 FEB [1913], leg. F.L. Stevens No. 389, ILL 28080 and 28081, PUR 3257; Villa Alba, 4 JAN [1913], leg. F.L. Stevens No. 527, ILL 28079, PUR 3259. **Holotype:** PUR 3258.

Uredo (Hemileia) gardeniae-thunbergiae P.C. Hennings in H. Baum. Kunene-Sambesi-Expedition, Berlin, p. 160. 1903. **Isotype:** Auf lebenden Blättern von *Gardenia thunbergia* L.f., [Südwest-Afrika], zwischen Kiteve und Humbe [s. Angola], alt. 1100 m, 1 JUN 1900, leg. H. Baum, Kunene-Sambesi-Expedition No. 956, ILL 27967.

Uredo globulosa J.C. Arthur. Mycologia 8:22–23. 1916. **Isotype:** On *Hypoxis decumbens* L. [= *H. hirsuta* (L.) Coville], Porto Rico [Puerto Rico], Las Marias, 10 JUL 1915, leg. F.L. Stevens No. 1127 [cited as J.C. Arthur No. 8127], ILL 27974. **Holotype:** PUR 13712. **Paratypes:** Bandera, 15 JUL 1915, leg. F.L. Stevens No. 1577 [cited as J.C. Arthur No. 8577], ILL 27973, PUR 13711; leg. F.L. Stevens No. 1630 [cited as J.C. Arthur No. 8630], ILL 27972, PUR 13710.

Uredo hameliae J.C. Arthur. Mycologia 8:23–24. 1916. **Isotype:** *Hamelia erecta* Jacquin [= *H. patens* Jacquin], Porto Rico [Puerto Rico], Lajas [misspelled "Lojos"], 17 JUN 1915, leg. F.L. Stevens No. 140 [cited as J.C. Arthur No. 7140], ILL 27985. **Holotype:** PUR 42939.

Uredo hawaiiensis J.C. Arthur in F.L. Stevens. Bernice P. Bishop Museum Bulletin 19:124. 1925. **Isolectotypes:** On *Carex wahuensis* C.A. Meyer [as *oahuensis*], Hawaii, [Island of] Hawaii, Kilauea, 17 JUL [1921], leg. F.L. Stevens No. 880, ILL 27986, BISH 500006. **Lectotype** [designated herein]: PUR F-8808.

Uredo jatrophicola J.C. Arthur. Mycologia 7:331. 1915. **Isotypes:** On Euphorbiaceae: *Jatropha curcas* L., Porto Rico [Puerto Rico], Hormigueros, 14 JAN [1913], leg. F.L. Stevens No. 220, ILL 28005, 28008, 28011 and 28012, BPI 70880. **Holotype:** PUR 42870. **Paratypes:** On *Jatropha gossypiifolia* L., Guayanilla, 13 NOV [1913], leg. F.L. Stevens No. 5866*bis*, ILL 28013; San Germán, 8 DEC [1913], leg. F.L. Stevens No.4790, ILL 28014; leg. F.L. Stevens No. 4113, ILL 28015.

Uredo kampuluvensis P.C. Hennings in H. Baum. Kunene-Sambesi-Expedition, Berlin, pp. 159–160. 1903. **Isosyntype:** Auf lederigen Blättern von *Baphia cornifolia* H.A.T. Harms, [Südwest-Afrika], am Kampuluvé [(Kuito) = Cuito River, s. Angola], alt. 1200 m, 7 APR 1900, leg. H. Baum, Kunene-Sambesi-Expedition No. 802, ILL 28020.

Uredo kansensis W.A. Kellerman & W.T. Swingle. Journal of Mycology 5:77–78. 1889. **Isosyntype:** On leaflets of *Amorpha fruticosa* L., Kansas, [Riley County], Manhattan, 21 JUN 1887, leg. W.A. Kellerman & W.T. Swingle No. 905 [ex Kansas Fungi], ILL 28021.

Uredo lutea J.C. Arthur. Mycologia 7:321–322. 1915. **Residual isosyntypes** [PUR numbers represent residual syntypes]: On *Cassia quinquangulata* sensu auct. non L.C. Richard [= *Senna* cf. *nitida* (L.C. Richard) Irwin & Barneby], Porto Rico [Puerto Rico, Maricao], 4 APR [1913], leg. F.L. Stevens No. 704, ILL 28026 and 28027, PUR 42816; Jajome Alto [as Jayome Alto], 3 DEC [1913], leg. F.L. Stevens No. 5653, ILL 28029, PUR 42814; Preston's Ranch, near Naguabo, 31 DEC [1913], leg. F.L. Stevens No. 6762, ILL 28028, PUR 42819. **Assumed lectotype:** PUR 42815 [for F.L. Stevens No. 404*bis*, which is not represented at ILL].

Uredo mauriae H. Sydow. Annales Mycologici 23:325. 1925. **Isotype:** Hab. in foliis *Mauriae glaucae* [= *Mauria glauca* J. Donnell Smith], Costa Rica, La Caja pr. San José, 14 FEB 1925, leg. H. Sydow, Fungi in Itinere Costaricensi Collecti No. 10, ILL 28036.

Uredo myopori G.B. Cummins. Bulletin of the Torrey Botanical Club 79:232. 1952. **Isotypes:** On leaves of *Myoporum sandwicense* (A. DC.) A. Gray, Hawaii, [Island of] Hawaii, N. Hilo District, Humuula Road, Parker Ranch, alt. 5500 ft. [ca. 1830 m], 26 DEC 1946, leg. D.P. Rogers No. 3183, ILL 32566, NY. **Holotype:** PUR F-14529.

Uredo ocfemiana F.L. Stevens. Philippine Agriculturist 20:87–89, fig. 2. 1931. **Holotype:** On *Pisonia alba* Span., Philippines, [Luzon], Tayabas, Sariaya, 14 SEP 1930 [as 10 SEP on the packet], leg. G.O. Ocfemia, as F.L. Stevens No. 595 [on the same leaves with *Aecidium ocfemianum* F.L. Stevens], ILL 28043. **Isotypes:** ILL 28041, BPI 70884.

Uredo proximella J.C. Arthur. Mycologia 7:324. 1915. **Isotype:** On *Lactuca intybacea* Jacquin [= *Launaea intybacea* (Jacquin) Beauvois], Porto Rico [Puerto Rico], Sabana Grande, 30 MAR [1913], leg. F.L. Stevens No. 318, ILL 28332. **Holotype:** PUR.

Uredo rubescens J.C. Arthur. Mycologia 7:327. 1915. **Isotype:** On Moraceae: *Dorstenia contrajerva* L., Porto Rico [Puerto Rico], Camuy, 22 NOV [1913], leg. F.L. Stevens No. 5011, ILL 28338. **Holotype:** PUR 42749.

Uredo sabiceicola J.C. Arthur. Mycologia 7:323–324. 1915. **Isotype:** On *Sabicea aspera* sensu auct. non Aublet [= *S.* cf. *villosa* Willd. ex Roemer & J.A. Schultes], Porto Rico [Puerto Rico], Mayagüez, 1 MAY [1913], leg. F.L. Stevens No. 1047, ILL 28340. **Holotype:** PUR 42932.

Uredo sauvagesiae J.C. Arthur. Mycologia 8:23. 1916. **Isotype:** On *Sauvagesia erecta* L., Porto Rico [Puerto Rico], Jajome Alto [as Jejome Alto], 17 JUL 1915, leg. F.L. Stevens No. 1376 [cited as J.C. Arthur No. 8376], ILL 28342. **Holotype:** PUR 42878.

Uredo spirostachydis J.C. Arthur. Bulletin of the Torrey Botanical Club 37:576. 1910. **Isotype:** On *Spirostachys occidentalis* S. Watson [= *Allenrolfea occidentalis* (S. Watson) O. Kuntze—the latter name on the packet], Arizona, north of Yuma, 26 APR 1906, leg. M.E. Jones No. 7815, ILL 31415. **Holotype:** PUR.

Uredo stevensiana J.C. Arthur. Mycologia 7:326. 1915. **Paratypes:** On *Axonopus compressus* (Sw.) Beauvois [as *Axonophus* (syn. cited: *Paspalum compressum* (Sw.) Rasp.)], Porto Rico [Puerto Rico], Mayagüez, 8 FEB [1913], leg. F.L. Stevens No. 280, ILL 2834; 9 JAN [1913], leg. F.L. Stevens No. 237, ILL 28351; on *Paspalum paniculatum* L., Vega Baja, 22 JAN [1913], leg. F.L. Stevens No. 373, ILL 28350; on *Paspalum ?helleri* Nash [= *P.* cf. *laxum* Lam.], Mayagüez, 30 APR [1913], leg. F.L. Stevens No. 932, ILL 28349. **Holotype:** PUR 18411.

Uredo stevensii J.C. Arthur in F.L. Stevens. Bernice P. Bishop Museum Bulletin 19:124. 1925. **Isolectotypes:** On *Euphorbia clusiaefolia* Hooker & Arnott [= *Chamaesyce clusiifolia* (Hooker & Arnott) J.C. Arthur, Hawaii, Kauai, pipe trail in Waimea Canyon, 15 JUN [1921], leg. F.L. Stevens No. 428, ILL 28359, BISH 500005. **Lectotype:** PUR F-8939 [designated by J.F. Hennen & C.S. Hodges, Mycologia 73:1122. 1981]. **Paratypes:** On *Euphorbia* sp., Oahu, Ahren's ditch trail, 8 JUN [1921], leg. F.L. Stevens No. 278, ILL 28358, PUR F-8940.

Uredo venustula J.C. Arthur. Mycologia 8:21. 1916. **Isotype:** On *Andropogon brevifolius* Sw., Porto Rico [Puerto Rico], Las Marias, 10 JUL 1915, leg. F.L. Stevens No. 1147 [cited as J.C. Arthur No. 8147], ILL 28366. **Holotype:** PUR 18193.

Uredo vicina J.C. Arthur. Mycologia 7:325. 1915. **Paratypes:** On *Wedelia lanceolata* DC., Porto Rico [Puerto Rico], Guanica, 3 FEB [1913], leg. F.L. Stevens No. 365, ILL 28368, PUR 42957. **Holotype:** PUR 42958 [for F.L. Stevens No. 365*bis*, which is not represented at ILL].

Urocystis gei J.B. Ellis & B.M. Everhart. Bulletin of the Torrey Botanical Club 27:572. 1900. **Isotype:** On leaves of *Geum ciliatum* Pursh [= *G. triflorum* Pursh var. *ciliatum* (Pursh) Fassett], Washington, Waitsburg, 7 MAY 1900, leg. R.M. Horner No. 1430 [ex J.B. Ellis & B.M. Everhart, Fungi Columbiani No. 1595], ILL 17778. **Holotype:** NY.

Uromyces alyxiae J.C. Arthur in F.L. Stevens. Bernice P. Bishop Museum Bulletin 19:117. 1925. **Isotypes:** On *Alyxia oliviformis* Gaudich. [cited as *olivaeformis*], Hawaii, Kauai, Kalalau trail, 16 JUN [1921], leg. F.L. Stevens No. 520, ILL 25214, BISH 500358 and 500359. **Holotype:** PUR F-3476. **Paratypes:** Leg. F.L. Stevens No. 519, ILL 25213; [Island of] Hawaii, Kona, Keauhou, Bishop Estate Road, 25 JUL [1921], leg. F.L. Stevens No. 1011, ILL 25216, PUR F-3477; Hamakua, upper ditch trail, 28 JUL [1921], leg. F.L. Stevens No. 1026, ILL 25215, PUR F-3475; 31 JUL [1921], leg. F.L. Stevens No. 1081, ILL 25211, PUR F-3474; Alakai Swamp, 22 AUG [1921], leg. O.H. Swezey, as F.L. Stevens No. 1167, ILL 25212.

Uromyces costaricensis H. Sydow. Annales Mycologici 23:312–313. 1925. **Isotype:** Hab. in foliis *Panici altissimi* [= *Panicum altissimum* G. Meyer], Costa Rica, Grecia, 19 JAN 1925, leg. H. Sydow, Fungi in Itinere Costaricensi Collecti No. 178, ILL 25647.

Uromyces densus J.C. Arthur. Mycologia 7:196. 1915. **Isotype:** On *Bidens pilosa* L., Porto Rico [Puerto Rico], Ponce, 8 NOV [1913], leg. F.L. Stevens No. 4266, ILL 25342. **Holotype:** PUR 38764.

Uromyces eleocharidis J.C. Arthur. Bulletin of the Torrey Botanical Club 33:514–515. 1906. **Paratypes:** On *Eleocharis palustris* (L.) Roemer & J.A. Schultes var. *glaucescens* (Willd.) A. Gray, Kansas, Stockton, 7 MAR 1906, leg. E. Bartholomew [ex Fungi Columbiani No. 2293], ILL 25360, PUR 12459. **Holotype:** PUR 12469.

Uromyces graminicola T.J. Burrill. Botanical Gazette 9:188. DEC 1884; Bulletin of the State Laboratory of Natural History 2(3):170. AUG 1885. **Isolectotype:** On *Panicum virgatum* L., McLean County, [Hudson], 20 JUL 1881, leg. A.B. Seymour [No. 2347], ILL 25972. **Lectotype:** PUR 11694. [The lectotype was designated by P. Ramachar and G.B. Cummins (Mycopathologia et Mycologia Applicata 19:55. 1963) and confirmed in G.B. Cummins (1971:456), a decision accepted herein.]. **Residual syntype:** On *Elymus virginicus* L. [the host plant may be misidentified], Illinois, Piatt County, [Mansfield], 10 AUG 1881, leg. A.B. Seymour [No. 1001], ILL 25988. ≡ *Nigredo graminicola* (T.J. Burrill) J.C. Arthur. Résultats Scientifiques du Congrès Botanique Vienne 1905. p. 343. 1906.

Uromyces koae J.C. Arthur in F.L. Stevens. Bernice P. Bishop Museum Bulletin 19:118. 1925. **Isotypes:** On *Acacia koa* A. Gray, Hawaii, Oahu, Tantalus, leg. North, as H.L. Lyon No. ?, ILL 25711, BISH 500004. **Holotype:** PUR F-2888. **Paratypes:** [Island of] Hawaii, 10 OCT 1913, leg. R.H. Hosmer, as H.L. Lyon No. 416, ILL 25709; Kauai, leg. O.H. Swezey s.n., ILL 25712; Maui, Honokahua, 17 JUL 1913, leg. R.H. Hosmer, as H.L. Lyon No. 359, ILL 25706; Pogue's ditch trail, 6 SEP [1921], leg. F.L. Stevens No. 1158, ILL 25708; Oahu, 19 JUL 1919, leg. H.L. Lyon No. 4, ILL 25710; Wahiawa, Ahren's ditch trail, 8 JUN [1921], leg. F.L. Stevens No. 291, ILL 25707.

Uromyces oenotherae T.J. Burrill. Botanical Gazette 9:187–188. DEC 1884; Bulletin of the Illinois State Laboratory of Natural History 2(3):162. AUG 1885. **Lectotype** [designated herein]: On *Oenothera linifolia* Nuttall, Illinois, [Jackson County], Makanda, 27 APR 1882, leg. A.B. Seymour [No. 4342], ILL 26258a [this particular packet marked "type"]. **Isolectotypes:** ILL 26258b, PUR 16801. **Residual syntype:** 28 APR 1882, leg. A.B. Seymour [No. 4359], ILL 26259.

Uromyces purus (H. Sydow) G.B. Cummins, 1977. **See basionym:** *Argomycetella pura* H. Sydow, 1925.

Uromyces rudbeckiae J.C. Arthur & E.W.D. Holway in T.J. Burrill. Bulletin of the Illinois State Laboratory of Natural History 2(3):163. 1885. **Residual isosyntypes** [residual syntypes at PUR]: On *Rudbeckia laciniata* L., Illinois, McHenry [County], 24 AUG [1881], leg. A.B. Seymour No. 1273, ILL 26614; Stephenson [County, Freeport], 21 SEP 1882, leg. A.B. Seymour No. 6084, ILL 26612. **Lectotype** [designated by G.B. Cummins (1978:203)]: PUR 38748.

Uromyces scirpi T.J. Burrill. Botanical Gazette 9:188. DEC 1884; Bulletin of the Illinois State Laboratory of Natural History 2(3):168–169. AUG 1885. **Holotype:** On *Scirpus fluviatilis* (Torrey) A. Gray, Illinois, Champaign [County, Urbana], 13 AUG 1881, leg. A.B. Seymour [No. 1031], ILL 26635. **Isotype:** PUR 12628.

Uropyxis holwayi (J.C. Arthur) J.C. Arthur, 1934. **See basionym:** *Calliospora holwayi* J.C. Arthur, 1905.

Ustilago montaniensis J.B. Ellis & B.M. Everhart. Journal of Mycology 6:119. 1891 [attributed to J.B. Ellis & E.W.D. Holway on the packet]. **Isotype:** On *Muhlenbergia glomerata* (Willd.) Trin., Montana, Sand Coulee, DEC 1887 [as JUL 1888 on the packet], leg. W.F. Anderson [ex J.B. Ellis & B.M. Everhart, North American Fungi No. 2263], ILL 17047.

Ustilago sieglingiae P.L. Ricker. Journal of Mycology 11:112. 1905. **Isotype:** On *Sieglingia purpurea* (Walter) O. Kuntze [= *Triplasis purpurea* (Walter) Chapman], Florida, Punta Rassa, AUG 1900, leg. A.S. Hitchcock s.n., ILL 17217.

Ustilago sphaerogena T.J. Burrill ex J.B. Ellis & B.M. Everhart. North American Fungi, Century 19, No. 1892. Anno 1887, nom. nud. **Isotype:** In ovaries of *Panicum crus-galli* L. [= *Echinochloa crus-galli* (L.) Beauvois], Illinois, Osborne, leg. A.B. Seymour No. 32678, ILL 17246.

V

Valsa farinosa J.B. Ellis. Bulletin of the Torrey Botanical Club 9:99. 1882. **Isotype:** [On dead shoots of *Quercus alba* L., New Jersey, Newfield, APR 1882, leg. J.B. Ellis, ex J.B. Ellis & B.M. Everhart, North American Fungi No. 1572], ILL 10449. **Holotype:** NY [no type citation in the protologue].

Valsa thujae C.H. Peck var. *foliicola* J.B. Ellis & B.M. Everhart. North American Fungi, Century 26, No. 2519. Anno 1891, nom. nud. **Isotype:** On dead leaves of *Thuja* sp., Canada, [Ontario], London, JAN 1890, leg. J. Dearness, ILL 10486. **Holotype:** NY.

Vararia ochroleuca (H. Bourdot & A. Galzin) M.A. Donk, 1930. **See basionym**: *Asterostromella ochroleuca* H. Bourdot & M. Galzin, 1911.

Vermicularia caricina J.B. Ellis & B.M. Everhart. North American Fungi, Century 28, No. 2779. Anno 1892, nom. nud. **Isotype**: On leaves of *Carex lurida* Wahlenb., Canada, [Ontario], London, AUG 1890, leg. J. Dearness, ILL 13059.

Vertixore atronitidum V.A.M. Miller & L. Bonar. University of California Publications in Botany 19:406–407, pl. 67, figs. 2–4; pl. 68, fig. 1; pl. 70, fig. 2. 1941. **Paratype**: On *Photinia arbutifolia* Lindley [= *Heteromeles arbutifolia* (Lindley) M. Roemer], California, Sonoma County, between Knights and Alexander Valley, 20 APR 1931, leg. H.E. Parks No. 3621 [ex California Fungi No. 749], ILL 30670.

Virgatospora echinofibrosa D.E. Finley. Mycologia 59:538–541, figs. 1–18. 1967. **Isolectotype**: On dead twig, Panama, Canal Zone, Barro Colorado Island, Pierson trail, 3 AUG 1964, leg. E.F. Morris & J.W. Strain No. 780, ILL 33543. **Lectotype** [designated by A. Rossman (1983:61)]: BPI.

Volutella uredinophila H. Sydow. Annales Mycologici 25:156–158. 1927. **Isotype**: Hab. parasitici in soris *Uredinis ramonensis* [= *Uredo ramonensis* H. Sydow], ad folia *Cassiae bacillaris* [*Cassia bacillaris* L.f. = *Senna bacillaris* (L.f.) Irwin & Barneby], Costa Rica, Cerro de San Isidro, pr. San Ramón, 9 FEB 1925, leg. H. Sydow, Fungi in Itinere Costaricensi Collecti No. 106, ILL 16583.

W

Wageria portoricensis F.L. Stevens & N.E. Dalbey. Mycologia 11(1):7–8, pl. 3, figs. 11–12. DEC 1918. **Holotype**: On *Gonzalagunia spicata* (Lam.) M. Gómez, Porto Rico [Puerto Rico], Jajome Alto, [17 JUL 1915], leg. F.L. Stevens No. 8407, ILL 9582. **Paratype**: El Alto de la Bandera, [15 JUL 1915], leg. F.L. Stevens No. 7636, ILL 9581.

X

Xenasma ludibundum D.P. Rogers & A.E. Liberta. Mycologia 52:902–903, fig. 12. (1960) 1961. **Holotype**: On bark of *Quercus* sp., [on the ground], Massachusetts, [Canton], Blue Hill, 31 NOV 1935, leg. D.P. Rogers No. 4, ILL 32579. **Isotypes:** ILL 32579a, FH, TRTC, and A.E. Liberta, pers. herb.

Xenolachne flagellifera D.P. Rogers. Mycologia 39:562–563, text fig. 1947. **Holotype:** Parasitic on the hymenium of *Hyaloscypha atomaria* (K. Starbäck) J.A.F. Nannfeldt, growing on the lower side of damp logs of *Libocedrus decurrens* Torrey [= *Calocedrus decurrens* (Torrey) Florin], Oregon, Umpqua Valley, east of Reedsport, 22 OCT 1938, leg. A.M. & D.P. Rogers No. 487, ILL 32810.

Xenolophium leve H. Sydow in F.L. Stevens. Bernice P. Bishop Museum Bulletin 19:97, text fig. 24a. 1925. **Holotype:** On dead bark of *Metrosideros* sp., Hawaii, [Island of] Hawaii, Kona, Keauhou, Bishop Estate Road, 23 JUL [1921], leg. F.L. Stevens No. 953, ILL 9660.

Xenolophium verrucosum H. Sydow in F.L. Stevens. Bernice P. Bishop Museum Bulletin 19:97–98, text fig. 24b. 1925. **Holotype:** On rotten wood (*Metrosideros* sp.), Hawaii, [Island of] Hawaii, Kona, Keauhou, Bishop Estate Road, 23 JUL [1921], leg. F.L. Stevens No. 955, ILL 9661.

Y

Yoshinagella japonica F. v. Höhnel. Sitzungsberichte der Kaiserl. Akademie der Wissenschaften in Wien. Mathematisch-Naturwissenschaftliche Klasse. Abt. 1, 122: 293–294. 1913. **Isotype:** Auf der Oberseite der Blätter von *Quercus glauca* Thunb., Japan, Tosa, Yoshino, Kamoda-mura und Noodzu-mura, 27 JUL 1912, leg. Torama Yoshinaga s.n., ILL 9288.

Yoshinagella nuda F.L. Stevens. Bernice P. Bishop Museum Bulletin 19:16–17, text fig. 1e, pl. I(D, H). 1925. **Syntypes** [additional ILL number and BISH numbers cited are isosyntypes]: On *Cibotium chamissoi* Kaulfuss, Hawaii, Oahu, Wahiawa, 31 MAY [1921], leg. F.L. Stevens No. 151, ILL 9298 and 9302, BISH 146466 and 500351; leg. F.L. Stevens No. 155, ILL 9292, BISH 146465 and 500348; Ahren's ditch trail, 8 JUN [1921], leg. F.L. Stevens No. 286, ILL 9301, BISH 146459 and 500347; Mt. Olympus, 10 JUN [1921], leg. F.L. Stevens No. 305, ILL 9290, BISH 146462 and 500346; leg. F.L. Stevens No. 307, ILL 9299, BISH 146460 and 500344; Waiahole ditch trail, 12 JUN [1921], leg. F.L. Stevens No. 388, ILL 9300, BISH 500343 and 500352; Tantalus, 20 JUN [1921], leg. F.L. Stevens No. 591, ILL 9297, BISH 146467 and 500350; 22 JUN [1921], leg. F.L. Stevens No. 656, ILL 9295, BISH

146468 and 500349; Olympus, 24 JUN [1921], leg. F.L. Stevens No. 664, ILL 9296, BISH 146461 and 500345; leg. F.L. Stevens No. 701, ILL 9294; [11 MAY 1913], leg. H.L. Lyon No. 331, ILL 9293, BISH 146464; leg. H.L. Lyon No. 419, ILL 9291, BISH 146463.

Yoshinagella polymorpha H.L. Lyon ex F.L. Stevens. Bernice P. Bishop Museum Bulletin 19:14–16, text fig. 1a–d, pl. I(A, B, E, F, G). 1925. **Holotype:** On living leaves of *Cibotium menziesii* Hooker [= *C. chamissoi* Kaulf.], Hawaii, Oahu, Olympus, 24 JUN [1921], leg. F.L. Stevens No. 694, ILL 9304. **Isotype:** BISH 500353. **Paratypes:** As Castle trail, [Konahuanu trail, 3 NOV] 1912, leg. H.L. Lyon No. 165, ILL 9312; Palolo Valley [as Palolo Ridge on the packet, 28 JAN] 1912, leg. H.L. Lyon No. 142, ILL 9311; leg. H.L. Lyon No. 433, ILL 9315; [as 1918 on the packet], leg. H.L. Lyon No. 468, ILL 9314; [Island of] Hawaii, Pahala, [20 APR] 1919, leg. H.L. Lyon No. 480, ILL 9313; Kauai, upper ditch trail, 31 JUL [1921?, no date on the packets], leg. F.L. Stevens No. 1161, ILL 9303 and 9310.

Yoshinagella polymorpha H.L. Lyon ex F.L. Stevens var. *pauciseta* F.L. Stevens. Bernice P. Bishop Museum Bulletin 19:16, pl. I(C). 1925. **Syntypes** [BISH numbers cited are isosyntypes]: On *Cibotium chamissoi* Kaulf., Hawaii, [Island of] Hawaii, Hamakua, upper ditch trail, 31 JUL [1921], leg. F.L. Stevens No. 1066, ILL 9309, BISH 146470 and 500356; leg. F.L. Stevens No. 1077, ILL 9307, BISH 500354 and 500355; Maui, Pogue's ditch trail, 6 SEP [1921], leg. F.L. Stevens No. 1156, ILL 9306, BISH 500357.

Z

Zignoella algaphila F.L. Stevens. Botanical Gazette 69:256. 1920. **Holotype:** On *Cephaleuros virescens* Kunze on *Artocarpus incisa* L.f., Porto Rico [Puerto Rico], Mayagüez, [24 JUL 1915], leg. F.L. Stevens No. 51, ILL 9583.

Zignoella (Trematostoma) nobilis H. Rehm. Leaflets of Philippine Botany 8:2950. 1916. **Assumed isotype:** Ad emortuum corticatum *Citrum nobilem* [= *Citrus nobilis* Lour.], Philippines, [Luzon, Laguna Prov.], Los Baños, MAY 1914, leg. C.F. Baker No. 3229 [ex Fungi Malayana No. 200], ILL 9584.

Zygodesmus albidus J.B. Ellis & B.D. Halsted ex B.D. Halsted. Bulletin of the Torrey Botanical Club 17:152. 1890. **Isotype:** On cultivated violet [= *Viola* sp.], New Jersey, New Brunswick, 26 MAR 1890, leg. B.D. Halsted s.n., ILL 32635, ex NY. **Holotype:** NY.

Zygodesmus hydnoideus M.J. Berkeley & M.A. Curtis in M.J. Berkeley. Grevillea 3:112. 1875. **Lectotype**: On rotten wood, Pennsylvania, leg. Michener No. 4335, ex M.A. Curtis herb., ILL 33549, ex NYS [lectotype designated by D.P. Rogers on the packet label and published herein].

Zygodesmus indigoferus J.B. Ellis & B.M. Everhart. Journal of Mycology 1:149. 1885. **Isotype**: On very rotten wood, Pennsylvania, West Chester, SEP 1885, leg. B.M. Everhart s.n., ILL 32545, ex NY. **Holotype**: NY.

Zygodesmus limoniisporus J.B. Ellis & B.M. Everhart. Proceedings of the Academy of Natural Sciences of Philadelphia 1891:87. 1891. **Isotype**: On rotten maple [= wood of *Acer* sp.], Canada, [Ontario], London, 19 OCT 1889, leg. J. Dearness No. 957, ILL 32638. **Holotype**: NY.

List of Lectotypes Located at ILL

[Fungus names printed in bold face indicate lectotypifications designated or first published herein]

Aecidium crotonopsidis **T.J. Burrill**, 1884: A.B. Seymour No. 4701, ILL 27716 [this particular packet marked "type"].
Aecidium diodiae **T.J. Burrill**, 1884: A.B. Seymour No. 4661, ILL 27474 [this particular packet marked "type"].
Aecidium myosotidis **T.J. Burrill**, 1884: A.B. Seymour No. 4029, ILL 27258 [we chose this specimen from two packets of No. 4029, both of which were marked "type." Isolectotype: ILL 27264].
***Asterina clermontiae* F.L. Stevens & R.W. Ryan in F.L. Stevens**, 1925: F.L. Stevens No. 1154 [as No. 1154a on the packet], ILL 6881.
Cercospora agerati F.L. Stevens, 1925 ≡ *Ragnhildiana agerati* (F.L. Stevens) F.L. Stevens & W.G. Solheim in W.G. Solheim & F.L. Stevens, 1931: F.L. Stevens No. 944, ILL 16296 [lectotype designated by C. Chupp (1953)].
Cercospora caseariae F.L. Stevens, 1917: F.L. Stevens No 99, ILL 14954 [lectotype designated by C. Chupp (1953)].
Cercospora pipturi F.L. Stevens & P.A. Glick in F.L. Stevens, 1925: F.L. Stevens No. 538, ILL 15471 [lectotype designated by C. Chupp (1953)].
***Corynelia clavata* (L.) P.A. Saccardo var. *portoricensis* F.L. Stevens**, 1917 ≡

Corynelia portoricensis (F.L. Stevens) H.M. Fitzpatrick, 1920: F.L. Stevens No. 784, ILL 9632 [lectotypification necessary because Stevens (1917) did not indicate a holotype, and the collection was subdivided].

Echidnodella melastomatacearum R.W. Ryan, 1924: F.L. Stevens No. 8160, ILL 6930.

Gloniella rubra F.L. Stevens, 1920: F.L. Stevens No. 4363, ILL 7605a.

Hyalosphaera miconiae F.L. Stevens, 1917: F.L. Stevens No. 207, ILL 8287 [two packets at ILL with the same accession number, the lectotype designated and annotated by A.Y. Rossman (1987:55)].

Irene escharoides H. Sydow, 1926 ≡ *Irenina escharoides* (H. Sydow) F.L. Stevens, 1927 ≡ *Meliola escharoides* (H. Sydow) R. Ciferri, 1954 ≡ *Meliola tabernaemontanae* C.L. Spegazzini var. *escharoides* (H. Sydow) C.G. Hansford, 1961: H. Sydow, Fungi in Itinere Costaricensi Collecti No. 393a, ILL 3736 [lectotypification by C.G. Hansford (1961:554) is assumed on the strength of his annotation of ILL 3736 as "type" and citation of F.L.S. herb. (ILL) for location of the type, although he erroneously cited the collection number as 293a, instead of 393a].

Irenina dalechampiae F.L. Stevens, 1927: F.L. Stevens No. 49, ILL 4084 [perhaps the holotype (?); citation of this collection as "type" by C.G. Hansford (1961:206) is interpreted as lectotypification].

Leptodiscus terrestris J.W. Gerdemann, 1953 ≡ *Mycoleptodiscus terrestris* (J.W. Gerdemann) S.A. Ostazeski, 1968: J.W. Gerdemann s.n., ILL 31238.

Marasmius nucicola W.B. McDougall, 1925: W.B. McDougall s.n., ILL 31106 [lectotype designated by D.E. Desjardin, in litt., 1992, and accepted herein].

Meliola annonae F.L. Stevens, 1928: F.L. Stevens No. 342, ILL 4584 [citation of specimen in the F.L.S. herb. (ILL) as "type" by C.G. Hansford (1961:39) is interpreted as lectotypification].

Meliola ardisiae H. Sydow, 1925: A.D.E. Elmer No. 17327, ILL 4623 [citation of specimen in F.L.S. herb. (ILL) as "type" by C.G. Hansford (1961:513) is interpreted as lectotypification].

Meliola bicornis H.G. Winter var. *calopogonii* F.L. Stevens, 1916 ≡ *Meliola scabriseta* C.G. Hansford & F.C. Deighton var. *calopogonii* (F.L. Stevens) C.G. Hansford, 1961: F.L. Stevens No. 8060, ILL 4707 [choice of this collection as "type" by C.G. Hansford (1961:277) is interpreted as lectotypification with the specimen at ILL, even though Stevens (1928:189) cited No. 3492 under type locality (the latter number is misquoted, it should be 3942). No choice of type is indicated on either of two packets at ILL.].

Meliola borneensis H. Sydow, 1923: M. Ramos No. 2138, ILL 4766 [citation of specimen in F.L.S. herb. (ILL) as "type" by C.G. Hansford (1961:35) is interpreted as lectotypification].

Meliola conica F.L. Stevens, 1928: F.L. Stevens No. 787, ILL 4956 [this is the specimen illustrated in the original publication, and confirmation of No. 787 as "type" by C.G. Hansford (1961:266) is interpreted as lectotypification].

Meliola crescentiae F.L. Stevens, 1928: F.L. Stevens No. 940, ILL 4992 [this collection is cited in the caption of fig. 42 in the original publication; C.G. Hansford's (1961:673) confirmation of No. 940 as "type" is accepted as lectotypification with the specimen at ILL].

Meliola didymopanacis P.C. Hennings var. *stevensii* **C.G. Hansford,** 1955: F.L. Stevens No. 7647, ILL 5066.

Meliola drepanochaeta H. Sydow, 1926: H. Sydow, Fungi in Itinere Costaricensi Collecti No. 163, ILL 5023 [citation of specimen in the F.L.S. herb. (ILL) as "type" by C.G. Hansford (1961:51) is interpreted as lectotypification. The packet at ILL has the original Sydow label, and a microscopic preparation of ILL 5023 by Hansford is marked "type"].

Meliola duggenae F.L. Stevens var. *panamensis* F.L. Stevens, 1928: F.L. Stevens No. 1314, ILL 5020 [citation of this collection as "type" by C.G. Hansford (1961:587) is accepted as lectotypification with the specimen at ILL].

Meliola euopla H. Sydow ex F.L. Stevens, 1928, nom. nov. **Based on**: *Meliola vicina* H. Sydow, 1926 (nom. illegit.), non H. Sydow, 1923: H. Sydow, Fungi in Itinere Costaricensi Collecti No. 133, ILL 6421.

Meliola exilis H. & P. Sydow, 1904 ≡ *Irene exilis* (H. & P. Sydow) F.L. Stevens, 1925 ≡ *Irenina exilis* (H. & P. Sydow) F.L. Stevens, 1927 ≡ *Asteridiella exilis* (H. & P. Sydow) C.G. Hansford ex C.G. Hansford, 1961: F.W. Neger s.n., ILL 4086 [citation of specimen in the F.L.S. herb. (ILL) as "type" by C.G. Hansford (1961:493) is interpreted as lectotypification. The packet was labeled and marked "type" in what appears to be H. Sydow's handwriting.].

Meliola holigarnae F.L. Stevens, 1928: L.J. Sedgwick, as F.L. Stevens No. 1981, ILL 5265 [lectotypification by F.L. Stevens in a second reference].

Meliola inocarpi F.L. Stevens, 1928: C.F. Baker, Fungi Malayana No. 459, ILL 5281 [as two microscopic preparations, the lectotype tentatively designated herein].

Meliola megalospora C.L. Spegazzini, 1881 ≡ *Irene megalospora* (C.L. Spegazzini) F. Theissen & H. Sydow, 1918 ≡ *Meliolina megalospora* (C.L. Spegazzini) F.L. Stevens, 1927 ≡ *Asteridiella megalospora* (C.L. Spegazzini) C.G. Hansford, 1961: Leg. ?, comm. C.L. Spegazzini, ILL 6501 [citation of specimen in F.L.S. herb. (ILL) as "type" by C.G. Hansford (1961:360) was interpreted as lectotypification by S.J. Hughes (1993:217) and is accepted herein].

Meliola ocoteicola F.L. Stevens, 1916: F.L. Stevens No. 7560, ILL 5709. [The author inadvertently indicated both No. 7560 and No. 4731 as "type" in the protologue; however, in his monograph (F.L. Stevens, 1928:279), he cited only No. 7560 under the type locality. This is interpreted as lectotypification, especially since the decision was upheld by C.G. Hansford (1961:57).].

***Meliola paulliniae* F.L. Stevens var. *dentata* F.L. Stevens**, 1928: F.L. Stevens No. 97, ILL 5868 [selected from 14 syntypes cited in the protologue. C.G. Hansford (1961: 432) erroneously cited the holotype of *Meliola serjaniae* F.L. Stevens var. *dentata* F.L. Stevens as "type"of this variety.].

Meliola pinicola J. Dearness, 1926 ≡ *Asteridiella pinicola* (J. Dearness) C.G. Hansford: G.G. Hedgcock No. 24394, comm. J. Dearness No. 5878, ILL 4219 [citation of specimen in the F.L.S. herb. (ILL) as "type" by C.G. Hansford (1961:751) is interpreted as lectotypification].

***Meliola schizolobii* H. & P. Sydow**, 1916: E.H.G. Ule No. 3495, ILL 6228 [part of the type collection, the holotype is presumed lost or destroyed].

Meliola tapirirae F.L. Stevens & L.R. Tehon, 1926: F.L. Stevens No. 330, ILL 6337 [no type is indicated in the protologue but both F.L. Stevens (1928:182) and C.G. Hansford (1961:458) selected No. 330. We accept this as lectotypification.].

***Meliola vaccinii* F.L. Stevens**, 1925: F.L. Stevens No. 739, ILL 6416.

Meliola vicina H. Sydow, 1923: E.D. Merrill No. 8886, ILL 6422 [annotation of this specimen and citation of F.L.S. herb. (ILL) for location of "type" by C.G. Hansford (1961:601) is interpreted as lectotypification].

***Meliola vicina* H. Sydow**, 1926 (**nom. illegit.**), non H. Sydow, 1923. ≡ *M. euopla* H. Sydow ex F.L. Stevens, 1928. nom. nov.: H. Sydow, Fungi in Itinere Costaricensi Collecti No. 133, ILL 6421.

Meliola xylopiae F.L. Stevens, 1928: F.L. Stevens No. 1102, ILL 6441 [citation of this collection as "type" by C.G. Hansford (1961:34) is accepted as lectotypfication with the specimen at ILL].

Meliolina sydowiana F.L. Stevens, 1925: F.L. Stevens No. 976, ILL 6519. [lectotype designated by S.J. Hughes, 1993].

***Morenoella miconiicola* R.W. Ryan**, 1924: F.L. Stevens No. 7451, ILL 7347a.

***Septoria clermontiae* F.L. Stevens & P.A. Young in F.L. Stevens**, 1925: F.L. Stevens No. 659, ILL 12109 [marked "type" by F.L. Stevens on the packet; accepted and published as lectotypification herein].

Septoria guettardae P. Garman, 1915: F.L. Stevens No. 4741 [as No. 9759 in the protologue; probably in error since the packet of No. 4741 is clearly labeled type], ILL 12273.

***Sphaerulina cibotii* F.L. Stevens & E.F. Guba in F.L. Stevens**, 1925: F.L. Stevens No. 545, ILL 9917 [text fig. 27b in the protologue represents this specimen of Stevens No. 545].

***Synchytrium sambuci* M.T. Cook,** 1949: M.T. Cook & W.J. Dickson s.n., ILL 21267.

***Uromyces oenotherae* T.J. Burrill,** 1884: A.B. Seymour [No. 4342], ILL 26258a [this particular packet marked "type"].

***Zygodesmus hydnoideus* M.J. Berkeley & M.A. Curtis in M.J. Berkeley,** 1875: Michener No. 4335, ex M.A. Curtis herb., ILL 33549, ex NYS [lectotype designated by D.P. Rogers on the packet label and published herein].

List of Undistributed Fungus Exsiccati Holdings of the University of Illinois Mycological Collections

Arthur, J.C. and E.W.D. Holway. Uredineae Exsiccatae et Icones. Fascicles I–IV, Nos. 1–60. 1894–1902.

Bartholomew, E. North American Uredinales. Centuries I–XXXV, Nos. 1–3500. 1911–1926.

Beck, G. und A. Zahlbrückner. Kryptogamae Exsiccatae, Editae a Museo Paletino Vindobonensi. Centuries 1–12.. 1894–1905; 14–24. 1906–1916. Supplement Nos. 1–100; 1115–2126. 1906–1916.

Berkeley, M.J. British Fungi: Consisting of Dried Specimens of the Species, Described in Vol. V, Part II of the English Flora: Together With Such as May Hereafter be Discovered Indigenous to Britain. Fascicles 2, 3, Nos. 61–120, 121–240. 1836–1837.

Briosi, G., F.Cavara and G. Pollacci. I Funghi Parassiti Delle Piante Coltivate Od Utili Essiccati Delineati E Descritti. Fascicles I–XIX, Nos. 1–475. 1888–1926.

Ciferri, R. Mycoflora Domingensis Exsiccata. Centuries I–III, Nos. 1–300. 1930–1932.

Ellis, J.B. North American Fungi. Series I. Centuries I–XV, Nos. 1–1500. 1878–1885. Series II. Centuries XVI–XXXVI, Nos. 1501–3600. 1886–1898.

Ellis, J.B. and B.M. Everhart. Fungi Columbiani. Centuries I–LI, Nos. 1–5100. 1893–1917.

Herpell, G. Sammlung Präparierter Hutpilze. Lieferungen 1–6, Nos. 1–135. 1880–1882.

Jaap, O. Fungi Selecti Exsiccati. Series 1, Nos. 1–25. 1903; Series 7, Nos. 151–175. 1906; Series 19–39, Nos. 451–959 and Supplement. 1911–1928.

Kellerman, W.A. Ohio Fungi. Fascicles I–X, Nos. 1–200. 1901–1905.

Krieger, K.W. Schädliche Pilze Unserer Kulturgewächse, Gesammelt und Herausgegeben. Fascicle 1, Nos. 47–50. 1896; Fascicle 5, Nos. 201–250. 1917.

Krieger, K.W. Fungi Saxonici Exsiccati. Die Pilze Sachsen's. Fascicles 47–50 and Supplement. 1915–1919.

Kunze, J. Fungi Selecti Exsiccati. Centuries 1, 2, Nos. 1–199. 1876–1877; Century 3, Nos. 301–400. 1880; Century 5, Nos. 501–600. ?1880.

Migula, E.F.A.W. Kryptogamae Germaniae, Austriae et Helvetiae Exsiccatae. Fascicle 3, Nos. 1–25. 1902; Fascicle 6, Nos. 26–50. 1903; Fascicles 13, 14, Nos. 51–75, 76–100. 1904; Fascicle 19, Nos. 101–125. 1904; Fascicle 23, Nos. 126–150. 1904; Fascicle 24, Nos. 151–175. 1904; Fascicles 33, 34, Nos. 176–200, 201–225. 1907; Fascicles 39, 40, Nos. 226–250, 251–275. 1922; Fascicles 43, 44, Nos. 276–300, 301–325. 1927; Fascicles 50, 51, Nos. 326–350, 351–375. 1931; Fascicles 56, 57, Nos. 376–400, 401–425. 1933.

Petrak, F. Mycotheca Generalis. Lieferungen 1, 2, Nos. 1–100. 1928.

Rabenhorst, G.L. Fungi Europaei Exsiccati. Centuries 1–40, Nos. 1–4000. 1859–1893.

Raciborski, M. Mycotheca Polonica. Fascicles 2, 3, Nos. 51–100, 101–150. 1959.

Ravenel, H.W. Fungi Caroliniani Exsiccati. Fascicle IV, Nos. 1–100. 1855; Fascicle V, Nos. 1–100. 1860.

Ravenel, H.W. and M.C. Cook. Fungi Americani Exsiccati. Centuries I–VIII, Nos. 1–800. 1878–1882.

Rehm, H. Ascomyceten. Fascicle XVI, Nos. 751–800. 1884; Fascicles LVI–LVII, Nos. 2126–2175. 1918.

Roumeguére, C. Fungi Selecti Gallici Exsiccati. Centuries I–LXXIV, Nos. 1–7400. 1879–1899.

Seymour, A.B. and F.S. Earle. Economic Fungi. I–XI. Nos. 1–550. 1890–1899. Supplement C, Nos. C1–C100. 1903; Nos. C101–C150. 1905.

Shear, C.L. New York Fungi. Centuries I–III, Nos. 1–400. 1893–1896.

Solheim, W.G. Mycoflora Saximontanensis Exsiccata. Centum I–IV, Nos. 1–400. 1934–1943.

Sydow, H. Fungi Exotici Exsiccati. Fascicles I–IX, Nos. 1–450. 1912–1915; Fascicles XII–XIX, Nos. 551–950. 1927–1934.

Sydow, H. and P. Sydow. Mycotheca Germanica. Fascicles 1–64, Nos. 1–3200. 1903–1938.

Sydow, P. Uredineen. Fascicles I–LV, Nos. 1–2750. 1888–1916.

Sydow, P. Ustilagineen. Fascicles I–XIII, Nos. 1–500. 1894–1915.
Sydow, P. Phycomyceten et Protomyceten. Fascicle IX, Nos. 326–350. 1916.
Thümen, F. von. Fungi Austriaci Exsiccati. Centuries 7–11, Nos. 601–1100. 1873–1874.
Thümen, F. von. Herbarium Mycologicum Oeconomicum. Fascicles I–XV, Nos. 1–750. 1872–1880.
Thümen, F. von. Mycotheca Universalis. Centuries I–XXIII, Nos. 1–2300. 1875–1884.
Ule, E. Mycotheca Brasiliensis. Century I, Nos. 1–100. 1905.
Vasudeva, R. Herbarium Cryptogamicorum. Indiae Orientalis. Series 2. Indian Ustilaginales, Fascicle 1, Nos. 1–50. 1954. Series 3. Indian *Cercosporae,* Fascicle 1, Nos. 1–50. 1955.
Weiss, J.E. Herbarium Pathologicum. Lieferungen 1–8, Nos. 1–200. 1916–1921. Decorative, Nos. 1–25. Orchard and Fruit, Nos. 1–25. Farm Crops, Nos. 1–25. [Publication dates unknown].

Literature Cited

Arthur, J.C. 1934. Manual of the rusts in United States and Canada. Illustrations by G.B. Cummins. Purdue Research Foundation, Lafayette, IN. 438 pp.

Backer, C.A. and R.C. Bakhuizen van den Brink. 1963–1968. Flora of Java (Spermatophytes only). N.V.P. Noordhoff, Groningen, the Netherlands. Three volumes.

Bailey, L.H. and E.Z. Bailey (initial compilers), and staff of the Liberty Hyde Bailey Hortorium. 1976. Hortus Third. Rev. and enl. MacMillan Publishing Co., Inc., New York, and Collier MacMillan Publishers, London. 1290 pp.

Bond, P. and P. Goldblatt. 1984. Plants of the Cape Flora. A descriptive Catalogue. Journal of South African Botany. Supplementary Volume 13. 455 pp.

Bridson, G.D.R. and E.R. Smith. 1991. Botanico-Periodicum-Huntianum/Supplementum. Hunt Institute for Botanical Documentation. Carnegie Mellon University. Pittsburg. 1068 pp.

Britton, N.L. and P. Wilson. 1923–1924. New York Academy of Sciences scientific survey of Porto Rico and the Virgin Islands. Botany of Porto Rico and the Virgin Islands. Published by the Academy. Vol. 5 (4 parts): 626 pp.

Britton, N.L. and P. Wilson. 1925–1930. New York Academy of Sciences scientific survey of Porto Rico and the Virgin Islands. Botany of Porto Rico and the Virgin Islands. Published by the Academy. Vol. 6 (4 parts): 663 pp.

Brown, P. and G.B. Stratton (editors). 1963. The world list of scientific periodicals published in the years 1900–1960. William Clowes and Sons. London and Beccles. Three volumes.

Brummitt, R.K. 1992. Vascular plant families and genera. Royal Botanic Gardens, Kew. Printed by Whitstable Litho Ltd. Whitstable, Kent, Great Britain. 804 pp.

Brummitt, R.K. and C.E. Powell, editors. 1992. Authors of plant names. Royal Botanic Gardens, Kew. Printed by Whitstable Litho Ltd. Whitstable, Kent, Great Britain. 732 pp.

Burrill, T.J. 1884. New species of Uredineae. Botanical Gazette 9:187–191.

Burrill, T.J. 1885. Parasitic fungi of Illinois. Part I. Bulletin of the Illinois State Laboratory of Natural History 2(3):141–255.

Burrill, T.J. 1887. Parasitic fungi of Illinois. Part II. Bulletin of the Illinois State Laboratory of Natural History 2(6):387–432.

Chupp, C. 1953. A monograph of the genus *Cercospora*. Cornell University. Ithaca, New York. 667 pp.

Correll, D.S. and H.B. Correll. 1982. Flora of the Bahama Archipelago (Including the Turks and Caicos Islands). J. Cramer, Vaduz. 1692 pp.

Cummins, G.B. 1971. The rust fungi of cereals, grasses and bamboos. Springer Verlag, New York, etc. 570 pp., illustr.

Cummins, G.B. 1978. Rust fungi on legumes and composites in North America. University of Arizona Press, Tucson. 424 pp., illustr.

Deighton, F.C. (1967) 1968. Nomenclatural notes on some Meliolineae. Some comments on the fungus names cited in Hansford's monograph (Sydowia, Beiheft 2, 1961) and in his supplement (Sydowia 16 [1962]:302–323, 1963). Sydowia Annales Mycologici 21:183–187.

Doidge, E.M. 1917. South African Perisporiales. I. Perisporiaceae. Transactions of the Royal Society of South Africa 5:713–750 (with plates LVII–LXVI).

Doidge, E.M. 1919. South African Perisporiaceae. II. Revisional notes. Transactions of the Royal Society of South Africa 7:193–197.

Doidge, E.M. 1921. South African Ascomycetes in the National Herbarium. Bothalia 1(1):5–32.

Doidge, E.M. 1922. South African Ascomycetes in the National Herbarium. Part II. Bothalia 1(2):65–82.

Farr, E.R., J.A. Leussink, and F.A. Stafleu. 1979. Index nominum genericorum (plantarum). Bohn, Scheltema and Holkema, Utrecht. dr. W. Junk b.v., Publishers. The Hague. Three volumes.

Farr, E.R., J.A. Leusink and G. Zijlstra. 1986. Index nominum genericorum (plantarum). Supplementum 1. Bohn, Scheltema & Holkema, Utrecht/Antwerpen. dr. W. Junk b.v., Publishers. The Hague/Boston. 126 pp.

Goos, R.D. and D.P. Gowing. 1992. Type specimens of fungi maintained at Herbarium Pacificum, Bernice P. Bishop Museum, Honolulu. Mycotaxon 43:177–198.

Greuter, W., F.R. Barrie, H.M. Burdet, W.G. Chaloner, V. Demoulin, D.L. Hawksworth, P.M. Jørgensen, D.H. Nicolson, P.C. Silva, P. Trehane, and J. McNeill (Editorial Commitee). 1994. International code of botanical nomenclature (Tokyo Code). Adopted by the Fifteenth International Botanical Congress, Yokohama, August–September 1993. Koeltz Scientific Books, Königstein, Germany. Regnum vegetabile Vol. 118:1–388.

Hansford, C.G. 1946. Contributions toward the fungus flora of Uganda.— VIII. New records (continued). Proceedings of the Linnean Society of London 157:138–212.

Hansford, C.G. 1961. The Meliolineae, a monograph. Beihefte zur Sydowia Annales Mycologici 2:1–806.

Hansford, C.G. 1963. The Meliolineae supplement. Sydowia Annales Mycologici 16:302-323.

Hennen, J.F. and C.S. Hodges, Jr. 1981. Hawaiian forest fungi. II. Species of *Puccinia* and *Uredo* on *Euphorbia*. Mycologia 73(6):1116–1122.

Holmgren, P.K., N.H. Holmgren, and L.C. Barnett. 1990. Index herbariorum. Part I: The herbaria of the world. 8th Edition. New York Botanical Garden. Bronx, New York. Regnum Vegetabile Vol. 120:1–693.

Hughes, S.J. 1953. Fungi from the Gold Coast. II. Mycological Papers. Commonwealth Mycological Institute 50:1–104.

Hughes, S.J. 1993. *Meliolina* and its excluded species. Mycological Papers. International Mycological Institute 166:1–255.

Index kewensis plantarum phanerogamarum. Nomina et synonyma omnium generum et specierum a Linnaeo usque ad annum MDCCCLXXXV complectens nomine recepto auctore patria unicuique plantae subjectis. 1895–present. Clarendon Press. Oxford, England. Two volumes & 18 supplements.

Kartesz, J.T. 1994. A synonymized checklist of the vascular flora of the United States, Canada, and Greenland. 2nd Edition. Timber Press. Portland. Vol. 1— Checklist, 622 pp. Vol. 2— Thesaurus, 816 pp.

Lawrence, G.H.M., A.F.G. Buchheim, G.S. Daniels and H. Dolezal. 1968. Botanico-Periodicum-Huntianum. Hunt Botanical Library. Pittsburgh. 1063 pp.

Li, Hui-Lin, et al., (editors). 1975–1979. Flora of Taiwan. 6 volumes. Taipei:Epoch.

Merrill, E.D. 1912. A Flora of Manila. Philippine Bureau of Printing. Manila. 490 pp.

Merrill, E.D. 1923–1926. An enumeration of Philippine flowering plants.

Philippine Bureau of Printing. Manila. Four volumes: Vol. 1(1, 2):1–240. 1923; Vol. 1(3):241–368. 1924; Vol. 1(4):369–463. 1925; Vol.2:1–530. 1923; Vol. 3:1–628. 1923; Vol. 4:1–515. 1926.

Nicolson, D.H. 1974. Orthography of names and epithets: Latinization of personal names. Taxon 23(4):549-561.

Nicolson, D.H. 1986. Species epithets and gender information. Taxon 35(2):323–328.

Nicolson, D.H. 1987. Species epithets ending in -*cola*, a retraction concerning -*colus*, -*colum*. Taxon 36(4):742–745.

Nicolson, D.H. 1991. Flora of Dominica. Part 2: Dicotyledoneae. Smithsonian Contributions to Botany 77:1–274.

Nicolson, D.H. and R.A. Brooks. 1974. Orthography of names and epithets: stems and compound words. Taxon 23(1):163–177.

Porter, K.I. and C.J. Koster (editors). 1970. The world list of scientific periodicals. New periodical titles 1960–1968. Butterworths, London. 603 pp. & indices.

Rossman, A.Y. 1983. The phragmosporous species of *Nectria* and related genera. Mycological Papers. International Mycological Institute 150:1–164.

Rossman, A.Y. 1987. The Tubeufiaceae and similar Loculoascomycetes. Mycological Papers. International Mycological Institute 157:1–71.

Ryan, R.W. 1924. The Microthyriaceae of Porto Rico. Mycologia 16:177–196.

Smith, A.C. 1979–1991. Flora Vitiensis Nova. A new flora of Fiji (Spermatophytes only). S.B. Printers, Inc. Printed for the National Tropical Botanical Garden, Lawai, Kauai, Honolulu, Hawaii. Five volumes.

Spegazzini, C.L. 1923 (1924). Algunos Honguitos Portoriqueños. Boletín de la Academia Nacional de Ciencias en Córdoba 26:335–368 (Preprint— journal part issued in 1924).

Stafleu, F.A. and R.S. Cowan. 1976–1988. Taxonomic literature. A selective guide to botanical publications and collections with dates, commentaries, and types. Second edition. Bohn, Scheltema & Holkema, Utrecht/Antwerpen. dr. W. Junk b.v., Publishers. The Hague/Boston. Seven volumes.

Stafleu, F.A. and E.A. Mennega. 1992–1995. Taxonomic literature. A selective guide to botanical publications and collections with dates, commentaries and types. Supplement 1: A–Ba. Regnum vegetabile Vol. 125:1–453. 1992. Supplement 2: Be–Bo. Regnum vegetabile Vol. 130:1–464. 1993. Supplement 3: Br–Ca. Regnum vegetabile Vol. 132:1–555. 1995.

Standley, P.C. 1937–1938. Flora of Costa Rica. Field Museum of Natural History. Botanical Series Vol. 18 (in four parts). Part 1:1–398. 1937; Part 2:399–780. 1937; Part 3:781–1133. 1938; Part 4:1135–1616. 1938.

Stearn, W.T. 1992. Botanical Latin. Fourth Edition. David & Charles, Brunel House, Newton Abbot, Devon, Great Britain. 546 pp.

Stevens, F.L. 1916. The genus *Meliola* in Porto Rico. Illinois Biological Monographs 2(4):469–554, with five plates. Reprint pagination:1–86.

Stevens, F.L. 1918. Some meliolicolous parasites and commensals from Porto Rico. Botanical Gazette 65:227–249.

Stevens, F.L. 1920. Dothidiaceous and other Porto Rican fungi. Botanical Gazette 69:248–257, with three text figures and two plates.

Stevens, F.L. (1923) 1924. Parasitic fungi from British Guiana and Trinidad. Illinois Biological Monographs 8(3):169–242, with 19 plates and a map. Reprint pagination:1–76.

Stevens, F.L. 1925. Hawaiian fungi. Bernice P. Bishop Museum Bulletin 19:1–189, with 35 text figures and ten plates.

Stevens, F.L. 1927a. The Meliolineae I. Annales Mycologici 25:409–469, with two plates.

Stevens, F.L. 1927b. Fungi from Costa Rica and Panama. Illinois Biological Monographs 11(2):153–254, with 18 plates and a map. Reprint pagination:1–102.

Stevens, F.L. 1928. The Meliolineae II. Annales Mycologici 26:165–383, with six plates.

Stevens, F.L. and E.F. Roldan. 1935. Philippine Meliolineae. Philippine Journal of Science 56:47–80.

Stevens, F.L. and L.R. Tehon. 1926. Species of *Meliola* and *Irene* from British Guiana and Trinidad. Mycologia 18:1–22, with 2 plates.

Tutin, T.G., V.H. Heywood, N.A. Burgess, D.M. Moore, D.H. Valentine, S.M. Walters, and D.A. Webb (editors). 1964–1980. Flora Europaea. Cambridge University Press. London, etc. Five volumes. Vol. 1 (1964):464 pp.; Vol. 2 (1968):455 pp.; Vol. 3 (1972):370 pp.; Vol. 4 (1976):505 pp.; Vol. 5 (1980):452 pp.

Wagner, W.L., D.R. Herbst and S.H. Sohmer. 1990. Manual of the Flora of Hawaii. Bishop Museum Special Publication 83. University of Hawaii Press; Bishop Museum Press. Two volumes: 1853 pp.

Wehmeyer, L.E. 1961. A World Monograph of the Genus *Pleospora* and its Segregates. The University of Michigan Press, Ann Arbor. 451 pp.

Young, E. 1915. Studies in Porto Rican parasitic fungi. Mycologia 7:143–150.